Polariton Chemistry

Polariton Chemistry

Molecules in Cavities

Edited by

Joel Yuen-Zhou
University of California San Diego

Noel C. Giebink
University of Michigan

Raphael F. Ribeiro
Emory University

WILEY

Published by John Wiley & Sons, Inc., Hoboken, New Jersey.

Library of Congress Cataloging-in-Publication Data has been applied for:

Hardback: 9781119783299
ePDF: 9781119783305
epub: 9781119783312
oBook: 9781119783329

Cover Design: Wiley
Cover Image: Courtesy of Arghadip Koner

Contents

Preface

Light and material excitations can strongly couple to form hybrid modes known as polaritons. Interest in the hybridization of photonic modes with crystalline solid excitations finds its origin in the classic works of Tolpygo [1] and Huang [2] on phonon-polaritons, and of Agranovich [3] and Hopfield [4] on exciton-polaritons; it is this last paper that coins the term "polariton." However, it was not until 1992 when Weisbuch et al. [5] demonstrated significant enhancement of these phenomena in inorganic quantum wells using Fabry–Pérot optical microcavities. In 1998, Lidzey et al. [6] replicated this achievement with excitons in a disordered organic film. In contrast to the cryogenic temperatures required for strong coupling with inorganic quantum well excitons, the organic version was readily observable at room-temperature owing to the large exciton binding energy of organic semiconductors. The demonstration of room-temperature molecular polaritons in systems that on their own right, i.e., in the absence of the cavity, would not feature strong coupling (owing to an insufficient dielectric contrast with the surrounding medium), ignited a small, but steady stream of research activity in the early 2000s in the organic photophysics community.

Early work studying the photophysics of molecular polaritons by Agranovich, Lidzey, La Rocca, and co-workers [7–10] revealed the prominent role of both static and dynamic disorder inherent to molecular excitations (the latter manifested in strong vibronic couplings) in the ensuing polariton relaxation dynamics. These studies played an important role in pursuing milestones of the polaritonics community as a whole, such as polariton lasing [11] and condensation [12, 13], but with the convenience of operating at room-temperature using molecular films [14–16]. These topics are still of much interest and are at the heart of current developments in molecular polaritonics. Naturally, throughout the course of these efforts, researchers inquired whether the photophysical kinetics of the organic molecules themselves could be modified by strong coupling; however, early work by Kéna-Cohen and Forrest [17] showed no detectable change in the kinetics of intersystem-crossing for an organic phosphor in the linear excitation regime. Yet, in the past 12 years, spearheaded notably by the Ebbesen group in Strasbourg, a shift occurred, indicating that polariton formation could in principle trigger a broad array of phenomena, including alterations in electronic excited-state [18] and even ground-state [19, 20] chemical kinetics and dynamics, phase transition hysteresis curves [21], and other chemical phenomena. These results sparked a wave of excitement in multiple disciplines, leading to the development of a new field of research broadly termed *polariton chemistry* [22–25], where the goal is to control the physicochemical properties and processes of matter using strong light–matter coupling. The technological significance of this field lies in its potential to bypass time-consuming and expensive conventional synthetic methods to drive chemical change with selective new processes that rely on confining the respective molecular systems inside a properly designed photonic architecture. On a fundamental level, the overarching goal is to understand the extent to which tailoring the optical environment of a material can alter its physical and/or chemical properties beyond what is expected from classical optics.

Given the aforementioned possibilities, *polariton chemistry* is currently a very active field, with researchers contributing from a wide range of disciplines, including physical chemistry, physical-organic chemistry, condensed matter physics, photonic engineering, plasmonics, and organic opto-electronics. Scientific meetings on the topic are vibrant, with interdisciplinary exchange among chemists, physicists, and engineers who have developed different tools to attack various facets of the problem. However, due to this same interdisciplinarity, language barriers are commonplace, and some difficulties also emerge in the discussion and assessment of results and techniques to reach certain conclusions. Furthermore, many of the mechanisms underlying strong coupling modification of material properties and chemical processes still remain opaque and controversial, with a recent realization that polaritons serve as trivial optical filters in a wide class of scenarios [CITE][26].

Although there are many good reviews on this evolving field, there is not, as of yet, a good starting point for incoming students and researchers to enter this fascinating and, by now, already vast field. The purpose of this book is to fill this important gap. It aims to present a timely overview of polariton chemistry, liberally defined to encompass any current efforts in molecular polaritonics, including its more traditional aspects involving spectroscopy and the exploration of collective phenomena (lasing, condensation, etc.). The chapters in this volume focus on both theoretical and experimental aspects of polariton chemistry, with the implicit aim of incentivizing stronger theory-experiment collaborative efforts that will be important for the field moving forward. Owing to the monographic nature of this volume and the different priorities and backgrounds of the broad readership, it is likely that a first reading will be selective and not sequential. Thus, in the next pages, we provide a diagram that is intended to help the reader navigate through the book (Figure 1).

This volume is organized into four parts. Part I (Basic Concepts) aims to provide the reader with an introduction to the subject, namely, the essential concepts and tools to understand molecular polaritons while simultaneously showcasing its direct relevance to current research. It begins with a pedagogical introduction to the molecular collective strong-coupling regime in Chapter 1 by del Po and Scholes, who discuss the prominent role of dark states in the interpretation of transient absorption and two-dimensional electronic spectra of these hybrid light–matter systems. These authors also discuss photophysical processes assisted by polaritons, such as intersystem crossing, triplet–triplet annihilation, and energy transfer. The contribution from Barnes in Chapter 2 connects the quantum and classical optics perspectives of strong light–matter coupling by formulating an equivalent description of the light–matter coupling strength in each case, which helps bridge the

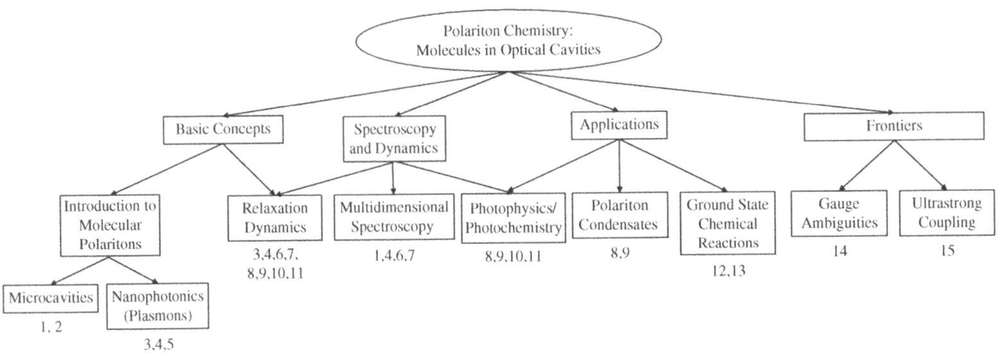

Figure 1 Organization of the book.

gap between quantum theory and experimentally measurable material properties. Most chapters in the volume discuss polaritons emerging from optical microcavity photon modes; however, the demonstration of the onset of strong coupling using fewer number of molecules can be achieved with plasmonic nanoparticles that feature nanoconfined photon (technically, plasmon) modes. Chapters 3 to 5 (and a small discussion in Chapter 15) address this very timely effort. Chapter 3 by Kotov and Shegai describes a variety of light–matter architectures supporting field confinement and plasmon-exciton polariton (plexciton) formation in configurations suitable for integration in devices. Highlights of this chapter are a discussion of the currently available architectures to achieve single-molecule strong coupling with these plexciton systems, as well as a description of micro-cavity assembly control via manipulation of the mesoscopic forces between mirrors. Chapter 4 by Finkelstein-Shapiro comprehensively describes the linear optical properties as well as dynamics of plexcitons. In these systems, the band structure, electron–phonon coupling, and other properties of the host material (metallic nanoparticles) play a key role as modulators of plexciton-assisted photophysics and photochemistry. Finally, in Chapter 5, Chickkaraddy and Baumberg go in-depth to describe the strategies to achieve single-molecule strong coupling using plasmonic nanocavities. Their focus on the design and implementation of nanoparticle-on-mirror systems provides a convenient approach to achieve extremely small mode volumes, and it stands as one of the few that has enabled single-molecule strong coupling to be realized experimentally.

Part II (Spectroscopy and Dynamics) places emphasis on the description of the relaxation processes afforded by the various types of molecular polaritons and the spectroscopic tools used to detect them. In Chapter 6, Dunkelberger et al. provide an experimental viewpoint on the nonlinear spectroscopy of vibrational polaritons (polaritons emerging from the strong coupling of high-frequency vibrational modes of molecules with infrared optical cavities). They show that, ultimately, a wide range of spectroscopic observations of these systems can be simply explained by considering changes in optical constants due to the nonstationarity of the optically prepared molecular excitations, thus demystifying the role of the cavity in the aforementioned experiments. Chapter 7 by Zhang offers a theoretical counterpart to the previous chapter, by providing a general theoretical formalism to model the multidimensional spectroscopy of molecular polaritons, both in the vibrational and electronic domain, based on density matrices and quantum Langevin equations. This formalism is useful because the standard nonlinear spectroscopy approach for bare molecular systems does not account for the multiple photon–matter exchanges that a microcavity facilitates in the strong light–matter regime. Chapter 8 by Groenhof et al. explains and illustrates a class of quantum-classical molecular dynamics algorithms designed for modeling organic exciton-polariton transport and photochemistry. Their computational models incorporate microscopic features of both the molecular system and the electromagnetic field, providing valuable insights into recent experiments in transport while highlighting the limitations of concepts such as a single polariton potential energy surface determining the outcome of polariton effects in photochemistry. Chapter 9 by Hagenmuller et al. focuses on the intriguing role of disorder in polaritonic systems. In their contribution, they clearly outline how disorder, usually seen as harmful to efficient energy transport and long-range entanglement in solid-state systems, can enhance exciton delocalization and intermolecular vibrational-electronic correlations in coherent strongly coupled systems modeled by the basic theories of many-body quantum optics.

Part III (Applications) aims to highlight two important applications of molecular polaritons. The first one is condensation, namely, the macroscopic occupation of a polariton mode which, as discussed above, occurs at room temperature for organic microcavities. In Chapter 10, Satapathy and Menon provide an overview of polariton condensation phenomena as well as a discussion of the challenges that must be overcome to streamline organic polariton condensation with a wider class of molecules; efforts to optimize the molecular materials and their nonlinearities are mentioned as important next steps. Related to this, in Chapter 11, Ishii et al. summarize the theoretical models and experimental verification of polariton radiative and nonradiative relaxation dynamics, a topic that is crucial for understanding how to design new materials and architectures that lower the threshold for polariton lasing and condensation. These two chapters are intimately related to the relaxation processes discussed in Chapter 8. The second highlighted application is vibropolaritonic chemistry, a fascinating yet poorly understood phenomenon whereby, in the putative absence of optical pumping (electronic), ground-state chemical reactions are reported to proceed with different kinetics (and maybe also thermodynamics) when substrates feature vibrational strong coupling. Chapter 12 by Michon and Simpkins describes an analytical method for monitoring the rate of chemical reactions in planar microcavities under vibrational strong coupling conditions. Their discussion is particularly helpful for new researchers entering the field because it highlights experimental pitfalls to avoid and best practices for obtaining reliable reaction rate data. Chapter 13 by Hirai goes on to summarize some of the most prominent experimental reports of vibropolaritonic chemistry, including cooperative solute–solvent effects, enzymatic reactions, and the impact of vibrational strong coupling on molecular self-assembly and crystallization.

Finally, Part IV (Frontiers) consists of two pedagogical contributions that clear up longstanding confusions in the literature concerning the proper description of light–matter coupling. These considerations become crucial when addressing the phenomenology of polaritonic systems in the regime of ultrastrong coupling, where heuristic approximations can lead to physically unsound conclusions. Chapter 14 by Stokes and Nazir provides an instructive quantum mechanical treatment of elementary light–matter interaction processes in optical microcavities using the multipolar and Coulomb gauges. Their chapter clearly illustrates the distinctive partitioning of light and matter degrees of freedom in each gauge and the consequences of such on observable processes. Finally, in Chapter 15, Rabl discusses the inadequacy of the quantum Rabi model in capturing the equilibrium physics of polaritonic systems under ultrastrong coupling conditions, as well as the controversies pertaining the putative superradiant phase transitions associated with such description. He goes on to provide an appropriate prescription to derive approximate effective Hamiltonians that can predict the behavior of these extreme systems and the corresponding phases they exhibit.

It is clear that the division of the volume into different parts is at best a rough roadmap for the reader and has only been provided as a pedagogical device, but shall not be regarded as a strict classification of contributions. In fact, most chapters deal in one way or another with basic concepts, spectroscopy and dynamics, and applications. Furthermore, a finite volume will necessarily miss important developments and contributions. For polariton chemistry, these include timely problems of cavity-induced phase transitions, heat transport effects in cavities, dynamical Casimir effects, and so forth. However, we hope this volume will provide the reader with a sufficient background to fearlessly explore the literature and inspire them to pursue creative new directions in this fascinating field of research. All the chapters have been written with the intent of providing valuable insights that have long been missing in the literature and which should help both incoming and experienced researchers in the field.

July 2024

Joel Yuen-Zhou, La Jolla
Raphael F. Ribeiro, Atlanta
Chris Giebink, Ann Harbor

References

1 Tolpygo, K. B. Physical properties of a rock salt lattice made up of deformable ions. *Zh. Eksp. Teor. Fiz.* **1950**, *20*(6), 497–509.
2 Huang. K. U. N. Lattice vibrations and optical waves in ionic crystals. *Nature* **1951**, *167*(4254), 779–780.
3 Agranovich, V. M. The influence of reabsorption upon the duration of the fluorescence of molecular crystals.1957.
4 Hopfield, J. J. Theory of the contribution of excitons to the complex dielectric constant of crystals. *Phys. Rev.* **1958**, *112*, 1555–1567. https://doi.org/10.1103/PhysRev.112.1555.
5 Weisbuch, C., Nishioka, M., Ishikawa, A., Arakawa, Y. Observation of the coupled exciton-photon mode splitting in a semiconductor quantum microcavity. *Phys. Rev. Lett.* **1992**, *69*, 3314–3317. https://doi.org/10.1103/PhysRevLett.69.3314.
6 Lidzey, D. G., Bradley, D. D. C., Skolnick, M. S., Virgili, T., Walker, S., Whittaker, D. M. Strong exciton–photon coupling in an organic semiconductor microcavity. *Nature* **1998**, *395*(6697): 53–55.
7 Agranovich, V. M., Litinskaia, M., Lidzey, D. G. Cavity polaritons in microcavities containing disordered organic semiconductors. *Phys. Rev. B* **2003**, *67*(8), 085311.
8 Litinskaya, M., Reineker, P., Agranovich, V. M. Fast polariton relaxation in strongly coupled organic microcavities. *J. Lumin.* **2004**, *110*(4), 364–372.
9 Michetti, P., La Rocca, G. C. Polariton states in disordered organic microcavities. *Phys. Rev. B* **2005**, *71*(11), 115320.
10 Michetti, P., La Rocca, G. C. Polariton-polariton scattering in organic microcavities at high excitation densities. *Phys. Rev. B* **2010**, *82*(11), 115327.
11 Deng, H., Weihs, G., Snoke, D., Bloch, J., Yamamoto, Y. Polariton lasing vs. photon lasing in a semiconductor microcavity. *Proc. Natl. Acad. Sci. USA* **2003**, *100*(26), 15318–15323.
12 Deng, H., Weihs, G., Santori, C., Bloch, J., Yamamoto, Y. Condensation of semiconductor microcavity exciton polaritons. *Science* **2002**, *298*(5591), 199–202.
13 Carusotto, I., Ciuti, C. Quantum fluids of light. *Rev. Mod. Phys.* **2013**, *85*, 299–366. https://doi.org/10.1103/RevModPhys.85.299.
14 Kéna-Cohen, S., Forrest, S. R. Room-temperature polariton lasing in an organic single-crystal microcavity. *Nat. Photon.* **2010**, *4*(6), 371–375.
15 La Rocca, G. C. Polariton lasing. *Nat. Photon.* **2010**, *4*(6), 343–345.
16 Daskalakis, K. S., Maier, S. A., Murray, R., Kéna-Cohen, S. Stéphane. Nonlinear interactions in an organic polariton condensate. *Nat. Mater.* **2014**, *13*(3), 271–278.
17 Kéna-Cohen, S., Forrest, S. R. Green polariton photoluminescence using the red-emitting phosphor ptoep. *Phys. Rev. B* **2007**, *76*(7), 075202.
18 Hutchison, J. A., Schwartz, T., Genet, C., Devaux, E., Ebbesen, T. W. Modifying chemical landscapes by coupling to vacuum fields. *Angew. Chem. Int. Ed.* **2012**, *51*(7), 1592–1596.
19 Thomas, A., George, J., Shalabney, A., Dryzhakov, M., Varma, S. J., Moran, J. et al. Ground-state chemical reactivity under vibrational coupling to the vacuum electromagnetic field. *Angew. Chem.* **2016**, *128*(38), 11634–11638.
20 Thomas, A., Lethuillier-Karl, L., Nagarajan, K., Vergauwe, R. M., George, J. et al. Tilting a ground-state reactivity landscape by vibrational strong coupling. *Science* **2019**, *363*(6427), 615–619.
21 Wang, S., Mika, A., Hutchison, J. A., Genet, C., Jouaiti, A., Hosseini, M. W et al. Phase transition of a perovskite strongly coupled to the vacuum field. *Nanoscale* **2014**, *6*(13), 7243–7248.

22 Ribeiro, R. F., Martínez-Martínez, L. A., Du, M., Campos-Gonzalez-Angulo, J., Yuen-Zhou, J. Polariton chemistry: controlling molecular dynamics with optical cavities. *Chem. Sci.* **2018**, *9*(30), 6325–6339.

23 Feist, J., Galego, J., Garcia-Vidal, F. J. Polaritonic chemistry with organic molecules. *ACS Photonics* **2018**, *5*(1), 205–216.

24 Garcia-Vidal, F. J., Ciuti, C., Ebbesen, T. W. Manipulating matter by strong coupling to vacuum fields. *Science* **2021**, *373*(6551), eabd0336.

25 Ebbesen, T. W., Rubio, A., Scholes, G. D. Introduction: polaritonic chemistry. *Chem. Rev.* **2023**, *123*(21), 12037–12038.

26 Schwennicke, K., Koner, A., Pérez-Sánchez, J. B., Xiong, W., Giebink, N. C., Weichman, M. L., Yuen-Zhou, J. When do molecular polaritons behave like optical filters?. *Chem. Soc. Rev.* **2025**. https://doi.org/10.1039/D4CS01024H.

Acknowledgments

We thank Ms. Elizabeth Amaladoss and Michael Leventhal for their support in the preparation of this volume.

Part I

Basic Concepts

1

Ultrafast Dynamics Under Electronic Strong Light–Matter Coupling

Courtney DelPo and Gregory Scholes

Department of Chemistry, Princeton University, Princeton, NJ, USA

1.1 Introduction: Energy Levels – Central to Science

The interaction of light with matter is ubiquitous and critical to the function of biological systems, the rate and likelihood of chemical reactions, and the performance of optoelectronic devices, to name a few examples. Photosynthesis begins with the absorption of light by a light-harvesting chromophore followed by the careful funnel of energy down to the reaction center where a series of intricately designed reactions can occur, ultimately leading to the vital production of oxygen [1]. In photoredox reactions, which are imperative to drug development in the pharmaceutical industry, a photocatalyst must absorb light and be promoted to its excited state for a reaction to proceed successfully [2]. A solar cell relies on the absorption of sunlight by active layer materials in order to produce electricity [3]. Many of the processes central to chemistry, biology, and materials science depend on the relationship between light and matter and, more specifically, on the energy levels of states within a molecule. The energy levels of the states of the various components of a light-harvesting complex control the funnel of energy to oxygen production. The energy levels within an excited photocatalyst dictate the steps that lead to the desired product. The active layer materials in a solar cell must be designed to absorb the maximal coverage of the solar spectrum. In the pursuit of an ability to control the energy levels of a system and thus control the function and reactivity of a molecule, electronic strong coupling induces a strong interaction between light and matter to engineer states at new energy levels, which allows for both energy level and photophysical pathway manipulation. Electronic strong coupling is a quantum phenomenon that can be used in some cases to surpass the capabilities of synthetic design of molecules.

Electronic strong coupling refers to the coupling of an electronic transition of a molecule to a light mode of an optical cavity [4–8]. An optical cavity consists of two reflective surfaces or metal mirrors. As light bounces back and forth between the mirrors, a standing wave of light of a particular wavelength or energy is formed (Figure 1.1). The wavelength or energy of the light mode at a given angle is determined by the refractive index of the material between the mirrors, n, and the length of the cavity, L, according to the following equation:

$$\lambda = 2nL \tag{1.1}$$

Polariton Chemistry: Molecules in Cavities, First Edition. Edited by Joel Yuen-Zhou, Noel C. Giebink and Raphael F. Ribeiro. Published 2026 by John Wiley & Sons, Inc.

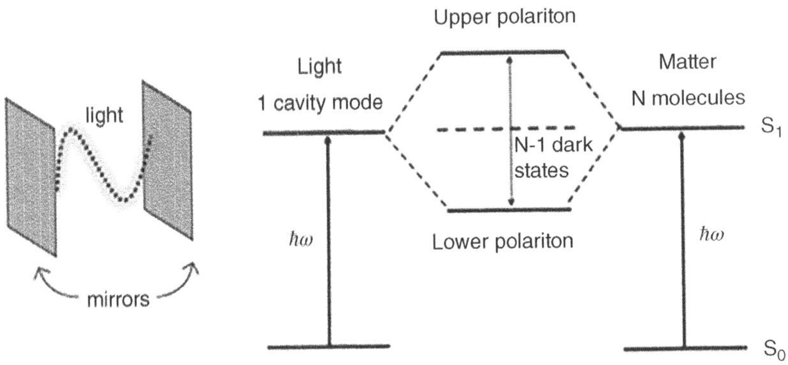

Figure 1.1 Structure and energy level representation of an optical cavity coupled to the lowest energy electronic transition, S_1, of a molecule.

The length of the cavity is tuned to bring the wavelength or energy of the light mode into resonance with the energy of an electronic transition of the molecules placed inside the optical cavity. The coupling of electronic transitions in the visible range leads to cavity lengths on the order of tens to hundreds of nanometers. This is often achieved by depositing a thin layer of the molecules dispersed in an inert polymer matrix between the two mirrors.

Under resonant conditions of strong coupling to one electronic transition, two new hybrid light–matter states called polaritons are formed (see Figure 1.1). There are N number of molecules coupled to 1 cavity mode; therefore, to conserve the total number of states, there must be $N + 1$ states in a cavity. Two of these states are the polariton states: an upper polariton, which is higher in energy, and a lower polariton, which is lower in energy than the energy of the coupled transition, but there is also a manifold of $N - 1$ states that remain at the energy of the coupled electronic transition that are referred to as the dark states. The transition from the ground state to the dark states is no longer optically allowed inside the cavity; however, the dark states can be populated through transitions with other excited states. The precise energy levels of the polariton states can be tuned by increasing or decreasing the coupling strength. The strength of the collective light–matter coupling is quantified by the energy difference between the upper and lower polariton states at resonance, a term called the Rabi splitting, Ω_R.

The magnitude of the Rabi splitting, Ω_R, is related to various properties of the system according to the following equation:

$$\hbar\Omega_R = 2\hbar g = 2\mu\sqrt{N}\sqrt{\frac{\hbar\omega}{2\varepsilon_0 v}} \tag{1.2}$$

where g is the strength of the coupling between the molecules and the cavity, μ is the magnitude of the transition dipole moment of the material, N is the number of molecules coupled to the cavity, ω is the frequency of the transition, ε_0 is the vacuum permittivity constant, and v is the cavity mode volume [5]. The strength of coupling for a system is typically increased by increasing the concentration of molecules ($\sqrt{N}$) or by controlling the orientation of the molecules for maximal alignment of the transition dipole moment of the molecule with the electromagnetic field of the cavity. To enter the strong coupling regime, the coupling strength must be large enough to overcome losses or dissipation from the cavity. This is accomplished by ensuring the magnitude of the Rabi splitting is greater than both the linewidth of the cavity and the linewidth of the molecular absorption. The coupling can shift from "strong" to an "ultrastrong" designation if the magnitude of the Rabi splitting is equal to at least 10–20% of the energy of the coupled electronic transition [9]. The properties of the polariton states are dependent on the coupling regime.

The energy of a polariton state is also anticipated to change with a change in the angle of the system. The energy of the cavity mode is angle-dependent. Within a resonant cavity, there is one angle, typically around 0°, where the energy of the cavity mode is resonant with the energy of the coupled electronic transition. As the magnitude of the angle of the cavity is increased or decreased from the resonant angle condition, the energy of the cavity mode shifts out of resonance with the energy of the coupled electronic transition, leading to a detuning of the cavity. As the cavity is detuned, the coupling changes and both the energy and light–matter fractional components of the polariton states change. Using the simplest model, the coupled oscillator model, the light–matter fractional components of a polariton can be quantified through Hopfield coefficients. The Hopfield coefficients, α and β, are calculated from the coupled oscillator Hamiltonian shown below [10]:

$$\begin{pmatrix} E_C(\theta) & \frac{\hbar\Omega_R}{2} \\ \frac{\hbar\Omega}{2} & E_X \end{pmatrix}\begin{pmatrix} \alpha \\ \beta \end{pmatrix} = E(\theta)\begin{pmatrix} \alpha \\ \beta \end{pmatrix} \tag{1.3}$$

where E_C is the cavity photon energy, E_X is the exciton energy, and θ is the angle.

The angle-dependent changes in the polariton features can be subtle or more drastic depending on the nature of the system. In some systems, particularly those involving broad-absorption organic semiconductors, the hallmark angle-dependent character of the polaritons might not be observed with traditional reflection measurements. The reflection measurements often employ unpolarized light sources to probe the polariton states. The energy of the polariton states can also be influenced by the polarization of light, and in some cases, the angle-dependent trends with s- and p-polarized light are opposite but equal in magnitude, leading to an observation of angle-independence with unpolarized light [11–13].

With the presence of polariton states at new energy levels in a system, the most intuitive application is to use these new energy levels to engineer energetically favorable processes. A process that is endothermic or uphill in energy can become exothermic or downhill in energy using the energy of the lower polariton state. The manipulation of energy levels can be used to manipulate photophysical processes because the rate or likelihood of a transition between two states depends, among other factors, on the energy difference between the two states. There are a few significant rules governing the photophysics of molecules, including the energy gap law and Fermi's golden rule. The energy gap law describes the rate of a nonradiative transition between states as:

$$k_{nr} \sim e^{\frac{-\Delta E}{k_B T}} \tag{1.4}$$

where ΔE is the energy gap between the initial and final states, k_B is the Boltzmann constant, and T is the temperature [14–17]. According to Fermi's golden rule, the rate of a transition between an initial and a final state, $k_{i\to f}$, is as follows:

$$k_{i\to f} = \frac{2\pi}{\hbar}|\langle f, H', i\rangle|^2 \rho(E_f) \tag{1.5}$$

where $|\langle f, H', i\rangle|^2$ is the matrix element of the perturbation H' between the initial and final states, and $\rho(E_f)$ is the density of states of the final state [18–20]. According to these rules, the likelihood of a transition between two states thus depends on the energy difference between the states, the coupling between the initial and final states, and the density of states of the final state. One other rule worth noting is Kasha's rule, which states that the emission of a system will occur from the lowest energy excited state, meaning internal conversion and vibrational relaxation to the lowest energy state will occur before emission [21]. Internal conversion is the nonradiative transition from a higher energy electronic state to a lower energy electronic state of the same spin.

As a nonradiative transition, internal conversion has a rate dependent on the energy difference between the two states as outlined in Eq. 1.4, but this rate, k_{IC}, also depends on a few other factors according to the following equation:

$$k_{IC} = \frac{C^2\sqrt{\pi}}{\hbar\sqrt{\hbar\omega_\upsilon \Delta E}} \exp\left\{-\Delta E/\hbar\omega_\upsilon \left[\log\left(\frac{2\Delta E}{\hbar\omega_\upsilon \Delta_\upsilon^2}\right) - 1\right]\right\} \tag{1.6}$$

where ΔE is the energy gap, C is the nonadiabatic coupling between the electronic states, ω_υ is the vibrational frequency involved in the transition, and Δ_υ is the dimensionless displacement between the states along the vibration [22]. It has been demonstrated that in a cavity system, k_{IC} cannot be accurately modeled on the basis of Eq. 1.6. In the cavity system, neither the energy gap nor the other factors of Eq. 1.6 related to structural changes in the internal conversion process determined k_{IC}, but rather the rate was controlled by the magnitude of the Rabi splitting [22]. In this example, the presence of the polariton states changed the rate of a process between nonpolariton states, supporting the claim that the degree of complexity in a cavity system is greater than can be explained by traditional models.

Transitions between polariton and nonpolariton states are also more complicated than the transitions between nonpolariton states in a noncavity system governed by the rules described above. As a result of their hybrid nature, polaritons exhibit the unique property of exceptional delocalization [23–26]. The delocalization of a polariton state can slow the anticipated rate of a transition between a polariton and nonpolariton state, which can introduce complications in using the energy of the polariton state to manipulate a photophysical pathway [27]. One photophysical pathway that has significant interest for manipulation through electronic strong coupling is the process of reverse intersystem crossing, namely the transition from a lower energy triplet to a higher energy singlet state. According to Hund's rule, the state of higher spin will have lower energy; therefore, energy of a triplet in a molecule will always be lower than the energy of a singlet. The introduction of a lower polariton state at an energy lower than the triplet state provides a pathway to turn this inherently endothermic process into an exothermic one. It has been demonstrated that polaritons can be used as intended in some instances for improved reverse intersystem crossing rates [10, 28]; however, for other systems it has been demonstrated that the polariton cannot be used, and the rate remains unchanged [29, 30].

The complex nature of the polariton state introduces factors that are not accounted for in the rules governing traditional molecular photophysics. These factors make it difficult to predict what will happen to a polariton in the excited state and consequently make it more difficult to use the polaritons to engineer a photophysical pathway or direct the function of a molecule. While polaritons in the electronic strong coupling regime have been successfully implemented into a variety of applications, including but not limited to: changing the work function of a material [31], enhanced and long-range energy transfer [32–35], up-conversion [36], barrier-free reverse intersystem crossing [10], and increased photoconductivity [37], in order to build upon polariton applications, we have to first enhance our fundamental understanding of polaritons. This requires a careful study of polaritons on a range of timescales. In this chapter, we survey what is known about polaritons, from the earliest timescales – tens of femtoseconds – through to some of the steady-state applications we observe with polaritons, and describe some of the more surprising results obtained with polaritons. While there are many ways to form polaritons and many architectures are implemented in cavity design, here, we specifically focus on cavities formed in the Fabry–Pérot type architecture.

1.2 Electronic Strong Coupling in Transient Absorption and Reflection Spectroscopy

1.2.1 Description of Transient Absorption and Reflection Spectroscopy

Transient absorption spectroscopy is a powerful tool to probe polariton dynamics on timescales of hundreds of femtoseconds to nanoseconds. The transient absorption setup consists of a narrowband pump used to excite a system at a particular energy and a broadband probe controlled by a delay stage to provide the time resolution from hundreds of femtoseconds through nanoseconds. A chopper is placed in the delay line of the pump to generate a pump-on and pump-off spectrum at each time delay (Figure 1.2). The output of this experiment is a change in absorbance, ΔA, spectrum recorded as a function of both time and wavelength. The change in absorbance is reported by subtracting the absorption spectrum with the pump off from the spectrum with the pump on at each time delay. The characteristic features that appear in a ΔA spectrum are an excited state absorption (ESA), a ground state bleach (GSB), and stimulated emission (SE). An ESA is a positive feature that signals a transition from one excited state to another excited state. A GSB is a negative feature that represents the depletion of molecules from the ground state. The GSB persists as long as there are molecules in the excited state and is always anticipated to appear where the sample absorbs in the steady-state absorption spectrum. However, overlapping ESA signals with the GSB can interfere with and, in some cases, erase the signatures of the GSB. An SE feature also appears as a negative feature in the ΔA spectrum. For a fully reflective sample, transient absorption spectroscopy is performed in the reflection geometry and the output is a ΔR spectrum, but the features mentioned above remain the same.

1.2.2 Polariton Signatures in Transient Absorption and Reflection Spectroscopy

Polaritons often exhibit derivative-like spectral features in the transient absorption or reflection spectra [6, 38–42]. The origin of the derivative-like feature has been explained in a few ways. One of the most common explanations is that the derivative-like feature arises from a contraction of the Rabi splitting induced by a depletion of molecules in the ground state. The pump pulse excites the

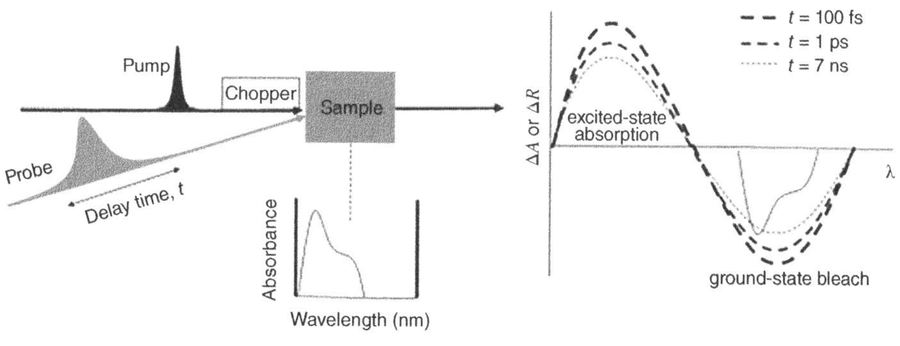

Figure 1.2 Simplified schematic of a pump-probe transient absorption spectroscopy setup.

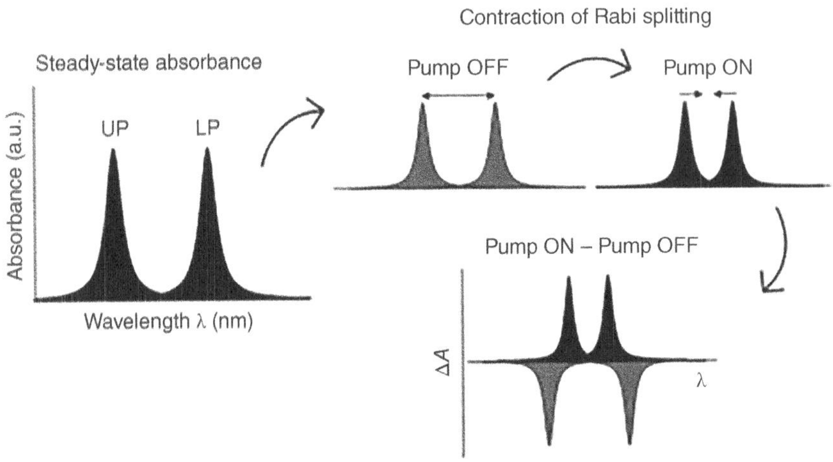

Figure 1.3 The derivative-like polariton feature in the ΔA spectrum explained through the contraction of Rabi splitting. The Rabi splitting decreases as molecules are depleted from the ground state resulting in a contracted polariton absorbance spectrum with the pump on. When the pump-off spectrum is subtracted from the pump-off, a derivative-like feature appears.

sample, resulting in a population of the molecules reaching the excited state and leaving behind a small deficit of ground state molecules. Because the Rabi splitting is dependent on the concentration of molecules in the ground state (Eq. 1.2), the Rabi splitting decreases with the pump on. The resulting ΔA (ΔR) or change in absorbance (reflection) spectrum from the pump-on spectrum subtracted from the pump-off spectrum thus shows a derivative-like feature at the energies of the polariton states (Figure 1.3). In this explanation, the entire derivative-like feature, with both a positive and negative component, is a GSB of the polariton.

In an alternative explanation, the ability for polaritons to exist in both a one- and two-particle manifold is introduced [29, 43]. The two-particle manifold for polaritons is similar to the biexciton phenomenon observed in some inorganic semiconductor systems. The biexciton phenomenon refers to the sequential excitation of two closely spaced electron–hole pairs that leads to an excitation near twice the energy of a single exciton excitation [44–47]. Within the cavity system, the same type of sequential excitation of two polariton species can occur, leading to the formation of states near twice the energy of the upper and lower polaritons. The negative part of the derivative-like polariton feature arises from a standard GSB of the polariton, while the positive part of the feature arises from an ESA. In the case of an upper polariton, the ESA originates from a transition from the upper polariton to a state slightly lower in energy than twice the energy of the upper polariton. In the case of the lower polariton, the ESA originates from a transition from the lower polariton to a state slightly higher in energy than twice the energy of the lower polariton. The result of the combined GSB and ESA is a derivative-like feature at the energy of the upper and lower polaritons. The two-particle manifold has also been invoked to explain the presence of a strong ESA overshadowing the presumed derivative-like feature at the energy of the lower polariton in some cavity systems. Within the two-particle manifold, there are not only states at approximately two times the energy of the upper and lower polaritons but also other additional states, including a crucial state at two times the energy of the cavity mode, 2ω (Figure 1.4).

As a result of the symmetric Rabi splitting with respect to the energy of the cavity mode in a resonant system, the energy of the transition from the upper polariton to 2ω is equal in energy to the lower polariton and appears as an ESA at this energy. Likewise, the energy of the transition

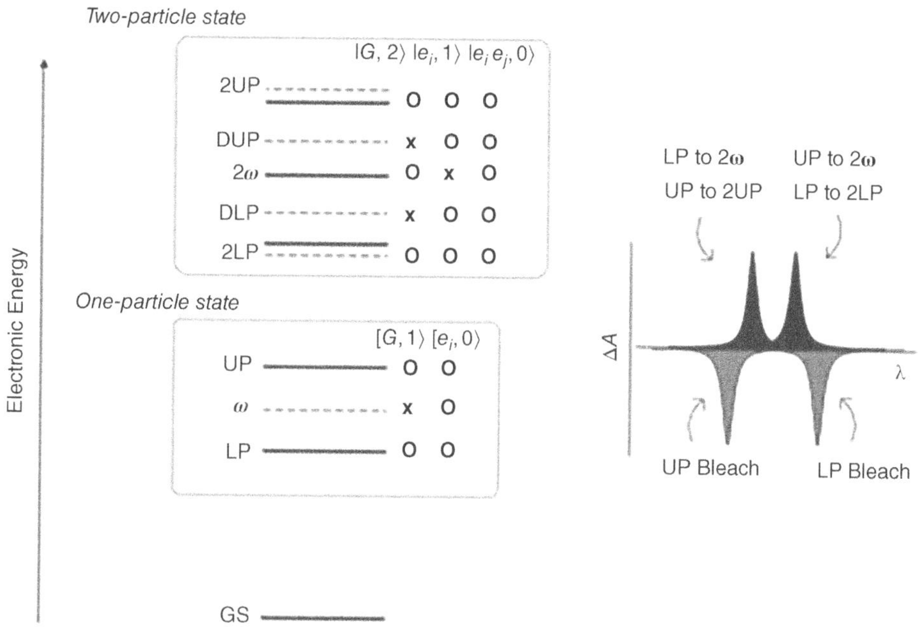

Figure 1.4 The two-particle manifold model used to explain the derivative-like polariton features in the ΔA spectrum. The ESA component in the derivative-like feature arises from a transition to a polariton in the two-particle manifold while the GSB component is a standard GSB [29].

from the lower polariton to 2ω is equal in energy to the upper polariton and appears as an ESA at this energy. There are, therefore, ESA features at the energy of the upper and lower polaritons that originate from the lower and upper polaritons, respectively. These transitions are enabled by the nature of the 2ω state, which is itself a polariton state. In some cavity systems, the ESA at the energy of the lower polariton is strong enough to hide the underlying derivative-like feature.

For a system with strong enough Rabi splitting and thus spectrally separated polariton and molecular states, the narrowband pump can be used to selectively populate the upper polariton and the lower polariton, or excite the system off-resonantly. A comparison of the dynamics when selectively exciting different states of a cavity system can provide selective information about the interactions between polaritons and nonpolariton states. One of the most surprising results observed through these transitions was the population of the dark states of a system on a picosecond timescale through a lower polariton nearly 1 eV lower in energy than the dark states. This type of uphill energy transfer is not possible for purely molecular states; however, in some cavity systems, the lower polariton has low enough entropy to reorder the states in terms of free energy (ΔG). The surprising reactivity is enabled by leveraging the difference in entropy between the ground state and the lower polariton, which is large enough in magnitude to overcome the large difference in electronic energy (ΔE) between the states (Figure 1.5). The change in entropy between two molecular states is normally negligible; therefore, the electronic energy is typically equal to the free energy and is an accurate determinant of the reactivity of a purely molecular system.

Another significant and surprising transition enabled by polariton states is a direct transition between the upper and lower polaritons that avoids intermediate relaxation to the dark state manifold. While this can be explained in the same context of the free energy argument presented above (Figure 1.5), it has also been experimentally observed in both transient absorption and reflection studies as well as two-dimensional electronic spectroscopy studies discussed in the next section.

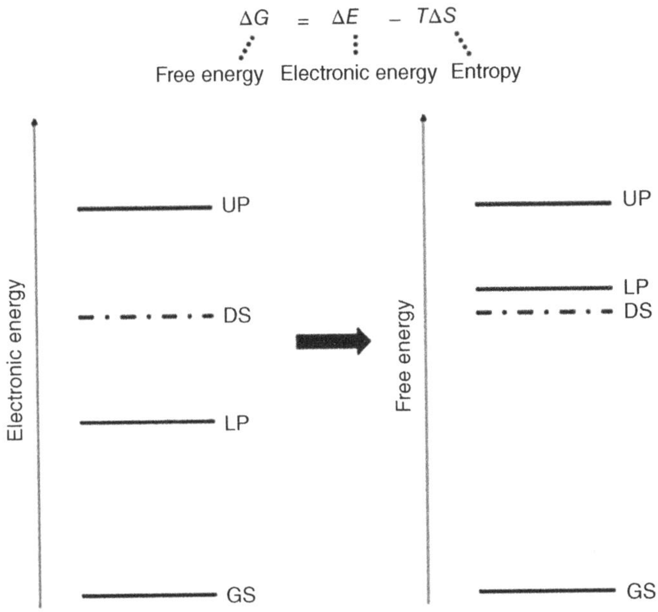

Figure 1.5 Entropy reorders polariton states. The large magnitude of the change in entropy from the ground state to the lower polariton state is enough to reorder the states in terms of free energy as compared to the ordering of states in terms of electronic energy.

1.3 Electronic Strong Coupling in Broadband and Two-dimensional Electronic Spectroscopy

The sub-hundred femtosecond timescale is of interest to polariton states because the lifetime of the photon inside an empty cavity is on this timescale. The photon lifetime is determined by the Q-factor (quality factor) of the cavity. The Q-factor depends primarily on the reflectivity of the mirrors that form the cavity and is calculated according to the following equation:

$$Q = \frac{\omega_{\text{cav}}}{\text{FWHM}_{\text{cav}}} \tag{1.7}$$

where ω_{cav} is the energy of the cavity mode and FWHM_{cav} is the full width at half maximum of the cavity mode [4]. For Fabry–Pérot type cavity systems with metallic mirrors, the Q-factor is typically around 10–20 and the corresponding photon lifetime is approximately 5 fs. To increase the lifetime of the photon in the cavity, distributed Bragg reflector (DBR) mirrors can be used to achieve Q-factors on the order of hundreds to thousands and photon lifetimes on the order of hundreds of femtoseconds. The spectroscopic techniques introduced to study the cavity systems on tens of femtosecond timescales are broadband transient absorption spectroscopy and two-dimensional electronic spectroscopy (2DES) [48–50]. Both of these techniques employ broadband pump pulses, which can enable a time resolution down to tens of femtoseconds. As a result of the time-bandwidth product, a broadband pulse with better time resolution sacrifices spectral resolution, removing the ability to selectively excite the polariton states in broadband transient absorption spectroscopy. The design of 2DES, however, introduces a third pulse and provides additional excitation frequency-dependent data. In 2DES, two broadband pulses are used to excite the sample with a time delay, τ, between these pulses. A Fourier transformation of this time delay gives the excitation frequency

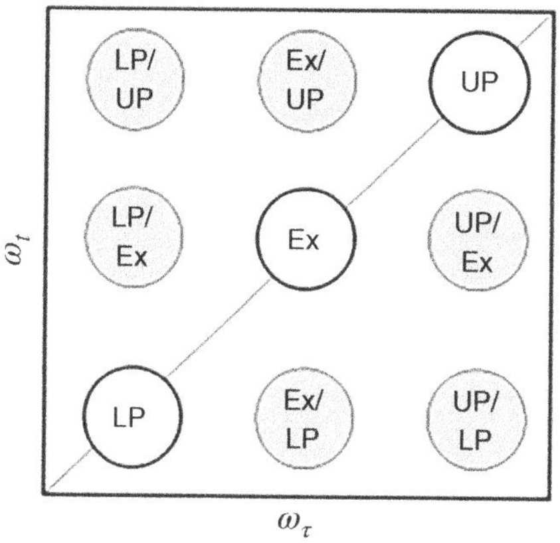

Figure 1.6 Representation of the on-diagonal and off-diagonal peaks in a 2D spectrum of a cavity system. E_x is the exciton or molecular signal, LP is the lower polariton, and UP is the upper polariton. *Source*: Adapted from Ref. [49].

axis, ω_τ. A third pulse is used to detect the transient response after a delay time, T, which is used as the detection frequency axis, ω_t. The information from a transient absorption measurement including the ESA, GSB, and SE signals can also be extracted from the 2DES data by taking a vertical slice of the spectrum at a particular ω_τ (Figure 1.6). The peaks along the diagonal in a 2D spectrum (Figure 1.6) approximate the linear absorption of the system with line shapes that provide information about the surrounding environment of the molecules. The off-diagonal peaks are the cross peaks which contain additional information regarding the coupling between two states and energy relaxation between two states. A pathway of direct energy relaxation from the upper polariton to the lower polariton was observed in a concurrent decay of the upper polariton signal and rise in the upper/lower polariton cross peak [49].

1.4 Electronic Strong Coupling in Applications

In the previous sections, we have discussed some of the surprising properties of polaritons and unintended transitions that occur within the states of a cavity system. Here we introduce a few examples of instances where polaritons have been successfully implemented in exactly the way intended, namely, as new energy levels to control photophysical pathways in a system. The lower polariton state has been demonstrated to interact with various triplet states of a molecule as shown in Figure 1.7 [10, 36, 51]. The polariton states also offer an alternative pathway for energy transfer in a donor–acceptor (D–A) cavity system as shown in Figure 1.8. It is worth noting that while the examples discussed below have been successfully demonstrated in cavity systems, there are also examples of attempts at the same applications with different molecules that have been unsuccessful as a result of the surprising properties of polaritons discussed in the sections above. The examples below are representations of possibilities with polariton states, with the caveat of the selection of the right molecule to couple to the cavity mode to achieve the desired result.

The first case is the case of barrier-free reverse intersystem crossing. Reverse intersystem crossing (RISC) is typically a thermally activated process with a rate determined by the energy difference between the singlet and triplet states of a molecule, the so-called singlet–triplet energy gap, which is inherently a positive value. Molecules are synthetically designed with small singlet–triplet energy

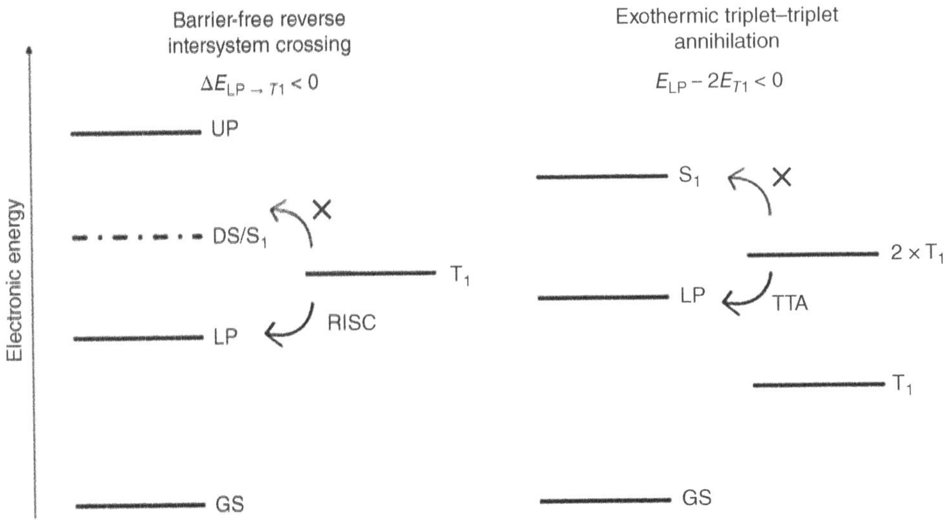

Figure 1.7 The energy of the lower polariton state is used to turn an endothermic process into an exothermic one. In the first case, the lower polariton is lower in energy than the triplet state and is used in the reverse intersystem crossing transition from the triplet to the lower polariton. Reverse intersystem crossing traditionally is a thermally activated process from a triplet to a higher energy singlet state. In the second case, the lower polariton is lower in energy than twice the energy of the triplet state. In triplet–triplet annihilation for upconversion, two triplets transition to a singlet if the energy of the singlet is lower in energy than twice the energy of the triplet. The lower polariton meets this energy condition in a system where the singlet does not.

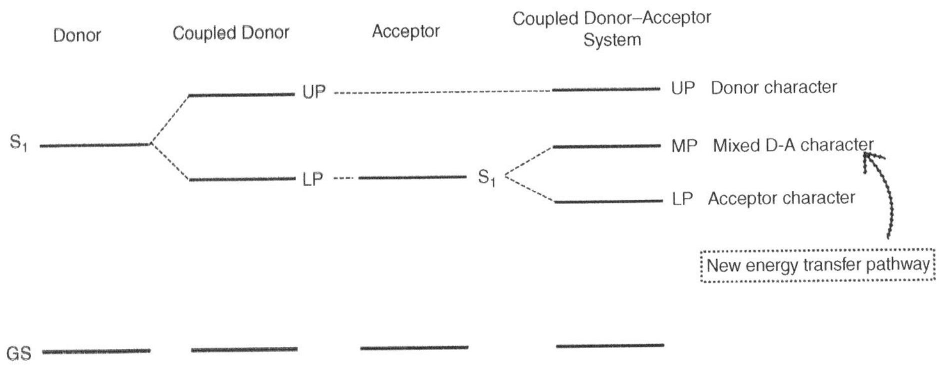

Figure 1.8 In a donor–acceptor polariton system, the polariton states (upper: UP, middle: MP, and lower: LP) offer a distance-independent and efficient pathway for energy transfer from the donor to the acceptor.

gaps to enable fast and efficient RISC for applications in organic light-emitting diodes (OLEDs). OLEDs operate on the principle of electroluminescence, whereby an electrical excitation leads to radiative decay or emission of light from an emitter. As a result of spin-statistics, the singlet and triplet states of the emitter are populated at a ratio of 1 : 3 under electrical excitation. In most organic molecules, the emission from the triplet state is not an allowed or efficient process. The process of RISC allows for triplet excitons to be harvested to the singlet state, where emission can occur in a spin-allowed and efficient process. Inside a cavity, the lower polariton state provides a lower energy and barrier-free pathway to spin-allowed emission.

The second case is the case of triplet–triplet annihilation (TTA), which is an energy transfer process by which two low-energy triplet states are converted to one high-energy singlet state and one ground state of a molecule. The upconversion of low-energy states to a higher energy state is a particularly useful concept for solar cells, where the less-efficient low-energy photons of the sun could be transformed into a higher energy state to properly excite a solar cell absorber [52]. TTA to form a singlet state can only occur if the energy of the singlet state is lower than twice the energy of the triplet state. It was demonstrated that this process could be enabled in molecules that do not meet the proper energy level conditions (i.e., the singlet state is too high in energy) by coupling the singlet to a resonant cavity mode to form the lower polariton state, which meets the energy level requirement for the observation of TTA.

Energy transfer in D–A systems with organic semiconductors can be governed by Forster-type energy transfer which occurs over short distances. The energy transfer rate is proportional to $1/R_{\mathbf{DA}^6}$ where R_{DA} is the D–A distance. The donor and acceptor must be designed to be in close proximity (approximately 10 nm or less) within a blend to achieve a reasonable and observable rate of energy transfer. The introduction of electronic strong coupling offers an alternative energy transfer pathway through the polariton states as shown above (Figure 1.8). The polariton energy transfer occurs as a result of the coupling to the cavity and is thus independent of D–A distance. This method begins with the coupling of the donor electronic transition to a resonant cavity mode, which leads to the formation of upper and lower polaritons. The acceptor is carefully selected so that the absorption of the acceptor overlaps with the lower polariton absorption leading to a second coupling and the formation of three total polariton states: an upper polariton, a middle polariton, and a lower polariton. The upper and lower polaritons predominantly have donor and acceptor character, respectively, while the middle polariton has mixed D–A character. Energy is thus funneled from the upper polariton and donor to the lower polariton and acceptor through the middle polariton state. This method of energy transfer has been demonstrated to lead to both an enhanced rate of energy transfer and distance-independent energy transfer.

1.5 Future Outlook of Ultrafast Dynamics in Electronic Strong Coupling

The ability to control the function and activity of a molecule and manipulate photophysical pathways in an excited state remains the ultimate goal with electronic strong coupling. In the current state of literature, we are unable to predict or explain why some polaritons undergo surprising transitions while others can be used exactly as intended in the manipulation of photophysical pathways. Spectroscopy, with the ability to probe all timescales from tens of femtoseconds to steady-state fluorescence, is the most useful tool for exploring the polariton properties unique to each cavity system. A thorough understanding of the interplay between properties of the molecule and the observed polariton properties will be necessary to achieve the goal of engineering polariton properties for applications.

References

1 Mirkovic, T., Ostroumov, E. E., Anna, J. M., Van Grondelle, R., Govindjee, Scholes, G. D. Light absorption and energy transfer in the antenna complexes of photosynthetic organisms. *Chem. Rev.* **2017**, *117*(2), 249–293.

2 Romero, N. A., Nicewicz, D. A. Organic photoredox catalysis. *Chem. Rev.* **2016**, *116*(17), 10075–10166.

3 Upama, M. B., Mahmud, M. A., Conibeer, G., Uddin, A. Trendsetters in high-efficiency organic solar cells: toward 20% power conversion efficiency. *Sol. RRL* **2020**, *4*(1), 1900342.

4 Hertzog, M., Wang, M., Mony, J., Börjesson, K. Strong light–matter interactions: a new direction within chemistry. *Chem. Soc. Rev.* **2019**, *48*(3), 937–961.

5 Ebbesen, T. W. Hybrid light–matter states in a molecular and material science perspective. *Acc. Chem. Res.* **2016**, *49*(11), 2403–2412.

6 George, J., Wang, S., Chervy, T., Canaguier-Durand, A., Schaeffer, G., Lehn, J. et al. Ultra-strong coupling of molecular materials: spectroscopy and dynamics. *Faraday Discuss.* **2015**, *178*, 281–294.

7 Wang, H., Wang, H., Sun, H., Cerea, A., Toma, A., De Angelis, F. et al. Dynamics of strongly coupled hybrid states by transient absorption spectroscopy. *Adv. Funct. Mater.* **2018**, *28*(48), 1801761.

8 Garcia-Vidal, F. J., Ciuti, C., Ebbesen, T. W. Manipulating matter by strong coupling to vacuum fields. *Science* **2021**, *373*(6551), eabd0336.

9 Frisk Kockum, A., Miranowicz, A., De Liberato, S., Savasta, S., Nori, F. Ultrastrong coupling between light and matter. *Nat. Rev. Phys.* **2019**, *1*(1), 19–40.

10 Yu, Y., Mallick, S., Wang, M., Börjesson, K. Barrier-free reverse-intersystem crossing in organic molecules by strong light-matter coupling. *Nat. Commun.* **2021**, *12*(1), 3255.

11 Bujalance, C., Esteso, V., Caliò, L., Lavarda, G., Torres, T., Feist, J. et al. Ultrastrong exciton–photon coupling in broadband solar absorbers. *J. Phys. Chem. Lett.* **2021**, *12*(43), 10706–10712.

12 Stelitano, S., De Luca, G., Savasta, S., Patané, S. Polarized emission from high quality microcavity based on active organic layered domains. *Appl. Phys. Lett.* **2008**, *93*(19), 193302.

13 Kéna-Cohen, S., Maier, S. A., Bradley, D. D. C. Ultrastrongly coupled exciton-polaritons in metal-clad organic semiconductor microcavities. *Adv. Opt. Mater.* **2013**, *1*(11), 827–833.

14 Jang, S. J. A. Simple generalization of the energy gap law for nonradiative processes. *J. Chem. Phys.* **2021**, *155*(16), 164106.

15 Fischer, S. Correlation function approach to radiationless transitions. *J. Chem. Phys.* **1970**, *53*(8), 3195–3207. https://doi.org/10.1063/1.1674470.

16 Freed, K. F. Radiationless transitions in molecules. *Acc. Chem. Res.* **1978**, *11*(2), 74–80.

17 Englman, R., Jortner, J. The energy gap law for radiationless transitions in large molecules. *Mol. Phys.* **1970**, *18*(2), 145–164.

18 Ribeiro, R. F., Martínez-Martínez, L. A., Du, M., Campos-Gonzalez-Angulo, J., Yuen-Zhou, J. Polariton chemistry: controlling molecular dynamics with optical cavities. *Chem. Sci.* **2018**, *9*(30), 6325–6339.

19 Dirac, P. A. M. The quantum theory of the emission and absorption of radiation. *R. Soc.* **1927**, *114*, 767.

20 Bransden, B. H., Joachain, C. J. *Quantum Mechanics*, 2nd ed. Pearson Education: London, 1999.

21 Kasha, M. Characterization of electronic transitions in complex molecules. *Discuss. Faraday Soc.* **1950**, *9*, 14.

22 Avramenko, A. G., Rury, A. S. Quantum control of ultrafast internal conversion using nanoconfined virtual photons. *J. Phys. Chem. Lett.* **2020**, *11*(3), 1013–1021.

23 Schachenmayer, J., Genes, C., Tignone, E., Pupillo, G. Cavity-enhanced transport of excitons. *Phys. Rev. Lett.* **2015**, *114*(19), 196403.

24 Feist, J., Garcia-Vidal, F. J. Extraordinary exciton conductance induced by strong coupling. *Phys. Rev. Lett.* **2015**, *114*(19), 196402.

25 Scholes, G. D., DelPo, C. A., Kudisch, B. Entropy reorders polariton states. *J. Phys. Chem. Lett.* **2020**, *11*, 6389.

26 Scholes, G. D. Polaritons and excitons: Hamiltonian design for enhanced coherence. *Proc. R. Soc. A* **2020**, *476*(2242), 20200278.

27 DelPo, C. A., Khan, S. U. Z., Park, K. H., Kudisch, B., Rand, B. P., Scholes, G. D. Polariton decay in donor–acceptor cavity systems. *J. Phys. Chem. Lett.* **2021**, *12*(40), 9774–9782.

28 Stranius, K., Hertzog, M., Börjesson, K. Selective manipulation of electronically excited states through strong light–matter interactions. *Nat. Commun.* **2018**, *9*(1), 2273.

29 DelPo, C. A., Kudisch, B., Park, K. H., Khan, S. U. Z., Fassioli, F., Fausti, D. et al. Polariton transitions in femtosecond transient absorption studies of ultrastrong light–molecule coupling. *J. Phys. Chem. Lett.* **2020**, *11*(7), 2667–2674.

30 Eizner, E., Martínez-Martínez, L. A., Yuen-Zhou, J., Kéna-Cohen, S. Inverting singlet and triplet excited states using strong light-matter coupling. *Sci. Adv.* **2019**, *5*(12), eaax4482.

31 Hutchison, J. A., Liscio, A., Schwartz, T., Canaguier-Durand, A., Genet, C., Palermo, V. et al. Tuning the work-function via strong coupling. *Adv. Mater.* **2013**, *25*(17), 2481–2485. https://doi.org/10.1002/adma.201203682.

32 Coles, D. M., Somaschi, N., Michetti, P., Clark, C., Lagoudakis, P. G., Savvidis, P. G. et al. Polariton-mediated energy transfer between organic dyes in a strongly coupled optical microcavity. *Nat. Mater.* **2014**, *13*(7), 712–719.

33 Zhong, X., Chervy, T., Zhang, L., Thomas, A., George, J., Genet, C. et al. Energy transfer between spatially separated entangled molecules. *Angew. Chem. Int. Ed.* **2017**, *56*(31), 9034–9038.

34 Tian, M., Li, X., Li, Z., Zhong, X. Analysis of the forward and reverse strongly coupled states on the nonradiative energy transfer effect. *J. Phys. Chem. Lett.* **2021**, *12*(20), 4944–4950.

35 Zhong, X., Chervy, T., Wang, S., George, J., Thomas, A., Hutchison, J. A. et al. Non-radiative energy transfer mediated by hybrid light-matter states. *Angew. Chem. Int. Ed.* **2016**, *55*(21), 6202–6206.

36 Ye, C., Mallick, S., Hertzog, M., Kowalewski, M., Borjesson, K. Direct transition from triplet excitons to hybrid light–matter states via triplet–triplet annihilation. *J. Am. Chem. Soc.* **2021**, *143*(19), 7501–7508.

37 Nagarajan, K., George, J., Thomas, A., Devaux, E., Chervy, T., Azzini, S. et al. Conductivity and photoconductivity of a P-type organic semiconductor under ultrastrong coupling. *ACS Nano* **2020**, *14*(8), 10219–10225.

38 Schwartz, T., Hutchison, J. A., Léonard, J., Genet, C., Haacke, S., Ebbesen, T. W. Polariton dynamics under strong light-molecule coupling. *Chem. Phys. Chem.* **2013**, *14*(1), 125–131.

39 Xiang, B., Ribeiro, R. F., Chen, L., Wang, J., Du, M., Yuen-Zhou, J. et al. State-selective polariton to dark state relaxation dynamics. *J. Phys. Chem. A* **2019**, *123*(28), 5918–5927.

40 Dunkelberger, A. D., Spann, B. T., Fears, K. P., Simpkins, B. S., Owrutsky, J. C. Modified relaxation dynamics and coherent energy exchange in coupled vibration-cavity polaritons. *Nat. Commun.* **2016**, *7*(1), 13504.

41 Xiang, B., Ribeiro, R. F., Li, Y., Dunkelberger, A. D., Simpkins, B. B., Yuen-Zhou, J. et al. Manipulating optical nonlinearities of molecular polaritons by delocalization. *Sci. Adv.* **2019**, *5*(9), eaax5196.

42 Virgili, T., Coles, D., Adawi, A. M., Clark, C., Michetti, P., Rajendran, S. K. et al. Ultrafast polariton relaxation dynamics in an organic semiconductor microcavity. *Phys. Rev. B* **2011**, *83*(24), 245309.

43 Fassioli, F., Park, K. H., Bard, S. E., Scholes, G. D. Femtosecond photophysics of molecular polaritons. *J. Phys. Chem. Lett.* **2021**, *12*(46), 11444–11459.

44 Styers-Barnett, D. J., Ellison, S. P., Mehl, B. P., Westlake, B. C., House, R. L., Park, C. et al. Exciton dynamics and biexciton formation in single-walled carbon nanotubes studied with femtosecond transient absorption spectroscopy. *J. Phys. Chem. C* **2008**, 112(12), 4507–4516.

45 Chemla, D. S., Shah, J. Many-body and correlation effects in semiconductors. *Nature* **2001**, *411*(6837), 549–557.

46 Binder, R., Koch, S. W. Nonequilibrium semiconductor dynamics. *Prog. Quantum Electron.* **1995**, *19*(4–5), 307–462.

47 Schäfer, W., Wegener, M. *Semiconductor Optics and Transport Phenomena*. Springer: Berlin, London, **2011**.

48 Renken, S., Pandya, R., Georgiou, K., Jayaprakash, R., Gai, L., Shen, Z. et al. Untargeted effects in organic exciton–polariton transient spectroscopy: a cautionary tale. *J. Chem. Phys.* **2021**, *155*(15), 154701.

49 Mewes, L., Wang, M., Ingle, R. A., Börjesson, K., Chergui, M. Energy relaxation pathways between light-matter states revealed by coherent two-dimensional spectroscopy. *Commun. Phys.* **2020**, *3*(1), 157.

50 Finkelstein-Shapiro, D., Mante, P. -A., Sarisozen, S., Wittenbecher, L., Minda, I., Balci, S. et al. Understanding radiative transitions and relaxation pathways in plexcitons. *Chem* **2021**, *7*(4), 1092–1107.

51 Polak, D., Jayaprakash, R., Lyons, T. P., Martínez-Martínez, L. Á., Leventis, A., Fallon, H. et al. Manipulating molecules with strong coupling: harvesting triplet excitons in organic exciton microcavities. *Chem. Sci.* **2020**, *11*(2), 343–354.

52 Gray, V., Dzebo, D., Abrahamsson, M., Albinsson, B., Moth-Poulsen, K. Triplet–triplet annihilation photon-upconversion: towards solar energy applications. *Phys. Chem. Chem. Phys.* **2014**, *16*(22), 10345–10352.

2

Molecular Strong Coupling: The Quantum to Classical Transition

William L. Barnes

Department of Physics and Astronomy, University of Exeter, Exeter, United Kingdom

2.1 Introduction

Strong coupling between atoms and photons is a striking phenomenon that is rooted in the vacuum fluctuations of the electromagnetic field; it is the foundation of atom optics [1], a key element in cavity quantum electrodynamics [2], and a resource in emerging quantum technologies [3]. Strong coupling involves the coherent interaction of an atomic/molecular resonance with an optical cavity mode. If the exchange rate of energy between the molecule and the cavity mode is greater than the dissipative rates, then the molecules and the cavity mode lose their separate identities and two new hybrid polariton modes emerge: part matter and part light. This is the strong coupling regime. Molecular strong coupling exploits an $\sqrt{N}$ scaling of this phenomenon when the coherent interaction involves N molecular resonators and a cavity mode. Molecular strong coupling may involve $>10^{12}$ molecules and has potential applications in modifying material properties and opening up new ways to control some chemical reactions [4–6].

Many of the phenomena associated with molecular strong coupling can successfully be predicted using entirely classical models, notably the frequency-dependent permittivity; there seems to be no role, or indeed need for vacuum fluctuations in the classical model, and yet these fluctuations seem essential from the quantum viewpoint. Whilst the primary aim of the present contribution is to provide a simple derivation of the strong coupling interaction strength in terms of parameters that can be directly related to bulk optical properties, thus facilitating the evaluation of new materials for strong coupling. A secondary aim is to compare the quantum/classical descriptions with the goal of identifying where in the classical model the vacuum fluctuations may lie hidden. The results reported here are not in general rigorous; they are intended to provide an intuitive guide and a number of approximations and conceptual simplifications will be made without discussion. The interested reader is referred to the thought-provoking article by Tserkizis et al. [7].

We will consider three different "systems," illustrated in Figure 2.1. The first is a single atom in a high-Q cavity (the realm of atom optics [8]), the second is an ensemble of molecules in a planar microcavity (the realm of polaritonic chemistry [9]), and the third is "simply" the optical response of a bulk material [10].

Polariton Chemistry: Molecules in Cavities, First Edition. Edited by Joel Yuen-Zhou, Noel C. Giebink and Raphael F. Ribeiro. Published 2026 by John Wiley & Sons, Inc.

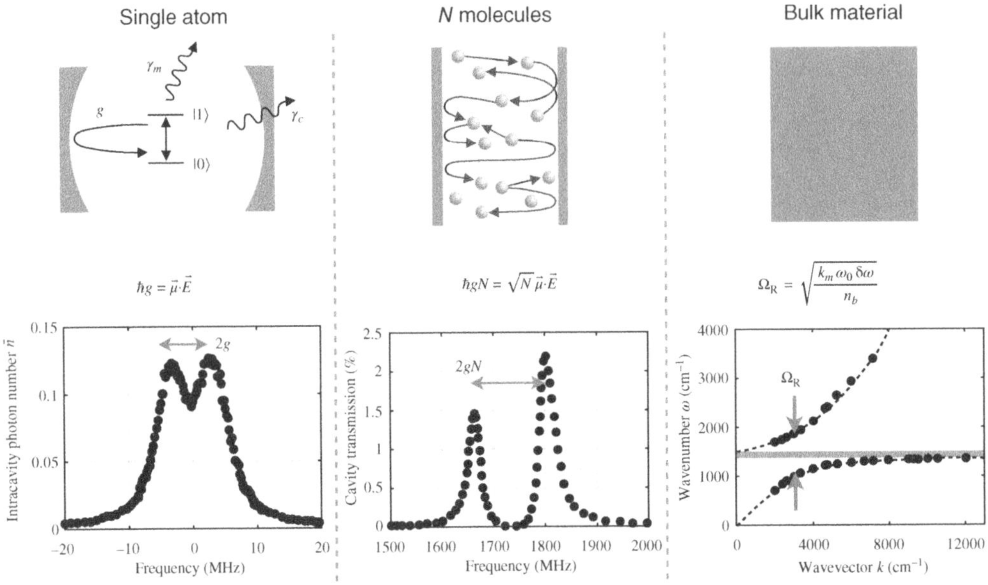

Figure 2.1 Strong coupling systems. *Left column:* strong coupling in atom optics. A single atom resides in a high-quality optical cavity (top). In the coupling regime, the strength of the coupling g is greater than the dissipation rates of the atom γ_m and cavity γ_c, and the atom and cavity mode become hybridized leading to a splitting in the cavity transmission. The data in the lower panel are adapted from the work of Thompson et al. [11]. The extent of the Rabi splitting $\Omega_R = 2g$ is indicated. *Middle column:* strong coupling for an ensemble of molecules in a planar microcavity. *N* molecules are located in the cavity and interact with the cavity mode. The strength of the interaction scales with $\sqrt{N}$. Again, a splitting in the cavity transmission is seen. The lower panel shows data adapted from the vibrational strong coupling experiment of Long et al. [12]. The extent of the Rabi splitting, now $\Omega_R = 2g_N$, is indicated. *Right column:* the bulk optical response of a material incorporating a resonance. Here, in the lower panel, the dispersion of the bulk polariton modes is indicated. Data were adapted from the work of Barra-Burillo et al. [10], and correspond to the bulk phonon-polariton modes in hBN. The extent of the splitting is now Ω_R. The shaded region is the Rhestralen band, where the permittivity is negative.

For the atom optics system and the planar optical microcavity system, the light–matter coupling strength, g, is determined by the interaction energy between the dipole moment μ of the atomic/molecular transition involved, and the vacuum electric field, E_{vac}, associated with a single cavity mode. For the atom optics case, and taking the free space value for the vacuum field strength [13], $E_{vac} = \sqrt{\hbar\omega_0/2V\varepsilon_0}$, we find [14],

$$g = \frac{1}{\hbar}\,\mu.E_{vac} = \frac{\mu}{\hbar}\sqrt{\frac{\hbar\omega_0}{2V\varepsilon_0}} = \mu\sqrt{\frac{\omega_0}{2\hbar\, V\varepsilon_0}} \tag{2.1}$$

where ω_0 is the (angular) frequency of the atomic transition, V is the volume of the cavity mode, and ε_0 is the permittivity of free space. In the strong coupling regime, the rate at which the atom and cavity exchange energy, g, exceeds any of the dissipation rates, specifically the atomic dephasing rate γ_m, and the cavity dephasing rate γ_c. In this strong coupling regime, the original identity of the atomic/molecular and cavity modes is lost, and two new hybrid modes emerge: part light and part matter. The emergence of these states is nicely seen in the transmission data shown at the bottom of Figure 2.1.

For an ensemble of molecules in a planar cavity, we can make use of the fact that the interaction scales as the square root of the number of molecules involved to write [14],

$$g_N = \frac{\sqrt{N}}{\hbar}\,\mu.E_{\text{vac}} = \mu\sqrt{\frac{N}{V}}\sqrt{\frac{\omega_0}{2\,\hbar\,\varepsilon_0\,\varepsilon_b}} \tag{2.2}$$

where we see that the density (concentration) of the molecules in the mode volume is very important. This $\sqrt{N}$ scaling is often sought as evidence of strong coupling [15]. Note also that we have now moved away from thinking of the cavity being filled by free space; the electric field strength inside the cavity now depends on the background relative permittivity of the cavity-filling medium, ε_b, and this has been included in Eq. (2.2) [16].

2.2 Interaction Strength and the Bulk Material Response

In what follows, we make a connection between classical linear response theory on the one hand and the quantum picture outlined above on the other. In doing so we have taken much inspiration and insight from previous work, including [17–19].

We begin with the Lorentz oscillator model, a model which, together with the concept of oscillator strengths, allows a description of the optical properties of many materials. The Lorentz oscillator model has a long history, see for example [20, 21]; the enduring appeal of such models is their simple conceptual framework, the intuitive understanding they embody, and their lasting effectiveness under an extensive range of conditions when compared with a wide variety of experimental data. In the Lorentz oscillator model, the frequency-dependent permittivity is given by [22],

$$\varepsilon(\omega) = \varepsilon_b + \frac{f\omega_p^2}{\omega_0^2 - \omega^2 - i\omega\gamma} \tag{2.3}$$

where f is the oscillator strength, ε_b is the background permittivity and, as noted above, γ is the de-phasing rate, and ω_p is the plasma frequency, discussed further below. An alternative and convenient formulation of Eq. (2.3) in terms of the resonance frequency, rather than the plasma frequency, is given by [23],

$$\varepsilon(\omega) = \varepsilon_b + \frac{f'\omega_0^2}{\omega_0^2 - \omega^2 - i\omega\gamma} \tag{2.4}$$

where f' is sometimes referred to as the reduced oscillator strength. We will use this formulation below since it allows an easy way to characterize the system with a normalized frequency parameter. The plasma frequency is given by Ref. [24],

$$\omega_p^2 = \frac{Ne^2}{V\varepsilon_0\,m_e} \tag{2.5}$$

where N/V is the density of molecules, e is the electronic charge, and m_e is the electron mass. If the electrons were free, this ω_p would indeed be the plasma frequency. Here, the electrons are not free, but the nomenclature has stuck. We should simply treat ω_p as shorthand for the parameters given in Eq. (2.5). Note that ω_p contains the density of the molecules, a material-specific property.

The oscillator strength, f, is related to the dipole moment of the relevant molecular transition and determines the strength with which the molecular resonator couples to an electric field [25]. Typically, $0 < f < 1$; a value of $f > 1$ is possible, but is rare. For an electron executing simple harmonic motion with an amplitude equal to the Bohr radius, $f = 1$. The oscillator strength is given by Ref. [26],[1]

$$f = \frac{2\, m_e\, \omega_0}{3\, \hbar\, e^2} |\mu|^2 \tag{2.6}$$

It might be tempting to think that the $\hbar$ in Eq. (2.6) is where the vacuum fluctuations are "hidden" in the classical description. However, the presence of $\hbar$ simply sets an energy scale but does not necessarily require a quantum description, instead, it is chosen to provide consistency with the results of experiments, see Ref. [27], Section 8.12.

At this point, we can take a shortcut to the result we will derive on the basis of a material response below. We can substitute Eqs. (2.5) and (2.6) into (2.2) to find,

$$g_N = \frac{1}{2}\sqrt{\frac{f\omega_p^2}{\varepsilon_b}} \tag{2.7}$$

Note that in evaluating Eq. (2.7) we have assumed that the dipole moments are all aligned with the field so that the factor of 3 in the denominator of Eq. (2.6) can be dropped (Eq. 2.2 assumes all dipole moments are aligned with the field).

Next, we want to find a way to evaluate the classical coupling strength, i.e. to find an expression for the coupling strength based on Eq. (2.3). If we can do that, then we can compare it to the quantum versions given in Eqs. (2.1) and (2.2). One approach is to look at the round-trip phase that light experiences in a planar microcavity, i.e. to look at the constructive interference condition that leads to the presence of cavity modes. To do this, we need the refractive index associated with the molecular resonance, so we next evaluate that refractive index. We start by finding the real and imaginary parts of the permittivity, $\varepsilon(\omega) = \varepsilon'(\omega) + i\varepsilon''(\omega)$. From Eq. (2.3) these are found to be,

$$\varepsilon'(\omega) = \varepsilon_b + \frac{f\omega_p^2(\omega_0^2 - \omega^2)}{(\omega_0^2 - \omega^2)^2 + \omega^2\gamma^2} \tag{2.8}$$

$$\varepsilon''(\omega) = \frac{i\omega\gamma f\omega_p^2}{(\omega_0^2 - \omega^2)^2 + \omega^2\gamma^2} \tag{2.9}$$

and using $(n'(\omega) + i\kappa(\omega))^2 = \varepsilon'(\omega) + i\varepsilon''(\omega)$ where $n'(\omega) + i\kappa(\omega)$ is the complex index of refraction, we find,

$$n'(\omega) = \sqrt{\varepsilon_b} + \frac{f\omega_p^2(\omega_0^2 - \omega^2)}{2\sqrt{\varepsilon_b}[(\omega_0^2 - \omega^2)^2 + \omega^2\gamma^2]} \tag{2.10}$$

$$\kappa(\omega) = \frac{f\omega_p^2\omega\gamma}{2\sqrt{\varepsilon_b}[(\omega_0^2 - \omega^2)^2 + \omega^2\gamma^2]} \tag{2.11}$$

where we have made an approximation by ignoring terms $\mathcal{O}(f^2)$ and above. Figure 2.2 shows representative plots of the permittivity and refractive index in the LO model.

It is now useful to look at placing a material with this (complex) refractive index in a planar microcavity. To begin with, we consider an empty cavity. The condition for such a cavity to support a resonant mode, i.e. for multiple reflections within the cavity to lead to a transmittance maximum, is that the round-trip optical phase is equal to (a multiple of) 2π.

1 Note that this formula is based on the assumption that the dipole moments are randomly oriented in space.

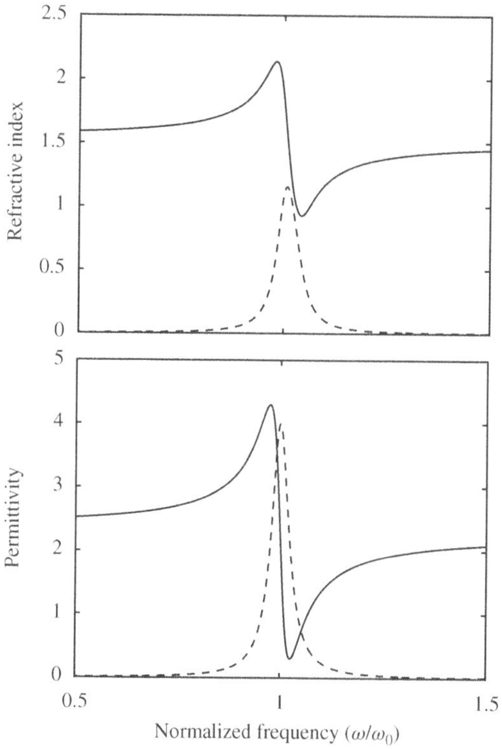

Figure 2.2 Refractive index and permittivity. The refractive index (left) and the permittivity (right) for the Lorentz oscillator model, Eq. (2.4). Both real (solid lines) and imaginary (dashed lines) parts are shown. The parameters used here were ε_b=2.25, f'=0.2, and $\gamma = \omega_0/20$.

Looking at the phase as a function of frequency in Figure 2.3 (upper), we can see that as the value of f' is increased, a transition takes place. For low values of f', there is just one crossing of the $\phi = 2\pi$ condition. However, as the value of f' is increased, the crossing point first becomes somewhat less well-defined (resulting in a spectrally broader transmission peak) and then changes character completely. There are now three crossing points, one at approximately the original frequency ω_0, and two new ones at ω_- and ω_+, see Figure 2.3 (upper). The frequencies ω_- and ω_+ correspond to the two new (hybrid) modes, usually referred to as the lower and upper polaritons. The original crossing point at $\sim \omega_0$ is not seen in the transmittance spectrum of the cavity, see Figure 2.1 (left and center); this is because the cavity at this frequency presents a significant impedance mismatch to incoming light; the light is thus reflected rather than transmitted. Note that the original crossing point is seen in other experimental configurations, e.g. where a surface plasmon mode takes the place of a planar cavity mode [29].

At this point, it is useful to connect the Lorentz oscillator model with the bulk polariton perspective. The dispersion relation of a bulk medium is associated with the propagation of electromagnetic waves in the medium. For each frequency ω the dispersion relation determines the allowed value for the wavevector, k, the dispersion relation is [30],

$$k = n(\omega)\frac{\omega}{c} \tag{2.12}$$

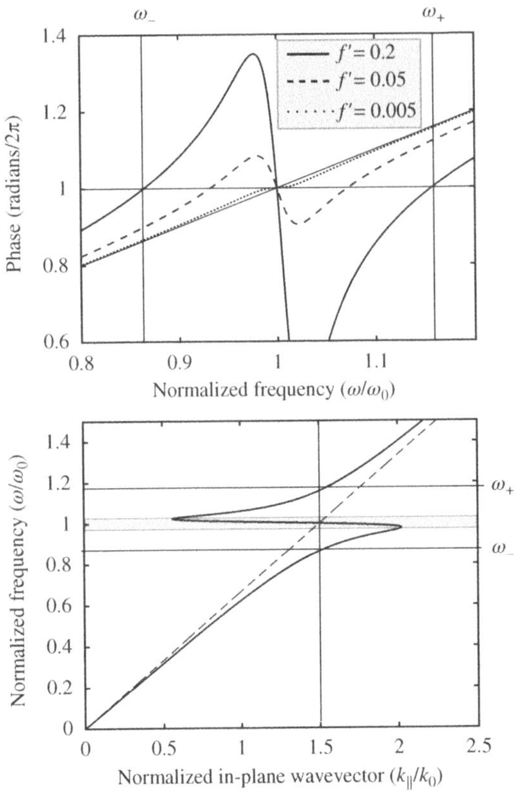

Figure 2.3 Phase and dispersion. Upper: Round-trip phase as a function of normalized frequency. The thin line is for an empty cavity. The dotted, dashed, and solid lines are for a cavity filled with materials whose optical response is given by the Lorentz oscillator model using Eq. (2.4), the reduced oscillator strength f' for each line is given in the inset. Lower: the bulk response of a material whose optical response is given by the Lorentz oscillator model in the upper panel with $f' = 0.2$. The other parameters are the same as those for Figure 2.2. The shaded region corresponds to the frequency range in which there are anomalous dispersions, i.e. $dn/d\omega$ is negative, see for example Ref. [28].

where c is the speed of light in free space. Since $n(\omega)$ is complex, we can look at either a complex frequency or a complex wavevector. For the plot shown in Figure 2.3 (lower), we have adopted the common convention of taking the real part of the refractive index.

Let us now go back and consider the round-trip phase for the microcavity. On resonance, the round-trip phase is equal to 2π, that is $2kL = 2\pi$ where L is the cavity thickness. For a cavity filled with a Lorentz oscillator medium, but with the molecular resonance (Lorentz oscillator) turned off, i.e. for an oscillator strength of $f = 0$, we can use Eq. (2.12) to find the value of k as,

$$k_{f=0}(\omega_0) = \sqrt{\varepsilon_b}\frac{\omega_0}{c} \tag{2.13}$$

If we now turn on the molecular resonance, by setting $f \neq 0$, then the round-trip phase for $f \neq 0$ will be the same as the round-trip phase for $f = 0$ at the frequencies $\omega_\pm$, see Figure 2.3 (upper), for which the associated wavevector is,

$$k_{f\neq 0}(\omega_\pm) = n'(\omega_\pm)\frac{\omega_\pm}{c} = \sqrt{\varepsilon_b}\frac{\omega_0}{c} \tag{2.14}$$

We want to determine what values of $\omega_\pm$ satisfy Eq. (2.14) for a given value of $n''(\omega_\pm)$. To do so, we follow Zhu et al. [18] and simplify our expression for the real part of the refractive index (Eq. 2.10) by making the approximation that $\omega_0 \sim \omega$. We can then write,

$$(\omega_0^2 - \omega^2) = (\omega_0 - \omega)(\omega_0 + \omega) = 2\Delta\omega \tag{2.15}$$

where $\Delta = \omega_0 - \omega$, and where for convenience we have written $\omega_\pm = \omega$. We can then write Eq. (2.10) as,

$$n'(\omega) = \sqrt{\varepsilon_b} + \frac{f\omega_p^2 2\Delta\omega}{2\sqrt{\varepsilon_b}[4\Delta^2\omega^2 + \omega^2\gamma^2]} \tag{2.16}$$

$$n'(\omega) = \sqrt{\varepsilon_b} + \frac{f\omega_p^2}{\sqrt{\varepsilon_b}[4\Delta^2 + \gamma^2]} \cdot \frac{\Delta}{\omega} \tag{2.17}$$

We can then use the fact that $\omega_0 = \Delta + \omega$ and put Eq. (2.14) in a similar form by writing,

$$n'(\omega)\omega = \sqrt{\varepsilon_b}\omega_0 \tag{2.18}$$

and using $\Delta = \omega_0 - \omega$, we have,

$$n'(\omega) = \sqrt{\varepsilon_b}\left(1 + \frac{\Delta}{\omega}\right) \tag{2.19}$$

so that by comparing Eqs. (2.19) and (2.17) we find the phase condition to be,

$$\frac{f\omega_p^2}{\sqrt{\varepsilon_b}(4\Delta^2 + \gamma^2)} = \sqrt{\varepsilon_b} \tag{2.20}$$

so that,

$$\Delta = \frac{\pm 1}{2}\sqrt{\frac{f\omega_p^2}{\varepsilon_b} - \gamma^2} \tag{2.21}$$

If we now recall (see Figure 2.1) that $g_N = \Delta$, then we can find the N molecule coupling strength to be,

$$g_N = \frac{1}{2}\sqrt{\frac{f\omega_p^2}{\varepsilon_b} - \gamma^2} \tag{2.22}$$

There are three things to note at this stage.

First: although we have derived Eq. (2.22) using a planar optical cavity as our system, we could also have derived it directly from Eq. (2.12). The solutions for $\omega_\pm$ are accordingly also presented in Figure 2.3 (upper and lower). We can thus see that the cavity polaritons we have found using the round-trip phase condition correspond to the bulk polaritons, i.e. the propagating modes in the bulk medium [30].

Second: we have not only predicted the magnitude of the coupling strength, g_N, we have also found a condition for strong coupling to occur, i.e.

$$\frac{f\omega_p^2}{\varepsilon_b} > \gamma^2 \tag{2.23}$$

which is very reminiscent of the "polariton swing parameter" introduced by Lagendijk and Tiggelen [17].[2]

Third: for $f\omega_p^2/\varepsilon_b \gg \gamma^2$, we have,

$$g_N \approx \frac{1}{2}\sqrt{f\omega_p^2/\varepsilon_b} \tag{2.24}$$

This result is the same as that given by Canales et al., who derived it from a quantum perspective [19], and has a long history, see for example Ref. [31]. The quantities f and ω_P can be derived from routine laboratory measurements so that Eq. (2.24) can be used as the basis for predicting whether and under what circumstances a particular molecular transition might exhibit strong coupling. For example, a common quantity determined in photophysics and photochemistry measurements for a thin film of material is the extinction coefficient, i.e. the imaginary part of the (complex) refractive index, which can be related to the oscillator strength as follows [32, 33].

We can find a version of Eq. (2.24) that is more amenable to experimental measurements of material parameters by solving Eq. (2.11) for the case of $\omega = \omega_0$, i.e. on resonance. For this condition (on resonance) κ will take its maximum value, κ_m; we find from Eq. (2.11) that,

$$f\omega_p^2 = 2\,\kappa_m\,\omega_0\,\gamma\,n_b \tag{2.25}$$

So that using Eq. (2.24) we have,

$$g_N \approx \frac{1}{2}\sqrt{\frac{\kappa_m\,\omega_0\,\delta\omega}{n_b}} \tag{2.26}$$

where we have set the width of the extinction feature, $\delta\omega$, to be $\delta\omega = 2\gamma$, which is frequently found to be a good approximation [23]. We can at this point look at an example, that of vibrational coupling involving the C=O bond in polymers such as PMMA and PVA [12, 34]. The extinction coefficient associated with the 1735 cm^{-1} C=O stretch resonance is $\kappa_m = 0.62$, and the width of the extinction peak is $\delta\omega = 30$ cm^{-1} so that the (maximum) predicted coupling strength is ~ 150 cm^{-1}. This compares with the values measured by Shalbney et al. and Long et al. of 166 and 140 cm^{-1} respectively, remarkably close given the approximations we have made. Here, we have taken the value of $n_b = 1.46$. Further details can be found in Ref. [23]. Whilst this example has made use of the extinction coefficient, it is straightforward to re-write Eq. (2.26) in terms of the absorbance or the molar absorption coefficient, see [35].

2.3 Comparing Quantum and Classical

It is interesting to look at the relationship between this classical expression for the coupling strength and the quantum version. We can substitute Eqs. (2.6) and (2.5) into Eq. (2.24) to give,

$$g_N = \frac{\mu}{2}\sqrt{\frac{2\,m_e\,\omega_0}{3\hbar\,e^2 n_b}}\sqrt{\frac{Ne^2}{V\varepsilon_o\,\varepsilon_b\,m_e}} \tag{2.27}$$

---

2 Lagendijk and Tiggelen's polariton swing parameter was $f\omega_p^2/\omega_0\gamma$ rather than the $f\omega_p^2/\gamma^2$ found here. This is because their swing parameter is in terms of permittivity rather than refractive index.

so that, if we note that the vacuum field for a background permittivity of ε_b is $E_{\text{vac}} = \sqrt{\hbar\omega_0/2V\varepsilon_0\varepsilon_b}$ [14], then we find,

$$g_N = \sqrt{\frac{N}{3}\frac{\mu}{\hbar}} E_{\text{vac}} \tag{2.28}$$

i.e. we recover the quantum result in Eq. (2.2), to within a factor of $\sqrt{3}$. This difference of $\sqrt{3}$ is because our definition of the oscillator strength as in Eq. (2.6) is based on the dipole moments being isotropically oriented in space, whereas Eq. (2.2) assumes the dipole moments and the field are aligned.

So how did the vacuum field in Eq. (2.28) emerge from the classical expression (Eq. 2.24)? We noted earlier that it is not really the oscillator strength that "hides" the quantum nature of the interaction, we need to look elsewhere.[3] If we accept the very semiclassical analysis presented here as a valid description of our system, then one way to proceed is to look at Eq. (2.28) where we might note that the field E_{vac} need not be quantum mechanical in nature, we could instead consider Eq. (2.28) to imply the existence of a classical zero-point electric field. This is a topic with a long history and is a subject known as stochastic electromagnetism, see Refs. [36–41]. Conventional physical wisdom tells us that the "quantum description" is the correct description. The fact that there is a correspondence between classical and quantum pictures – as outlined here – is to be expected, but the analysis outlined above shows that if we want to see how the quantum and classical are related in a simple way, then we have to adopt into our classical description elements that we usually only think of as quantum, specifically a fluctuating vacuum field, an "element" of classical physics we are usually not aware of and consequently overlook. Nonetheless, whether one takes a quantum or classical view of vacuum fluctuations, their existence has been beautifully seen in experiments [42].

In summary, we have used linear dispersion theory to derive a polariton parameter that will be useful in connecting the material properties of liquids and solids being considered for polariton chemistry experiments, not to see whether their chemistry may be altered, but rather to establish in a simple manner whether any considered material is capable of exhibiting sufficient Rabi splitting to merit further consideration. Along the way, we have discussed the quantum-to-classical transition, and in particular, tried to look at where the vacuum fluctuations of the cavity electric field, fluctuations so central to the quantum description, may lie hidden in the classical viewpoint.

Acknowledgments

The author is grateful to Marie Rider, Christos Tserkezis and Felipe Herrera for valuable discussions, and to the European Research Council for funding Project Photmat (ERC-2016-AdG-742222). No new data were generated for this work.

3 Perhaps one way to look at what we have done is to re-examine the expression for the free space (quantum) vacuum field strength, $E_{\text{vac}} = \sqrt{\hbar\omega_0/2V\varepsilon_0}$, and to consider whether any insight can be gained by re-writing it as,

$$\varepsilon_0 = \frac{\hbar\omega_0}{2VE^2} \tag{2.29}$$

If we take this view, then the finite value for the permittivity of free space implies that there are zero-point fluctuations (E^2 should more properly be $\overline{E^2}$), an idea that has been discussed before [43, 44].

References

1 Haroche, S., Raimond, J.M. *Exploring the Quantum*. Oxford Graduate Texts. Oxford University Press: Oxford, **2006**. ISBN 978-0-19-968041-3.

2 Hinds, E.A. Pertubative cavity quantum electrodynamics. *Cavity Quantum Electrodynamics*. Academic Press Inc: Cambridge, MA, Chap 1, 1–56, **1994**.

3 Wallraff, A., Schuster, D.I., Blais, A., Frunzio, L., Huang, R.S., Majer, J. et al. *Nature* **2004**, *431*, 162–167.

4 Ebbesen, T.W. *Acc. Chem. Res.* **2016**, *49*, 2403–2412.

5 Herrera, F., Owrutsky, J. *J. Chem. Phys.* **2020**, *152*, 100902.

6 Hertzog, M., Wang, M., Mony, J., Börjesson, K. *Chem. Soc. Rev.* **2019**, *48*, 937–961.

7 Tserkezis, C., Fernández-Domínguez, A.I., Gonçalves, P.A.D., Todisco, F., Cox, J.D., Busch, K., et al. *Rep. Prog. Phys.* **2020**, *83*, 082401.

8 Kaluzny, Y., Goy, P., Gross, M., Raimond, J.M., Haroche, S. *Phys. Rev. Lett.* **1983**, *51*, 1175–1178.

9 Lidzey, D.G., Bradley, D.D.C., Skolnick, M.S., Virgil, T., Walker, S., Whittaker, D.M. *Nature* **1998**, *395*, 53–55.

10 Barra-Burillo, M., Muniain, U., Catalano, S., Autore, M., Casanova, F., Hueso, L.E., et al. *Nat. Commun.* **2021**, *12*, 6206.

11 Thompson, R.J., Rempe, G., Kimble, H.J. *Phys. Rev. Lett.* **1992**, *68*, 1132–1135.

12 Long, J.P., Simpkins, B.S. *ACS Photonics* **2015**, *2*, 130–136.

13 Loudon, R. *The Quantum Theory of Light* 2nd ed. Oxford, **1983**.

14 Rider, M.S., Barnes, W.L. *Contemp. Phys.* **2021**, *62*, 217–232.

15 Dintinger, J., Klein, S., Bustos, F., Barnes, W.L., Ebbesen, T.W. *Phys. Rev. B* **2005**, *71*, 035424.

16 Barnett, S.M., Huttner, B., Loudon, R., Matloob, R. *J. Phys. B: At, Mol. Opt. Phys.* **1996**, *29*, 3763–3781.

17 Lagendijk, A., van Tiggelen, B.A. *Phys. Rep.* **1996**, *270*, 143–215.

18 Zhu, Y., Gauthier, D.J., Morin, S.E., Wu, Q., Carmichael, H.J., Mossberg, T.W. *Phys. Rev. Lett.* **1990**, *64*, 2499–2502.

19 Canales, A., Baranov, D.G., Antosiewicz, T.J., Shegai, T. *J. Chem. Phys.* **2021**, *154*, 024701.

20 Longair, M. *Quantum Concepts in Physics*. Cambridge University Press: Cambridge, **2013**. ISBN 978-1-107-01709-2.

21 Smith, D., Graham, G. *J. Phys. Colloq.* **1980**, *41*, C6-80–C6-83.

22 Fox, M. *Optical Properties of Solids*, 2nd ed. Oxford University Press: Oxford, **2010**.

23 Barnes, W.L. *Phys. Scripta*. 99(6): 065560, **2023**.

24 Bohren, C.F., Huffman, D.R. *Absorption and Scattering of Light by Small Particles*. Wiley VCH: Weinheim, **2004**. ISBN 978-0-471-29340-8.

25 Hilborn, R.C. *Am. J. Phys.* **1982**, *50*, 982–986.

26 Novotny, L., Hecht, B. *Principles of Nano-optics*, 1st ed. Cambridge University Press: Cambridge, **2006**.

27 Milonni, P.W. *The Quantum Vacuum*. Academic Press: Cambridge, MA, **1993**.

28 Pockrand, I., Brillante, A., Möbius, D. *J. Chem. Phys.* **1982**, *77*, 6289–6295.

29 Menghrajani, K.S., Nash, G.R., Barnes, W.L. *ACS Photonics* **2019**, *6*, 2110–2116.

30 Mills, D.L., Burstein, E. *Rep. Prog. Phys.* **1974**, *37*, 817–926.

31 Hopfield, J.J. *Phys. Rev.* **1958**, *112*, 1555–1567.

32 Kuhn, H., Försterling, H.D., Waldeck, D.H. *Principles of Physical Chemistry*, 2nd ed. Wiley: Hoboken, NJ, **2009**. ISBN 978-0-470-08964-4.

33 Valeur, B., Berberan-Santos, M.N. *Molecular Fluorescence: Principles and Applications*, 2nd ed. Wiley VCH: Weinheim, **2012**.

34 Shalabney, A., George, J., Hutchison, J., Pupillo, G., Genet, C., Ebbesen, T.W. *Nat. Commun.* **2015**, *6*, 1–6.

35 Rider, M.S., Johnson, E.C., Bates, D., Wardley, W.P., et al. Strong coupling in molecular systems: a simple predictor employing routine optical measurements. Nanophotonics. **2024**,13(14): 2453–2567. https://doi.org/10.1515/nanoph-2023-0879.

36 Marshall, T.W. *Proc. R. Soc. A* **1963**, *A276*, 475–491.

37 Boyer, T.H. *Phys. Rev. D* **1975**, *11*, 790–808.

38 Ibison, M., Haisch, B. *Phys. Rev. A* **1996**, *54*, 2737–2744.

39 Boyer, T.H. *Atoms* **2019**, *7*, 29.

40 Boyer, T.H. *Eur. J. Phys.* **2021**, *43*, 025201.

41 Cetto, A.M., de la Peña, L. *Entropy* **2022**, *24*, 1717.

42 Riek, C., Seletskiy, D.V., Moskalenko, A.S., Schmidt, J.F., Krauspe, P., Eckart, E., et al. *Science* **2015**, *350*, 420–423.

43 Leuchs, G., Villar, A.S., Sánchez-Soto, L.L. *Appl. Phys. B* **2010**, *100*, 9–13.

44 Leuchs, G., Hawton, M., Sánchez-Soto, L.L. *Physics* **2023**, *5*, 179–192.

3

The Role of Cavity in Polaritonics: Plasmonic Nanoparticles, Self-hybridized Polaritons, and Casimir Self-assembly

Oleg V. Kotov and Timur O. Shegai

Department of Physics, Chalmers University of Technology, Gothenburg, Sweden

This chapter deals with the role of "cavity" in the formation of polaritonic states. We use the term cavity in the sense of an object that is able to provide confinement of electromagnetic radiation and support a set of optical modes (or quasinormal modes [QNMs]). Thus, our treatment includes both open and closed cavities. The optical modes of the cavity are assumed to interact with material excitations (of either electronic or vibrational origin), and provided the interaction is strong enough, to reach the polaritonic regime. In this chapter, we specifically consider plasmonic nanoparticles, planar Fabry–Pérot microcavities, and nanostructured resonant materials as contemporary ways to create polaritons. In all of these realizations, the role of the corresponding cavity is to provide confinement of electromagnetic radiation to enable its strong interaction with material excitations. In practice, it is extremely challenging to confine electromagnetic radiation ideally, for which reason at least a tiny connection with the external field usually remains. Through this connection, the cavity can exchange energy with the environment and thus be both excited and radiationally decayed. A wide range of potential implementations, in principle, satisfy such a view of a cavity. It is precisely because of such a diverse palette of possible realizations, that we pose questions about the role of the cavity in polariton formation. Furthermore, we are interested in such key cavity characteristics as quality factor, mode volume, as well as cavity openness, influencing this process.

In this chapter, we consider two specific examples of open cavities – plasmonic nanoparticles, and nanostructured resonant materials supporting self-hybridized polaritons without external mirrors. On the other hand, the last part of this chapter is devoted to planar Fabry–Pérot microcavities, which are more standard from the polaritonic and quantum optics point of view. Here, we are interested in them because of the recently discovered ways to construct this type of cavity by using the Casimir effect. Furthermore, we briefly discuss the relevance of the Casimir effect for polaritonics and possible ways in which polaritonics and Casimir communities can learn from each other. Finally, we note that this chapter is partially based on the recent review on the topic by Bhuyan et al. [1].

We start by very briefly recalling that a typical way to quantify the light–matter coupling strength, g, is through a scalar product between the transition dipole moment of the material excitation and the vacuum field of the cavity.

Polariton Chemistry: Molecules in Cavities, First Edition. Edited by Joel Yuen-Zhou, Noel C. Giebink and Raphael F. Ribeiro. Published 2026 by John Wiley & Sons, Inc.

$$g = \vec{\mu} \cdot \vec{E}_{vac} \sim \mu\sqrt{\frac{1}{V}} \tag{3.1}$$

Strong coupling is realized when $g > (\gamma_{cav} + \gamma_0)/2$, where γ_{cav}, γ_0 are dissipation rates of the cavity and the emitter, respectively. The strong-coupling condition is challenging to meet for a single quantum emitter, particularly at room temperature, because it requires a combination of high transition dipole moments (special types of transitions) with cavities that are able to strongly localize light (e.g. plasmonic nanoparticles). Hence, the typical solution is to realize strong coupling with many, $N >> 1$, emitters. In an assumption that *all* emitters are coupled to the cavity field with the same strength (which is very often not the case), the collective coupling strength, $g_{col} = \sqrt{N}g$. For sufficiently high N, g_{col} may get substantially greater than g and, hence, it may be more suitable for the realization of the strong- and ultrastrong-coupling regimes, as we shall see in multiple examples below.

3.1 Plasmonic Resonators

Plasmonic resonators, thanks to their ability to localize electromagnetic excitations on deep subwavelength scales, are promising for the realization of polaritons, especially with a relatively small number of emitters and at room temperature. However, before discussing their usefulness for strong light–matter interactions in more detail, it is important to consider several factors influencing their behavior as hosts for electromagnetic excitations. In particular, it is important to appreciate the fact that plasmonic resonators are lossy and open systems, which creates certain difficulties in their theoretical treatment from the point of view of quantum optics. Below, we briefly review several modern approaches toward resolving these issues, after which we provide a historical perspective of recent achievements in the experimental realization of polaritonic states based on plasmonic resonators, including individual plasmonic nanoantennas and ultimately the single-emitter limit. Furthermore, we briefly review photophysical and photochemical processes occurring in plasmon–exciton (plexciton) systems and outlook a few promising research directions.

3.1.1 Light–Matter Interactions Using Plasmonic Resonators and Their Arrays

The description of the light–matter coupling provided in the previous sections works well for low-loss, nonradiative, single-mode cavities, where one can unambiguously define the mode volume V and the number of the coupled emitters N (as Fabry–Pérot cavities typically are considered). In these, the mirrors provide defined boundaries for both the electromagnetic field and the molecular volume, resulting in a collective coupling strength (Eq. 2.1) that scales with the molecular concentration ($\sqrt{N/V}$). In other words, there is a complete basis of discrete normal modes with real eigenfrequencies and the effective mode volume is just related to the physical volume occupied by the mode. Small losses in such resonators are usually described by perturbation of an ideal system, which broadens the eigenstates with a Lorentzian line shape, but leaves unchanged the resonance frequencies and field distributions [2]. Although dielectric cavities (such as Fabry–Pérot) allow to achieve high Q-factors ($\sim 10^6$) their mode volume is basically limited by the diffraction limit ($V \sim \lambda^3$). This limit can be overcome by using another way of enhancing the electromagnetic field, namely by using a plasmonic resonator. Instead of relying on an optical standing wave as in a Fabry–Pérot cavity (diffraction limited), the oscillation of free charge carriers within a plasmonic resonator provides an enhancement of the electromagnetic field. The frequency and mode volume depend on the size and shape of the plasmonic resonator. Furthermore, arrays of resonators can be made where the resonance frequency of individual resonators is matched with the distance between resonators to achieve a constructive interference, and thus further enhancement of the electromagnetic field. Plasmonic resonators can achieve very small mode volumes, on the order of $V \sim 10^{-4}\lambda^3$, but this inevitably entails Ohmic losses and very low Q values ($\sim$10). Such lossy systems have no true eigenmodes, but resonances with a given linewidth embedded in the continuum. However, to describe light–matter interactions at a quantum level, one needs to quantize the electromagnetic modes. This is much more difficult for open lossy systems as compared with the well-defined Fabry–Pérot cavity. Besides, as plasmonic resonators are open, comparable in size with the emitters themselves, and lossy, they are not well-defined in terms of V and N. The proportionality $g_N \propto \sqrt{N/V}$ is, therefore, questionable [3], although *ab initio* calculations show, that as an order of magnitude estimation, Eq. (2.1) may work well even for plasmonic resonators [4, 5]. Nevertheless, such lossy systems can no longer be described by real-valued eigenmodes, but only with complex-valued frequencies $\tilde{\omega}_n = \omega_n - i\gamma_n$, where the imaginary part gives the leakage rate. The Q-factor is defined as $Q = -\mathrm{Re}(\tilde{\omega}_n)/2\mathrm{Im}(\tilde{\omega}_n)$, which is consistent with an energy balance consideration. However, the imaginary part of the eigenmodes values gives a spatial divergence of the fields outside of the resonator and hence a divergence of the mode volume integral [6], which makes the definition of V quite challenging. For similar reasons, the common Purcell factor expression is not applicable for open lossy systems, as it is derived under the assumptions that normal (lossless) modes can be defined for the system, so that the local density of states is written as a sum of them, and this sum is dominated by a single mode [7].

So, plasmonic resonators are open non-Hermitian systems with complex-valued eigenmodes and they cannot be reduced to a single-mode description within the Tavis–Cummings model. Instead, in general, one needs to work with macroscopic quantum electrodynamics (QED) Hamiltonian [8] for multiple emitters, and have the plasmonic system described by the classical dyadic Green's function $G(\vec{r}, \vec{r}', \omega)$. Such a Hamiltonian contains the frequency-dependent light–matter coupling expressed through the Green's function as: $g(\omega) \propto \omega\sqrt{\mathrm{Im}\{G(\vec{r}, \vec{r}', \omega)\}}$ [9]. However, the integral in this function goes from zero to infinity over the entire continuum of frequencies, which makes it too computationally costly to solve [10].

To avoid expensive numerical calculations, one needs to build a model system Hamiltonian represented in terms of discrete bosonic modes, which describes the original system with good accuracy. Nowadays, there are two main approaches to constructing an equivalent model Hamiltonian, which provides a few-mode quantized description of plasmonic resonators. One is based on finding the s QNMs [6, 11] of the resonator with complex-valued frequencies $\tilde{\omega}_n$. Then, the quantized modes are defined as superpositions of the bosonic field operators with coefficients determined by the QNMs. These can be obtained from the poles calculations of the electromagnetic scattering matrix in the complex frequency plane [12]. The resulting quantized modes are orthonormalized to obtain approximate discrete lossy modes. The negative sign of $\mathrm{Im}(\tilde{\omega}_n) = -\gamma_n$, which is demanded by the causality of electromagnetic fields, results in spatial divergence of the QNM fields outside of the resonator, thus, making it difficult to properly define their normalization value. When using the QNMs method, one has to overcome this fundamental obstacle with the help of regularization techniques [6, 13, 14]. More details about the QNMs approach can be found in [14–17].

An alternative approach [18–20] does not require calculation and explicit quantization of the QNMs. Instead, it is based on fitting the full spectral density of a system obtained through classical electromagnetic simulations $J(\omega) \propto \omega^2 \text{Im}\{G(\vec{r}, \vec{r}\,', \omega)\}$. A model system is constructed with a number of lossy and *interacting* modes, which are linearly coupled to the quantum emitter and to an independent Markovian (spectrally flat) background bath. The Markov approximation allows one to use the Lindblad master equation with a standard dissipation term. Furthermore, Fano diagonalization results in a simple analytical solution for the spectral density $J_{\text{mod}}(\omega)$. Thus, it is not just a Lorentzian fitting widely used to quantize plasmonic systems in the quasistatic approximation. Here, the fitting procedure includes the intermode interaction, which allows one to reproduce the Fano-like resonances in the spectral density function. Fano-like resonances are typical for hybrid metallodielectric systems [12] and usually correspond to the interference of localized plasmons with standing-wave modes. So this fitting yields not only the frequencies and dissipation of the discrete modes, and all coupling constants between them and the emitters, but also the intermode interaction constants. This procedure requires the dyadic Green's function calculations as the only input. This approach is a readily implementable and computationally efficient tool to construct a quantum description of light–matter interactions in plasmonic resonators. Note that both described approaches are also useful for modeling cavity-free polaritonic systems with self-hybridized modes, as we discuss in the next section [21].

Plasmonics pave the way toward the strong-coupling regime by confining light at the subwavelength scale using metallic nanoresonators [22, 23]. The main advantage of plasmonic cavities over Fabry–Pérot ones is that they possess small subwavelength mode volumes, which allow for achieving the strong-coupling regime even on a single-molecule level and at room temperature. However, the involvement of free carriers in the metallic components inevitably entails Ohmic losses and a strong reduction of the cavity quality factor.

In the early works preceding the observation of Rabi splitting in plasmonic systems, the coupling of molecular or excitonic resonances to plasmonic nanostructures was usually considered as a "coherent" coupling effect [24–29]. Such an effect in the weak interaction regime is characterized by a Fano-like profile of the scattering spectrum (an asymmetric line-shape caused by two resonances of different linewidth) [30, 31]. However, due to the presence of two peaks in the scattering, this regime can potentially be confused with the true Rabi splitting, which can be reliably recognized only when the splitting substantially exceeds the full-width at half maxima of polaritonic bands or by signs of polariton splitting in absorption or photoluminescence spectra [32–35]. In the coherent coupling regime, the scattering dip is a result of destructive interference between the plasmon and excitonic resonance in the far-field [36]. Unlike in the strong-coupling regime, when the two peaks corresponding to the polariton modes split at a value exceeding the polariton linewidth $(\gamma_0 + \gamma_{pl})/2$, in the coherent coupling regime, this splitting can still be much smaller than the plasmon linewidth $2\gamma_{pl}$. However, only strongly coupled systems support an ultrafast energy exchange between two polariton states (Rabi oscillations), which, for example, has been observed in real time for a plasmon–exciton (plexcitonic) system on a $\sim$ 10 fs timescale [37].

To the best of our knowledge, the first demonstration of plexcitonic strong coupling was performed at room temperature for an ensemble of organic molecules (J-aggregate) covering a silver hole array [38]. After that, a large number of both theoretical and experimental works reported similar values of Rabi splitting in different systems consisting of metallic and molecular components [39–45]. From the biochemistry perspective, it is quite interesting that plexcitonic strong coupling in light-harvesting complexes (namely in purple bacteria) is sensitive to the specific molecular organization and the protein coverage of the pigment molecules [46]. Among numerous manifestations of plexcitonic strong coupling, here, we mention several especially striking examples: ultrafast manipulation of strong coupling in molecular aggregate hybrid nanostructures [47, 48], ultrafast energy transfer between molecular assemblies and surface plasmons [49], and coherent emission from a disordered molecular system [50]. It is important to note that the plexcitonic strong coupling has been obtained in a great variety of systems with different geometries and materials of both excitonic and plasmonic components, including metallic nanoshells and core-shells [41, 51, 52], sharp tips and platelets [53, 54], nanostars [55, 56], and nanostructures of higher complexity [57]. Some works confirmed the strong-coupling regime by observing signs of polariton formation not only in the scattering spectra, but also in the photoluminescence [52, 58], which as discussed in this chapter provides additional evidence.

For any geometry, Ohmic losses reduce the cavity quality factor; however, arranging plasmonic nanoparticles into periodic arrays offers a way to increase the quality factor through the Fano interference of the localized plasmons with the diffractive modes. Owing to surface lattice resonances, plasmonic arrays combined with excitonic layers enable a boost of the excitonic emission to the level of lasing, which has been demonstrated by several groups in the near-infrared and visible ranges using various plasmonic arrays coupled to organic dyes [59–65] (more works can be found in reference [66]). Rabi splitting in plasmonic diffractive arrays coupled to various excitonic layers (organic dyes or quantum dots) has been observed in several configurations [67–71]. An attractive advantage of this type of plasmonic structures is the ability to change the effective mass and the composition of the polariton state by tuning the geometrical parameters of the lattice [68, 69]. In contrast to individual nanoantennas, exciton–polaritons in plasmonic arrays can demonstrate a delocalized spatial behavior, which can be used for coherent energy transfer on the scale of the lattice [72]. Owing to the strong coupling to surface lattice resonances, a high degree of spatial coherence can be obtained even for very exciton-like (80%) polaritons [73]. Such a sharing of the photonic coherence by the excitons is a key step to switch from ordinary lasing to Bose–Einstein condensation of exciton–polaritons [74, 75]. More details about plexcitonic strong coupling in the diffractive arrays can be found in references [76, 77].

3.1.2 Single Plasmonic Resonators

The observations of the plexcitonic strong coupling discussed above were realized in ensembles or periodic arrays of plasmonic nanoparticles. However, such ensembles are not very well suited for studying single molecules or quantum dots. In order to better control the coupling with single emitters and to reduce the contribution of various ensemble mechanisms to the plasmonic resonance broadening, researchers turned their interest toward the coupling of excitonic materials to *single* nanoantennas. The pioneering observations of plexcitonic strong coupling in single plasmonic nanoantennas interacting with J-aggregates were reported by Schlather et al. [78] using plasmonic dimers and Zengin et al. [79] using individual silver nanorods. Later, Zengin et al. [80] reached the strong-coupling regime using an individual silver nanoprism coupled to J-aggregates and estimated from the experimental data that the number of excitons involved in such a plexcitonic coupling is no more than ≈ 80. Furthermore, the nanoprism used in this work compresses the mode volume enough to achieve comparable to photonic crystals or microring resonator cavities strong-coupling figure-of-merit $Q/\sqrt{V}$. In later works by Wersäll et al. [81, 82] the plexcitonic strong coupling in a single plasmonic nanoprism embedded in J-aggregates was demonstrated not only by splitting in the dark-field scattering spectra but also by temperature-dependent photoluminescence spectroscopy.

Nanoparticles of several tens of nanometers in size, on the one hand, create a large electric field enhancement near the surface, but on the other hand, as the light penetration depth in the metal is comparable to nanoparticle size, most of the mode energy is concentrated inside the particle [7, 80]. That is why plasmonic dimers have become so popular – they are able not only to enhance the electric field in the gap region but also to reduce the mode volume by localizing the mode within the gap region both inside and outside of the metal [35, 78, 83]. The bowtie-shaped dimers are among the most effective and convenient gap nanoantennas from the experimental point of view. In the gap region, they are essentially metallic nanomirrors with a width of about the size of the gap and arranged perpendicular to the substrate. Such a configuration provides exceptionally high electric field enhancement in the gap while using normal incident light for excitation. Moreover, this configuration is highly selective to light polarization, with the light polarized along the bowtie axis providing the highest enhancement. Plasmonic nanocavities supporting gap plasmon modes can also be realized using the nanoparticle-on-mirror strategy. These can be implemented using various types of metallic particles, including nanocubes, nanorods, nanoprisms, nanospheres, etc., and will be discussed in detail below.

3.1.3 The Single-emitter Limit

Different realizations of plexcitonic strong coupling with single plasmonic nanoantennas paved the way to approach the single-emitter limit. Its achievement opens up new opportunities in the creation and control of quantum states at the single-photon level, which has great prospects for the creation of single-photon sources [90], quantum optical circuits [91], and ultralow-power lasers [92, 93]. In contrast to the molecular aggregates with a large number of emitters, single molecules have transition dipole moments of a few Debyes, which therefore requires even stronger compression of the field in a nanocavity to reach the strong-coupling regime. Quantum dots, usually having an order of magnitude larger transition dipole moments [77], in this sense, are more promising and convenient as a platform for the realization of strong coupling with single emitters.

The main experimental challenges in reaching the single-molecule strong-coupling limit are twofold – first, on the nanocavity side, one needs to ensure strong enough confinement of the electromagnetic field without substantially increasing the losses; second, on the molecule side, one needs a molecule with a strong enough transition dipole moment, which could physically fit inside the nanocavity and, ideally, be oriented along the direction of the maximal field. Moreover, the plasmonic resonance and the molecular resonance must be spectrally matched. Such a scenario is rather hard to fulfill but, on a general note, requires a combination of some plasmonic nanogap cavities (such as plasmonic dimers or particle-on-mirror) with strongly resonant organic chromophores. Remarkably, the conditions for reaching the single-molecule strong-coupling regime are similar to the optimal configuration of single-molecule surface-enhanced (resonant) Raman scattering (SERS) [94–96]. Therefore, it is probably not very surprising that single-molecule strong-coupling and single-molecule SERS were realized in similar setups, and in general, these disciplines are closely related and develop in parallel nowadays, although the latter is considerably more mature [97].

Following this line of arguments, probably the first work noticing the relevance of SERS to single-molecule plexcitonic strong coupling was the report by Itoh et al. [98], where Rhodamine-6G molecules in the nanogap of a silver nanoparticle dimer were studied by means of Raman and dark-field scattering. Later, record-high numbers in terms of the plasmonic gap miniaturization have been set in a series of works using nanoparticle-on-mirror geometry. First, for a gold nanosphere on a gold mirror forming a 0.9-nm gap with ultra-confined mode volume ($\approx 40\ \text{nm}^3$) Chikkaraddy et al. [99] demonstrated the plexcitonic strong coupling down to a single molecule of methylene blue placed in the gap and oriented normal to the metal surface using a barrel-shaped non-resonant host molecule. Notably, the dark-field scattering spectra clearly revealed the effect of the transition dipole moment orientation, as a pronounced Rabi splitting was observed only when molecules were oriented perpendicular to the metal surface. Furthermore, based on a similar nanoparticle-on-mirror geometry (Figure 3.1a), the formation and disassembling of atomic defects were demonstrated in the gold surfaces forming the 0.9-nm gap by sufficiently strong laser irradiation [100]. Later, plasmonic nanocavities with atomic defects or protrusions were shown

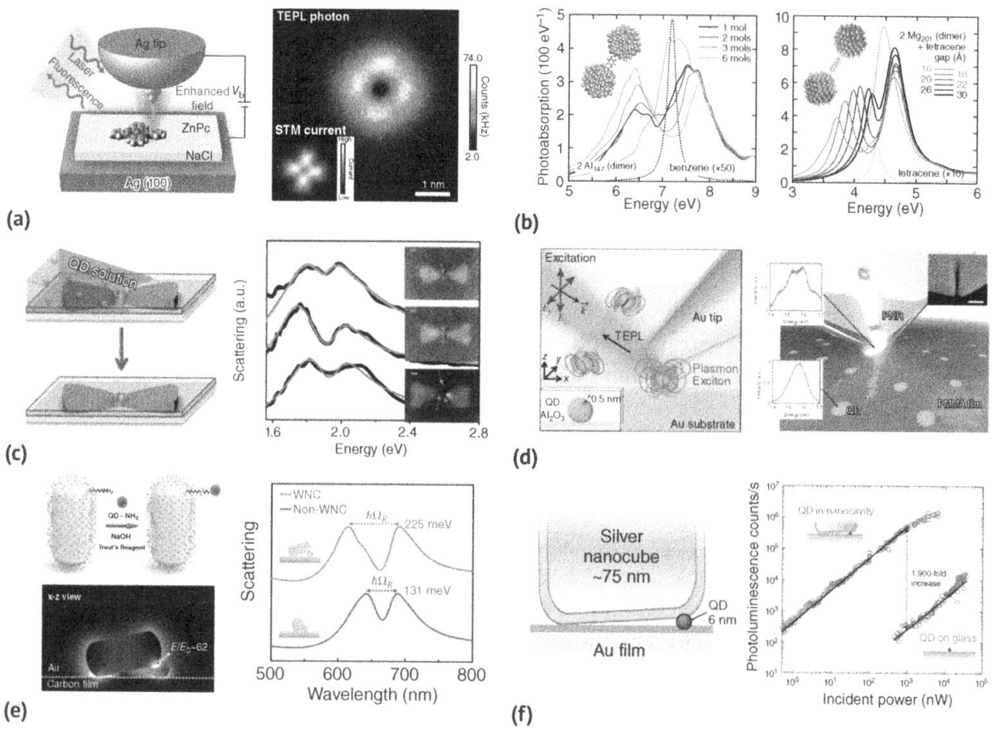

Figure 3.1 Single-emitter Rabi splitting at room temperature. (a) Schematic of scanning tunneling microscopy setup with the atomistic protrusion at the silver tip giving the enhancement of photoluminescence from a molecule under the tip (left). Simultaneously recorded tip-enhanced photoluminescence photon image and scanning tunneling microscope photon image (inset) of a single zinc phthalocyanine molecule on three-monolayer-thick NaCl/Ag(100). [84]/Springer Nature. (b) Ab initio calculations of photoabsorption spectra of Al_{147} dimers with a 10 Å gap coupled to a various number of benzene molecules showing the ultrastrong coupling even for the single molecule (left). The same for Mg_{201} dimers with a single benzene molecule for various dimer gaps (right). [5]/American Chemical Society/CC BY 4.0. (c) Schematic of the interfacial capillary force-assisted method for positioning quantum dots in the gap of bowtie structures (left). Scattering spectra of silver bowtie antennas with one, two, and three CdSe/ZnS quantum dots in the gap showing record values of Rabi splitting (right). Reproduced from [85]/Springer Nature/CC BY 4.0. (d) Schematics of a scanning probe microscopy setup to study quantum dots in the strong-coupling regime (left) and slit-like scanning probe combining the near-field spectroscopy technique with the ideas of gap nanoantennas (right). [86, 87]/American Association for the Advancement of Science - AAAS/CC by 4.0. (e) Schematic (top left) of the self-positioning of quantum dots at the sharper corners near the ends of gold nanorods, where the electric field (bottom left) and Rabi splitting (right) are maximized. [88]/American Chemical Society. (f) Schematic of a single quantum dot in the gap between a silver nanocube and a gold film (right). The location of the quantum dot near the corner of the nanocube, where the field enhancements are the largest, gives a 1900-fold increase in room-temperature emission (right). Adapted from [77].

to be useful for single-molecule photoluminescence sub-nanometer imaging. The tip-enhanced photoluminescence (TEPL) technique can provide a similar resolution (~ 8Å) as a scanning tunneling microscope can have (Figure 3.1a) [84].

As described above, the single-molecule strong coupling is challenging to realize experimentally. Thus, a considerable effort to better understand single-molecule plexcitonic systems is attempted theoretically through numerical simulations [4, 101, 102]. In particular, although single-molecule SERS and single-molecule strong coupling have been experimentally realized, reaching the *ultrastrong* coupling regime with a single molecule remains challenging. To fill this gap, recent time-dependent density functional theory calculations have shown that a single organic molecule can reach ultrastrong coupling with a plasmonic dimer consisting of a few hundred atoms [5] (Figure 3.1b). Experimental realization of such a strategy remains to be demonstrated and possibly could involve electron energy loss spectroscopy, as suggested previously [103].

In parallel with single-molecule strong coupling, great progress toward single-emitter plexcitonic strong coupling was achieved using individual colloidal quantum dots. As mentioned above, quantum dots possess substantially higher transition dipole moments in comparison to molecular emitters. Therefore, single-emitter strong coupling, in this case, does not necessarily require enormous compression of optical modes and alignment of the transition dipole moment with the electromagnetic field. Pioneering works reporting such a strategy include Santhosh et al. [85] and Hartsfield et al. [104]. Hartsfield et al. used a gold nanosphere on a glass substrate coupled to a single CdSe/ZnS quantum dot but did not obtain a pronounced Rabi splitting. Santhosh et al. managed to place a single CdSe/ZnS quantum dot in the 20-nm gap of a bowtie silver nanoantenna, using a polymeric mask defined by electron beam lithography and interfacial capillary force, and achieved a Rabi splitting of up to 240 meV (Figure 3.1c). To our knowledge, this result is a record one for single-emitter plexcitonic strong coupling to date. The quantum nature of the emission from this system was confirmed recently by second-order photon correlation measurements [105]. In addition, the contribution of dark plasmon modes in the plexcitonic strong coupling using this system was revealed recently by means of electron energy loss spectroscopy [106].

After these initial works, strong coupling between a single quantum dot and a gap plasmon has been realized in different configurations of nanogaps. Leng et al. [107] observed weak coupling (the Purcell effect), intermediate coupling (Fano interference), and strong coupling (Rabi splitting) at room temperature for a single-colloidal quantum dot in the gap between a gold nanoparticle and a silver film. Gross et al. [87] and Park et al. [86] combined the scanning probe microscopy technique with the nanogap formation between a scanning tip and a gold mirror, or at the apex of slit-like scanning probe, which enabled better control and imaging of the strong coupling between the gap plasmons and single quantum dots (Figure 3.1d). Li et al. [88] used the surface functionalization of gold nanorods with cetyltrimethylammonium bromide (CTAB) molecules to provide better control of the position, spacing, and quantity of the quantum dots involved in the coupling. The advantage of this approach is in the self-positioning of quantum dots at the sharper corners near the ends of the nanorods, where the electric field and coupling strength are maximized. It is based on the fact that the fewest CTAB molecules attach at the sharper corners near the ends of the nanorods resulting in higher possibilities for binding the surface-functionalized quantum dots on these sharper corners (Figure 3.1e). Similar nanogap structures were also used to enhance the brightness of single-photon emission from nitrogen-vacancy centers in a diamond [108] or to increase room-temperature single-photon emission from quantum dots [89] (Figure 3.1f). More details on plexcitonic strong-coupling applications and the most complete list of different plexcitonic systems can be found in recent reviews [109–113].

3.1.4 Plexcitonic Photophysics and Photochemistry

Strong coupling of molecules with light in a cavity can be used to modify photochemical reactions. Pioneering experiments in this field used planar metallic cavities since they are easier to fabricate, and strong coupling with molecules at room temperature does not require a high-Q cavity mode [119, 120]. Later on, to reach the single-molecule limit, or at least to work with a much smaller number of molecules, single plasmonic nanoantennas were used [80, 99, 100, 114], allowing on the one hand, to achieve a smaller mode volume required for the strong coupling with a small number of molecules, and on the other hand, providing greater access to monitor and control the system. Furthermore, this potentially allows for the resolution of the large N problem, which is typical for planar Fabry–Pérot microcavities and which may hamper observation of photophysical and photochemical processes [121–123]. Some plasmonic nanoantennas provide an additional advantage over microcavities or plasmonic nanoparticle arrays: they can be chemically synthesized in large numbers and exist as colloidal suspensions in a solution [124, 125].

It should be noted that contemporary experiments in this field are aimed at proving the principle of polariton photochemistry or photophysics, and not necessarily at obtaining the best characteristics of the corresponding phenomena. In this regard, the use of plasmonic nanoparticles may influence various photochemical effects even in the weak coupling regime through a range of mechanisms. For instance, if one needs to speed up a reaction, plasmonic nanoparticles can be used as sources of hot electrons (see [126]). The way chemical reactions can be slowed down is highly dependent on their mechanisms. For example, the suppression of photobleaching of organic chromophores observed in [114] in the strong-coupling regime can be alternatively achieved by several other methods, ranging from straightforward use of oxygen scavenger reagents [127] to quenching of long-lived triplet states [128–130] and surface-enhanced spontaneous emission [131–133]. For the sake of clarity, we stress that in this section, we will focus specifically on the role of plexcitonic strong coupling in photophysics and photochemistry.

One of the early experiments in the field of plexciton-polariton photochemistry was carried out by Munkhbat et al. [114]. The experiment demonstrated a beneficial impact of plexcitonic strong coupling on one of the most important classes of photochemical processes – the photo-oxidation (photobleaching) reaction (Figure 3.2a). Its mechanism involves photodynamic interactions between the excited triplet state of dye molecules and atmospheric triplet oxygen ($^3O_2$). Due to the long lifetime of the triplet state, the excited molecules have a higher probability of interacting with environmental oxygen and thus undergo photobleaching. The interaction with a cavity mode can significantly change the relaxation pathways and, accordingly, photobleaching both in weak coupling (Purcell regime) and strong coupling (Rabi splitting). However, owing to the delocalized and coherent nature of polaritons, only in the strong-coupling regime the effect becomes collective and the relaxation pathways can be modified to a very large extent. The formation of polaritons opens up a new relaxation pathway that avoids falling from the excited singlet state into the long-lived triplet state. Due to the partial plasmonic nature of polariton states, their lifetimes are much shorter (~10 fs) than the typical singlet-triplet transition rate (γ_{ISC} from ~10 ns to ms), which provides a fast relaxation from the excited to the ground state, bypassing the triplet (Figure 3.2b). To make this mechanism most effective, one needs, on the one hand, to maximize the relaxation rates from the upper polariton to the ground state (γ_{UP}) and from the incoherent state (exciton reservoir) to the lower polariton ($\gamma_{D\to\mathrm{LP}}$), but on the other hand, one needs to minimize the relaxation from the upper polariton to the exciton reservoir ($\gamma_{\mathrm{UP}\to D}$) and the resonant excitation of the upper polariton itself (γ_{exc}). Based on this, one can expect that red-detuned particles, owing to the reduced plasmonic component of the upper polariton, give lower γ_{exc} and higher $\gamma_{D\to\mathrm{LP}}$, while blue-detuned

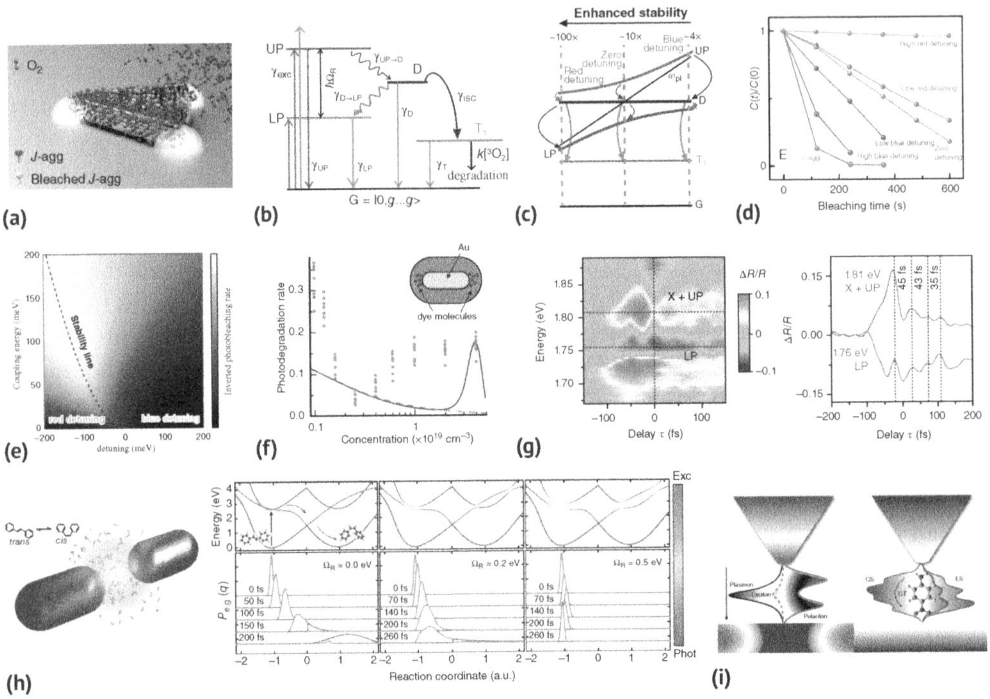

Figure 3.2 Plexcitonic photochemistry and photophysics. (a) Schematic of the photobleaching reaction of *J*-aggregates strongly coupled to a silver nanoprism. (b) Photobleaching mechanism of a strongly coupled system and transitions between various states: polaritons provide a fast relaxation from the excited to the ground state, bypassing the long-lived triplet. (c) Schematic of the photobleaching process as a function of plasmon–exciton detuning. The transparency of arrows shows high/low probability of the corresponding transition, demonstrating that red-detuned particles are more stable than blue-detuned ones. (d) The relative change in the concentration of active organic molecules for various plasmon–exciton detuning, compared with the uncoupled bare *J*-aggregates. [140]/American Association for the Advancement of Science - AAAS. (e) Inverted population of the triplet state shows the optimal balance between the detuning and coupling energy for which the photobleaching suppression is maximized (the stability line). [115]/American Chemical Society. (f) The photodegradation rate of dye molecules vs. molecular concentration. The dots are experimental data obtained from different samples. The solid and dashed lines are full and simplified theoretical curves, respectively. Reproduced from [116]/American Chemical Society. (g) Coherent ultrafast manipulation of the coupling energy in strongly coupled *J*-aggregate/metal hybrid nanostructures by controlling the exciton density on a 10 fs timescale. [37]/Springer Nature. (h) The left panel shows a schematic of molecules coupled to a plasmonic nanocavity and a sketch of the photoisomerization reaction between a trans- and cis-isomer. The right block of six panels shows a calculation of the photoisomerization suppression under strong coupling for single molecules. The top row in it shows how the formation of polaritons provides new minima in the potential energy surfaces and even barriers at larger coupling strength. The bottom row shows the corresponding trapping of the nuclear wavepacket, which means that the photochemical reaction is slowed down. [117]/Springer Nature/CC BY 4.0. (i) Coupled-cluster calculation of plexcitonic strong coupling reveals the modification of the ground-state (GS) electronic density caused by the molecule–plasmon interaction. The direction of charge transfer (CT) is along the arrow on the right panel. [118]/American Chemical Society.

particles, *vice versa*, provide higher γ_{exc} and lower $\gamma_{D \rightarrow LP}$ (Figure 3.2c). Note that this effect does not require minimization of the plasmonic losses. On the contrary, due to them, the lifetime of the excited state is significantly reduced, which makes it possible to avoid falling into the triplet state and the corresponding degradation of the molecules. However, too high plasmon losses are also undesirable, since instead of the strong-coupling regime, they will quench the photoluminescence of dyes into the metal.

The described mechanism of dye stabilization was demonstrated by using a set of separate plasmonic silver nanoprisms providing different initial values of strong coupling to excitons in a *J*-aggregate (Figure 3.2a). A 100-fold stabilization of organic dyes was shown for the red-detuned nanoparticles (Figure 3.2d). Based on the observation that strongly coupled molecules are significantly more stable than uncoupled ones the authors concluded that the role of the hot-electron effects in this case is negligible. Indeed, hot electrons should additionally destabilize the molecules, and for the relatively large size of nanoparticles used in this experiment (~100 nm) hot electrons are more likely to thermalize inside the metal particle before they reach the surface.

A theoretical description of the observed effect on the basis of pioneering works by Agranovich et al. [134, 135] and Litinskaya et al. [136] was also presented. In a later theoretical study [115], the optimal balance between the values of red detuning and Rabi frequency for which the photobleaching suppression is most pronounced was found (see the stability line in Figure 3.2e). Subsequently, similar effects were demonstrated experimentally with plasmonic nanoparticles and Fabry–Pérot cavities strongly coupled to various excitonic systems. A reduction of photodegradation of an organic semiconductor polymer sandwiched between silver mirrors, providing a giant Rabi splitting of ~1.0 eV, was observed [137] for a very important material for organic photovoltaics – 2,5-poly(3-hexylthiophene) (P3HT). Normalizing the reduction in the oxygen flow caused by the top silver mirror, the authors obtained a three-fold reduction of the photodegradation rate [137]. Furthermore, enhanced delayed fluorescence was observed in tetracene single crystal strongly coupled to plasmonic nanoparticle arrays [138]. The resonance behavior (Figure 3.2f) with the dye molecule concentration was observed for the photobleaching rate of the dye molecules strongly coupled to gold nanorods [116].

From the photophysics side, Vasa et al. [37] for the first time observed real-time ultrafast Rabi oscillations in a plexcitonic strong-coupling system and demonstrated coherent manipulation of the coupling energy by controlling the exciton density on a 10 fs timescale (Figure 3.2g). Ultrafast pump-probe studies reported a femtosecond energy transfer between excitons and plasmons [124], and the relaxation pathways in plexcitons [139–141]. Particularly, Finkelstein-Shapiro et al. [139] showed that the plexcitonic dynamics beyond a few femtoseconds must be considered in terms of hot-electron distributions instead of lower and upper polariton branches. Using femtosecond-transient absorption spectroscopy, Park *et al.* [141] demonstrated that due to the additional energy transfer channels both the upper and lower polaritons have shorter lifetimes than the bare excitons in a plexcitonic system consisting of a perovskite and a plasmonic nanoparticle lattice. A similar result was observed recently for the molecular photoswitches strongly coupled to anisotropic plasmonic nanoantennas [142].

Strong coupling can also provide protection of emitters from quenching even in close proximity to a metal nanoparticle [143], giving rise to the "quenching of quenching." Unlike isolated nanoparticles, which usually quench emitters by their decay at distances less than 10 nm, plasmonic nanocavities, owing to the mode hybridization, suppress the emitter's decay into nonradiative channels and promote the re-emission of its energy.

Regarding the interpretation of the photoluminescence spectra of the polariton system, there were several suggestions about the nature of the high-energy emission peak. In some works, it was concluded that it corresponds to the upper polariton emission [58], while others argued that it arises from the uncoupled molecules [134, 136] or dark polaritons [144]. For the case of plexcitons, the works by Wersäll et al. [81, 82] contribute to this discussion by performing correlative dark-field and photoluminescence spectroscopy measurements on the same individual plasmon-molecule hybrid nanostructure. The unusual temperature dependence of polaritonic spectra helped the authors to attribute the lower energy photoluminescence peak to the lower polariton, whereas the higher energy peak was assigned to uncoupled molecules and incoherent states [82]. This is in agreement with observations made in organic microcavity polaritons [145].

We now turn our attention to the theoretical progress in organic polaritons and specifically plexcitons. First theoretical studies of exciton-polaritons did not address specifically the regime of plexcitonic strong coupling but rather focused on more general models, mostly Jaynes–Cummings or Tavis–Cummings, considering single-mode lossless cavities and treating organic molecules as two-level systems [146]. Such an approach suffices to estimate the Rabi splitting but does not allow calculating photochemical processes that require taking into account the internal molecular degrees of freedom. The first attempts to account for the rovibrational molecular states were based on the usage of effective decay and dephasing rates [147, 148]. Subsequently, a number of works [117, 118, 149–164] based on microscopic theories of organic molecules and their strong coupling to cavity modes that contain almost full descriptions of nuclear, electronic, and photonic degrees of freedom came out. Some of these works predict not only a modification of molecular structure in the strong-coupling regime [149–151] but also became the first works explaining the influence of strong-coupling on photochemical reactions [117, 152, 153].

In particular, Galego et al. [117] predicted the suppression of photoisomerization reaction for molecules strongly coupled to quantized light fields and were probably the first who considered it for plasmonic nanoantennas as well. They showed that the formation of polaritons provides new minima in the potential energy surfaces in which the excited nuclear wavepackets can be trapped, such that the photochemical reaction will be slowed down. When the coupling becomes even stronger, a barrier, much larger than the thermal energy, appears and the photoisomerization process becomes almost completely suppressed (Figure 3.2h). Moreover, this potential energy barrier becomes even higher with the increased number of molecules coupled to the light mode, thus, giving rise to a "collective protection" effect. The collective nature of polaritons leads to the formation of a polaritonic "supermolecule" involving the degrees of freedom of all molecules and allowing to trigger a many-molecule reaction by just a single photon [165], which circumvents the second law of photochemistry (also known as the Stark–Einstein law). However, only in their later work, a more accurate model for plasmonic nanoantennas was built, which accounts for the plasmon-induced dissipative processes [166]. The most complete to-date theory of plexcitonic strong coupling, which nonperturbatively treats the molecular electronic density relaxation upon polariton formation together with plasmon-molecule correlation, has appeared only recently [118]. It extends the QED-coupled-cluster method [167, 168] already applied to optical cavities, to realistic plasmonic nanocavities, thus, providing the full quantum description of both molecule and plasmonic subsystems. Particularly, the modification of the ground-state electronic density caused by the molecule–plasmon interaction was demonstrated for the realistic molecule and plasmonic nanocavity (Figure 3.2i), which is beyond the standard Jaynes–Cummings picture.

In concluding this section, we would like to bring the reader's attention to several promising examples of future directions in the field of plexcitonic photophysics and photochemistry. A recent proposal [169] and its experimental realizations [170, 171] of molecular frequency optomechanical upconversion in plasmonic nanocavities can be potentially improved by reaching the vibrational strong-coupling regime in these systems (or alternatively phonon-polaritons in solid-state systems). Furthermore, recent studies of strong coupling of coherent phonons to excitons in different semiconductors [172, 173] can be enriched by coupling to plasmonic systems. Additionally, the influence of strong coupling between a plasmonic resonator and organic molecules on hot carrier generation is an important task, as usually these two processes are considered as separate contributions, and often, to demonstrate the role of strong coupling, researchers strive to minimize the hot carrier contribution. The first theoretical studies on this subject have just begun to appear [174]. In connection with the development of quantum technologies, it could be interesting to observe the influence of the strong-coupling regime on the photon entanglement in plasmonic lattices [175]. Finally, let us note a strong-coupling-induced symmetry breaking which can boost nonlinear optical processes [176, 177].

3.2 Self-hybridized Polaritons

Among the open non-Hermitian systems, including plasmonic resonators considered above, the *self-hybridized* cavities have recently attracted special attention. Thanks to the proper size and sufficiently high contrast of the refractive index with the environment, such systems enable the hybridization of their own material resonance(s) and volumetric optical modes, thus, combining resonator and emitter subsystems within the same object. In this case, self-hybridized or cavity-free polaritons arise, with the Rabi splitting approaching (but not exceeding) the so-called bulk polariton splitting limit. The latter was pioneered by Hopfield in 1958 who studied bulk polaritons in 3D continuous semiconductors [178] (see also Mills and Burstein [179]). The bulk polariton splitting is a function of the material resonance only (in the Lorentzian description determined in terms of oscillator strength – f, plasma frequency – ω_p, and background permittivity – ε_∞) and does not depend on the parameters of the cavity, $g_b = \frac{\omega_p}{2}\sqrt{\frac{f}{\varepsilon_\infty}}$. When the resonant material is confined in space, the splitting and detuning start to depend on the geometrical parameters. The calculation of such systems requires QNMs solutions obtained from the poles of the scattering matrix in the complex frequency plane [21], as we discussed above.

Self-hybridized polaritons were first considered theoretically for spherical resonant materials [180]. Around the same time, the first experimental observation of such polaritons was reported by Takazawa et al. [181] using the exciton–polaritons in organic nanofibers. This topic was further developed by Liao et al. [182] who may have been the first to demonstrate room-temperature exciton–polariton lasing from self-assembled organic nanowires. However, in these works, the authors did not focus on the self-hybridized nature of the polaritons. Only in recent years, have open systems attracted the attention of researchers and many different implementations of self-hybridized polaritons have appeared within different geometries, material platforms, and frequency ranges. Among a large number of such works, we note realizations based on transition metal dichalcogenides [183–185], perovskites [186, 187], and organic molecules [188–191] realizing exciton–polaritons in the visible and near-infrared range, as well as polar crystals such as hexagonal boron nitride (hBN) with phonon-polaritons in the mid-infrared range [192] (Figure 3.3a), organic waveguide gratings with vibro-polaritons in the THz range [193], and even in plasmonic systems with plasmon–interband interactions in the ultraviolet range [194]. Moreover, 3D crystals of plasmonic nanoparticles can reach the deep strong-coupling regime (the coupling strength exceeds the plasmon frequency) when the plasmonic nanoparticles are ~10 times larger than the interparticle gaps [195] (Figure 3.3b).

One interesting example of self-hybridized polaritonic systems existing in natural environments is vibrational Mie self-polaritons in ~1–10 μm water droplets, which were recently shown theoretically and experimentally to reach the ultrastrong-coupling regime [21, 196] (Figure 3.3c). Interestingly, such droplet sizes are encountered in natural fogs, mists, and clouds and could therefore be present in atmospheric science measurements. Indeed, Arnott et al. [197] have measured

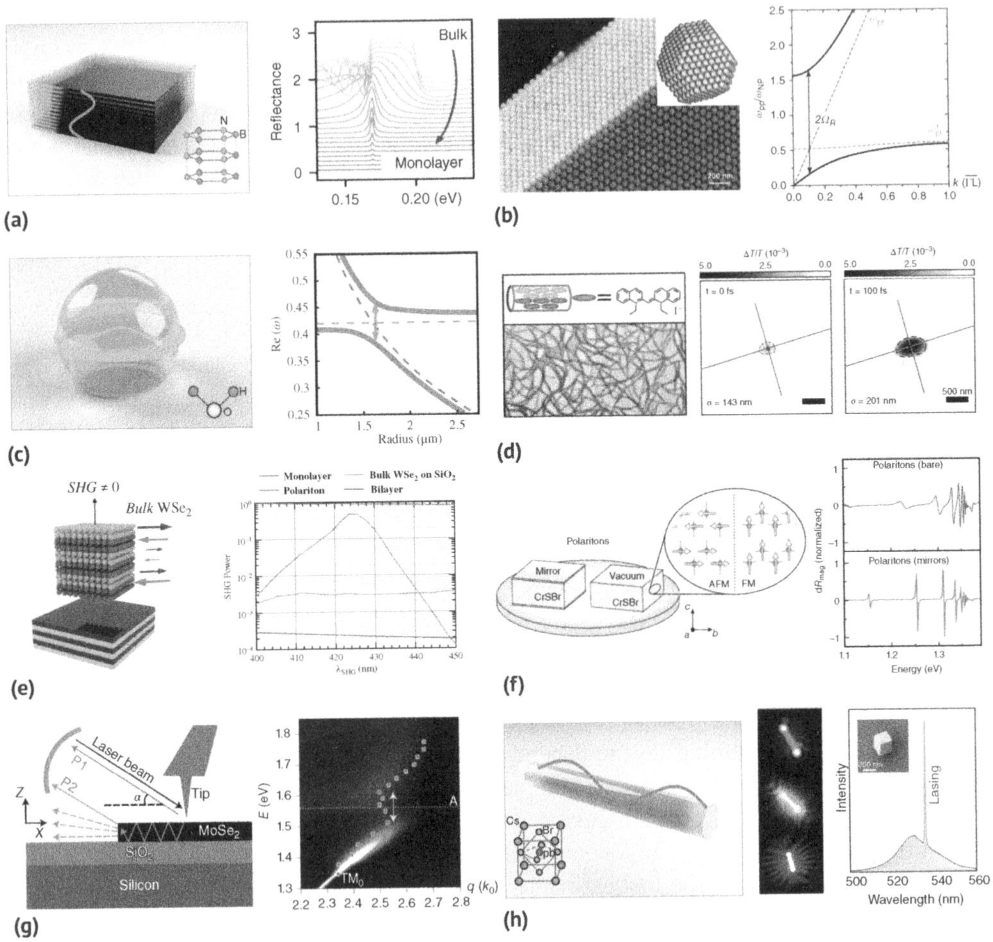

Figure 3.3 Self-hybridized polaritons in open cavities. (a) Schematic of an hBN slab with phononic self-polaritons (left) and reflectance spectra of an optically thick hBN on a polystyrene substrate. The traces for different thicknesses are offset for clarity (right) Left - [21]/AIP Publishing/CC BY 4.0. Right - [192]/American Chemical Society. (b) Edge view and schematic of a face-centered cubic crystal consisting of gold nanoparticles (left) and the plasmon–polariton dispersion of it with a large Rabi frequency ($\Omega_R = 1.4\,\omega_{pl}$) corresponding to the deep strong-coupling regime (right). [195]/Springer Nature. (c) Schematic of a water droplet with vibrational Mie self-polaritons (left) and the splitting of Mie TM-modes for a water sphere depending on its radius (right). [21]/AIP Publishing/CC BY 4.0. (d) Schematic and electron-microscopy images (scale bar 100 nm) of self-assembled nanotubular *J*-aggregates supporting self-hybridized exciton-polaritons (left) and femtosecond pump spot spatiotemporal dynamics showing the typical scale of enhanced exciton transport (right), here the dotted line shows the radial Gaussian standard deviation (σ). [190]/Springer Nature/CC BY 4.0. (e) Schematic of the resonant uncompensated bulk second-harmonic generation (SHG) in WSe_2 due to an asymmetric distribution of nonlinear polarization caused by the coupling to a photonic cavity (left) and the corresponding SHG signal for 140 nm-thick WSe_2 compared to its monolayer and bilayer (right) [210]/American Chemical Society. (f) Schematic of CrSBr magnetic crystals with and without highly reflective mirrors (left) and calculations of the magnon-induced reflectance changes in the bare and closed-cavity cases. (g) Schematic of the experimental setup (left) and calculated dispersion color map with squares indicating experimental points (right) from the observation of self-hybridized exciton-polaritons in $MoSe_2$ slabs beyond the light line. [212]/Springer Nature. (h) Schematic of a perovskite $CsPbBr_3$ nanowire with excitonic self-polaritons (left), optical images in the lasing mode operation of $CH_3NH_3PbX_3$ [X = I (red), Br (green), Cl (blue)] perovskite nanowires (center), emission spectra for a subwavelength $CsPbBr_3$ nanocube (right) [21, 186, 213, 214]/AIP Publishing/CC BY 4.0/Springer Nature/American Chemical Society.

infrared spectra of laboratory clouds and observed peculiar features in the 3000–3700 cm^{-1} range (corresponding to symmetric and antisymmetric O–H stretching of liquid water), although without explicitly interpreting these peculiarities in terms of self-hybridized Mie polaritons in water droplets. Nevertheless, such types of droplets could potentially find use as polaritonic platforms or as cavities for reaching strong coupling with various types of admixed molecules or dissolved ions, since higher order whispering gallery modes (WGM) in these water droplets possess exceptionally high Q-factors, as has been recently demonstrated [198, 199].

One may ask the question whether strong coupling in self-hybridized open cavities could potentially be promising in terms of modification of material properties, such as polaritonic chemistry and exciton transport [200]. In this regard, it has recently been proposed that polaritonic chemistry effects should be referenced with respect to bulk polaritons, since such polaritons exist even in the absence of any confinement [201]. It has also been shown that open cavities (slabs) do not reach the same level of vibrational density of states as mirror-based Fabry–Pérot microcavity polaritons do (but rather remain close to the level of bulk polaritons), suggesting that their effect on polaritonic chemistry would be minor [201]. Thus, open cavities could potentially influence polaritonic chemistry only if they support high Q-factor modes.

In terms of exciton transport, open cavities could provide an additional benefit in comparison to ordinary mirror-based cavities. In particular, mirror-based Fabry–Pérot cavities, due to the highly reflective external mirrors, achieve weakly-broadened optical modes and therefore reach the strong-coupling regime easier than their cavity-free counterparts. However, the presence of external mirrors, in addition to complicating the fabrication, causes difficulties with charge injection and light out-coupling in prospective polaritonic devices. Open cavities with moderately confined photonic states, on the one hand, enable the realization of strong coupling and, on the other hand, help overcome these obstacles. This makes them potentially interesting for future optoelectronic device applications. For instance, it has been predicted that exciton transport in organic materials can be enhanced when the molecules are strongly coupled to an electromagnetic mode [202, 203]. The delocalization of the exciton–polariton modes helps them bypass the disorder and other imperfections in organic semiconductors and thus overcome exponential suppression of the transmission properties. Remarkably, this prediction was confirmed for organic layers not only in an artificial cavity with external mirrors [204] or in plasmonic nanoparticle arrays [205], but even in bare nanocylinders of J-aggregates with self-hybridized polaritonic states [190] (Figure 3.3d). In the latter experiment, when the refractive index of the organic material and the external optical environment were matched with oil, no ultrafast exciton–polariton transport was observed. That confirms the crucial role of polaritons in the exciton transport enhancement, as for their formation sufficiently large optical contrast is needed while oil cannot affect the exciton transport. Furthermore, the excitonic components of exciton–polaritons can be manipulated by an external electric field, which due to strong exciton–photon coupling enables the realization of asymmetric light propagation in organic nanowire waveguides [206].

Exciton transport enhancement in bare organic layers is interesting for device applications. In such systems, the polariton propagation length is typically on the order of hundreds of nanometers (Figure 3.3d). However, by coupling to Bloch surface waves supported by distributed Bragg reflector (DBR) substrates under the excitonic layer, the polariton propagation can be enhanced by up to three orders of magnitude both for disordered organic layers [207, 208] and crystalline semiconductors [209]. This is due to the high Q-factor, long lifetime, and large group velocity of the Bloch surface waves. A polaritonic mode with 80% photonic weight entirely overcomes the disorder and shows a ballistic propagation with distances ~ 100 μm and at a speed of two-thirds of the speed of light.

Furthermore, self-hybridized exciton–polaritons were shown to relax the crystal symmetry selection rules that govern second-order nonlinearities [210]. Usually, there is no second harmonic generation (SHG) from bulk crystals with inversion symmetry because the macroscopic nonlinear polarization vanishes for all even-order nonlinear processes. For example, a giant SHG signal from a MoS_2 monolayer is almost completely suppressed in its bulk 2H-phase. However, self-hybridized polaritons, can provide an asymmetric distribution of nonlinear polarization thus giving rise to the resonant uncompensated bulk SHG signal. For the bulk WSe_2, films of 140-nm thickness, the largest SHG enhancement (comparable to its monolayer SHG) was demonstrated for the DBR substrate combined with an air spacer (Figure 3.3e) [210]. Recently, self-hybridized exciton–polaritons tunable by external magnetic field and temperature have been demonstrated in bare crystals of magnetic semiconductor CrSBr [211]. The exciton peak in the optical spectrum of this material can be significantly shifted by moderately strong magnetic fields that switch the equilibrium antiferromagnetic order into a ferromagnetic one below the Néel temperature, Figure 3.3f.

In addition to self-hybridized polaritons above the light line discussed in previous paragraphs, one can also realize strong coupling with guided modes beyond the light line (Figure 3.3g). This has been done with excitons in organic molecules [215] and transition metal dichalcogenides [212, 216, 217], and phonons in hBN that were intermixed with vibrations of organic molecules [218]. We will not consider this type of polaritons further; however, for additional references, we refer the interested reader to reviews by Basov et al. [219] and Low et al. [220].

Finally, it is worth mentioning that guided modes in perovskite nanowires and nanoplates were also shown to self-hybridize strongly with excitons [186, 187]. Remarkably, by varying the composition of perovskite nanowires, it is possible to tune the emission wavelength throughout the entire visible spectrum (Figure 3.3h). Furthermore, lasing, including that in the polaritonic regime, has been reported in such structures [213, 221, 222]. A more complete list of observations of lasing and self-hybridized polaritons in different perovskite-based systems can be found in the reviews in references [223–225].

3.3 Casimir Microcavities

Although the previous sections were devoted to plasmonic and open self-coupled systems, we feel it is important to discuss the developments in planar microcavities as well. In particular, we would like to bring the reader's attention to the geometry of a planar microcavity system. Interestingly, it closely resembles that of the famous Casimir effect [226], – the effect of attraction between two uncharged perfect electric conductor plates in vacuum. This effect is known to produce a sizable pressure (force per unit area), which is given only by fundamental constants and the distance between the plates, a, in accordance with Eq. (3.2).

$$F_C = \frac{\hbar c\pi^2}{240a^4}. \tag{3.2}$$

The Casimir effect is a significant effect in QED and is directly linked to zero-point fluctuations of quantum vacuum [227, 228]. An apparent similarity between the Casimir problem and the planar microcavity problem suggests that polaritons in planar microcavities can potentially harness the Casimir effect for their assembly. In addition, it may imply a deeper connection between the zero-point energy Casimir problem and findings related to ground-state energy corrections found in the strong- and ultrastrong-coupling regimes [229].

Specifically, the Casimir effect may contribute to the developments in fabrication methods for Fabry–Pérot microcavities. With this in mind, it is worth mentioning an intriguing platform based on Casimir self-assembly, which promises to become a possible future direction for polaritonics. This approach relies on harnessing the quantum electrodynamics Casimir–Lifshitz forces to assemble microcavities in a liquid environment at room temperature. It has first been suggested theoretically [234, 235] and later realized experimentally using a delicate balance between attractive and repulsive Casimir–Lifshitz [236] forces in a system comprising a few micron gold nanoflakes in ethanol solution trapped above an interface consisting of a continuous gold mirror and a Teflon spacer [230] (see Figure 3.4a). Later, a similar idea was proposed theoretically, where a glass spacer and glycerol were used for the Casimir repulsion and the flotation of the top mirror played a role of attraction instead of the attracting Casimir force [231] (see Figure 3.4b).

Afterwards, a similar approach was used to create self-assembled Fabry–Pérot microcavities in an aqueous solution, where the equilibrium distance (~100 – 200 nm) between the gold nanoflakes (which were playing the role of microscopic mirrors in the cavity) was controlled by balancing the electrostatic repulsion and Casimir–Lifshitz attraction (see Figure 3.4c) [232]. Furthermore, excitons in a few-layer WSe_2 flake positioned between the mirrors were used to reach the strong-coupling regime with the optical mode of the self-assembled microcavity. Such an arrangement allows to dynamically modulate the microcavity resonance by radiation pressure, the salinity of the solution, and temperature and in doing so tune the system *in* and *out* of the strong-coupling

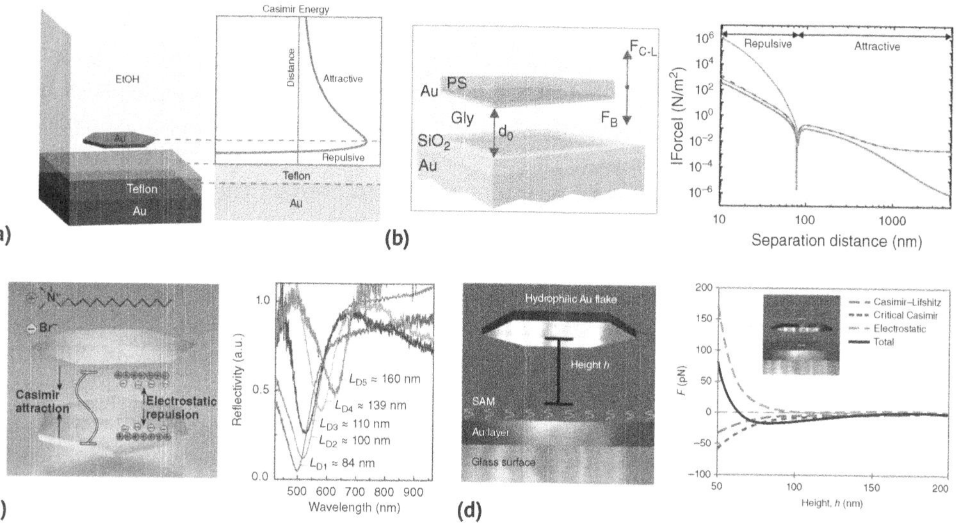

Figure 3.4 Self-assembled Casimir microcavities. (a) Schematic of experimental demonstration of stable Casimir equilibrium formed by repulsive Casimir force at short distances and attractive one at long distances. The refractive index of Teflon is lower than that of ethanol, which gives rise to the repulsive Casimir effect. [230]/American Association for the Advancement of Science - AAAS. (b) Theoretical proposal of stable Casimir equilibrium formed by flotation and repulsive Casimir forces. Here, the repulsive component is given by the appropriate refractive indices of the glycerol-glass combination. [231]/American Chemical Society. (c) Schematic of a self-assembled microcavity formed by the joint action of the attractive Casimir and the repulsive electrostatic forces (left). A spacer layer is not needed here and the electrostatic repulsion is tuned by the concentration of a ligand (CTAB). The reflection spectra for different CTAB concentrations show high tunability of the cavity in the visible range (right). [232]/Springer Nature. (d) Similar to the setup in (c) but with a critical binary liquid mixture instead of water, which gives rise to the additional highly tunable critical Casimir force. The attractive/repulsive character of this force depends on the chemical functionalization of the gold flake and the surface. The magnitude of the effect can be tuned by the temperature. [233]/CC BY 4.0/Springer Nature.

regime [232]. The latter may be advantageous for a direct comparison between the weak and strong-coupling regimes in one and the same material system, without the need to prepare several samples and therefore suffering from experimental inhomogeneities. This platform could be further enriched by introducing so-called critical Casimir forces into the self-assembly process, which, as was demonstrated recently, can even counteract Casimir–Lifshitz attraction (see Figure 3.4d) [233].

While this field of research is still in its infancy, looking ahead, we foresee that the Casimir self-assembly platform can be exploited for further studies in the fields of molecular polaritonics and polaritonic chemistry. Specifically, this platform allows for significant advantages, including, extension of the effect to other spectra ranges (in particular near- and mid-infrared), extension of material platforms, and combination with micro- and nanofluidics. The latter promises to scale up the platform, for example, through a combination of top-down lithography with colloidal science (initial steps towards this vision have recently been taken [237]). Such scaling can be realized in both horizontal and vertical directions, potentially leading to Casimir-driven metamaterials. Furthermore, it allows for a facile exchange of chemicals by microfluidics means and a possibility for active control through tunable electrostatic interactions or radiation pressure. This platform may hence bridge the gap between QED, polaritons, self-assembly, colloidal science, and microfluidics.

3.4 Conclusions and Outlook

In this chapter, we discussed various polaritonic platforms and the cavity's role in these realizations. We also discussed the potential use of these cavities in polaritonic photophysical and photochemical processes. Briefly repeating the main points here: (i) Plasmonic nanoparticles are useful for compact polaritonic platforms down to the single quantum emitter level (with e.g. organic chromophores and colloidal quantum dots) even at room temperature. Their main advantage – subwavelength mode volumes – is pushing the limits of the coupling strength with an individual emitter. Their main disadvantages are losses and low-quality factors. Therefore, they are less useful in situations where long lifetimes of polaritonic states are required. (ii) In the case of self-hybridized polaritons, we showed that optical modes sustained by the material itself may in some cases be enough to reach the strong-coupling regime with matter excitations within the very same material. This regime not only simplifies the way to create polaritons, but also provides fundamental insights into the limits of strong light–matter coupling. For instance, it states that the maximum level of coupling strength that is possible to reach with a given material is not dependent on the cavity but instead fully determined by material characteristics (such as the oscillator strength, volumetric density of oscillators, and background refractive index) and is called the bulk Rabi splitting limit. This regime challenges the way we think about cavities and their optical modes. It also poses questions about the extent to which the quality factors of such cavities are important and does the mismatch in refractive indices at the boundary of the material with its environment, such as air, solvent, or substrate provides enough contrast to make these polaritonic states useful. It is possible that at least in some cases, the answer is positive. Furthermore, we believe that even if future research shows that these states turn out to be impractical, their mere existence deserves the attention of the polaritonic community. The main disadvantage of these self-coupled states, in our opinion, is that the cavity subpart(s) and the emitter subpart(s) cannot be manipulated independently. (iii) Casimir self-assembly as a means to construct optical Fabry–Pérot microcavities is an exciting experimental development. It has been already demonstrated to lead to tunable polaritonic platforms. In the future, it may prove useful for new types of polaritonic applications and functionalities, e.g. nanofluidic and/or optomechanical systems. Furthermore, these tunable Casimir microcavities and polaritons pose fundamental questions about the relation between the Casimir effect and strong light–matter coupling. The main disadvantage of this system, in our opinion, is a relatively narrow range of parameters where Casimir self-assembly provides stable Fabry–Pérot microcavities, as well as the sensitivity of these microcavities to intense laser excitation, which likely renders it incompatible with high laser power applications such as e.g. polaritonic lasing.

This chapter is clearly not comprehensive, but we hope that in combination with other chapters in this book, as well as recent reviews on the topic (e.g. [1, 20, 238]), it will provide enough material for the interested reader. Among the future directions in the entire molecular polaritonics field, we foresee (photo)chemical experiments on a few-molecules and ultimately on the single-molecule level in both electronic and vibrational strong-coupling regimes. Furthermore, vibrational polariton condensation is an excellent non-equilibrium problem, which may be interesting for manipulating chemical processes too. We also envision prospects in studying systems combining polaritons and optical nonlinearities, for example, involving polar crystals supporting surface phonon-polaritons. We believe that plasmonic, self-hybridized, and Casimir cavities discussed in this Chapter will play an important role in advancing these studies.

References

1 Bhuyan, R., Mony, J., Kotov, O., Castellanos, G. W., Gómez Rivas, J., et al. The rise and current status of polaritonic photochemistry and photophysics. *Chem. Rev.* **2023**, *123*(18), 10877–10919.

2 Chang, R. K., Campillo, A. J. *Optical Processes in Microcavities*. World Scientific. **1996**. https://doi.org/10.1142/2828.

3 Tserkezis, C., Fernández-Domínguez, A. I., Gonçalves, P. D., Todisco, F., Cox, J. D., et al. On the applicability of quantum-optical concepts in strong-coupling nanophotonics. *Rep. Prog. Phys.* **2020**, *83*(8), 082401.

4 Rossi, T. P., Shegai, T., Erhart, P., Antosiewicz, T. J., Strong plasmon-molecule coupling at the nanoscale revealed by first-principles modeling. *Nat. Commun.* **2019**, *10*(1), 3336.

5 Kuisma, M., Rousseaux, B., Czajkowski, K. M., Rossi, T. P., Shegai, T., Erhart, P. et al. Ultrastrong coupling of a single molecule to a plasmonic nanocavity: a first-principles study. *ACS Photonics.* **2022**, *9*(3), 1065–1077.

6 Sauvan, C., Hugonin, J. P., Maksymov, I. S., Lalanne, P. Theory of the spontaneous optical emission of nanosize photonic and plasmon resonators. *Phys. Rev. Lett.* **2013**, *110*, 237401.

7 Koenderink, A. F. On the use of Purcell factors for plasmon antennas. *Opt. Lett.* **2010**, *35*, 4208–4210.

8 Buhmann, S. Y., Welsch, D. G. Dispersion forces in macroscopic quantum electrodynamics. *Prog. Quant. Elect.* **2007**, *31*(2), 51–130.

9 Buhmann, S. Y., Welsch, D. G. Casimir-polder forces on excited atoms in the strong atom-field coupling regime. *Phys. Rev. A.* **2008**, 77:012110.

10 Del Pino, J., Schröder, F. A., Chin, A. W., Feist, J., Garcia-Vidal, F. J. Tensor network simulation of non-Markovian dynamics in organic polaritons. *Phys. Rev. Lett.* **2018**, *121*, 227401.

11 Ge, R. C., Kristensen, P. T., Young, J. F., Hughes, S. Quasinormal mode approach to modelling light-emission and propagation in nanoplasmonics. *New J. Phys.* **2014**, *16*(11), 113048.

12 Franke, S., Hughes, S., Dezfouli, M. K., Kristensen, P. T., Busch, K., Knorr, A. et al. Quantization of quasinormal modes for open cavities and plasmonic cavity quantum electrodynamics. *Phys. Rev. Lett.* **2019**, *122*, 213901.

13 Kristensen, P. T., Ge, R. C., Hughes, S. Normalization of quasinormal modes in leaky optical cavities and plasmonic resonators. *Phys. Rev. A*. **2015**, *92*, 053810.

14 Lalanne, P., Yan, W., Vynck, K., Sauvan, C., Hugonin, J. P. Light interaction with photonic and plasmonic resonances. *Laser Photon. Rev*. **2018**, *12*(5), 1700113.

15 Kristensen, P. T., Hughes, S. Modes and mode volumes of leaky optical cavities and plasmonic nanoresonators. *ACS Photonics*. **2014**, *1*(1), 2–10.

16 Kristensen, P. T., Herrmann, K., Intravaia, F., Busch, K. Modeling electromagnetic resonators using quasinormal modes. **2020**, *Adv. Opt. Photonics*, *12*(3), 612–708.

17 Sauvan, C., Wu, T., Zarouf, R., Muljarov, E. A., Lalanne, P. Normalization, orthogonality, and completeness of quasinormal modes of open systems: the case of electromagnetism. *Opt. Express*. **2022**, *30*(5), 6846–6885.

18 Medina, I., García-Vidal, F. J., Fernández-Domínguez, A. I., Feist, J. Few-mode field quantization of arbitrary electromagnetic spectral densities. *Phys. Rev. Lett*. **2021**, *126*, 093601.

19 Sánchez-Barquilla, M., García-Vidal, F. J., Fernández-Domínguez, A. I., Feist, J. Few-mode field quantization for multiple emitters. *Nanophotonics*. **2022**, *11*(19), 4363–4374.

20 Sánchez-Barquilla, M., Fernández-Domínguez, A. I., Feist, J., García-Vidal, F. J. A theoretical perspective on molecular polaritonics. *ACS Photonics*. **2022**, *9*(6), 1830–1841.

21 Canales, A., Baranov, D. G., Antosiewicz, T. J., Shegai, T. Abundance of cavity-free polaritonic states in resonant materials and nanostructures. *J. Chem. Phys*. **2021**, *154*(2), 024701.

22 Pelton, M., Aizpurua, J., Bryant. G. Metal nanoparticle plasmonics. *Laser Photon. Rev*. **2008**, *2*(3), 136–159.

23 Schuller, J. A., Barnard, E. S., Cai, W., Jun, Y. C., White, J. S., Brongersma, M. L. Plasmonics for extreme light concentration and manipulation. *Nat. Mater*. **2010**, *9*(3), 193–204.

24 Wiederrecht, G. P., Wurtz, G. A., Hranisavljevic, J. Coherent coupling of molecular excitons to electronic polarizations of noble metal nanoparticles. *Nano Lett*. **2004**, *4*, 2121.

25 Govorov, A. O., Bryant, G. W., Zhang, W., Skeini, T., Lee, J., Kotov, N. A. et al. Exciton-plasmon interaction and hybrid excitons in semiconductor-metal nanoparticle assemblies. *Nano Lett*. **2006**, *6*(5), 984–994.

26 Liu, G. L., Long, Y. T., Choi, Y., Kang, T., Lee, L. P. Quantized plasmon quenching dips nanospectroscopy via plasmon resonance energy transfer. *Nat. Methods*. **2007**, *4*, 1015.

27 Ni, W., Yang, Z., Chen, H., Li, L., Wang, J. Coupling between molecular and plasmonic resonances in freestanding dye-gold nanorod hybrid nanostructures. *J. Am. Chem. Soc*. **2008**, *130*(21), 6692–6693.

28 Achermann, M. Exciton-plasmon interactions in metal-semiconductor nanostructures. *J. Phys. Chem. Lett*. **2010**, *1*(19), 2837–2843.

29 Ni, W., Ambjornsson, T., Apell, S. P., Chen, H., Wang, J. Observing plasmonic-molecular resonance coupling on single gold nanorods. *Nano Lett*. **2010**, *10*, 77.

30 Chen, X. W., Sandoghdar, V., Agio, M. Coherent interaction of light with a metallic structure coupled to a single quantum emitter: from superabsorption to cloaking. *Phys. Rev. Lett*. **Apr 2013**, *110*, 153605.

31 Zhang, Y., Meng, Q. S., Zhang, L., Luo, Y., Yu, Y. J., Yang, B. et al. Sub-nanometre control of the coherent interaction between a single molecule and a plasmonic nanocavity. *Nat. Commun*. **2017**, *8*, 15225.

32 Khitrova, G., Gibbs, H. M., Kira, M., Koch, S. W., Scherer, A. Vacuum Rabi splitting in semiconductors. *Nat. Phys*. **2006**, *2*(2), 81–90.

33 Antosiewicz, T. J., Apell, S., Shegai, T. Plasmon-exciton interactions in a core-shell geometry: from enhanced absorption to strong coupling. *ACS Photonics*. **2014**, *1*, 454–463.

34 Murata, N., Hata, R., Ishihara, H. Crossover between energy transparency resonance and Rabi splitting in antenna-molecule coupled systems. *J. Phys. Chem. C.* **2015**, *119*, 25493.

35 Yang, Z., Antosiewicz, T. J., Shegai, T. Role of material loss and mode volume of plasmonic nanocavities for strong plasmon-exciton interactions. *Opt. Express.* **2016**, *24*(18), 20374–20381.

36 Wu, X., Gray, S. K., Pelton, M. Quantum-dot-induced transparency in a nanoscale plasmonic resonator. *Opt. Express.* **2010**, *18*(23), 23633–23645.

37 Vasa, P., Wang, W., Pomraenke, R., Lammers, M., Maiuri, M., Manzoni, C. et al. Real-time observation of ultrafast Rabi oscillations between excitons and plasmons in metal nanostructures with J-aggregates. *Nat. Photon.* **2013**, *7*(2), 128–132.

38 Dintinger, J., Klein, S., Bustos, F., Barnes, W. L., Ebbesen, T. W. Strong coupling between surface plasmon-polaritons and organic molecules in subwavelength hole arrays. *Phys. Rev. B.* **2005**, *71*(3), 035424.

39 Sugawara, Y., Kelf, T. A., Baumberg, J. J., Abdelsalam, M. E., Bartlett, P. N. Strong coupling between localized plasmons and organic excitons in metal nanovoids. *Phys. Rev. Lett.* **2006**, *97*(26), 266808.

40 Wurtz, G. A., Evans, P. R., Hendren, W., Atkinson, R., Dickson, W., Pollard, R. J. et al. Molecular plasmonics with tunable exciton-plasmon coupling strength in J-aggregate hybridized au nanorod assemblies. *Nano Lett.* **2007**, 7: 1297.

41 Fofang, N. T., Park, T. H., Neumann, O., Mirin, N. A., Nordlander, P., Halas, N. J. Plexcitonic nanoparticles: Plasmon-exciton coupling in nanoshell-J-aggregate complexes. *Nano Lett.* **2008**, *8*(10), 3481–3487.

42 Hakala, T. K., Toppari, J. J., Kuzyk, A., Pettersson, M., Tikkanen, H. et al. Vacuum Rabi splitting and strong-coupling dynamics for surface-plasmon polaritons and rhodamine 6G molecules. *Phys. Rev. Lett.* **2009**, *103*(5), 053602.

43 Djoumessi Lekeufack, D., Brioude, A., Coleman, A. W., Miele, P., Bellessa, J., De Zeng, L. et al. Core-shell gold J-aggregate nanoparticles for highly efficient strong coupling applications. *Appl. Phys. Lett.* **2010**, *96*(25), 2010–2013.

44 Peruffo, N., Gil, G., Corni, S., Mancin, F., Collini, E. Selective switching of multiple plexcitons in colloidal materials: directing the energy flow at the nanoscale. *Nanoscale.* **2021**, *13*, 6005–6015.

45 Peruffo, N., Parolin, G., Collini, E., Corni, S., Mancin, F. Engineering the aggregation of dyes on ligand-shell protected gold nanoparticles to promote plexcitons formation. *Nanomaterials.* **2022**, *12*, 1180.

46 Tsargorodska, A., Cartron, M. L., Vasilev, C., Kodali, G., Mass, O. A., Baumberg, J. J. et al. Strong coupling of localized surface plasmons to excitons in light-harvesting complexes. *Nano Lett.* **2016**, *16*(11), 6850–6856.

47 Vasa, P., Pomraenke, R., Cirmi, G., Re, E. D., Wang, W., Schwieger, S. et al. Ultrafast manipulation of strong coupling in metal-molecular aggregate hybrid nanostructures. *ACS Nano.* **2010**, *4*(12), 7559–7565.

48 Peruffo, N., Mancin, F., Collini, E. Ultrafast dynamics of multiple plexcitons in colloidal nanomaterials. *J. Phys. Chem. Lett.* **2022**, *13*, 6412–6419.

49 Sukharev, M., Seideman, T., Gordon, R. J., Salomon, A., Prior, Y. Ultrafast energy transfer between molecular assemblies and surface plasmons in the strong coupling regime. *ACS Nano.* **2014**, *8*(1), 807–817.

50 Aberra Guebrou, S., Symonds, C., Homeyer, E., Plenet, J. C., Gartstein, Y. N., Agranovich, V. M. et al. Coherent emission from a disordered organic semiconductor induced by strong coupling with surface plasmons. *Phys. Rev. Lett.* **2012**, *108*, 066401.

51 Fofang, N. T., Grady, N. K., Fan, Z., Govorov, A. O., Halas, N. J. Plexciton dynamics: exciton-plasmon coupling in a J-aggregate-Au nanoshell complex provides a mechanism for nonlinearity. *Nano Lett.* **2011**, *11*(4), 1556–1560.

52 Zhou, N., Yuan, M., Gao, Y., Li, D., Yang, D. Silver nanoshell plasmonically controlled emission of semiconductor quantum dots in the strong coupling regime. *ACS Nano*. **2016**, *10*(4), 4154–4163.
53 D'Agostino, S., Alpeggiani, F., Andreani, L. C. Strong coupling between a dipole emitter and localized plasmons: Enhancement by sharp silver tips. *Opt. Express*. **2013**, *21*(23), 27602–27610.
54 DeLacy, B. G., Miller, O. D., Hsu, C. W., Zander, Z., Lacey, S., Yagloski, R. et al. Coherent plasmon-exciton coupling in silver platelet-J-aggregate nanocomposites. *Nano Lett*. **2015**, *15*(4), 2588–2593.
55 Melnikau, D., Savateeva, D., Susha, A., Rogach, A. L., Rakovich, Y. P. Strong plasmon-exciton coupling in a hybrid system of gold nanostars and J-aggregates. *Nanoscale Res. Lett*. **2013**, *8*(1), 134.
56 Peruffo, N., Mancin, F., Collini, E. Plexcitonic nanohybrids based on gold nanourchins: the role of the capping layer. *J. Phys. Chem. C*. **2021**, *125*, 19897–19905.
57 Benz, A., Campione, S., Klem, J. F., Sinclair, M. B., Brener, I. Control of strong light-matter coupling using the capacitance of metamaterial nanocavities. *Nano Lett*. **2015**, *15*(3), 1959–1966.
58 Melnikau, D., Esteban, R., Savateeva, D., Sánchez-Iglesias, A., Grzelczak, M., Schmidt, M. K. et al. Rabi splitting in photoluminescence spectra of hybrid systems of gold nanorods and J-aggregates. *J. Phys. Chem. Lett*. **2016**, *7*(2), 354–362.
59 Suh, J. Y., Kim, C. H., Zhou, W., Huntington, M. D., Co, D. T., Wasielewski, M. R. et al. Odom. Plasmonic bowtie nanolaser arrays. *Nano Lett*. **2012**, *12*, 5769.
60 Zhou, W., Dridi, M., Yong Suh, J., Kim, C. H., Co, D. T., Wasielewski, M. R. et al. Lasing action in strongly coupled plasmonic nanocavity arrays. *Nat. Nanotechnol*. **2013**, *8*(7), 506–511.
61 van Beijnum, F., van Veldhoven, P. J., Geluk, E. J., de Dood, M. J., 't Hooft, G. W., van Exter, M. P. Surface plasmon lasing observed in metal hole arrays. *Phys. Rev. Lett*. **2013**, *110*, 206802.
62 Yang, A., Hoang, T. B., Dridi, M., Deeb, C., Mikkelsen, M. H., Schatz, G. C. et al. Real-time tunable lasing from plasmonic nanocavity arrays. *Nat. Commun*. **2015**, *6*(1), 6939.
63 Hakala, T. K., Rekola, H. T., Väkeväinen, A. I., Martikainen, J.-P., Nečada, M., Moilanen, A. J. et al. Lasing in dark and bright modes of a finite-sized plasmonic lattice. *Nat. Commun*. **2017**, *8*, 13687.
64 Ramezani, M., Halpin, A., Fernández-Domínguez, A. I., Feist, J., Rodriguez, S. R. K., Garcia-Vidal, F. J. et al. Plasmon-exciton-polariton lasing. *Optica*. **2017**, *4*(1), 31–37.
65 Pourjamal, S., Hakala, T. K., Nečada, M., Freire-Fernández, F., Kataja, M., Rekola, H. et al. Lasing in Ni nanodisk arrays. *ACS Nano*. **2019**, *13*(5), 5686–5692.
66 Wang, D., Wang, W., Knudson, M. P., Schatz, G. C., Odom, T. W. Structural engineering in plasmon nanolasers. *Chem. Rev*. **2017**, *118*(6), 2865–2881.
67 Bellessa, J. Symonds, C. Vynck, K. Lemaitre, A. Brioude, A. Beaur, L. et al. Giant Rabi splitting between localized mixed plasmon-exciton states in a two-dimensional array of nanosize metallic disks in an organic semiconductor. *Phys. Rev*. **2009**, *B*, *80*, 033303.
68 Baudrion, A. L., Perron, A., Veltri, A., Bouhelier, A., Michel Adam, P., Bachelot. R. et al. Reversible strong coupling in silver nanoparticle arrays using photochromic molecules. *Nano Lett*. **2013**, *13*, 282–286.
69 Rodriguez, S. R. K., Gómez Rivas, J. Surface lattice resonances strongly coupled to rhodamine 6G excitons: tuning the plasmon-exciton-polariton mass and composition. *Opt. Express*. **2013**, *21*(22), 27411–27421.
70 Väkeväinen, A. I., Moerland, R. J., Rekola, H. T., Eskelinen, A.-P., Martikainen, J.-P., Kim, D.-H. et al. Plasmonic surface lattice resonances at the strong coupling regime. *Nano Lett*. **2014**, *14*(4), 1721–1727.
71 Yadav, R. K., Bourgeois, M. R., Cherqui, C., Juarez, X. G., Wang, W. et al. Room temperature weak-to-strong coupling and the emergence of collective emission from quantum dots coupled to plasmonic arrays. *ACS Nano*. **2020**, *14*(6), 7347–7357.

72 Yadav, R. K., Otten, M., Wang, W., Cortes, C. L., Gosztola, D. J., Wiederrecht, G. P. et al. Strongly coupled exciton–surface lattice resonances engineer long-range energy propagation. *Nano Lett.* **2020**, *20*(7), 5043–5049.

73 Shi, L., Hakala, T. K., Rekola, H. T., Martikainen, J.-P., Moerland, R. J., Törmä, P. Spatial coherence properties of organic molecules coupled to plasmonic surface lattice resonances in the weak and strong coupling regimes. *Phys. Rev. Lett.* **2014**, *112*, 153002.

74 Rodriguez, S. R K., Feist, J., Verschuuren, M. A., Garcia Vidal, F.J., Gómez Rivas, J. Thermalization and cooling of plasmon-exciton polaritons: towards quantum condensation. *Phys. Rev. Lett.* **2013**, *111*(16), 166802.

75 Hakala, T. K., Moilanen, A. J., Väkeväinen, A. I., Guo, R., Martikainen, J. P., Daskalakis, K. S. et al. Bose–einstein condensation in a plasmonic lattice. *Nat. Phys.* **2018**, *14*(7), 739–744.

76 Törmä, P., Barnes, W. L. Strong coupling between surface plasmon polaritons and emitters: a review. *Rep. Prog. Phys.* **2015**, *78*(1), 013901.

77 Baranov, D. G., Wersall, M., Cuadra, J., Antosiewicz, T. J., Shegai, T. Novel nanostructures and materials for strong light–matter interactions. *ACS Photonics.* **2018**, *5*(1), 24–42.

78 Schlather, A. E., Large, N., Urban, A. S., Nordlander, P., Halas, N. J. Near-field mediated plexcitonic coupling and giant Rabi splitting in individual metallic dimers. *Nano Lett.* **2013**, *13*(7), 3281–3286.

79 Zengin, G., Johansson, G., Johansson, P., Antosiewicz, T. J., Käll, M., Shegai, T. Approaching the strong coupling limit in single plasmonic nanorods interacting with J-aggregates. *Sci. Rep.* **2013**, *3*, 3074.

80 Zengin, G., Wersäll, M., Nilsson, S., Antosiewicz, T. J., Käll, M., Shegai, T. Realizing strong light-matter interactions between single-nanoparticle plasmons and molecular excitons at ambient conditions. *Phys. Rev. Lett.* **2015**, *114*(15), 157401.

81 Wersaell, M., Cuadra, J., Antosiewicz, T. J., Balci, S., Shegai, T. Observation of mode splitting in photoluminescence of individual plasmonic nanoparticles strongly coupled to molecular excitons. *Nano Lett.* **2017**, *17*(1), 551–558.

82 Wersaell, M., Munkhbat, B., Baranov, D. G., Herrera, F., Cao, J., Antosiewicz, T. J. et al. Correlative dark-field and photoluminescence spectroscopy of individual plasmon–molecule hybrid nanostructures in a strong coupling regime. *ACS Photonics.* **2019**, *6*(10), 2570–2576.

83 Savasta, S., Saija, R., Ridolfo, A., Stefano, O. D., Denti, P., Borghese, F. Nanopolaritons: vacuum Rabi splitting with a single quantum dot in the center of a dimer nanoantenna. *ACS Nano.* **2010**, *4*(11), 6369–6376.

84 Yang, B., Chen, G., Ghafoor, A., Zhang, Y., Zhang, Y., Zhang, Y. et al. Sub-nanometre resolution in single-molecule photoluminescence imaging. *Nat. Photon.* **2020**, *14*(11), 693–699.

85 Santhosh, K., Bitton, O., Chuntonov, L., Haran, G. Vacuum Rabi splitting in a plasmonic cavity at the single quantum emitter limit. *Nat. Commun.* **2016**, *7*, 11823.

86 Park, K. D., May, M. A., Leng, H., Wang, J., Kropp, J. A., Gougousi, T. et al. Tip-enhanced strong coupling spectroscopy, imaging, and control of a single quantum emitter. *Sci. Adv.* **2019**, *5*(7), eaav5931.

87 Gross, H., Hamm, J. M., Tufarelli, T., Hess, O., Hecht, B. Near-field strong coupling of single quantum dots. *Sci. Adv.* **2018**, *4*(3), eaar4906.

88 Li, J. Y., Li, W., Liu, J., Zhong, J., Liu, R., Chen, H. et al. Room-temperature strong coupling between a single quantum dot and a single plasmonic nanoparticle. *Nano Lett.* **2022**, *22*(12), 4686–4693.

89 Hoang, T. B., Akselrod, G. M., Mikkelsen, M. H. Ultrafast room-temperature single photon emission from quantum dots coupled to plasmonic nanocavities. *Nano Lett.* **2016**, *16*(1), 270–275.

90 Lounis, B., Orrit, M. Single-photon sources. *Rep. Prog. Phys.* 68(5). **1129**, 2005.

91 Elshaari, A. W., Pernice, W., Srinivasan, K., Benson, O., Zwiller, V. Hybrid integrated quantum photonic circuits. *Nat. Photon.* **2020**, *14*(5), 285–298.

92 McKeever, J., Boca, A., Boozer, A. D., Buck, J. R., Kimble, H. J. Experimental realization of a one-atom laser in the regime of strong coupling. *Nature*. **2003**, *425*(6955), 268–271.

93 Nomura, M., Kumagai, N., Iwamoto, S., Ota, Y., Arakawa, Y. Laser oscillation in a strongly coupled single-quantum-dot–nanocavity system. *Nat. Phys.* **2010**, *6*(4), 279–283.

94 Kneipp, K., Wang, Y., Kneipp, H., Perelman, L. T., Itzkan, I., Dasari, R. R. et al. Single molecule detection using surface-enhanced Raman scattering (sers). *Phys. Rev. Lett.* **1997**, *78*(9), 1667.

95 Nie, S., Emory, S. R. Probing single molecules and single nanoparticles by surface-enhanced Raman scattering. *Science*. **1997**, *275*(5303), 1102–1106.

96 Xu, H., Bjerneld, E. J., Käll, M., Börjesson, L. Spectroscopy of single hemoglobin molecules by surface enhanced Raman scattering. *Phys. Rev. Lett.* **1999**, *83*(21), 4357.

97 Langer, J., Jimenez de Aberasturi, D., Aizpurua, J., Alvarez-Puebla, R. A., Auguié, B., Baumberg, J. J. et al. Present and future of surface-enhanced Raman scattering. *ACS Nano*. **2019**, *14*(1), 28–117.

98 Itoh, T., Yamamoto, Y. S., Tamaru, H., Biju, V., Wakida, S. I., Ozaki, Y. et al. Single-molecular surface-enhanced resonance Raman scattering as a quantitative probe of local electromagnetic field: The case of strong coupling between plasmonic and excitonic resonance. *Phys. Rev. B*. **May 2014**, *89*, 195436.

99 Chikkaraddy, R., De Nijs, B., Benz, F., Barrow, S. J., Scherman, O. A., Rosta, E. et al. Single-molecule strong coupling at room temperature in plasmonic nanocavities. *Nature*. **2016**, *535*(7610), 127–130.

100 Benz, F., Schmidt, M. K., Dreismann, A., Chikkaraddy, R., Zhang, Y., Demetriadou, A. et al. Single-molecule optomechanics in "picocavities". *Science*. **2016**, *354*(6313), 726–729.

101 Neuman, T., Esteban, R., Casanova, D., García-Vidal, F. J., Aizpurua, J. Coupling of molecular emitters and plasmonic cavities beyond the point-dipole approximation. *Nano Lett.* **2018**, *18*(4), 2358–2364.

102 Jestädt, R., Ruggenthaler, M., Oliveira, J. T., Rubio, A., Appel, H. Light-matter interactions within the Ehrenfest–Maxwell–Pauli–Kohn–Sham framework: fundamentals, implementation, and nano-optical applications. *Adv. Phys.* **2019**, *68*(4), 225–333.

103 Konečná, A., Neuman, T., Aizpurua, J., Hillenbrand, R. Surface-enhanced molecular electron energy loss spectroscopy. *ACS Nano*. **2018**, *12*(5), 4775–4786.

104 Hartsfield, T., Chang, W. S., Yang, S.C., Ma, T., Shi, J., Sun, L. et al. Single quantum dot controls a plasmonic cavity's scattering and anisotropy. *Proc. Natl. Acad. Sci. U. S. A.* **2015**, *112*, 12288–12292.

105 Gupta, S. N., Bitton, O., Neuman, T., Esteban, R., Chuntonov, L., Aizpurua, J. et al. Complex plasmon-exciton dynamics revealed through quantum dot light emission in a nanocavity. *Nat. Commun.* **2021**, *12*(1), 487.

106 Bitton, O., Gupta, S. N., Houben, L., Kvapil, M., Křápek, V., Šikola, T. et al. Vacuum Rabi splitting of a dark plasmonic cavity mode revealed by fast electrons. *Nat. Commun.* **2020**, *11*(1), 487.

107 Leng, H., Szychowski, B., Daniel, M. C., Pelton, M. Strong coupling and induced transparency at room temperature with single quantum dots and gap plasmons. *Nat. Commun.* **2018**, *9*(1), 4012.

108 Andersen, S. K., Kumar, S., Bozhevolnyi, S. I. Ultrabright linearly polarized photon generation from a nitrogen vacancy center in a nanocube dimer antenna. *Nano Lett.* **2017**, *17*(6), 3889–3895.

109 Flick, J., Rivera, N., Narang, P. Strong light-matter coupling in quantum chemistry and quantum photonics. *Nanophotonics*. **2018**, *7*(9), 1479–1501.

110 Wei, H., Yan, X., Niu, Y., Li, Q., Jia, Z., Xu, H. Plasmon–exciton interactions: spontaneous emission and strong coupling. *Adv. Funct. Mater.* **2021**, *31*(51), 2100889.

111 Anantharaman, S. B., Jo, K., Jariwala, D. Exciton–photonics: from fundamental science to applications. *ACS Nano*. **2021**, *15*(8), 12628–12654.

112 Xiong, X., Kongsuwan, N., Lai, Y., Png, C. E., Wu, L., Hess, O. Room-temperature plexcitonic strong coupling: Ultrafast dynamics for quantum applications. *Appl. Phys. Lett.* **2021**, *118*(13), 130501.

113 Zhao, Q., Zhou, W. J., Deng, Y. H., Zheng, Y. Q., Shi, Z. H., Ang, L. K. et al. Plexcitonic strong coupling: unique features, applications, and challenges. *J. Phys. D: Appl. Phys.* **2022**, *55*(20), 203002.

114 Munkhbat, B., Wersäll, M., Baranov, D. G., Antosiewicz, T. J., Shegai, T. Suppression of photo-oxidation of organic chromophores by strong coupling to plasmonic nanoantennas. *Sci. Adv.* **2018**, *4*(7), eaas9552.

115 Nefedkin, N., Andrianov, E., Vinogradov, A. The role of strong coupling in the process of photobleaching suppression. *J. Phys. Chem. C.* **2020**, *124*(33), 18234–18242.

116 Doronin, I. V., Kalmykov, A. S., Zyablovsky, A. A., Andrianov, E. S., Khlebtsov, B. N., Melentiev, P. N. et al. Resonant concentration-driven control of dye molecule photodegradation via strong optical coupling to plasmonic nanoparticles. *Nano Lett.* **2021**, *22*(1), 105–110.

117 Galego, J., Garcia-Vidal, F. J., Feist, J. Suppressing photochemical reactions with quantized light fields. *Nat. Commun.* **2016**, *7*(1), 13841.

118 Fregoni, J., Haugland, T. S., Pipolo, S., Giovannini, T., Koch, H., Corni, S. et al. Strong coupling between localized surface plasmons and molecules by coupled cluster theory. *Nano Lett.* **2021**, *21*(15), 6664–6670.

119 Schwartz, T., Hutchison, J. A., Genet, C., Ebbesen, T. W. Reversible switching of ultrastrong light-molecule coupling. *Phys. Rev. Lett.* **2011**, *106*(19), 196405.

120 Hutchison, J. A., Schwartz, T., Genet, C., Devaux, E., Ebbesen, T. W. Modifying chemical landscapes by coupling to vacuum fields. *Angew. Chem. Int. Ed.* **2012**, *51*(7), 1592–1596.

121 Ribeiro, R. F., Martínez-Martínez, L. A., Du, M., Campos-Gonzalez-Angulo, J., Yuen-Zhou, J. Polariton chemistry: controlling molecular dynamics with optical cavities. *Chem. Sci.* **2018**, *9*(30), 6325–6339.

122 Feist, J., Galego, J., Garcia-Vidal, F. J. Polaritonic chemistry with organic molecules. *ACS Photonics*. **2018**, *5*(1), 205–216.

123 Li, T. E., Cui, B., Subotnik, J. E., Nitzan, A. Molecular polaritonics: chemical dynamics under strong light–matter coupling. *Ann. Rev. Phys. Chem.* **2022**, *73*, 43–71.

124 Balci, S., Kocabas, C., Kücüköz, B., Karatay, A., Akhüseyin, E., Yaglioglu, H. G. et al, Probing ultrafast energy transfer between excitons and plasmons in the ultrastrong coupling regime. *Appl. Phys. Lett.* **2014**, *105*(5), 051105.

125 Balci, S., Kucukoz, B., Balci, O., Karatay, A., Kocabas, C., Yaglioglu, G. et al, Tunable plexcitonic nanoparticles: a model system for studying plasmon–exciton interaction from the weak to the ultrastrong coupling regime. *ACS Photonics*. **2016**, *3*(11), 2010–2016.

126 Zhang, Y., He, S., Guo, W., Hu, Y., Huang, J., Mulcahy, J. R. et al. Surface-plasmon-driven hot electron photochemistry. *Chem. Rev.* **2017**, *118*(6), 2927–2954.

127 Rasnik, I., McKinney, S. A., Ha, T. Nonblinking and long-lasting single-molecule fluorescence imaging. *Nat. Methods*. **2006**, *3*(11), 891–893.

128 Hale, G. D., Jackson, J. B., Shmakova, O. E., Lee, T. R., Halas, N. J. Enhancing the active lifetime of luminescent semiconducting polymers via doping with metal nanoshells. *Appl. Phys. Lett.* **2001**, *78*(11), 1502–1504.

129 Kéna-Cohen, S., Wiener, A., Sivan, Y., Stavrinou, P. N., Bradley, D. D., Horsfield, A. et al. Plasmonic sinks for the selective removal of long-lived states. *ACS Nano*. **2011**, *5*(12), 9958–9965.

130 Manjavacas, A., Fenollosa, R., Rodriguez, I., Jiménez, M. C., Miranda, M. A., Meseguer, F. Magnetic light and forbidden photochemistry: the case of singlet oxygen. *J. Mater. Chem. C.* **2017**, *5*(45), 11824–11831.

131 Weiss, A., Haran, G. Time-dependent single-molecule raman scattering as a probe of surface dynamics. *J. Phys. Chem. B.* **2001**, *105*(49), 12348–12354.

132 Cang, H., Liu, Y., Wang, Y., Yin, X., Zhang, X. Giant suppression of photobleaching for single molecule detection via the purcell effect. *Nano Lett.* **2013**, *13*(12), 5949–5953.

133 Yuan, H., Khatua, S., Zijlstra, P., Yorulmaz, M., Orrit, M. Thousand-fold enhancement of single-molecule fluorescence near a single gold nanorod. *Angew. Chem.* **2013**, *125*(4), 1255–1259.

134 Agranovich, V. M., Litinskaia, M., Lidzey, D. G Cavity polaritons in microcavities containing disordered organic semiconductors. *Phys. Rev. B.* **2003**, *67*(8), 085311.

135 Agranovich, V. M., Gartstein, Y. N., Litinskaya, M. Hybrid resonant organic–inorganic nanostructures for optoelectronic applications. *Chem. Rev.* **2011**, *111*(9), 5179–5214.

136 Litinskaya, M., Reineker, P., Agranovich, V. M. Fast polariton relaxation in strongly coupled organic microcavities. *J. Lumin.* **2004**, *110*(4), 364–372.

137 Peters, V. N., Faruk, M. O., Asane, J., Alexander, R., Peters, D. A. A., Prayakarao, S. et al. Effect of strong coupling on photodegradation of the semiconducting polymer P3HT. *Optica.* **2019**, *6*(3), 318–325.

138 Berghuis, A. M., Halpin, A., Le-Van, Q., Ramezani, M., Wang, S., Murai, S. et al. Enhanced delayed fluorescence in tetracene crystals by strong light-matter coupling. *Adv. Funct. Mater.* **2019**, *29*(36), 1901317.

139 Finkelstein-Shapiro, D., Mante, P. A., Sarisozen, S., Wittenbecher, L., Minda, I., Balci, S. et al. Understanding radiative transitions and relaxation pathways in plexcitons. *Chem.* **2021**, *7*(4), 1092–1107.

140 Sample, A. D., Guan, J., Hu, J., Reese, T., Cherqui, C. R. et al. Strong coupling between plasmons and molecular excitons in metal–organic frameworks. *Nano Lett.* **2021**, *21*(18), 7775–7780.

141 Park, J. E., López-Arteaga, R., Sample, A. D., Cherqui, C. R., Spanopoulos, I., Guan, J. et al. Polariton dynamics in two-dimensional ruddlesden–popper perovskites strongly coupled with plasmonic lattices. *ACS Nano.* **2022**, *16*(3), 3917–3925.

142 Kuttruff, J., Romanelli, M., Pedrueza-Villalmanzo, E., Allerbeck, J., Fregoni, J., Saavedra-Becerril, V. et al. Sub-picosecond collapse of molecular polaritons to pure molecular transition in plasmonic photoswitch-nanoantennas. *Nat. Commun.* **2023**, *14*(1), 3875.

143 Kongsuwan, N., Demetriadou, A., Chikkaraddy, R., Benz, F., Turek, V. A., Keyser, U. F. et al. Suppressed quenching and strong-coupling of purcell-enhanced single-molecule emission in plasmonic nanocavities. *ACS Photonics.* **2018**, *5*(1), 186–191.

144 Herrera, F., Spano, F. C. Dark vibronic polaritons and the spectroscopy of organic microcavities. *Phys. Rev. Lett.* **2017**, *118*(22), 223601.

145 Coles, D. M., Michetti, P., Clark, C., Tsoi, W. C., Adawi, A. M., Kim, J. S. et al. Vibrationally assisted polariton-relaxation processes in strongly coupled organic-semiconductor microcavities. *Adv. Funct. Mater.* **2011**, *21*(19), 3691–3696.

146 Agranovich, V., Benisty, H., Weisbuch, C. Organic and inorganic quantum wells in a microcavity: Frenkel-Wannier-Mott excitons hybridization and energy transformation. *Solid State Commun.* **1997**, *102*(8), 631–636.

147 González-Tudela, A., Huidobro, P. A., Martín-Moreno, L., Tejedor, C., García-Vidal, F. J. Theory of strong coupling between quantum emitters and propagating surface plasmons. *Phys. Rev. Lett.* **2013**, *110*(12), 126801.

148 Mazza, L., Kéna-Cohen, S., Michetti, P., La Rocca, G. C. Microscopic theory of polariton lasing via vibronically assisted scattering. *Phys. Rev. B.* **2013**, *88*(7), 075321.

149 Spano, F. C. Optical microcavities enhance the exciton coherence length and eliminate vibronic coupling in J-aggregates. *J. Chem. Phys.* **2015**, *142*(18), 184707.

150 Galego, J., Garcia-Vidal, F. J., Feist, J. Cavity-induced modifications of molecular structure in the strong-coupling regime. *Phys. Rev. X.* **2015**, *5*(4), 041022.

151 Ćwik, J. A., Kirton, P., Liberato, S. D., Keeling, J. Excitonic spectral features in strongly coupled organic polaritons. *Phys. Rev. A.* **2016**, *93*(3), 033840.

152 Herrera, F., Spano, F. C. Cavity-controlled chemistry in molecular ensembles. *Phys. Rev. Lett.* **2016**, *116*(23), 238301.

153 Kowalewski, M., Bennett, K., Mukamel, S. Non-adiabatic dynamics of molecules in optical cavities. *J. Chem. Phys.* **2016**, *144*(5), 054309.

154 Flick, J., Ruggenthaler, M., Appel, H., Rubio, A. Atoms and molecules in cavities, from weak to strong coupling in quantum-electrodynamics (qed) chemistry. *Proc. Natl. Acad. Sci. U. S. A.* **2017**, *114*(12), 3026–3034.

155 Luk, H. L., Feist, J., Toppari, J. J., Groenhof, G. Multiscale molecular dynamics simulations of polaritonic chemistry. *J. Chem. Theory Comput.* **2017**, *13*(9), 4324–4335.

156 Schäfer, C., Ruggenthaler, M., Rubio, A. Ab initio nonrelativistic quantum electrodynamics: bridging quantum chemistry and quantum optics from weak to strong coupling. *Phys. Rev. A.* **2018**, *98*(4), 043801.

157 Vendrell, O. Coherent dynamics in cavity femtochemistry: Application of the multi-configuration time-dependent hartree method. *Chem. Phys.* **2018**, *509*, 55–65.

158 Szidarovszky, T., Halász, G. J., Császár, A. G., Cederbaum, L. S., Vibók, Á. Conical intersections induced by quantum light: Field-dressed spectra from the weak to the ultrastrong coupling regimes. *J. Phys. Chem. Lett.* **2018**, *9*(21), 6215–6223.

159 Fregoni, J., Granucci, G., Coccia, E., Persico, M., Corni, S. Manipulating azobenzene photoisomerization through strong light–molecule coupling. *Nat. Commun.* **2018**, *9*(1), 4688. ISSN 2041-1723.

160 Campos-Gonzalez-Angulo, J. A., Ribeiro, R. F., Yuen-Zhou, J. Resonant catalysis of thermally activated chemical reactions with vibrational polaritons. *Nat. Commun.* **2019**, *10*(1), 4685.

161 Ulusoy, I. S., Gomez, J. A., Vendrell, O. Modifying the nonradiative decay dynamics through conical intersections via collective coupling to a cavity mode. *J. Phys. Chem. A.* **2019**, *123*(41), 8832–8844.

162 Fregoni, J., Corni, S., Persico, M., Granucci, G. Photochemistry in the strong coupling regime: a trajectory surface hopping scheme. *J. Comput. Chem.* **2020**, *41*(23), 2033–2044.

163 Fregoni, J., Granucci, G., Persico, M., Corni, S. Strong coupling with light enhances the photoisomerization quantum yield of azobenzene. *Chem.* **2020**, *6*(1), 250–265.

164 Coccia, E., Fregoni, J., Guido, C. A., Marsili, M., Pipolo, S., Corni, S. et al. Hybrid theoretical models for molecular nanoplasmonics. *J. Chem. Phys.* **2020**, *153*(20), 200901.

165 Galego, J., Garcia-Vidal, F. J., Feist, J. Many-molecule reaction triggered by a single photon in polaritonic chemistry. *Phys. Rev. Lett.* **2017**, *119*(13), 136001.

166 Felicetti, S., Fregoni, J., Schnappinger, T., Reiter, S., de Vivie-Riedle, R., Feist, J. Photoprotecting uracil by coupling with lossy nanocavities. *J. Phys. Chem. Lett.* **2020**, *11*(20), 8810–8818.

167 Haugland, T. S., Ronca, E., KjØnstad, E. F., Rubio, A., Koch, H. Coupled cluster theory for molecular polaritons: Changing ground and excited states. *Phys. Rev. X.* **2020**, *10*, 041043.

168 Haugland, T. S., Ronca, E., KjØnstad, E. F., Rubio, A., Koch, H. Intermolecular interactions in optical cavities: an ab initio QED study. *J. Chem. Phys.* **2021**, *154*(9), 094113.

169 Roelli, P., Martin-Cano, D., Kippenberg, T. J, Galland, C. Molecular platform for frequency upconversion at the single-photon level. *Phys. Rev. X.* **2020**, *10*(3), 031057.

170 Chen, W., Roelli, P., Hu, H., Verlekar, S., Amirtharaj, S. P., Barreda, A. I. et al. Continuous-wave frequency upconversion with a molecular optomechanical nanocavity. *Science.* **2021**, *374*(6572), 1264–1267.

171 Xomalis, A., Zheng, X., Chikkaraddy, R., Koczor-Benda, Z., Miele, E., Rosta, E. et al. Detecting mid-infrared light by molecular frequency upconversion in dual-wavelength nanoantennas. *Science*. **2021**, *374*(6572), 1268–1271.

172 Nguyen, X. T., Winte, K., Timmer, D., Rakita, Y., Ceratti, D. R., Aharon, S. et al. Phonon-driven intra-exciton Rabi oscillations in cspbbr3 halide perovskites. *Nat. Commun.* **2023**, *14*(1), 1047.

173 Sayers, C. J., Genco, A., Trovatello, C., Conte, S. D., Khaustov, V., Cervantes-Villanueva, J. et al. Strong coupling of coherent phonons to excitons in semiconducting monolayer $MoTe_2$. *arXiv preprint arXiv:2302.07561*, **2023**.

174 Kluczyk-Korch, K., Antosiewicz, T. J. Hot carrier generation in a strongly coupled molecule–plasmonic nanoparticle system. *Nanophotonics*. **2023**, *12*(9), 1711–1722.

175 Wang, Z., Li, L., Wei, S., Shi, X., Xiao, J., Guo, Z. et al. Manipulating light–matter interaction into strong coupling regime for photon entanglement in plasmonic lattices. *J. Appl. Phys.* **2023**, *133*(6), 063101.

176 Welakuh, D. M., Narang, P. Tunable nonlinearity and efficient harmonic generation from a strongly coupled light–matter system. *ACS Photonics*. **2023**, *10*(2):383–393.

177 Welakuh, D. M., Narang, P. Nonlinear optical processes in centrosymmetric systems by cavity-induced symmetry breaking. *ACS Photonics*. **2024**, *11*(2), 369–377.

178 Hopfield, J. J. Theory of the contribution of excitons to the complex dielectric constant of crystals. *Phys. Rev.* **1958**, *112*, 1555–1567.

179 Mills, D L., Burstein, E. Polaritons: the electromagnetic modes of media. *Rep. Prog. Phys.* **1974**, *37*(7), 817.

180 Platts, C. E., Kaliteevski, M. A., Brand, S., Abram, R. A., Iorsh, I. V., and Kavokin, A. V. Whispering-gallery exciton polaritons in submicron spheres. *Phys. Rev. B*. **2009**, *79*(24), 245322. doi: 10.1103/PhysRevB.79.245322.

181 Takazawa, K., Inoue, J. I., Mitsuishi, K., Takamasu, T. Fraction of a millimeter propagation of exciton polaritons in photoexcited nanofibers of organic dye. *Phys. Rev. Lett.* **2010**, *105*(6), 067401.

182 Liao, Q., Xu, Z., Zhong, X., Dang, W., Shi, Q., Zhang, C. et al. An organic nanowire waveguide exciton–polariton sub-microlaser and its photonic application. *J. Mater. Chem. C*. **2014**, *2*(15), 2773–2778.

183 Wang, Q., Sun, L., Zhang, B., Chen, C., Shen, X., Lu, W. Direct observation of strong light-exciton coupling in thin Ws_2 flakes. *Opt. Express*. **2016**, *24*(7), 7151–7157.

184 Munkhbat, B., Baranov, D. G., Stührenberg, M., Wersäll, M., Bisht, A., Shegai, T. Self-hybridized exciton-polaritons in multilayers of transition metal dichalcogenides for efficient light absorption. *ACS Photonics*. **2019**, *6*, 139–147.

185 Verre, R., Baranov, D. G., Munkhbat, B., Cuadra, J., Käll, M., Shegai, T. Transition metal dichalcogenide nanodisks as high-index dielectric Mie nanoresonators. *Nat. Nanotechnol.* **2019**, *14*, 679–684.

186 Tiguntseva, E. Y., Baranov, D. G., Pushkarev, A. P., Munkhbat, B., Komissarenko, F., Franckevicius, M. et al. Tunable hybrid fano resonances in halide perovskite nanoparticles. *Nano Lett.* **2018**, *18*(9), 5522–5529.

187 Zhang, S., Shang, Q., Du, W., Shi, J., Wu, Z., Mi, Y. et al. Strong exciton–photon coupling in hybrid inorganic–organic perovskite micro/nanowires. *Adv. Opt. Mater.* **2018**, *6*(2), 1701032.

188 Chen, P. Z., Weng, Y. X., Niu, L. Y., Chen, Y. Z., Wu, L. Z., Tung, C. H. et al. Light-harvesting systems based on organic nanocrystals to mimic chlorosomes. *Angew. Chem. Int. Ed.* **2016**, *55*(8), 2759–2763.

189 Georgiou, K., Jayaprakash, R., Lidzey, D. G. Strong coupling of organic dyes located at the surface of a dielectric slab microcavity. *J. Phys. Chem. Lett.* **2020**, *11*(22), 9893–9900.

190 Pandya, R., Chen, R Y., Gu, Q., Sung, J., Schnedermann, C., Ojambati, O. S. et al. Microcavity-like exciton-polaritons can be the primary photoexcitation in bare organic semiconductors. *Nat. Commun.*, *12*(1):6519, 2021.

191 Thomas, P. A., Menghrajani, K. S., Barnes, W. L. Cavity-free ultrastrong light-matter coupling. *J. Phys. Chem. Lett.* **2021**, *12*(29), 6914–6918.

192 Ma, E. Y., Hu, J., Waldecker, L., Watanabe, K., Taniguchi, T., Liu, F. et al. The reststrahlen effect in the optically thin limit: A framework for resonant response in thin media. *Nano Lett.* **2022**, *22*(20), 8389–8393.

193 Kaeek, M., Damari, R., Roth, M., Fleischer, S., Schwartz, T. Strong coupling in a self-coupled terahertz photonic crystal. *ACS Photonics.* **2021**, *8*(7), 1881–1888.

194 Pirzadeh, Z., Pakizeh, T., Miljkovic, V., Langhammer, C., Dmitriev, A. Plasmon–interband coupling in nickel nanoantennas. *ACS Photonics.* **2014**, *1*(3), 158–162.

195 Mueller, N. S., Okamura, Y., Vieira, B. G., Juergensen, S., Lange, H., Barros, E. B. et al. Deep strong light–matter coupling in plasmonic nanoparticle crystals. *Nature.* **2020**, *583*(7818), 780–784.

196 Canales, A., Kotov, O. V., Küçüköz, B., Shegai, T. O. Self-hybridized vibrational-Mie polaritons in water droplets. *Phys. Rev. Lett.* **2024**, *132*(19), 193804.

197 Arnott, W. P., Schmitt, C., Liu, Y., Hallett, J. Droplet size spectra and water-vapor concentration of laboratory water clouds: inversion of fourier transform infrared (500–5000 cm^{-1}) optical-depth measurement. *Appl. Opt.* **1997**, *36*(21), 5205–5216.

198 Maayani, S., Martin, L. L., Carmon, T. Water-walled microfluidics for high-optical finesse cavities. *Nat. Commun.* **2016**, *7*(1), 10435.

199 Giorgini, A., Avino, S., Malara, P., De Natale, P., Gagliardi, G. Fundamental limits in high-q droplet microresonators. *Sci. Rep.* **2017**, *7*(1), 41997.

200 Garcia-Vidal, F. J., Ciuti, C., Ebbesen, T. W. Manipulating matter by strong coupling to vacuum fields. *Science.* **2021**, *373*(6551), eabd0336.

201 Vurgaftman, I., Simpkins, B. S., Dunkelberger, A. D., Owrutsky, J. C. Comparative analysis of polaritons in bulk, dielectric slabs, and planar cavities with implications for cavity-modified reactivity. *J. Chem. Phys.* **2022**, *156*(3), 034110.

202 Feist, J., Garcia-Vidal, F. J. Extraordinary exciton conductance induced by strong coupling. *Phys. Rev. Lett.* **2015**, *114*(19), 196402.

203 Schachenmayer, J., Genes, C., Tignone, E., Pupillo, G. Cavity-enhanced transport of excitons. *Phys. Rev. Lett.* **2015**, *114*, 196403.

204 Rozenman, G. G., Akulov, K., Golombek, A., Schwartz, T. Long-range transport of organic exciton-polaritons revealed by ultrafast microscopy. *ACS Photonics.* **2018**, *5*(1), 105–110.

205 Berghuis, A. M., Tichauer, R. H., de Jong, L. M., Sokolovskii, I., Bai, P., Ramezani, M. et al. Controlling exciton propagation in organic crystals through strong coupling to plasmonic nanoparticle arrays. *ACS Photonics*, 2022.

206 Cui, Q. H., Peng, Q., Luo, Y., Jiang, Y., Yan, Y., Wei, C. et al. Asymmetric photon transport in organic semiconductor nanowires through electrically controlled exciton diffusion. *Sci. Adv.* **2018**, *4*(3), eaap9861.

207 Hou, S., Khatoniar, M., Ding, K., Qu, Y., Napolov, A., Menon, V. M. et al. Ultralong-range energy transport in a disordered organic semiconductor at room temperature via coherent exciton-polariton propagation. *Adv. Mater.* **2020**, *32*(28), 2002127.

208 Balasubrahmaniyam, M., Simkhovich, A., Golombek, A., Sandik, G., Ankonina, G., Schwartz, T. et al, From enhanced diffusion to ultrafast ballistic motion of hybrid light–matter excitations. *Nat. Mater.* **2023**, *22*(3), 338–344.

209 Guo, Q., Wu, B., Du, R., Ji, J., Wu, K., Li, Y. et al. Boosting exciton transport in WSe_2 by engineering its photonic substrate. *ACS Photonics*. **2022**, *9*(8), 2817–2824.

210 Khatoniar, M., Bushati, R., Mekawy, A., Dirnberger, F., Alú, A., Menon, V. M. Relaxing symmetry rules for nonlinear optical interactions in van der waals materials via strong light–matter coupling. *ACS Photonics*. **2022**, *9*(2), 503–510.

211 Dirnberger, F., Quan, J., Bushati, R., Diederich, G. M., Florian, M., Klein, J. et al. Magneto-optics in a van der waals magnet tuned by self-hybridized polaritons. *Nature*. **Aug 2023**, *620*(7974), 533–537.

212 Hu, F., Luan, Y., Scott, M. E., Yan, J., Mandrus, D. G., Xu, X. et al. Imaging exciton–polariton transport in $MoSe_2$ waveguides. *Nat. Photon*. **2017**, *11*(6), 356–360.

213 Zhu, H., Fu, Y., Meng, F., Wu, X., Gong, Z., Ding, Q. et al. Lead halide perovskite nanowire lasers with low lasing thresholds and high quality factors. *Nat. Mater*. **2015**, *14*(6), 636–642.

214 URL: https://jin.chem.wisc.edu.

215 Menghrajani, K. S., Barnes, W. L. Strong coupling beyond the light-line. *ACS Photonics*. **2020**, *7*(9), 2448–2459.

216 Hu, F., Luan, Y., Speltz, J., Zhong, D., Liu, C. H., Yan, J. et al. Imaging propagative exciton polaritons in atomically thin wse_2 waveguides. *Phys. Rev. B*. **2019**, *100*(12), 121301.

217 Canales, A., Kotov, O., Shegai, T. O. Perfect absorption and strong coupling in supported MoS_2 multilayers. *ACS Nano*. **2023**, *17*(4), 3401–3411.

218 Bylinkin, A., Schnell, M., Calavalle, F., Li, P., Taboada-Gutièrrez, J., Liu, S. et al. Real-space observation of vibrational strong coupling between propagating phonon polaritons and organic molecules. *Nat. Photon*. **2021**, *15*(3), 197–202.

219 Basov, D. N., Fogler, M. M., García de Abajo, F. J. Polaritons in van der waals materials. *Science*. **2016**, *354*(6309), aag1992.

220 Low, T., Chaves, A., Caldwell, J. D., Kumar, A., Fang, N. X., Avouris, P. et al. Polaritons in layered two-dimensional materials. *Nat. Mater*. **2017**, *16*(2), 182–194.

221 Evans, T. J., Schlaus, A., Fu, Y., Zhong, X., Atallah, T. L., Spencer, M. S. et al. Continuous-wave lasing in cesium lead bromide perovskite nanowires. *Adv. Opt. Mater*. **2018**, *6*(2), 1700982.

222 Zhang, Q., Su, R., Liu, X., Xing, J., Sum, T. C., Xiong, Q. High-quality whispering-gallery-mode lasing from cesium lead halide perovskite nanoplatelets. *Adv. Funct. Mater*. **2016**, *26*(34), 6238–6245.

223 Fu, Y., Zhu, H., Chen, J., Hautzinger, M. P., Zhu, X.-Y., Jin, S. Metal halide perovskite nanostructures for optoelectronic applications and the study of physical properties. *Nat. Rev. Mater*. **2019**, *4*(3), 169–188.

224 Makarov, S., Furasova, A., Tiguntseva, E., Hemmetter, A., Berestennikov, A., Pushkarev, A. et al. Halide-perovskite resonant nanophotonics. *Adv. Opt. Mater*. **2019**, *7*(1), 1800784.

225 Su, R., Fieramosca, A., Zhang, Q., Nguyen, H. S., Deleporte, E., Chen, Z. et al. Perovskite semiconductors for room-temperature exciton-polaritonics. *Nat. Mater*. **2021**, *20*(10), 1315–1324.

226 Casimir, H. B. On the attraction between two perfectly conducting plates. In *Proc. Kon. Ned. Akad. Wet*. 1948, 51, 793.

227 Milonni, P. W. *The Quantum Vacuum: An Introduction to Quantum Electrodynamics*. Academic Press, 2013.

228 Sernelius, B. E. *Fundamentals of van der Waals and Casimir Interactions*, 102. Springer, 2018.

229 Kockum, A. F., Miranowicz, A., Liberato, S. D., Savasta, S., Nori, F. Ultrastrong coupling between light and matter. *Nat. Rev. Phys*. **2019**, *1*(1), 19–40.

230 Zhao, R., Li, L., Yang, S., Bao, W., Xia, Y., Ashby, P. et al. Stable casimir equilibria and quantum trapping. *Science*. **2019**, *364*(6444), 984–987.

231 Esteso, V., Carretero-Palacios, S., Míguez, H. Casimir–Lifshitz force based optical resonators. *J. Phys. Chem. Lett.* **2019**, *10*(19), 5856–5860.

232 Munkhbat, B., Canales, A., Küçüköz, B., Baranov, D. G., Shegai, T. O. Tunable self-assembled Casimir microcavities and polaritons. *Nature.* **2021**, *597*(7875), 214–219.

233 Schmidt, F., Callegari, A., Daddi-Moussa-Ider, A., Munkhbat, B., Verre, R., Shegai. T. et al. Tunable critical Casimir forces counteract Casimir–Lifshitz attraction. *Nat. Phys.* **2023**, *19*(2), 271–278.

234 Zhao, R., Koschny, T., Economou, E. N., Soukoulis, C. M. Repulsive Casimir forces with finite-thickness slabs. *Phys. Rev. B.* **2011**, *83*, 075108.

235 Dou, M., Lou, F., Boström, M., Brevik, I., Persson, C. Casimir quantum levitation tuned by means of material properties and geometries. *Phys. Rev. B.* **2014**, *89*, 201407.

236 Munday, J. N., Capasso, F., Parsegian, V. A. Measured long-range repulsive Casimir–Lifshitz forces. *Nature.* **2009**, *457*(7226), 170–173.

237 Küçüköz, B., Kotov, O. V., Canales, A., Polyakov, Y. A., Agrawal, A. V., Antosiewicz, T. J. et al. Quantum trapping and rotational self-alignment in triangular Casimir microcavities. *Sci. Adv.* **2024**, *10*(17), eadn1825.

238 Ruggenthaler, M., Sidler, D., Rubio, A. Understanding polaritonic chemistry from ab initio quantum electrodynamics. *Chem. Rev.* **2023**, *123*(19), 11191–11229.

4

Plexciton Photophysics

Daniel Finkelstein-Shapiro

Instituto de Química, Universidad Nacional Autonoma de México, Ciudad de México, México

4.1 Goal of this Chapter

Plexcitons are hybrid light–matter states formed from the coupling of an exciton and a plasmon. They are a type of polariton in which the light field is confined within a plasmonic nanoparticle. Their sub-wavelength dimensions, ability to reach strong coupling with a single or few emitters [1], and possibility to interact with other molecules in a solvent make them especially interesting. This attention has resulted in excellent reviews on the general aspects [2,3], the design principles [4], the coupling regimes [5], their applications [6], their coherent dynamics [7], emission [8], and non-linear optical response [9]. This chapter focuses on their ultrafast dynamics, from the very earliest times accessible by femtosecond pulses until the return to the ground state. There are many additional processes that take place in plexcitons compared to microcavity polaritons, and much fewer ultrafast studies [10–23]. We provide an overview of the known dissipation mechanisms known so far to form a framework for future experiments. The discussion is heavily centered on plexcitons formed from organic molecules and metallic nanoparticles or surface lattice resonances, since most ultrafast studies have been done on these systems, and discuss to a lesser extent quantum dots and two-dimensional materials.

Ultimately, polaritons based on microcavities, nanocavities, open cavities, or plasmonic colloidal nanoparticles will have different applications, and these can be informed in part by their ultrafast dynamics. We start by discussing the energy landscape of plexcitons. We detail their synthesis and resulting structures, followed by a discussion of their ultrafast dynamics, the models to simulate them, and possible pitfalls when analyzing the data. We conclude with an outlook on the possible applications.

4.2 What is a Plexciton and How Is It Different from a Cavity Polariton

A plexciton is a hybrid composed of a plasmon and one or several emitters. In practice, however, there can be up to four important components: the plasmon, the emitters, a capping ligand layer for the nanoparticle, and the solvent if in colloidal suspension (including acids, bases, and

Polariton Chemistry: Molecules in Cavities, First Edition. Edited by Joel Yuen-Zhou, Noel C. Giebink and Raphael F. Ribeiro. Published 2026 by John Wiley & Sons, Inc.

spectator ions). The plasmon excitations are supported by metallic colloidal nanoparticles, metallic nanoparticles on a surface, or metallic arrays that form surface lattice resonances [24]. The excitonic or emitter part comes from organic molecules [4], quantum dots, or two-dimensional materials [25]. The coupling between plasmon and emitter arises most typically from a dipolar interaction [26]. To obtain a strong enough coupling, a large number of emitters N_e are used, with the coupling strength scaling as $\sqrt{N_e}$ as in microcavities [27]. Unlike a cavity, the close proximity of the plasmon and matter components makes it possible for higher order multipoles of the metal to play a role in the coupling [28]. We define any given plexciton by the emitter and plasmon frequencies, ω_e and ω_p, the dephasing rates of the optical coherences γ_p an γ_e, and the coupling strength between them g. We also characterize a given system by the number of emitters N_e and plasmon resonances N_p involved, which will help to define the likely photophysics as well as the appropriate spectroscopies to probe them. As for microcavity polaritonic states, there are two parameters that label different phenomenological regimes. The strength $S = \frac{2g}{(\gamma_p - \gamma_e)}$ separates the weak coupling ($S < 1$) from the strong coupling ($S > 1$) while the cooperativity $C = \frac{g^2}{\gamma_p \gamma_e}$ indicates where interferences are important ($C > 1$) [5]. Some of the systems investigated using ultrafast spectroscopy are in the weak coupling regime. It is not easy to distinguish them from the strong coupling regime based on peak separation alone because interference dips can look like Rabi splitting [5, 29, 30]. They also show interesting photophysics, and we will include them in the discussion. We will refer to both as plexcitons and specify where necessary if they are in the weak or strong coupling regime.

Plexcitons differ fundamentally from microcavity polaritons in three key points: (i) the emitters can be labile when in solution, reorienting and detaching from the metallic nanoparticle [31, 32], (ii) the number of emitters can reach the single or few-molecule limit [1, 33], and (iii) the plasmonic cavity has other dissipation mechanisms and additional interactions at the metal–molecule interface than are possible with a microcavity [10, 18, 34, 35]. To a first approximation, we understand the energy landscape of microcavity polaritons arising from N_e emitters coupled to a cavity, as two separate bright modes, the upper and lower polariton, and $N_e - 1$ dark states in between [36, 37]. The replacement of a cavity by a metal nanoparticle induces a more complicated energy structure, partly because there is significant structural disorder in the arrangement, and partly because of the metallic band structure. We show a typical example in Figure 4.1. We consider N_J *J*-aggregates consisting of a bright mode and a manifold of dark states (DS1) coupling to a plasmon transition in a nanoprism. The coupling of the plasmon with N_J aggregates results in additional $N_J - 1$ dark modes (DS2) at a different energy than DS1. Furthermore, those *J*-aggregates that fall outside of the hot spots do not couple strongly to the plasmon [38]. They constitute a third type of dark state that can participate in exciton or charge transfer (DS3). On the plasmonic cavity side, the plasmon–electron coupling provides a pathway for plasmon decay into hot electron–hole pairs [39]. The proximity of the molecules to the surface also opens nonradiative interactions between molecular dark states and the metal, which can result in energy quenching and charge transfer. The plasmon decay into hot electron–hole states, as opposed to photons exiting the system, provides an additional layer of many-body physics that influences the late-time dynamics in ways that could be useful for chemistry [40–42].

The detailed photophysics and photochemistry will depend on the energy landscape of Figure 4.1, as well as the geometry and type of nanoparticles, the emitters, their orientation with respect to the surface, the capping ligands, and the solvent. All of these factors vary depending on the synthesis and conditions of study.

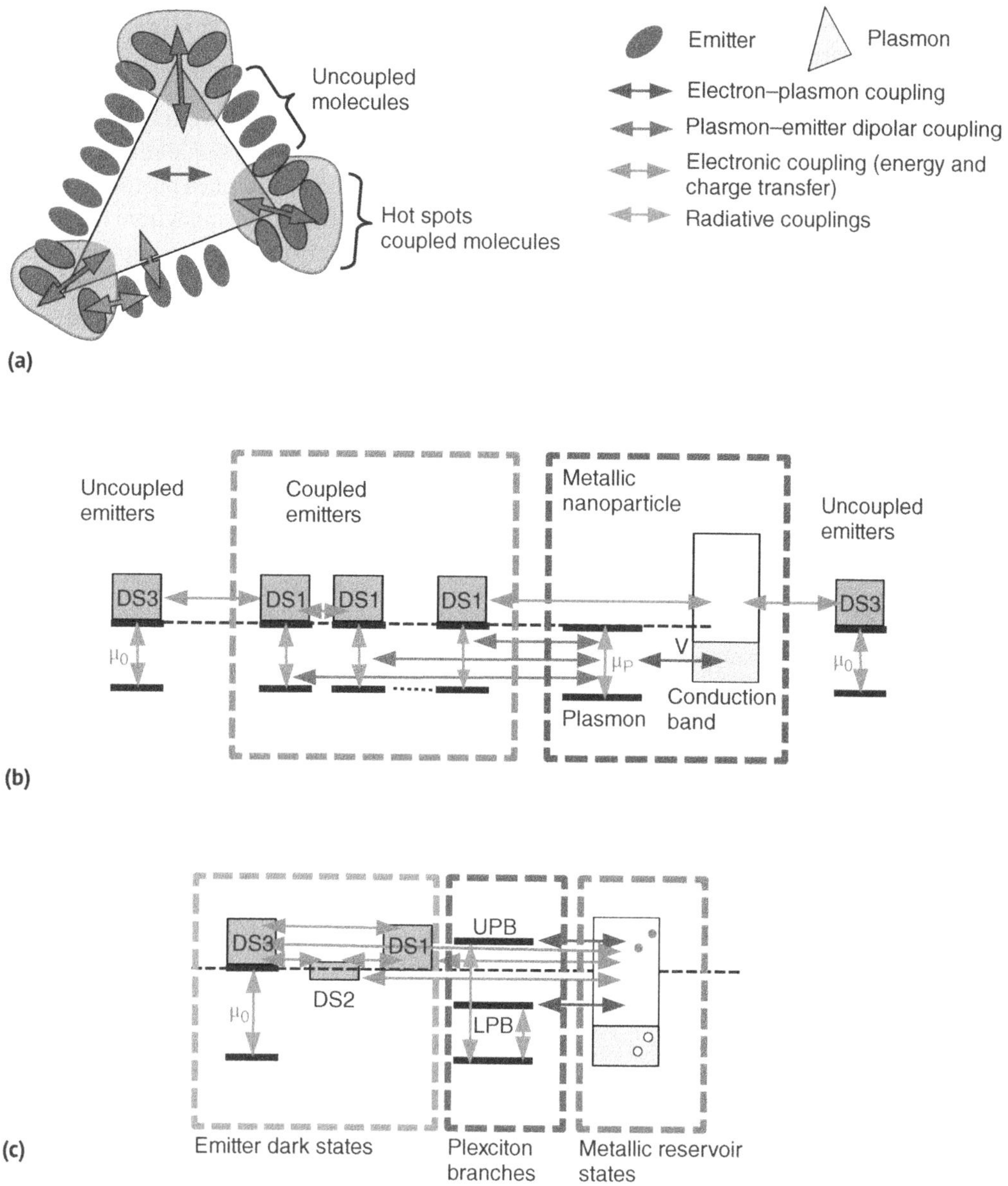

Figure 4.1 (a) Physical system consisting of emitters (in this case *J*-aggregates) coupled to a metallic nanoparticle. The sharp edges of the metal generate hot spots that provide preferential coupling to only a subset of molecules. (b) Energy structure in the site picture showing components and couplings, including dipolar coupling between the bright states of *J*-aggregates and the plasmon, the electronic coupling due to physical proximity between molecules and metal, the electron-plasmon coupling generating Landau damping. (c) Plexcitonic picture showing the two bright plexciton branches and different types of dark state manifolds, including excitations of the metal.

4.3 Synthesis of Plexcitons and Their Structure: Influence and Consequence on the Photophysics

In pairing a number N_e of emitters to a number N_p of plasmonic nanoparticles, we are left to choose from a large library of molecules, semiconducting nanoparticles, two-dimensional materials, and metallic nanoparticles. Each one provides different advantages, structures, excited potential energy surfaces, and excited state dynamics. It is not possible to infer the ultrafast dynamics from the linear optical absorption alone. The legitimate intention to strive for stronger and stronger couplings leads to a race for larger peak separations, which can leave one blind to the fact that for each successful system, a thorough structural characterization – which is in most of the cases lacking – as well as an appropriate study of its dynamics after excitation is necessary to understand the possible application the system can be used for.

Plexcitons made from the same starting materials can still produce different combinations (N_p, N_e). Low N_e are important for reducing the emitter dark state manifold density of states and occur, for example, in quantum dots in plasmonic bow-tie arrangements on surfaces [33, 43]. They are more easily studied using time-resolved fluorescence and in the future with time-resolved photoelectron spectroscopy. Large N_e (100–1000s of molecules) are more typical of colloidal systems [14, 31, 32] or surface lattice resonances, which have enough optical density to be probed easily by transient absorption [13]. Surface lattice resonance arrays are crucial for ultrafast studies in the strong coupling regime, as the slower dephasing of the collective plasmon resonance allows multiple Rabi oscillations to be measured.

The number of emitters N_e participating in the coupling dictates the size of the manifold of dark states – scaling as $N_e - 1$ – and when present tends to dominate both the excited state dynamics and the optical nonlinearity of the plexciton. As shown in Figure 4.1, structural details such as whether molecules couple to the plasmon as monomers, *J*-aggregates or H-aggregates give us information on the level alignment of dark states with respect to the polaritonic branches. The transition from any given initial bright state to dark states depends on the density of states at that energy and, for example, is more favorable for DS1 than for DS2.

4.3.1 Colloidal Systems-based on Organic Molecules

They consist of metallic nanoparticles covered by a chromophore. Ag and Au nanoparticles remain the most common plasmonic component because of their ease of synthesis, variety of shapes, and strong resonance. The types of emitters that can be used are more extensive, although cyanines are the most common, with 5,5,6,6-tetrachloro-di(4-sulfobutyl) benzimidazolocarbocyanine (TDBC) being the workhorse of polaritonic and plexcitonic systems [4]. To obtain the strongest couplings, the shapes have favored the inclusion of sharp edges, such as in prisms, disks, donuts, and urchins. Unfortunately, the advantage afforded by sharp edges can be transient as often times they either reconstruct over time, or break off from the core after functionalization [18, 32]. Some systems, however, can also last for several months without noticeable changes in their absorption spectrum, with negligible dye degradation, even with prolonged laser exposure. This enhanced stability has been explained by the opening of relaxation pathways that suppress the formation of oxidative oxygen species [44].

The capping layer of the nanoparticle, the chromophore, its concentration relative to the nanoparticles, and the pH have been shown to change whether the molecules adsorb as monomers, *J*-aggregates, *H*-aggregates, and in which geometry [31]. Studies of porphyrin derivatives on small and large Au nanoparticles showed that in basic conditions (pH = 11) porphyrins form

H-aggregates, while in acidic conditions (pH = 2) they tend to form *J*-aggregates. The surface coverage tunes the fraction of monomer or aggregate, while the particle size further determines the relative orientation with respect to the surface. Porphyrins on small nanoparticles are guided by the capping layer to adopt a flat geometry on the surface, while porphyrins on large nanoparticles prefer a quasi-perpendicular orientiation. Moreover, symmetric binding groups on a porphyrin were shown to create dynamical nanocavities of colloidal Au nanoparticles with increased hot-spots [31].

4.3.2 Open Cavities

Open cavities plexcitons are made by depositing chromophores on plasmon lattices. Au gratings have played a very important role in the study of early time dynamics (< 100 fs) [13]. The chromophores entering the grooves of the grating are strongly coupled, while the outermost layer does not couple to the plasmon. A first advantage of open cavities is that the collective plasmon mode slows down its dephasing significantly [7, 13, 16], and another one is that the cavity does not filter any wavelengths of light. As opposed to metallic nanoparticles that fix the plasmon resonance, a full dispersion curve can be obtained. Other important semi-open cavities were TDBC on aluminum nanodisks [45, 46], and WS_2 deposited on Ag arrays [47]. There is less variation in the structure of these plexcitons that has been reported compared to colloidal systems, mainly because of the absence of a solvent and the formation of a rigid film.

4.3.3 Surface Nanocavities

Nanocavities are made from two nanoparticles or a nanoparticle and a surface, separated by a gap where the field strength increases. The few emitter strong coupling limit can be achieved here, and so these are the family of systems that can provide an idea of the dynamics without a dark state manifold. The emitters can be transition metal dichalcogenides, molecules [1], or quantum dots [5, 33]. Each nanocavity is different, and these have been studied mainly with single-molecule fluorescence spectroscopy.

4.4 Photophysics of Plexcitons

There are several time-resolved methods that can be used to probe plexciton dynamics. They all consist of an initial pump pulse sequence that prepares the system in the excited state, followed by a detection sequence some waiting time (or population time) later. If the pump sequence consists of a pulse pair whose delay is scanned in time, then the experiment will give resolution of the excitation frequency after Fourier transforming over the delay. We can do an initial separation of the methods depending on the measured observable. Transient absorption and two-dimensional electronic photon echo spectroscopy (2DES) measure an induced polarization [48, 49]. The other family of methods probes an observable related to an induced population, for example, fluorescence [33], photocurrent [50], or photoelectron emission [51–53]. The methods differ in the sample requirements, time resolution, sensitivity, and the information that they provide. For example, transient absorption cannot directly measure the dynamics in (truly) dark states, whereas photoelectron spectroscopy can probe all electronic states. Transient absorption spectroscopy and 2DES require large optical densities (0.3–0.5) meaning a large amount of sample. It is challenging to apply them for single plasmon or single plexciton spectroscopy [54]; however, a big advantage is that they can be done in solution and so can study the dynamical structures we have discussed earlier [31].

They can also be used in films and open cavities as well. They provide the best time resolutions down to around 10–20 fs.

Fluorescence spectroscopy can be done at the single-particle level, and can be simpler to interpret. However, it strictly probes the population of emissive states. There are different variations of the method. The most common one is time-correlated single-photon counting (TCSPC) and has a resolution of a few tens of picoseconds at best. The statistics of the emissions can be measured and used to derive the number of emitters in the cavity [33]. Streak camera detection or upconversion-gated fluorescence can improve the time resolution of the experiment to the sub-picosecond regime [55], while two-dimensional fluorescence spectroscopy has both the same time resolution as transient absorption and provides excitation wavelength resolution in the second dimension [56], although it has not been applied to plexcitons.

Time-resolved photoelectron spectroscopy can be done at the single-particle level, and resolve the energy of every electron inside the metal, and hence measure dark states. It cannot be done yet in solution and requires careful analysis [52, 57]. It has yet to be applied to study time-resolved dynamics in plexcitons, although there are encouraging results in bare metals and atoms [51, 58]. Transient absorption or 2DES can be simpler to interpret in molecular systems with well-separated resonances because each transition results in well-separated signals (even if the lineshapes can be very complicated and carry a lot of physics). In systems with extended band structures like metals, the response of a probe photon is energy-averaged over all the electrons in the band structure, requiring more detailed calculations [59–61]. Photoelectron spectroscopies are likely to be a way forward for more discriminating signals.

Regardless of the spectroscopy used, the photophysics of plexcitons can be divided into an initial transient response to the laser fields followed by the coherent dynamics between plasmon and exciton (Rabi oscillations), ending by dynamics attributable to either molecular, metallic, or interfacial processes.

We start with a brief description of the dynamics known in emitters (molecules, semiconductors, and two-dimensional materials), followed by the photophysics in metals. We then describe the photophysics that has been reported in plexcitons. Compared to molecular or metallic photophysics, the field of plexciton photophysics is still emerging. Several of the assignments appearing in the literature are tentative and should be tested and taken as a guideline to interpret subsequent experiments. Ideally, we can follow in the footsteps of the study of their individual components and explore the temperature, particle size, pulse intensity, and exciting frequency dependence of the dynamics to delineate what is plexciton photophysics. We include the dependence on the pulse properties as much as possible, to suggest directions of interest for future experiments.

4.4.1 Emitters

4.4.1.1 Organic Molecules

Emitters can be chosen from a variety of organic molecules, quantum dots, and two-dimensional materials. Of the plexcitons built from organic molecules, around 80% are made with cyanine derivatives that couple to the plasmons as *J*-aggregates (see Figure 4.1, where DS_1 is the dark states of *J*-aggregates and DS_2 are the dark states formed by coupling several *J*-aggregates to the plasmon). There is an extremely strong body of literature investigating the photophysics of *J*-aggregates with transient absorption and two-dimensional electronic spectroscopy [62–65]. *J*-aggregates have narrow linewidths and small Stokes shifts, with correspondingly long coherence times (several tens of femtoseconds at room temperature). The first process after dephasing is a transfer to dark states lying above the bright transition. Although not energetically favorable, this rate is very fast

(~ 100 fs) given the large density of states [18]. Once in the molecular dark state (DS1), the excitations can live for very long times, during which they diffuse along the aggregate. When two excitations meet, they decay via exciton–exciton annihilation, a pump-pulse intensity-dependent process happening on the order of a few ps [62, 63]. Eventually, the excited state transitions back to the bright state and recombines radiatively on the order of a few ns. Ground-state bleach and stimulated emission overlap and the excited state absorption is usually blueshifted with respect to the ground-state absorption.

4.4.1.2 Inorganic Materials

Quantum dots have been used in plexcitons. They have large transition dipole moments, which make possible single or few-molecule strong coupling regimes [3, 5, 33]. Their dynamics vary but are similar to those we will describe for metals. After initial excitation and dephasing, the exciton dissociates into electron–hole pairs that undergo electron–electron scattering, electron–hole scattering to eventually thermalize to the band edge. From there, there are many possible transitions back to the ground state, either by photon emission, nonradiatively, or via Auger processes following different carrier density dependencies. In quantum dots, bi-excitons and trions are possible and common [66, 67]. Transition metal dichalcogenides have also been incorporated into nanocavities. Qualitatively, they have the same processes as quantum dots, although the magnitudes vary due to the difference in magnitude of the many-body interactions [68].

4.4.2 Metallic Nanoparticle

4.4.2.1 Dephasing of the Plasmonic Transition

The plasmon can confine the light field below the diffraction limit in a coherent superposition of the electromagnetic radiation and the conduction electrons [2]. In metals, there are two main mechanisms that dephase the plasmon: (i) radiation damping – whereby the plasmon decays radiatively – is dominant for sizes larger than 10 nm and (ii) Landau damping – whereby the plasmon decays into electron and hole excitations inside the band structure – is the dominant mechanism for small diameters < 4 nm [35, 39, 69, 70]. The typical dephasing rates of localized surface plasmon resonances range from a few to tens of femtoseconds with surface lattice resonances extending their coherence time upwards of 100 fs [24]. The short lifetimes are measured interferometrically [71, 72] or inferred from linewidth analysis [18]. As a caution, linewidth analysis serves as an upper bound for the dephasing rate since the heterogeneity also contributes to the linewidth [60, 61]. The shape of choice for the metal nanoparticles due to their strong hot spots are prisms, urchins, sponges or donuts [4]. Their dimensions fall both below and above the dominance threshold of radiation vs. Landau damping.

The generation of metallic excitations in the band structure can occur via plasmon decay (as in Ag) or, in addition, have contributions from direct excitation of interband transitions (as in Au), in both cases resulting in an out-of-equilibrium distribution of electrons in the metal $f(E)$ [34]. For Pt, the two resonances have even been shown to interfere and produce Fano lineshapes [73].

4.4.2.2 Dynamics of the Out-of-equilibrium Electron Distribution

After dephasing, the electron distribution of the metal $f(E)$ is out-of-equilibrium as a fraction of electrons has been promoted from filled to empty states. The electrons can subsequently scatter among themselves, and with phonons from the lattice, exchanging energy in both processes [34, 54, 57, 74–82]. Although they are not sequential, it is common to study them separately, assuming that electron–electron scattering is much faster than electron–phonon scattering. This separation of timescales allows us to treat the electrons and the lattice as two energy reservoirs in internal thermal equilibrium that slowly come to thermal equilibrium between themselves. It is an extremely useful picture to develop intuition and derive a dependence of these processes on the pulse intensity and frequency. This will ultimately help in solidifying the assignments of each rate process. We have from Fermi liquid theory that $\tau_{ee}^{-1} = K(E - E_f)^2$, where τ_{ee} is the e-e scattering time, E is the scattering electrons energy and E_F is the Fermi energy and K is a proportionality constant [75, 83]. Importantly, the scattering rate increases with the energy of the electrons. This electron energy is the plasmon energy if the excitations come from plasmon decay, or related to the frequency of the pump pulse if they come from an interband transition. τ_{ee} can be taken smaller than 500 fs in Ag and around 100 fs in Au. A quadratic dependence on the pulse frequency and intensity has been measured for Au [61]. Once the electrons are thermalized with each other, we begin to consider them as a thermal reservoir at temperature T_e with the heat capacity of an ideal Fermi gas $C_e = \gamma T_e$, with γ a constant. These electron temperatures can reach from a few tens of degrees above the lattice temperature up to $\sim$1500 K.

4.4.2.3 Thermalization of the Electron Reservoir with the Lattice Reservoir

The energy rate of exchange between electrons and phonons is approximately proportional to the energy difference between the two reservoirs, that is, the energy flow $\frac{dE_e}{dt}$ out of the electron reservoir is:

$$T_e\gamma\frac{\mathrm{d}T_e}{\mathrm{d}t} = -G_{eph}(T_e - T_l) \tag{4.1}$$

where T_e, T_l are the electron and lattice temperatures, G_{eph} is the electron–phonon coupling constant, and E_e is the energy of the electron reservoir [75]. Because of the temperature-dependent heat capacity, the temperature rise of the electron reservoir is not proportional to the pulse energy. For intense pulses, the temperature rise of the electron reservoir per pulse energy is smaller than for low-intensity pulses. The energy flow out of the electron reservoir is no longer proportional to the input energy, and at high pulse powers the two reservoirs take longer to thermalize. An effective electron–phonon scattering time was derived [75]:

$$\tau_{\text{e-ph}} \approx \frac{\gamma T_l}{G_{eph}}\left(1 + \frac{E_L}{2\gamma T_l^2}\right) \tag{4.2}$$

where E_L is the energy of the pump pulse, T_l is taken as the final electron temperature after equilibration. For large pulse energies, thermalization time with the lattice increases. This theory agrees very well with experiment [61, 75].

The last step of the return to equilibrium is the thermalization of the lattice with its environment. This is governed by the phonon–phonon scattering time $\tau_{\text{ph-ph}}$ is much slower than all of the other processes, and of the order hundreds of picoseconds to nanoseconds [54, 84, 85]. Under some conditions, the energy deposited in the lattice launches a coherent phonon mode of the entire nanoparticle (around 10–100 GHz) – a very useful process to diagnose the fate of the initial photon energy. In any system where coherent phonons are launched, there are two main mechanisms: the displacive mechanism whereby excited electron can exert enough pressure on the lattice that it kicks it outwards (analogous to having a pendulum and giving it an impulse that sets it in motion), or the indirect mechanism, whereby the energy can warm the lattice enough that its equilibrium position changes instantaneously on the timescale of the lattice vibrations (analogous to translating instantaneously the fixing point of the string of a pendulum laterally, thereby transforming the current position of the weight into the point of maximum displacement and causing it to move) [76, 80–82, 85]. These two mechanisms have different initial phases with respect to time zero. The displacive mechanism oscillates with as $\sin(\omega_0 t - \phi_D)$ where

$$\phi_D = \arctan\left(\frac{\omega_0}{1/\tau_{e-ph} - 1/\tau_h}\right) \tag{4.3}$$

while the indirect mechanism oscillates as $\cos(\omega_0 t - \phi_I)$, where

$$\phi_I = \arctan\left(\frac{1}{\omega_0 \tau_h} + \frac{\omega_0}{1/\tau_{e-ph} - 2/\tau_h}\right) \tag{4.4}$$

where τ_h is the homogeneous dephasing time of the nanoparticle phonon mode (which can be estimated for a perfect sphere of radius R of a material with sound velocity v_L as $\tau_h = \alpha R/v_L$ with $\alpha \sim 2.1 - 2.9$ and is upwards of 1 ps for a 5 nm radius Au NP and ω_0 is the phonon mode frequency) [85].

The resulting modulation of the optical absorption by the phonon mode is a strong oscillation of the plasmon frequency, via the change in volume of the nanoparticle. As we discuss in the next section, care should be taken not to confuse an excited state with a thermally hot ground state. If the heterogeneous broadening is large enough, or if the energy in the pump pulse is not strong enough to launch a visible phonon mode, we can still observe a shift in the spectrum resulting from the warming of the lattice. The dominance of displacive vs. indirect mechanism depends on the particle size. Below 10 nm, the displacive mechanism is important, while for larger sizes it is the indirect thermal effect that dominates. For nanoparticle sizes used in plexcitons, it is the latter effect that will be more common.

We have provided here a very abridged summary of the main processes that occur in metals. We have striven to emphasize the functional dependence on important experimental knobs, namely the pump pulse frequency and power, and in doing so have skipped a number of subtleties of the metallic dynamics that are studied in metals. These should be kept in mind as the interpretations become more refined [51].

4.4.3 Plexcitons

With a knowledge of the dynamics in the individual components, we need to understand what emerging dynamics occur from the coherent coupling of emitters and plasmons, which new pathways open due to interfacial processes, and how the previously described individual mechanisms are affected by the coupling of emitters and plasmons. An overview of the dynamics that have been measured in plexcitons is summarized in Figure 4.2, with time constants also reported in Table 4.1, with most contributions coming from studies done on plexctions built from organic molecules. A given system does not show all processes, and not all systems have been measured at all timescales.

4.4.3.1 Rabi Oscillations

Rabi oscillations are the hallmark of strong coupling. They have been measured both in surface lattice resonances [7, 13, 16] and more recently in colloidal nanoparticles, with both two-dimensional spectroscopy and transient absorption. Vasa et al. were the first ones to observe oscillations with a period of 35 fs on 2,2-dimethyl-8-phenyl-5,6,5,6-dibenzothiacarbocyanine chloride adsorbed on Au surface lattice resonance [13]. They noticed that the period of the oscillation decreased over time after excitation. This was attributed to the reduction in Rabi contraction due to radiation damping. As the pump pulse excites the molecules out of the ground state, the Rabi splitting is reduced on

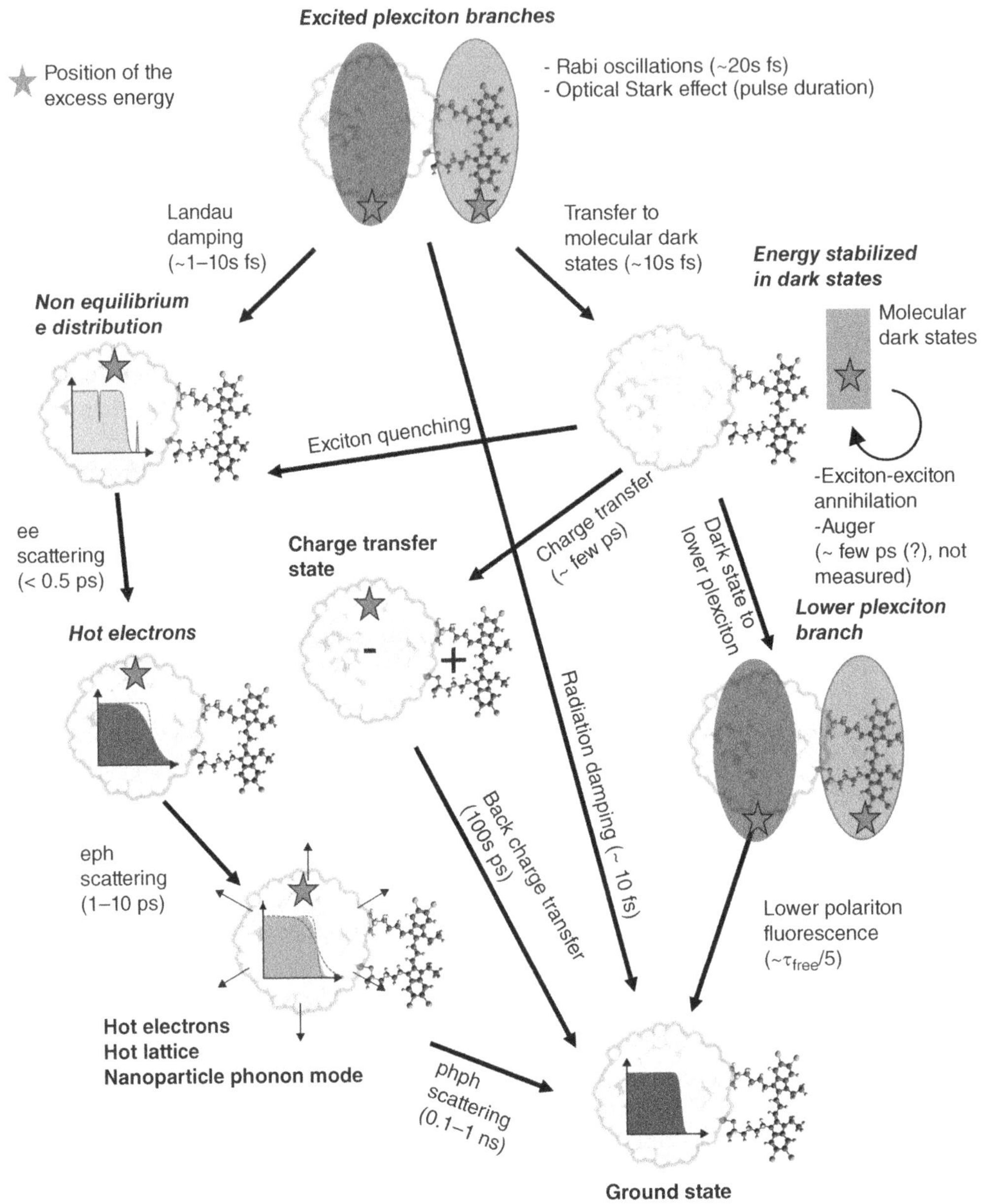

Figure 4.2 Relaxation pathways in plexcitons. The red star marks whether the energy resides in the nanoparticle or the emitter. The inset in the nanoparticles show the electron distribution in the conduction band.

account of having less molecules in the ground state (see Section 4.5). At very short times, radiation damping can return excitations back to the ground state, increasing the Rabi splitting and hence the observed oscillation's period. Rabi oscillations are in general not easy to see in colloidal nanoparticles where losses are much larger. Peruffo et al. have measured only the first cycle with period 23 fs for TDBC on Au nanourchins with a dephasing time of 16 fs [20, 22].

Table 4.1 Processes and time constants occurring in plexcitons made from molecular aggregates.

Process	Range	Intensity and wavelength dependence	Modification in plexcitons	References
Rabi oscillation	~ 30 fs			[13]
e-e scattering	$\sim 50 - 500$ fs	$\tau_{ee} \propto I_{\text{pulse}}^{-2}, \omega_{\text{pulse}}^{-2}$ [a]	Increases	[18, 34, 88]
e-ph scattering	~ 1 ps	τ_{e-ph} increases with increasing pump pulse power	Follows metallic behavior Longer	[61, 84, 89]
Exciton–exciton annihilation	1–10 ps	Timescale decreases with increasing pump power	–	[62, 63]
Electron injection into metal	5 ps		–	[12]
Electron recombination from metal	100–500 ps		–	[10, 12]
Lattice thermalization	> 300 ps		–	[18, 84]
Acoustic mode period	~ 10 ps	Visible at high powers	Period increases in plexcitons or is absent	[84]

a. In Au.

4.4.3.2 Plexciton Decay

The energy that exists in the polariton branches decays quickly, either to molecular reservoirs (i.e., molecular dark states), metallic reservoirs (Landau damping) or back into the far field (Radiation damping). The radiation damping timescale was measured in the open Au cavities at 20 fs [13]. The timescale for transfer to molecular dark states in TDBC-Ag plexcitons was measured at 40 fs, roughly twice as fast as in the *J*-aggregates alone, possibly on account of a more favorable energy level alignment [18]. Landau damping extracted from linewidth analysis was measured for the same system to occur on the timescale of a few femtoseconds [18].

4.4.3.3 Dynamics in the Molecular (or Emitter) Dark States

There is very little information that has been gleaned as to what happens with the molecular dark states. Exciton–exciton annihilation (or Auger processes in semiconductors) have clear pump pulse intensity dependencies and no convincing signatures of these processes have appeared yet. What has been observed is that in spite of the long lifetimes in bare molecules, in some systems, the association of molecules and plasmons tends to quench what would be expected to be a long-lived excitation in the molecules, and limit its lifetime to that of the longest many-body processes inside the metal [14, 18].

4.4.3.4 Dynamics in Metallic Band Structure

Plasmon–electron coupling generates hot electron–hole state in plexcitons as well. Theoretical calculations suggest that the hot electrons appear in resonance with the upper and lower plexciton branches [86]. From there, processes mirror those in metals, electron–electron and electron–phonon scattering takes place [18, 20, 22, 23, 45, 46, 87]. There are, however, conflicting reports.

Some report an increase in the lifetime of sub-ps processes which can be attributed to many-body scattering [14, 18, 20]. In contrast, some reports measure a decrease in τ_{e-ph} interpreted as an increase in interfacial scattering in the aggregates as well as a better electron–phonon spectral overlap [87]. Possibly again the difference lies in the preparation of the samples where different mechanisms can be at play. Pulse intensity-dependent measurements on Au plexcitons have shown that τ_{e-ph} increases with pulse intensity as expected for pure Au [20, 23]. These first studies are encouraging as to the possibility of modifying effectively the excitations in the metal as a result of the chromophore layer.

4.4.3.5 Interfacial Processes

Having emitters and metals in such close proximity opens many interfacial channels for energy and charge. We previously mentioned the quenching of the molecular dark states by the presence of the surface. A few studies in the weak coupling regime have reported an increase of the excitation lifetime resulting from electron injection from molecule to metal, with potentially hundreds of molecules injecting charge on a single nanoparticle [10, 12]. This mechanism was present on Ag but not Au, evidencing a need for level alignment [10]. A flat-face geometry was hypothesized in one study [10] but not the other [12]. The charge injection from molecule to metal was measured to be at 6 ps, while recombination took much longer and ranged from 100 to 300 ps, much longer than any metallic excited state process.

4.4.3.6 Coherent Phonons

The coherently launched nanoparticle phonon mode can be present or absent due to the influence of a molecular layer. Under similar excitation conditions (> 1 nJ/pulse), we observed nanoparticle oscillations in bare Ag prisms but not in plexcitons [18]. This was consistent with energy being dissipated outside of the nanoparticle. For increased pump powers similar to those used in transient absorption, we could recover the oscillations. Consistent with this, Kirshner et al. have found phonon modes in transient absorption spectra of bypiramids decorated with thiacarbocyanine dyes [90]. The presence of the dye slightly increases the oscillation period in plexcitons compared to plasmons (42.52 ps increasing to 43.61 ps), which is expected from adding mass on the surface of a resonator.

4.4.3.7 Fluorescence from the Lower Plexciton Branch

Static measurements of TDBC on Ag showed experimentally and theoretically that emission happens from the lower polariton branch as well as uncoupled *J*-aggregates. A last pathway to consider is the repopulation of the lower polariton by molecular dark states followed by emission [43, 91]. TCSPC experiments yielded a fast instrument response limited component and a slower component. On plexcitons built from quantum dots in Au bow-ties, Gupta et al. extracted the emission lifetimes by direct measurement as well as by the emission statistics (Hanburry-Brown–Twiss experiment) [33]. They observed that the quantum dot emission rate outside of the cavity at 24 ns was only slightly reduced (by a factor of 5) to around 5 ns inside the cavity, which is not a very large Purcell enhancement factor. Curiously, measurements in microcavities also yield this factor of 5 reduction in the lifetime [92]. While offering moderate radiative rate enhancements, it seems that the strong coupling regime in plexcitons does not make the radiative transition from the lower polariton branch an extremely efficient decay pathway, when compared to all the others [8]. Unfortunately, ultrafast studies are seldom done using both transient absorption spectroscopy and time-resolved fluorescence.

4.4.3.8 Charge Migration

There are two further important ultrafast dynamics phenomena worth mentioning. The delocalized nature of plasmons makes them good candidates for long-range energy migration [8]. This spatial migration was measured by ultrafast microscopy recently [21]. The second effect is vibrationally assisted plexciton condensation, which was measured in open cavities using pump pulses below and above the threshold. Above the threshold, a clear vibrational structure in the signal emerged [93]. Both of these are exciting new directions related to two possible plexciton applications.

4.5 Spectral Signatures

It is essential to know the spectral signatures of the different states described in the previous section (Figure 4.2). Given that the vast majority of experiments are related to transient absorption, we focus on those experiments. Fluorescence signatures are simpler to interpret, while photoelectron spectroscopy requires a different theory that has been very recently demonstrated experimentally and theoretically for bare metal [51] and which is not developed for plexcitons yet. Phenomenological models such as the two- and three-temperature models have proven useful for interpreting carrier dynamics in metals and could be adapted to describe relaxation processes in plexcitons [94].

We start with a brief summary of linear absorption. There are three approaches that can successfully reproduce the spectra: coupled mode theory [95], non-Hermitian Hamiltonians [26, 30], and Mie scattering for core–shell nanoparticles [96, 97]. It has been recognized throughout the literature that in the weak coupling regime the spectra of molecules on plasmons show interference patterns (called electromagnetically induced transparency or Fano interferences), while above it, they show a Rabi splitting [5, 29, 30]. Unfortunately, interferences can look very much like a Rabi splitting [30], and for some systems, it is not so obvious whether there is a Rabi splitting or an interference, in particular if there is heterogeneous broadening of the plasmon resonance so that the dephasing rate cannot be read from the width of the resonance in the linear absorption spectrum. The dynamics have not been studied in enough detail in both of these cases, with a variety of couplings and detunings to understand exactly the consequences in the ultrafast dynamics at early and long times under both weak and strong coupling.

Rigorously speaking, nonlinear spectroscopies of a given order should be modeled by calculating all of the double-sided Feynman diagrams [98] (Figure 4.3a), classified according to the contributions of ground-state absorption (GSB), representing the transitions that are bleached in the excited system, simulated emission (SE), which are the emitted photons of the system in the excited state, and excited state absorption (ESA) which corresponds to the new transitions to higher lying excited states from the first excited state manifold. Note that we represent the GSB and SE contributions as positive signals, while ESA appears as a negative signal. In general, coupled oscillator approach or Mie theory, which is a linear theory, should not be suitable for isolating a signal to a given order in the electric field. However, in plexcitons and polaritons, there is a drastic simplification: the excited state quickly transitions to dark states, and the response of the plexciton to the probe pulse is dominated by all of the remaining molecules in the ground state. Thus, there is no SE, and the ESA contribution is that of the ground state of a system that has less molecules in the ground state, that is less molecules that participate in the vacuum Rabi splitting. This is referred to as the Rabi contraction nonlinearity (see Figure 4.3b). The spectrum can be simulated by the substraction of two linear responses with varying numbers of molecules, and this is why it has also been simulated with very good agreement by input–output theory in infrared microcavities [36]. There is not much information concerning the charge transfer transient spectra, but from the available experiments

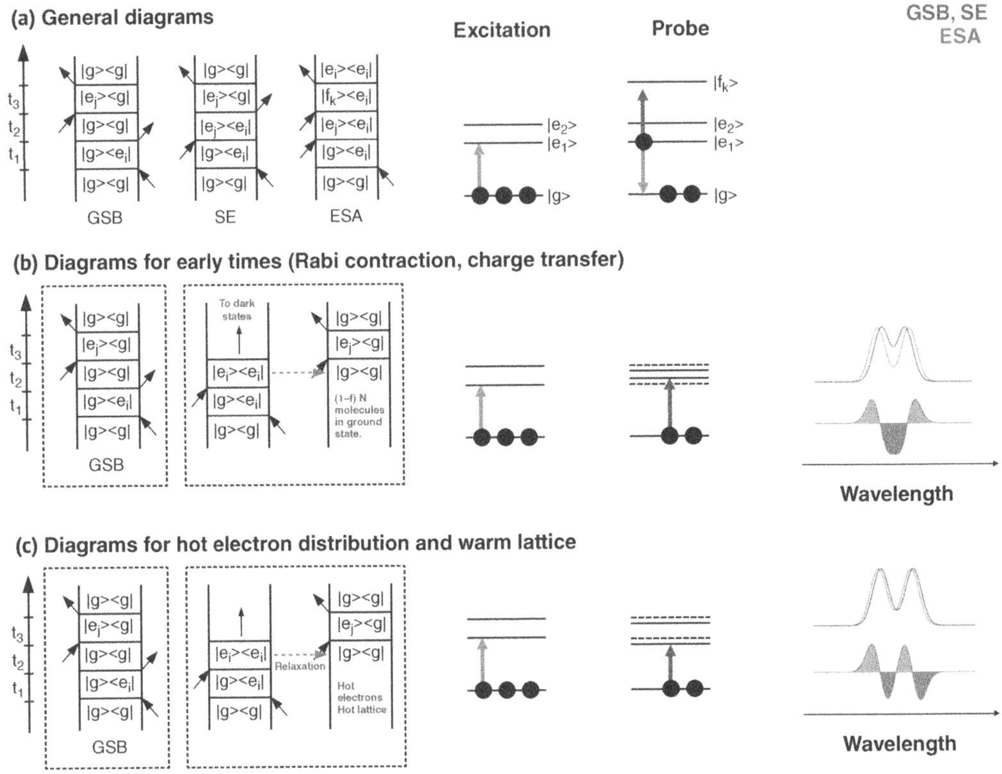

Figure 4.3 Feynman diagrams and corresponding transitions along with spectral signatures for (a) a general case, (b) a Rabi contraction, and (c) a hot electron distribution and a hot lattice.

we can see that it looks very similar to a Rabi contraction. Recent simulations with non-Hermitian Hamiltonians have suggested a symmetry change in the 2DES spectrum when going from the weak to the strong coupling [30]. Discussion of other nonlinearities is also ongoing [11, 19, 46].

The metallic response in the intermediate times that involves the evolution of the electron and lattice reservoirs is also luckily very similar. The metal with hot electron and lattice reservoirs can be modeled by a red-shifted and broadened Lorentzian [34], and the transient absorption or 2DES signal can be calculated by substracting to the linear absorption the signal of a hot equilibrium distribution (Figure 4.3c). We used this approach to simulate a hot plexciton ground state at late times [18]. Stete et al. adapted the Rosei model [99] coupled to a core–shell formula for the Mie scattering of plexcitons to successfully model the later times dynamics (> 200 fs) of a lattice with an increased temperature [23]. A good agreement with the spectral signatures and with the dependence on pump pulse intensity was found. In terms of many-body processes, the only region that has not been fit to experiments is the earlier times, where electron–electron scattering dominates (10–200 fs in Au). The broad form of the spectral signatures should be good support for discriminating whether molecular electronic excited states are still there (Rabi contraction) or if we are observing ground state dynamics. With these general guidelines already established, we can start to focus on the finer points of plexciton dynamics that are hiding in the spectra, in particular those occurring in the 10–200 fs timescale where plexcitons dephase.

There are two additional processes that have been measured and invite to caution as they can be mistaken for intrinsic dynamics. Lietard et al. measuring two-dimensional spectra for Au nanorods found a very pronounced transient grating signal for negative times and during pulse overlap [60]. Vasa et al. observed that a red detuned excitation dressed the plexciton, causing an Optical Stark Shift as derivative-like shape at very early times [16].

4.5.1 Suggestions for Approaching Transient Absorption Spectra of Plexcitons

- Carry out a good lineshape fit of the linear absorption to help determine the parameters of the Hamiltonian.
- Be weary of the very early times. Optical Stark effect or transient grating signals (in the case of 2DES) can dominate around pulse overlap but also misleadingly beyond.
- Hypothesize a nonlinearity. Currently, a Rabi contraction is a good guess, although others are being investigated.
- Can the models of a hot lattice explain intermediate times (0.2–5 ps)? Does the late time (100 ps) spectra look like a hot plexciton? Recent models such as the one from Stete et al. [23] should reproduce the spectra.
- Is there a charge transfer from molecule to metal? Tentatively, this can be seen as a very long-lived Rabi contraction signal (upwards of 100 ps).
- Is there a vibrational coherent mode of the nanoparticle? If so, the phase of the oscillation can help in ascertaining that the electron–phonon timescale is appropriately assigned.
- When there is still flexibility in performing the experiment, different pump intensities or having samples with Au and Ag go a long way in evidencing the electron–phonon scattering time and suppressing charge transfer.

4.6 Applications

More than anything, plexcitons are different from microcavities in that the interaction with solvent and solute molecules can occur. Long-range energy transfer will probably be helped by surface lattice resonances, and directional emission, given the flexibility in preparing the nanoparticles, should also be further explored.

4.6.1 Photostability

Detailed experiments on the photostability of plexcitons composed of cyanine dyes of Ag prisms were carried out by Munkbhat et al. [44]. They found that TDBC, which usually bleaches after minutes of irradiation, becomes quite stable under strong coupling. An extensive study of photodegradation rates and fluorescence yields confirmed a model for stabilization involving the modification of decay pathways. The interaction of oxygen with an adsorbed layer is just an example of an undesirable reaction that was quenched. However, further studies on the interaction with other molecules for other chemical transformations – photoprotection or otherwise – seem like a good application.

4.6.2 Hot Electron Hole Generation

Hot electron photochemistry is the process of catalyzing a reaction with hot electrons from a metal. We and others have found that the lifetime of the hot carriers in decorated metals can increase compared to the bare metals. A picture that awaits confirmation is that the molecular dark states stabilize the excitation and continuously feed the metal with energy via interfacial coupling.

This results in a longer lifetime compared to bare metal, where the hot electron distribution cannot be re-excited. The study of photophysics in the presence of possible reactants could shed light on the feasibility of this proposal. After all, hot electron generation via plasmon decay is predicted to be strongly affected under strong coupling, so that more than one mechanism could make the hot electrons more reactive.

4.6.3 Chiral Cavities and Phase Transitions

A promising avenue for further study is ultrafast plexciton condensation and directional emission, particularly given the flexibility of plasmonic nanoparticles. For instance, chiral Au nanoparticles have been synthesized and coupled strongly to molecules [100, 101]. This provides a platform for studying ultrafast chiral dynamics above and below plexciton condensation.

4.7 Conclusion

Plexcitons offer notable advantages over microcavities: miniaturization, few-molecule limit and stronger interaction possibilities with the environment. This comes at the cost of a large heterogeneity of coupling strengths, strong disorder, a sensitive structure to the synthesis conditions, a dynamical structure when in water, and a great deal of possible decay pathways for the plexciton branches. The fundamental knowledge of plexciton photophysics is still being pieced together. Reiterated measurements of dynamics in the metallic band have bolstered support for some of the assignments, notably the electron–phonon scattering, but still many details remain, in particular, surrounding the dynamics in the molecular dark states. The present day knowledge should provide a platform for investigating how exactly the strong coupling changes these parameters and how useful these can become. Contrasting different metals with or without interband transitions, multiple experimental conditions and novel spectroscopies should be able to steadily provide a picture of the modification due to strongly coupled emitters. It is hoped that gathering the known timescales and processes in this chapter will help to guide and frame future experiments, as well as ask the correct questions that will unleash possible plexctonic applications in physics and chemistry.

Acknowledgments

D.F.-S. acknowledges grant ECOS-CONAHCYT No. 321169 as well as funding from DGAPA-PAPIIT IA207223.

References

1 Chikkaraddy, R., De Nijs, B., Benz, F., Barrow, S. J., Scherman, O. A., Rosta, E. et al. Single-molecule strong coupling at room temperature in plasmonic nanocavities. *Nature* **2016**, *535*(7610), 127–130.

2 Bitton, O., Gupta, S. N., Haran, G. Quantum dot plasmonics: from weak to strong coupling. *Nanophotonics* **2019**, *8*(4), 559–575.

3 Bitton, O., Haran, G. Plasmonic cavities and individual quantum emitters in the strong coupling limit. *Acc. Chem. Res.* **2022**, *55*(12), 1659–1668.

4 Peruffo, N., Bruschi, M., Fresch, B., Mancin, F., Collini, E. Identification of design principles for the preparation of colloidal plexcitonic materials. *Langmuir* **2023**, *39*(36), 12793–12806.

5 Pelton, M., Storm, S. D., Leng, H. Strong coupling of emitters to single plasmonic nanoparticles: exciton-induced transparency and Rabi splitting. *Nanoscale* **2019**, *11*(31), 14540–14552

6 Manuel, A. P., Kirkey, A., Mahdi, N., Shankar, K. Plexcitonics – fundamental principles and optoelectronic applications. *J. Mater. Chem. C* **2019**, *7*, 1821–1853.

7 Vasa, P., Lienau. C. Strong light-matter interaction in quantum emitter/metal hybrid nanostructures. *ACS Photonics* **2018**, *5*(1), 2–23.

8 Wei, H., Yan, X., Niu, Y., Li, Q., Jia, Z., Xu. H. Plasmon–exciton interactions: spontaneous emission and strong coupling. *Adv. Funct. Mater.* **2021**, *31*(51), 2100889.

9 Chen, Y., Sun, M. Plexcitonics: plasmon–exciton coupling for enhancing spectroscopy, optical chirality, and nonlinearity. *Nanoscale* **2023**, *15*, 11834–11851.

10 Hranisavljevic, J., Dimitrijevic, N. M., Wurtz, G. A., Wiederrecht, G. P. Photoinduced charge separation reactions of J-aggregates coated on silver nanoparticles. *J. Am. Chem. Soc.* **2002**, *124*(17), 4536–4537.

11 Fofang, N. T., Grady, N. K., Fan, Z., Govorov, A. O., Halas, N. J. Plexciton dynamics: exciton-plasmon coupling in a J-aggregate-au nanoshell complex provides a mechanism for nonlinearity. *Nano Lett.* **2011**, *11*(4), 1556–1560.

12 Murphy, S., Huang, L., Kamat, P. V. Charge-transfer complexation and excited-state interactions in porphyrin-silver nanoparticle hybrid structures. *J. Phys. Chem. C* **2011**, *115*(46), 22761–22769.

13 Vasa, P., Wang, W., Pomraenke, R., Lammers, M., Maiuri, M., Manzoni, C. et al. Real-time observation of ultrafast rabi oscillations between excitons and plasmons in metal nanostructures with j-aggregates. *Nat. Photonics* **2013**, *7*(2), 128–132.

14 Balci, S., Kocabas, C., Kücüköz, B., Karatay, A., Akhüseyin, E., Gul Yaglioglu, H. et al. Probing ultrafast energy transfer between excitons and plasmons in the ultrastrong coupling regime. *Appl. Phys. Lett.* **2014**, *105*(5), 051105.

15 Wang, W., Vasa, P., Pomraenke, R., Vogelgesang, R., De Sio, A., Sommer, E. et al. Interplay between strong coupling and radiative damping of excitons and surface plasmon polaritons in hybrid nanostructures. *ACS Nano* **2014**, *8*(1), 1056–1064.

16 Vasa, P., Wang, W., Pomraenke, R., Maiuri, M., Manzoni, C., Cerullo, G. et al. Optical stark effects in *j*-aggregate–metal hybrid nanostructures exhibiting a strong exciton–surface-plasmon-polariton interaction. *Phys. Rev. Lett.* **2015**, *114*: 036802.

17 Zhong, J. H., Yi, J. M., Wang, D., Korte, A., Chimeh, A., Schaaf, P. et al. Ultrafast optical dynamics of a nonlinearly coupled au plasmon-zno exciton nanostructure. In: *The 22nd International Conference on Ultrafast Phenomena 2020*, Optica Publishing Group, **2020**, p. W2B.3.

18 Finkelstein-Shapiro, D., Mante, P. A., Sarisozen, S., Wittenbecher, L., Minda, I., Balci, S. et al. Understanding radiative transitions and relaxation pathways in plexcitons. *Chem* **2021**, *7*(4), 1092–1107.

19 Timmer, D., Zhong, J. H., Gittinger, M., Quenzel, T., Stephan, S., Zablocki, J. et al. Rabi oscillations in disguise: strong exciton-surface plasmon polariton coupling probed by two-dimensional electronic spectroscopy, In: *The International Conference on Ultrafast Phenomena (UP)*. Optica Publishing Group, **2022**, p. Th3A.5.

20 Peruffo, N., Mancin, F., Collini, E. Ultrafast dynamics of multiple plexcitons in colloidal nanomaterials: the mediating action of plasmon resonances and dark states. *J. Phys. Chem. Lett.* **2022**, *13*(28), 6412–6419.

21 Jin, L., Sample, A. D., Sun, D., Gao, Y., Deng, S., Li, R. et al. Enhanced two-dimensional exciton propagation via strong light-matter coupling with surface lattice plasmons. *ACS Photonics* **2023**, *10*(6), 1983–1991.

22 Peruffo, N., Mancin, F., Collini, E. Coherent dynamics in solutions of colloidal plexcitonic nanohybrids at room temperature. *Adv. Opt. Mater.* **2023**, *11*(15), 2203010.

23 Stete, F., Bargheer, M., Koopman, W. Ultrafast dynamics in plasmon–exciton core–shell systems: the role of heat. *Nanoscale* **2023**, *15*(40), 16307–16313.

24 Kravets, V. G., Kabashin, A. V., Barnes, W. L., Grigorenko, A. N. Plasmonic surface lattice resonances: a review of properties and applications. *Chem. Rev.* **2018**, *118*(12), 5912–5951.

25 Anantharaman, S. B., Jo, K., Jariwala, D. Exciton-photonics: from fundamental science to applications. *ACS Nano* **2021**, *15*(8), 12628–12654.

26 Delga, A., Feist, J., Bravo-Abad, J., Garcia-Vidal, F. J. Theory of strong coupling between quantum emitters and localized surface plasmons. *J. Optics* **2014**, *16*(11), 114018.

27 Törmä, P., Barnes, W. L. Strong coupling between surface plasmon polaritons and emitters: a review. *Rep. Prog. Phys.* **2015**, *78*(1), 013901.

28 Rousseaux, B., Baranov, D. G., Antosiewicz, T. J., Shegai, T., Johansson, G. Strong coupling as an interplay of quantum emitter hybridization with plasmonic dark and bright modes. *Phys. Rev. Res.* **2020**, *2*, 033056.

29 Faucheaux, J. A., Fu, J., Jain, P. K. Unified theoretical framework for realizing diverse regimes of strong coupling between plasmons and electronic transitions. *J. Phys. Chem. C* **2014**, *118*(5), 2710–2717.

30 Finkelstein-Shapiro, D., Mante, P. A., Balci, S., Zigmantas, D., Pullerits, T. Non-Hermitian Hamiltonians for linear and nonlinear optical response: a model for plexcitons. *J. Chem. Phys.* **2023**, *158*(10), 104104.

31 Peruffo, N., Parolin, G., Collini, E., Corni, S., Mancin, F. Engineering the aggregation of dyes on ligand-shell protected gold nanoparticles to promote plexcitons formation. *Nanomaterials* **2022**, *12*(7), 1180.

32 Peruffo, N., Mancin, F., Collini, E. Plexcitonic nanohybrids based on gold nanourchins: the role of the capping layer. *J. Phys. Chem. C* **2021**, *125*(36), 19897–19905.

33 Gupta, S. N., Bitton, O., Neuman, T., Esteban, R., Chuntonov, L., Aizpurua, J. Complex plasmon-exciton dynamics revealed through quantum dot light emission in a nanocavity. *Nat. Commun.* **2021**, *12*(1), 1310.

34 Voisin, C., Del Fatti, N., Christofilos, D., Vallée, F. Ultrafast electron dynamics and optical nonlinearities in metal nanoparticles. *J. Phys. Chem. B* **2001**, *105*(12), 2264–2280.

35 Khurgin, J. B. Hot carriers generated by plasmons: where are they generated and where do they go from there? *Faraday Discuss.* **2019**, *214*, 35–58.

36 Ribeiro, R. F., Martínez-Martínez, L. A., Du, M., Campos-Gonzalez-Angulo, J., Yuen-Zhou, J. Polariton chemistry: controlling molecular dynamics with optical cavities. *Chem. Sci.* **2018**, *9*, 6325–6339.

37 Ebbesen, T. W. Hybrid light–matter states in a molecular and material science perspective. *Acc. Chem. Res.* **2016**, *49*(11), 2403–2412.

38 Yankovich, A. B., Munkhbat, B., Baranov, D. G., Cuadra, J., Olsén, E., Lourenço-Martins, H. et al. Visualizing spatial variations of plasmon-exciton polaritons at the nanoscale using electron microscopy. *Nano Lett.* **2019**, *19*(11), 8171–8181.

39 Román Castellanos, L., Hess, O., Lischner, J. Single plasmon hot carrier generation in metallic nanoparticles. *Commun. Phys.* **2019**, *2*(1), 47.

40 Zhang, Z., Zhang, C., Zheng, H., Xu, H. Plasmon-driven catalysis on molecules and nanomaterials. *Acc. Chem. Res.* **2019**, *52*(9), 2506–2515.

41 Marchuk, K., Willets, K. A. Localized surface plasmons and hot electrons. *Chem. Phys.* **2014**, *445*, 95–104.

42 Hou, W., Cronin, S. B. A review of surface plasmon resonance-enhanced photocatalysis. *Adv. Funct. Mater.* **2013**, *23*(13), 1612–1619.

43 Wersäll, M., Munkhbat, B., Baranov, D. G., Herrera, F., Cao, J., Antosiewicz, T. J. et al. Correlative dark-field and photoluminescence spectroscopy of individual plasmon-molecule hybrid nanostructures in a strong coupling regime. *ACS Photonics* **2019**, *6*(10), 2570–2576.

44 Munkhbat, B., Wersäll, M., Baranov, D. G., Antosiewicz, T. J., Shegai, T. Suppression of photo-oxidation of organic chromophores by strong coupling to plasmonic nanoantennas. *Sci. Adv.* **2018**, *4*(7), eaas9552.

45 Eizner, E., Avayu, O., Ditcovski, R., Ellenbogen, T. Aluminum nanoantenna complexes for strong coupling between excitons and localized surface plasmons. *Nano Lett.* **2015**, *15*(9), 6215–6221.

46 Eizner, E., Akulov, K., Schwartz, T., Ellenbogen, T. Temporal dynamics of localized exciton-polaritons in composite organic-plasmonic metasurfaces. *Nano Lett.* **2017**, *17*(12), 7675–7683.

47 Tang, Y., Zhang, Y., Liu, Q., Wei, K., Cheng, X. A., Shi, L. et al. Interacting plexcitons for designed ultrafast optical nonlinearity in a monolayer semiconductor. *Light Sci. Appl.* **2022**, *11*(1), 94.

48 Zigmantas, D., Polívka, T., Persson, P., Sundström, V. Ultrafast laser spectroscopy uncovers mechanisms of light energy conversion in photosynthesis and sustainable energy materials. *Chem. Phys. Rev.* **2022**, *3*(4), 041303.11.

49 Jonas, M. Two-dimensional femtosecond spectroscopy. *Annu. Rev. Phys. Chem.* **2003**, *54*(1), 425–463.

50 Karki, K. J., Widom, J. R., Seibt, J., Moody, I., Lonergan, M. C., Pullerits, T. et al. Coherent two-dimensional photocurrent spectroscopy in a PbS quantum dot photocell. *Nat. Commun.* **2014**, *5*(1), 5869.

51 Pettine, J., Maioli, P., Vallée, F., Del Fatti, N., Nesbitt, D. J. Energy-resolved femtosecond hot electron dynamics in single plasmonic nanoparticles. *ACS Nano* **2023**, *17*(11), 10721–10732.

52 Wittenbecher, E., Viñas Boström, L., Vogelsang, J., Lehman, S., Dick, K. A., Verdozzi, C. et al. Unraveling the ultrafast hot electron dynamics in semiconductor nanowires. *ACS Nano* **2021**, *15*(1), 1133–1144.

53 Wittenbecher, L., Zigmantas, D. Correction of fabry-perot interference effects in phase and amplitude pulse shapers based on liquid crystal spatial light modulators. *Opt. Exp.* **2019**, *27*(16), 22970–22982.

54 Hartland, G. V. Optical studies of dynamics in noble metal nanostructures. *Chem. Rev.* **2011**, *111*(6), 3858–3887.

55 Guarin, C. A., Villabona-Monsalve, J. P., López-Arteaga, R., Peon, J. Dynamics of the higher lying excited states of cyanine dyes: an ultrafast fluorescence study. *J. Phys. Chem. B* **2013**, *117*(24), 7352–7362.

56 Widom, J. R., Johnson, N. P., von Hippel, P. H., Marcus, A. H. Solution conformation of 2-aminopurine dinucleotide determined by ultraviolet two-dimensional fluorescence spectroscopy. *N. J. Phys.* **2013**, *15*(2), 025028.

57 Ramasesha, K., Leone, S. R., Neumark, D. M. Real-time probing of electron dynamics using attosecond time-resolved spectroscopy. *Annu. Rev. Phys. Chem.* **2016**, *67*(1), 41–63.

58 Uhl, D., Bangert, U., Bruder, L., Stienkemeier, F. Coherent optical 2D photoelectron spectroscopy. *Optica* **2021**, *8*(10), 1316–1324.

59 Brown, A. M., Sundararaman, R., Narang, P., Schwartzberg, A. M., Goddard III, W. A. Experimental and ab initio ultrafast carrier dynamics in plasmonic nanoparticles. *Phys. Rev. Lett.* **2017**, *118*, 087401.

60 Lietard, A., Hsieh, C. S., Rhee, H., Cho, M. Electron heating and thermal relaxation of gold nanorods revealed by two-dimensional electronic spectroscopy. *Nat. Commun.* **2018**, *9*(1), 891.

61 Jeffries, W. R., Park, K., Vaia, R. A., Knappenberger Jr., K. L. Resolving electron-electron scattering in plasmonic nanorod ensembles using two-dimensional electronic spectroscopy. *Nano Lett.* **2020**, *20*(10), 7722–7727.

62 Lee, J. H., Min, C. K., Joo, T. Ultrafast optical dynamics of excitons in J-aggregates. *J. Chem. Phys.* **2001**, *114*(1), 377.

63 Sundström, V., Gillbro, T., Gadonas, R. A., Piskarskas, A. Annihilation of singlet excitons in j aggregates of pseudoisocyanine (pic) studied by pico- and subpicosecond spectroscopy. *J. Chem. Phys.* **1988**, *89*(5), 2754–2762.

64 Minoshima, K., Taiji, M., Misawa, K., Kobayashi, T. Femtosecond nonlinear optical dynamics of excitons in J-aggregates. *Chem. Phys. Lett.* **1994**, *218*(1), 67.

65 Milota, F., Prokhorenko, V. I., Mancal, T., von Berlepsch, H., Bixner, O., Kauffmann, H. F. et al. Vibronic and vibrational coherences in two-dimensional electronic spectra of supramolecular J-aggregates. *J. Phys. Chem. A* **2013**, *117*(29), 6007–6014.

66 Zidek, K., Zheng, K., Abdellah, M., Lenngren, N., Chábera, P., Pullerits, T. Ultrafast dynamics of multiple exciton harvesting in the CdSe-ZnO system: electron injection versus auger recombination. *Nano Lett.* **2012**, *12*(12), 6393–6399.

67 Honarfar, A., Mourad, H., Lin, W., Polukeev, A., Rahaman, A., Abdellah, M. et al. Photoexcitation dynamics in electrochemically charged cdse quantum dots: from hot carrier cooling to auger recombination of negative trions. *ACS Appl. Energy Mater.* **2020**, *3*(12), 12525–12531.

68 Li, Y., Jia, S., Yang, M., Xinyu, S., Haiyang, X., Xinfeng, L. Ultrafast carrier dynamics in two-dimensional transition metal dichalcogenides. *J. Mater. Chem. C* **2019**, *7*, 4304–4319.

69 Molina, R. A., Weinmann, D., Jalabert, R. A. Oscillatory size dependence of the surface plasmon linewidth in metallic nanoparticles. *Phys. Rev. B* **2002**. *65*, 155427.

70 Weick, G., Molina, R. A., Weinmann, D., Jalabert, R. A. Lifetime of the first and second collective excitations in metallic nanoparticles. *Phys. Rev. B* **2005**, *72*, 115410.

71 Anderson, A., Deryckx, K. S., Xu, X. G., Steinmeyer, G., Raschke, M. B. Few-femtosecond plasmon dephasing of a single metallic nanostructure from optical response function reconstruction by interferometric frequency resolved optical gating. *Nano Lett.* **2010**. *10*(7), 2519–2524.

72 Nishiyama, Y., Imaeda, K., Imura, K., Okamoto, H. Plasmon dephasing in single gold nanorods observed by ultrafast time-resolved near-field optical microscopy. *J. Phys. Chem. C* **2015**, *119*(28), 16215–16222.

73 Hao, Q., Juluri, B. K., Zheng, Y. B., Wang, B., Chiang, I. K., Jensen, L. Effects of intrinsic fano interference on surface enhanced raman spectroscopy: comparison between platinum and gold. *J. Phys. Chem. C* **2010**, *114*(42), 18059–18066.

74 Hamanaka, Y., Hayashi, N., Nakamura, A., Omi, S. Ultrafast relaxation dynamics of electrons in silver nanocrystals embedded in glass. *J. Lumin.* **1998**, 76, 221–225.

75 Hodak, J., Martini, I., Hartland, G. V. Ultrafast study of electron–phonon coupling in colloidal gold particles. *Chem. Phys. Lett.* **1998**, *284*(1), 135–141.

76 Del Fatti, N., Voisin, C., Christofilos, D., Vallée, F., Flytzanis, C. Acoustic vibration of metal films and nanoparticles. *J. Phys. Chem. A* **2000**. *104*(18), 4321–4326.

77 Del Fatti, N., Voisin, C., Achermann, M., Tzortzakis, S., Christofilos, D., Vallée, F. Nonequilibrium electron dynamics in noble metals. *Phys. Rev. B* **2000**, *61*, 16956–16966.

78 Voisin, C., Christofilos, D., Del Fatti, N., Vallée, F. Femtosecond surface plasmon resonance dynamics and electron-electron interactions in silver nanoparticles. *Eur. Phys. J. D Atom. Mol. Opt. Plasma Phys.* **2001**, *16*(1), 139–144.

79 Rethfeld, B., Kaiser, A., Vicanek, M., Simon, G. Ultrafast dynamics of nonequilibrium electrons in metals under femtosecond laser irradiation. *Phys. Rev. B* **2002**, *65*, 214303.

80 Hartland, G. V. Coherent vibrational motion in metal particles: determination of the vibrational amplitude and excitation mechanism. *J. Chem. Phys.* **2002**, *116*(18), 8048–8055.

81 Hu, M., Wang, X., Hartland, G. V., Mulvaney, P., Juste, J. P., Sader, J. E. Vibrational response of nanorods to ultrafast laser induced heating: theoretical and experimental analysis. *J. Am. Chem. Soc.* **2003**, *125*(48), 14925–14933.

82 Hartland, G. V. Coherent excitation of vibrational modes in metallic nanoparticles. *Annu. Rev. Phys. Chem.* **2006**, *57*(1), 403–430.

83 Pines, D., Nozières, P. *The Theory of Quantum Liquids: Normal Fermi Liquids*. Advanced book classics series. Routledge, 1966.

84 Hodak, J. H., Henglein, A., Hartland, G. V. Size dependent properties of Au particles: coherent excitation and dephasing of acoustic vibrational modes. *J. Chem. Phys.* **1999**, *111*(18), 8613–8621.

85 Voisin, C., Del Fatti, N., Christofilos, D., Vallée, F. Time-resolved investigation of the vibrational dynamics of metal nanoparticles. *Appl. Surf. Sci.* **2000**, *164*(1), 131–139.

86 Kluczyk-Korch, K., Antosiewicz, T. J. Hot carrier generation in a strongly coupled molecule–plasmonic nanoparticle system. *Nanophotonics* **2023**, *12*(9), 1711–1722.

87 Jain, P. K., Qian, W., El-Sayed, M. A. Ultrafast electron relaxation dynamics in coupled metal nanoparticles in aggregates. *J. Phys. Chem. B* **2006**, *110*(1), 136–142.

88 Voisin, C., Christofilos, D., Loukakos, P. A., Del Fatti, N., Vallée, F., Lermé, J. et al. Ultrafast electron-electron scattering and energy exchanges in noble-metal nanoparticles. *Phys. Rev. B* **2004**, *69*, 195416.

89 Arbouet, A., Voisin, C., Christofilos, D., Langot, P., Fatti, N. D., Vallée, F. Electron-phonon scattering in metal clusters. *Phys. Rev. Lett.* **2003**, *90*, 177401.

90 Kirschner, M. S., Ding, W., Li, Y., Chapman, C. T., Lei, A., Lin, X. M. Phonon-driven oscillatory plasmonic excitonic nanomaterials. *Nano Lett.* **2018**, *18*(1), 442–448.

91 Wersall, M., Cuadra, J., Antosiewicz, T. J., Balci, S., Shegai, T. Observation of mode splitting in photoluminescence of individual plasmonic nanoparticles strongly coupled to molecular excitons. *Nano Lett.* **2017**, *17*(1), 551–558.

92 Schwartz, T., Hutchison, J. A., Léonard, J., Genet, C., Haacke, S., Ebbesen, T. W. Polariton dynamics under strong light-molecule coupling. *ChemPhysChem* **2013**, *14*(1), 125–131.

93 Ramezani, M., Halpin, A., Wang, S., Berghuis, M., Rivas, J. G. Ultrafast dynamics of nonequilibrium organic exciton-polariton condensates. *Nano Lett.* **2019**, *19*(12), 8590–8596.

94 Schirato, A., Maiuri, M., Cerullo, G., Della Valle, G.. Ultrafast hot electron dynamics in plasmonic nanostructures: experiments, modelling, design. Nanophotonics. 2023;12(1):1–28. https://doi.org/https://doi.org/10.1515/nanoph-2022-0592.

95 Hu, H., Shi, Z., Zhang, S., Xu, H. Unified treatment of scattering, absorption, and luminescence spectra from a plasmon–exciton hybrid by temporal coupled-mode theory. *J. Chem. Phys.* **2021**, *155*(7), 074104.

96 Lebedev, V. S., Medvedev, A. S. Plasmon — exciton coupling effects in light absorption and scattering by metal/j-aggregate bilayer nanoparticles. *Quantum Electron.* **2012**, *42*(8), 701.

97 Stete, F., Koopman, W., Henkel, C., Benson, O., Kewes, G., Bargheer, M. Optical spectra of plasmon-exciton core-shell nanoparticles: a heuristic quantum approach. *ACS Photonics* **2023**, *10*(8), 2511–2520.

98 Mukamel, S. *Principles of Nonlinear Optical Spectroscopy*. Oxford University Press: Oxford, 1995.

99 Rosei, R., Antonangeli, F., Grassano, U. M. d bands position and width in gold from very low temperature thermomodulation measurements. *Surf. Sci.* **1973**, *37*, 689–699.

100 Wu, F., Guo, J., Huang, Y., Liang, K., Jin, L., Li, J. et al. Plexcitonic optical chirality: strong exciton-plasmon coupling in chiral j-aggregate-metal nanoparticle complexes. *ACS Nano* **2021**, *15*(2), 2292–2300.

101 Lee, H. E., Ahn, H. Y., Mun, J., Lee, Y. Y., Kim, M., Cho, N. H. Amino-acid- and peptide-directed synthesis of chiral plasmonic gold nanoparticles. *Nature* **2018**, *556*(7701), 360–365.

5

Coupling of Nanocavities to Molecules

Rohit Chikkaraddy[1] *and Jeremy Baumberg*[2]

[1] *School of Physics and Astronomy, University of Birmingham, Birmingham, UK*
[2] *NanoPhotonics Centre, Cavendish Laboratory, Department of Physics, University of Cambridge, Cambridge, UK*

5.1 Light and Molecules

The question of how to couple, enhance, and harness the coupling of light to single molecules has been a topic of interest since the 1950s when techniques were developed to trap single atoms. Optical detection of a single molecule was demonstrated in 1989. Later, it was revealed that light can be used not just to probe the natural states of atoms or molecules, but also to change the energy states such that new hybrid energy states will form with appealing quantum properties from both light and matter components [1–4]. Before discussing these hybrid states, we first consider the natural energy landscape of a molecule.

Unlike atoms, molecules have complex electronic, vibrational, and rotational energy states. The energy levels of a molecule can be described in terms of an energy landscape known as the Jablonsky diagram [5, 6] (Figure 5.1a). Most of the photophysical properties occur from electronic excitations in which the incident light of energy $\hbar\omega_i$ excites the molecule to higher electronic states. In most π-conjugated organic molecules, electronic transitions occur within the visible spectrum of light, ranging from 300 to 800 nm. Usually, the ground electronic level ($|S_0\rangle$) is filled with paired electron spins resulting in singlet states. The excited state is a spin-preserved singlet state ($|S_1\rangle$ or $|S_2\rangle$). Depending on how the different electronic and vibrational wave packets are coupled, the relaxation dynamics can be complicated. A typical relaxation process involves (1) fast nonradiative vibronic relaxation ($v_i^{|S_1\rangle} \rightarrow v_0^{|S_1\rangle}$) on the timescale of picoseconds (10^{-12} s), followed by (2) a spontaneous radiative decay from the lowest vibrational state in the higher electronic state to the ground electronic state ($|S_1\rangle^{v_0} \rightarrow |S_0\rangle^{v_i}$). This process typically happens on the timescale of nanoseconds (10^{-9} s) and is termed as fluorescence. Apart from this process, an electron in an excited state can undergo spin-flips, resulting in a transition from singlet-to-triplet state ($|S_1\rangle \rightarrow |T_1\rangle$). The lifetime of $|T_1\rangle$ is usually three to four orders of magnitude longer than the lifetime of $|S_1\rangle$. Therefore, the molecule is "dark" when it is in the triplet state. However, certain molecules are highly reactive with oxygen in their $|T_1\rangle$ state, resulting in the irreversible bleaching of the molecule. All these relaxation processes are also dependent on external environmental factors, such as temperature, solvent, and the local ionic environment, which makes them

Polariton Chemistry: Molecules in Cavities, First Edition. Edited by Joel Yuen-Zhou, Noel C. Giebink and Raphael F. Ribeiro. Published 2026 by John Wiley & Sons, Inc.

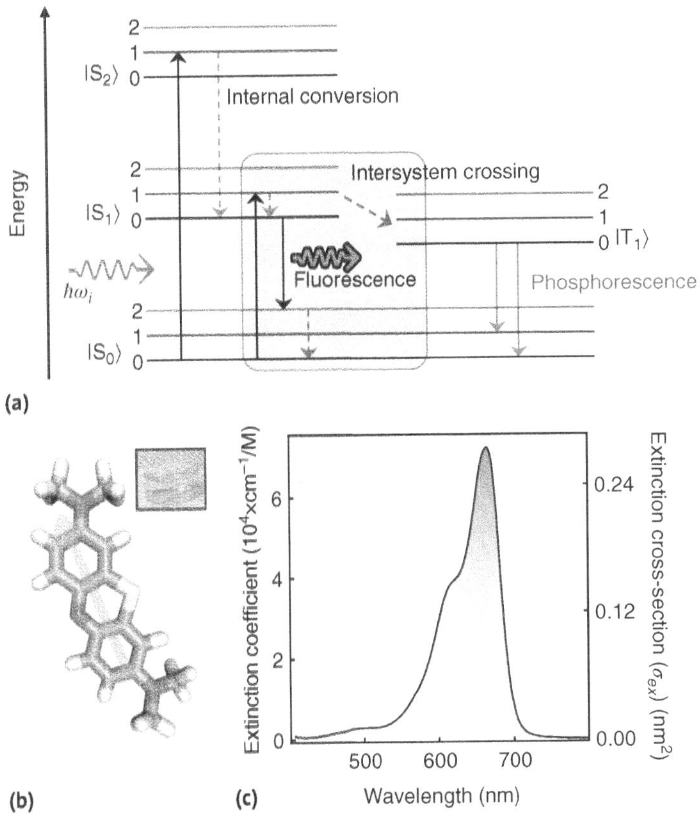

Figure 5.1 Energy level diagram for molecules. (a) Jablonsky diagram representing energy relaxation pathways after resonant excitation to higher electronic states. (b) Chemical structure of methylene blue (MB) molecule. Inset is an image of a 10 μM solution of MB in water. (c) Extinction spectrum for the MB sample shown in (b).

complicated to disentangle. In this chapter, we will discuss the first electronic excitation and its relaxation pathways (as highlighted in the gray box of Figure 5.1a).

The resulting allowed optical transitions can be probed via absorption and emission spectroscopy. At room temperature, the absorption of a dye molecule, such as MB (chemical structure shown in Figure 5.1b) is easily detectable by the human eye as a blue color (a 10 μM solution, Figure 5.1b inset). However, molecules also scatter/emit light and the probability of such resonant scattering events is proportional to the population in the excited state and the life time of the excited electronic state. In a simplistic classical description of light–molecule interactions (where the molecule is described as an oscillating point dipole), the scattering cross-section is given by the Abraham–Lorentz equation [7],

$$\sigma_{\text{scat}} = \sigma^0_{\text{scat}} \frac{\Gamma_s^2}{4\delta^2 + \Gamma_s^2} \tag{5.1}$$

where Γ_s is the total damping rate and δ is the detuning between the incident light and the oscillating dipole frequency. The scattering cross-section on resonance ($\sigma^0_{\text{scat}} = 3\lambda_0^2/2\pi$) depends only on the transition/resonant frequency (with wavelength λ_0) of the dipole and is independent of all the other atomic or molecular properties. Therefore, this is the maximum achievable scattering cross-section for a single quantum system. The integrated oscillator strength is constant, but the scattering cross-section of the transition is broadened under ambient conditions due to the increased thermal decoherence processes. Therefore, the peak of the scattering cross-section reduces to much smaller values. At room temperature, σ^0_{scat} is thus scaled down by approximately 10^{-6} ($\Gamma_{nat}/\Gamma_{hom}$), where Γ_{nat} is the natural (radiative) linewidth and Γ_{hom} is the homogenous broadening, which increases due to thermal dephasing, thus, limiting σ_{scat} to values <0.2 nm^2, much smaller than the geometric area of the molecule (Figure 5.1c).

The transition dipole moment characterizing this optical excitation is given by [8],

$$\mu = \left(\left(\frac{3}{4\pi^2}\right) \hbar\sigma_{scat}\gamma_e\epsilon_0\lambda_e\right)^{\frac{1}{2}} \tag{5.2}$$

where σ_{scat}, γ_e, and λ_e are obtained from the experimental extinction spectrum, respectively, corresponding to extinction cross-section, full-width at half maximum (FWHM) of optical resonance at λ_e, resulting in the dipole strength of 3.8 D (=1.27 C.m) for MB. In comparison to other strong light-emitting molecules with large conjugation lengths, such as cyanine-5 dye (Cy5) molecules, MB has a dipole moment that is 2.5 times weaker. For the special class of molecules termed *J*-aggregates in which the exciton is delocalized over many molecules (>15), the dipole strength can be as large as 20 D. However, in the context of this chapter, the useful figure of merit when comparing different exciton strengths is the dipole moment per unit volume of the electronic state (Table 5.1). Molecules have lower optical cross-section compared to inorganic materials, such as quantum wells and two-dimensional transition metal dichalcogenides (TMDs), but they are also much smaller in size. This low optical interaction of molecules with light makes the light–molecule coupling extremely weak at room temperature. Enhancing the light–matter interaction requires the design of innovative optical cavities with optical field volumes on the scale of the molecular size, which is discussed in the following section.

5.2 Optical Cavities

An optical cavity confines light in space, typically through multiple reflections. These cavities are essential to effectively couple light with matter (molecules, quantum dots (QDs), nitrogen-vacancy centers, etc.). An ideal cavity would confine light for infinitely long times in an infinitesimally small space. Realistic cavities are characterized by two parameters that determine their deviation from such ideal cavities. The first is the quality factor (Q), which is proportional to the confinement time

Table 5.1 Comparing optical transition strengths of different light-absorbing materials.

Optical system	Dipole moment (Debye)	Exciton size (nm^3)	Figure of merit (μ/V_e)
MB	3.8	1 (1 × 1 × 1)	3.80
Cy5	10.1	2.5 (2.5 × 1 × 1)	4.04
J-aggregates	20.5	100 (2 × 1 × 50)	0.21
Quantum dots (CdSe)	15	125 (5 × 5 × 5)	0.12
Quantum well (GaAs)	90	25 × 10^3 (50 × 50 × 10)	3.6 × 10^{-3}
TMD (WS_2)	50	200 (0.5 × 20 × 20)	0.25

in units of optical period and the second is the mode volume (V_m), which determines the spatial confinement.

Over the past two decades, many optical cavities have been designed to enhance the coupling of single emitters (Figure 5.2a), such as planar interfaces [9], near-field probes [10], photonic crystals [11, 12], microcavities [13], and metal nanostructures [14–16]. The rapid development in solid-state growth and fabrication techniques has allowed the design of optical cavities with $Q > 10^9$ and cavity volumes reaching the diffraction limit ($V_m/V_\lambda \sim 1$ at wavelength λ and refractive index n with $V_\lambda = (\lambda/n)^3$). Cavities have a variety of applications in modern technology (data storage, integrated optical chips, sensing, lasing, filters, etc.); however, each of these applications requires different cavity parameters. To choose the optimal optical cavity for a particular process, a landscape can be drawn with appropriate conditions (along with emitter properties) for the right niche. As

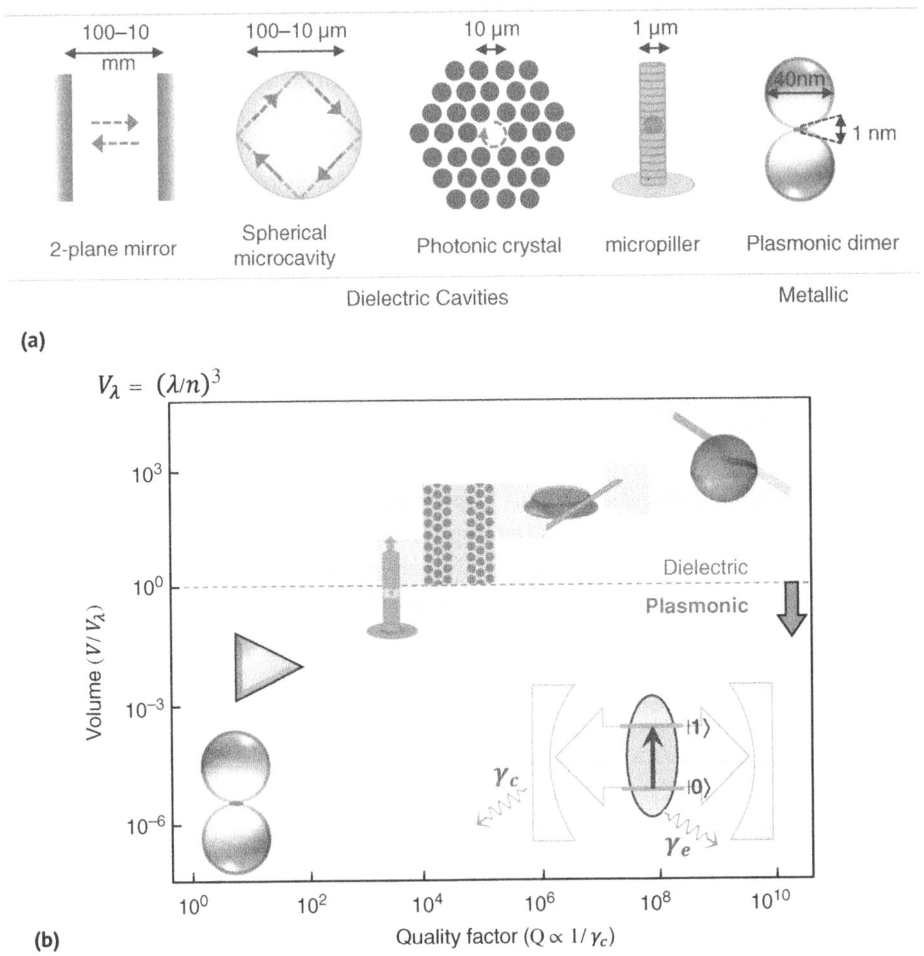

Figure 5.2 Optical cavities. (a) Different optical cavities arranged in order of decreasing cavity size. (b) Quality factor, Q, of each cavity is plotted against its effective volume (V/V_λ). The icons show realizations of each type of optical cavity: from whispering gallery spheres, microdisks, photonic crystals, micropillars, plasmonic nanotriangles, and plasmonic nanodimers. Inset shows the coupling of emitter to the cavity with different loss paths.

Table 5.2 Scaling of different optical processes.

Optical process	Scaling
Spontaneous emission	Q/V_m
Nonlinear threshold (lasing)	Q^2/V_m
Sensing	Q/V_m
Strong light–matter coupling	$Q/\sqrt{V_m}$

tabulated in Table 5.2, different aspects of light-emitter coupling have different exponential scaling of cavity parameters. All of these cavities occupy different positions in a log(Q) versus log(V_m) plot (Figure 5.2b). In this context, we can find the best niche for strong light–molecule coupling. We will focus on plasmonic cavities that are unique in comparison to other cavities. Optical field confinement in plasmonic cavities is far below the diffraction limit; however, Q is limited to the values <100. In the following sections, we will discuss how plasmons confine light below the diffraction limit, why Q is <100, and the rationale behind the choice of plasmonic cavity over dielectric cavity confinement.

5.3 Free-electron Model

The optical and conductive properties of a metal are dominated by the electrons in the conduction band. In metals, the conduction band is partially filled and the electrons are relatively free to move within it. The behavior of these free electrons in the conduction band and the interband excitations of electrons from energetically lower bands to the conduction band determine how the metal interacts with electromagnetic radiation. The properties of the macroscopic light–matter interaction can be described by a complex dielectric function $\varepsilon(\omega)$. The conduction band electrons can (for simple metals) be treated as a free-electron gas, which are displaced from their equilibrium position when they interact with light. This free-electron gas moves against a static background of positively charged ions. Upon excitation with light of the correct frequency and momentum, a charge oscillation will be induced in the free-electron gas. In a classical picture, this can be described as a harmonic oscillator [17],

$$m^* \frac{\mathrm{d}^2 x}{\mathrm{d}t^2} + m^* \gamma_c \left(\frac{\mathrm{d}x}{\mathrm{d}t}\right) = -eE \tag{5.3}$$

For an electric field E, this describes the response of a single electron with charge e, displacement x, and damping rate γ_c. The fact that electrons are moving in the crystal lattice is taken into account by the effective mass, m^*. The displacement of electrons against a positive background leads to a microscopic dipole moment $p = ex$. The macroscopic polarization P considering all the individual dipole moments for N electrons, assuming that x and E have the same time dependence is,

$$P = -ex\left(\frac{N}{V}\right) = -\frac{ne^2}{m^*\left(\omega^2 + i\gamma_c\omega\right)} E \tag{5.4}$$

Here, n describes the electron density (number of electrons N per unit volume V). The dielectric function $\varepsilon(\omega)$ can be obtained from P as,

$$\varepsilon(\omega) = 1 + \left(\frac{P}{\varepsilon_0 E}\right) = 1 - \frac{ne^2}{\varepsilon_0 m^*\left(\omega^2 + i\gamma_c\omega\right)} = 1 - \frac{\omega_p^2}{\omega^2 + i\gamma_c\omega} \tag{5.5}$$

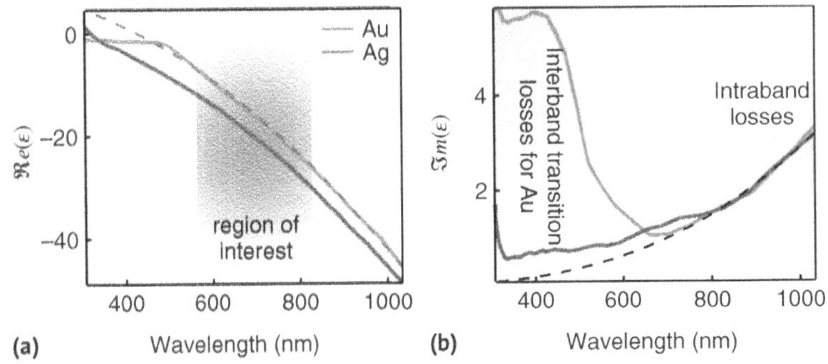

Figure 5.3 Dielectric function. (a) Real and (b) imaginary part of the dielectric function of for gold (solid light gray) and silver (solid dark gray), highlighting the wavelength of interest for this study (gray region in [a]). $\Re e(\varepsilon)$ for Au is fitted with a Drude model (dashed black line in [a]).

The plasma frequency ω_p describes the eigenfrequency of harmonic oscillation of a free-electron gas in a metal and is characteristic of each given metal. Also, considering the background contribution from the bound electrons can often be taken as constant across optical frequencies, the total dielectric function can be written as,

$$\varepsilon(\omega) = \varepsilon_\infty - \frac{\omega_p^2}{\omega^2 + i\gamma_c\omega} \tag{5.6}$$

The polarization of a metal at frequencies much larger than the plasma frequency is expressed in the parametric value of ε_∞, for example, due to filled d-bands below the conducting s-band. For noble metals, interband transitions (below ~500 nm for gold) result in increased absorption. This Drude–Lorentz model does not take into account interband transitions and hence is limited to longer wavelengths. Figure 5.3 shows a comparison of the real and imaginary parts of the dielectric function for gold from the experimental data taken by Johnson and Christy [18, 19] with a fit from the Drude–Lorentz model. For the wavelengths of interest in this chapter (highlighted region in Figure 5.3a), this model serves as a reasonable approximation.

The losses in the system are determined by $\Im m(\varepsilon)$. In the near-infrared wavelengths of interest, gold possesses the lowest interband and intraband losses (similar for silver). Due to the negative values of $\Re e(\varepsilon)$, light does not penetrate far into the metals and therefore they are highly reflective. This results in localized charge oscillations on the metal surface, which are discussed in the next section.

5.4 Introduction to Plasmons

Fundamental limits on the confinement of light given by the classical diffraction limit have been a huge limitation for conventional cavities, especially when the size of an emitter is 100–1000 times smaller than the round trip of photons in the cavity. However, over the past decade, it has become evident that evanescent fields supported by negative permittivity metals can overcome the diffraction limit and confine light fields on the nanometer scale. This is possible due to the collective oscillations of free electrons excited by light, termed surface plasmon polaritons (SPPs) (Figure 5.4). SPPs come under the wider class of surface modes in which the interaction between light and matter leads to the possibility of a bound surface mode, such as surface phonon polaritons [20] and surface exciton polaritons [21]. Here, we will focus on SPPs. There are two fundamental modes of excitations associated with SPPs: (1) propagating SPPs (Figure 5.4a–c), in which

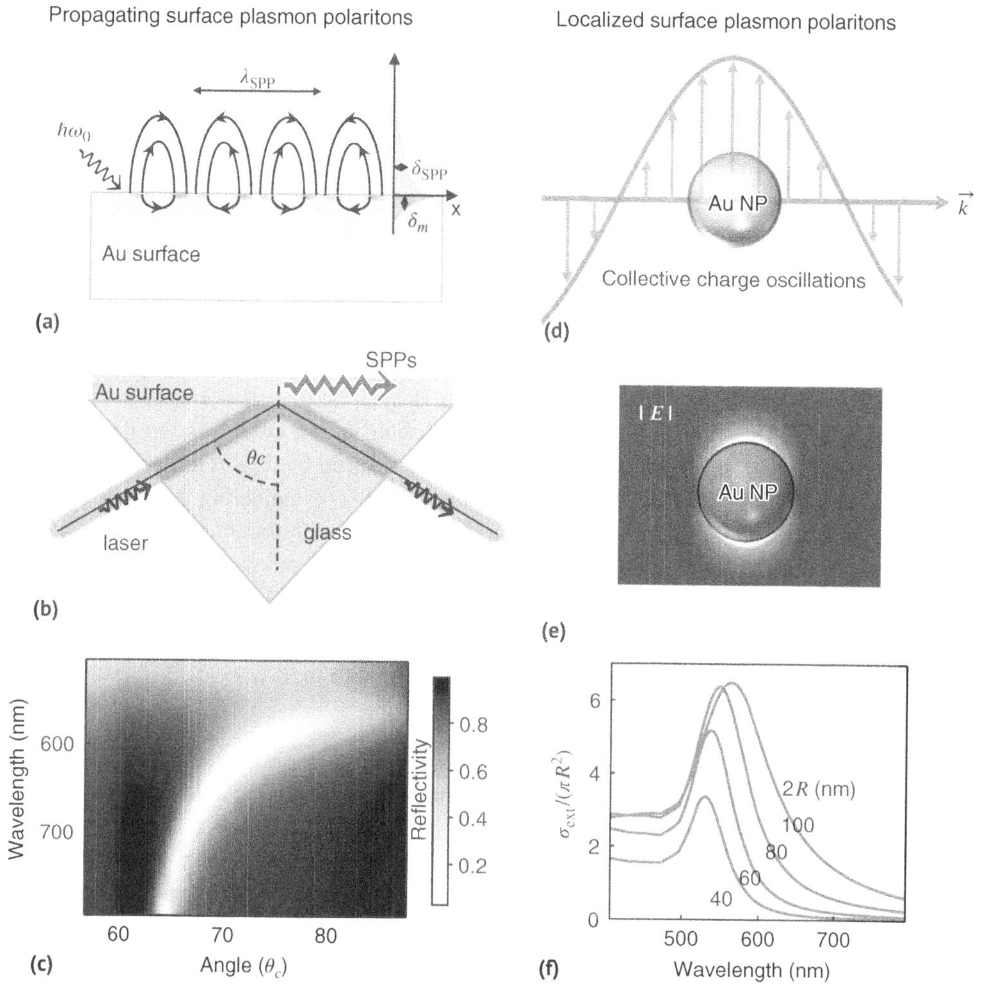

Figure 5.4 Localized and propagating modes in plasmonic nanostructures. (a) Schematic interaction of metal film with light and (b) launch and propagation of SPP modes on metal films. (c) Dispersion of SPPs on 50 nm Au film on glass substrate, with outside environment being water. Calculations are based on solving Fresnel equation for TM waves in Kretschmann configuration. (d) Schematic representation of metal nanosphere interacting with light and (e) the associated field distribution. The maximum field enhancement (E/E_0) near the interface is approximately eight. (f) Enhancement spectrum for different size gold nanoparticles (Au NPs) is normalized to geometric area of NP. Calculated from three-dimensional finite-domain time-difference (FDTD) simulations in water medium.

charge oscillations travel across a metal–dielectric interface (conventionally termed as SPPs) and (2) localized SPPs (LSP) where modes are specially confined (Figure 5.4d–f). This confinement is seen in metallic nanostructures of various shapes (spheres, rods, discs, prisms, cubes, etc.). Generally, localized modes are used for sensing purposes, whereas propagating modes find their utility in signal transport. Both these modes exhibit the following two properties, which make plasmon polaritons interesting for coupling with molecules:

1. Near field: confinement and enhancement
2. Far field: scattering and directional emission

5.4.1 Propagating Surface Plasmon Polaritons

SPPs are propagating optical fields bound to the metal–dielectric interface. However, the dispersion of these modes does not allow direct coupling with freely propagating light. The dispersion relation is given by [17],

$$K_{\mathrm{SPP}} = \frac{\omega}{c}\sqrt{\frac{\varepsilon_1\varepsilon_2}{\varepsilon_1 + \varepsilon_2}} \tag{5.7}$$

where ε_1 and ε_2 are the frequency-dependent relative permittivities of the two media, metal and dielectric, respectively. This mismatch in the wave-vector condition can be fulfilled by prism coupling in which evanescent fields at dielectric interfaces excite the plasmon polaritons. Various techniques have been developed to effectively excite SPPs, such as coupling with gratings, near-field coupling with fibers or emitters, and nonlinear coupling.

This dispersion of SPPs is generally mapped via the Kretschmann configuration (a common prism coupling technique), in which the evanescent field produced by total internal reflection is coupled to the metal film in contact above a prism (Figure 5.4b). By adjusting the angle of incidence, the in-plane wavevector of the evanescent field is matched with that of the SPP mode. The SPP excitation is clear from the dip in reflection and calculations for one such Au film system are shown in Figure 5.4c. The dispersion is similar to that of free-space light at low frequency, but at higher frequencies, the dispersion bends and reaches an asymptotic limit, indicating strong field confinement.

5.4.2 Localized Surface Plasmon Polaritons

Another mode of surface plasmon excitation is a confined field around a metallic nanostructure. The curvature of the nanostructure itself enables the direct excitation of this mode. Again, the near-field properties such as confinement and enhancement of optical fields are associated with the fact that plasmon modes are bound to the interface of metal–dielectric nanostructures. For example, the optical near-field distribution of a single gold nanosphere (Au NP) immersed in water is shown in Figure 5.4e. The optical resonances of this system are obtained from full-wave finite-domain time-difference (FDTD) simulations. The field strength decays exponentially into the surrounding media. The decay length into the dielectric medium for LSP modes is considerably smaller than for SPP modes [22]. The confinement of the optical field to the proximity of the surface is evident from the evanescent nature of the field. Field distributions around the nanostructure strongly depend on the geometry of the nanostructure.

The coupled oscillations of free electrons and light are generally treated as electromagnetic dipoles, which coherently radiate light at the frequency of charge oscillations, which gives rise to far-field scattering. The typical cross-sections are larger than the geometric area of nanostructure itself, and are orders of magnitude greater than the absorption for any known molecule. For simple spherical NPs, scattering and absorption cross-sections are given by [23],

$$\sigma_{\mathrm{scat}} = \frac{8\pi}{3}k^4R^6\left|\frac{\varepsilon_1 - \varepsilon_2}{\varepsilon_1 + 2\varepsilon_2}\right| = \frac{k^4}{6\pi}|\alpha_1|^2 \Rightarrow \sigma_{\mathrm{scat}} \propto \frac{R^6}{\lambda^4} \tag{5.8}$$

$$\sigma_{\mathrm{abs}} = 4\pi kR^3\,\Im m\left\{\frac{\varepsilon_1 - \varepsilon_2}{\varepsilon_1 + 2\varepsilon_2}\right\} = k\,\Im m\{\alpha_1\} \Rightarrow \sigma_{\mathrm{scat}} \propto \frac{R^3}{\lambda} \tag{5.9}$$

Here, α_1 is the dipolar polarizibility of a metallic NP of radius R, and ε_1, ε_2 are the frequency-dependent relative permittivities of the metal and the surrounding dielectric, respectively. Wavevector (k) is related to the frequency (ω) and the speed of light in vacuum (c) as $k = \omega/c$.

It is important to note that, for plasmonic NPs with dimensions greater than 20 nm, scattering cross-sections dominate, and their absorption cross-sections can be neglected, whereas for those with dimensions less than 5 nm, absorption cross-sections dominate and scattering is negligible.

Scattering by plasmonic nanostructures can be uniquely tuned and modified either by the geometry of the nanostructure and also by organizing nanostructures into the desired architecture. Some structures are designed to achieve omnidirectional light scattering. Such unique properties are studied under the concept of a "plasmonic antenna" [16], which has gained relevance in channeling single photons at desired angles for efficient collection and quantum processing.

5.5 Cavity Description for Plasmon Modes

A single nanostructure provides significant field enhancement ($E/E_0 \sim 10$) and confinement ($V_m \sim R^3, \frac{V_m}{V_\lambda} \sim 10^{-2} - 10^{-3}$) of fields surrounding the nanostructure. However, significant improvement in near-field intensities is obtained from NP dimers, where the plasmonic resonances from each NP hybridize to form a bonding dipolar mode (l_1), which confines fields into the gap between the nanostructures (Figure 5.5a). This in-phase dipole–dipole coupling results in the

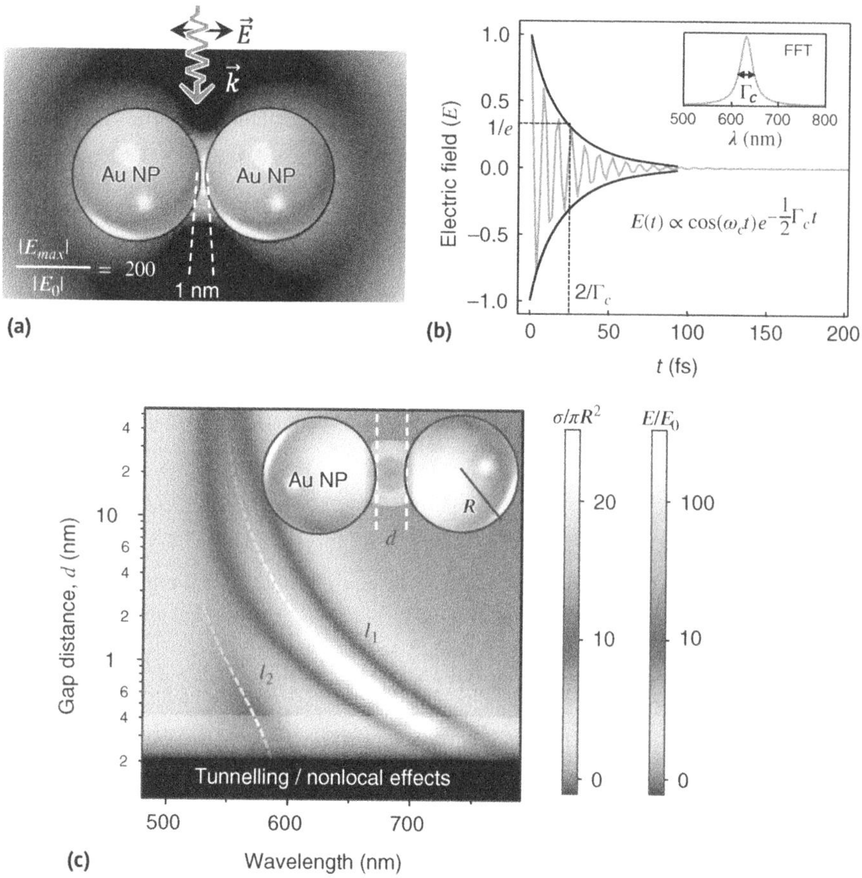

Figure 5.5 Plasmonic nanodimer cavity. (a) Spatial distribution of electric-field enhancement in Au plasmonic dimer system with 1-nm gap, obtained from full-wave FDTD simulation at bonding dipolar plasmon resonance (l_1) at 650 nm. (b) Temporal decay of the optical fields in the gap, which is used to estimate the quality factor. (c) Variation of plasmonic nanocavity modes for different gap size.

lowering of plasmon resonance energy, whereas out-of-phase coupling results in a higher energy mode that is generally "dark" and optically nonaccessible in the far field. Associated with this confinement is an enhancement in the strength of the electric field adjacent to the surface. With appropriate geometries of nanostructures, near-field enhancements (E/E_0) can be made as high as 10^2–10^4 with spatial confinements as small as 10 nm^3. How to extract and quantify cavity parameters Q and V_m is discussed in the following section.

5.5.1 Qualify Factor

The quality factor Q, in simple terms, measures the rate at which the optical energy decays from the cavity due to losses, such as absorption, scattering, or leakage. This can be quantitatively expressed as the ratio of energy stored inside the cavity to the energy loss per oscillation cycle.

$$Q = \frac{2\pi}{T} \frac{u(t)}{-\left[\frac{du(t)}{dt}\right]} = \frac{2\pi}{\Gamma_c \cdot T} = \frac{\omega_c}{\Gamma_c} \tag{5.10}$$

Here, $u(t) \propto e^{\left(-\frac{1}{2}\Gamma_c t\right)^2}$ is the energy density and Γ_c is the decay time, with $-\frac{du(t)}{dt} \propto \Gamma_c e^{\Gamma_c t}$ being the decay of energy density, and ω_c is the resonant frequency. For a plasmonic nanodimer cavity, these decay rates are dominated by the material absorption losses (Joule heating) and radiative scattering rates. They are estimated by monitoring fields in the gap using time domain simulations (Figure 5.5b). This time-domain response is equivalently represented in the frequency domain via Fourier transforms (Figure 5.5b, inset). Here, Q is experimentally extracted from the linewidths and resonance frequency is obtained through scattering spectroscopy. Note that this works only for isolated dipolar modes of the nanostructures; for coupled NPs with complex mode structures, one needs to consider several other factors.

The extinction spectrum for an Au dimer (with $2R = 40$ nm) is obtained from FDTD simulation for various gap sizes, d (Figure 5.5c). The cross-section (σ) of the coupled plasmon mode (l_1) becomes stronger for $d < 5$ nm indicating that the strong electrostatic coupling between the NP and Q increases up to values of approximately 30. This is also because the $\Im m(\varepsilon)$ of Au is at a minimum in the window of 600–800 nm. In fact, in this spectral range, both Au and Ag have relatively weak intraband and interband losses. Typically, $Q \sim 15$–30 for an Au nanodimer at the l_1 cavity resonance with $d < 5$ nm.

5.5.2 Mode Volume

In simple terms, the mode volume (V_m) of a cavity is the volume in which optical energy density is confined. The local energy density of the system is given by,

$$W(r, \omega) = \frac{1}{2}\left(\frac{\partial\left[\omega\,\varepsilon(r, w)\right]}{\partial\omega}\varepsilon_0 \left|E(r, \omega)\right|^2 + \mu_0 \left|H(r, \omega)\right|^2\right) \tag{5.11}$$

where $\varepsilon(r, \omega)$ is the permittivity at position r. To correctly account for the negative permittivity and dispersive character of plasmonic metals, one needs to use the term $\frac{\partial[\omega\,\varepsilon(r,w)]}{\partial\omega}$ for the electric energy density inside the metal. V_m is obtained by integrating the energy density over the volume of the nanoplasmonic structure, and normalizing it to the maximum value of $W(r)$ at each ω as,

$$V_m(\omega) = \frac{\int W(r, \omega)\, d^3 r}{max\left[W(r, \omega)\right]} \tag{5.12}$$

It is important to note that $W(r, \omega)$ is complex inside the metal, due to the complex ε, hence $V_m(\omega)$ is complex. Here, $\Re e(V_m)$ and $\Re e(W)$ are associated with the energy density in the system, and $\Im m(V_m)$ and $\Im m(W_m)$ with the energy dissipation in the metal. Since $\Im m(V_m)$ is commonly neglected, the equation reduces to the general definition of V_m as,

$$V_m(\omega) = \frac{\int \varepsilon(r)|E(r)|^2 \, d^3r}{max\left[\varepsilon(r)|E(r)|^2\right]} \tag{5.13}$$

Note that this is exactly applicable only to lossless dielectric cavities.

To obtain an accurate value for $\int W(r, \omega)\, d^3r$, the electric and magnetic fields should be calculated in a three-dimensional volume and followed by integration. Accurately calculating these values using FDTD requires high mesh density for metallic nanostructures, especially when the gaps are <2 nm. The mesh accuracy of the simulations defines the resolution of the confined fields in the gap, and therefore dramatically affects the value of $V_m(\omega)$. For gaps of 1 nm, calculations were converged with 0.3 nm meshing for NP dimers of size $2R = 40$ nm. The effective mode volume is minimum at $\lambda_c = 655$ nm giving rise to $V_m = 35.7 \pm 5$ nm^3.

5.6 Nanocavities

Reliable fabrication of nanogaps with such high degrees of repeatability is a problematic practical issue. Various techniques have been developed based on both bottom-up synthesis and top-down fabrication. Top-down techniques, including (Figure 5.6): (1) high-resolution electron beam lithography [24], (2) computer-programmed electromigration processes [25], and (3) electrochemical deposition/etching on AFM tips [26], which have all been used to make gaps <5 nm. However, these methods are not reliable for large-scale fabrication (e.g. making many thousand gaps with well-controlled geometry at the nanoscale). More importantly, defects and grain boundaries are hard to control in top-down processes.

On the other hand, bottom-up processes, such as colloidal synthesis, produce fairly uniform (±10%) nanostructures in large quantities (liter scale). However, methods to make nanogaps are still limited to binding these nanostructures with organic linker, such as dithiols [29], block co-polymers [27], or DNA linkers [30], which generally results in low yield (<30%) and "floppy" gaps. Separation techniques have been developed to increase the yield of nanogap dimers, which include gel-electrophoresis and differential centrifugation [28], in which the formed structures are generally "shell-isolated" with polymers for stabilization in high-ionic concentration.

To overcome these challenges, different constructs are better used, combining a solid surface with single NPs. Using this NP-on-mirror geometry (NPoM) gives optical properties similar to plasmonic nanodimers. The fabrication of this geometry is much simpler than fabrication methods required to create most other nanogap geometries.

5.6.1 Nanoparticle on Mirror

An equivalent system to the NP dimer is an NP placed over a metal surface with a small dielectric gap (Figure 5.7). Analogous to the coupling in an NP dimer, charge oscillations in the NP are coupled with image charges in the film for NPoMs, resulting in an effective dimer system.

The simple dipole coupling analogy does not provide a complete picture of mode coupling in the NPoM geometry, although the fundamental mode l_1 has a similar charge distribution to the plasmonic dimer. Electromagnetic properties of this system were studied already in the 1980s [31, 32]

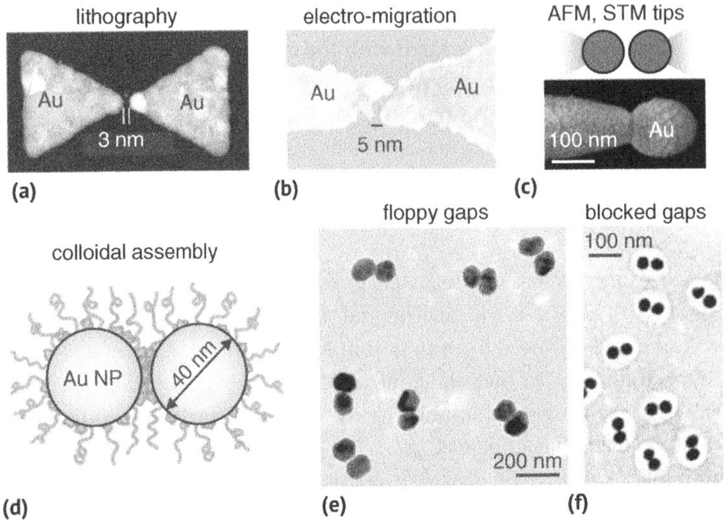

Figure 5.6 Plasmonic nanogaps. (a) Scanning transmission electron microscope (TEM) image of plasmonic bowtie antenna fabricated using e-beam lithography. *Source*: Reproduced from [24]/American Chemical Society. (b) Scanning electron microscope (SEM) image of nanogap formed from electro-migration. *Source*: Reproduced from [25]/Springer Nature. (c) SEM image of atomic force microscope tips coated with single plasmonic NP. *Source*: Reproduced from [26]/Springer Nature. (d) Colloidal assembly of nanodimer gaps with poly(ethylene glycol) (PEG) and poly(methyl methacrylate) (PMMA). *Source*: Reproduced from [27]/American Chemical Society and (e) SEM images of resulting dimers. *Source*: Reproduced from [27]/American Chemical Society. (f) TEM image of nanodimers purified using differential centrifugation after encapsulating with polystyrene-block-poly(acrylic acid). *Source*: Reproduced from [28]/American Chemical Society.

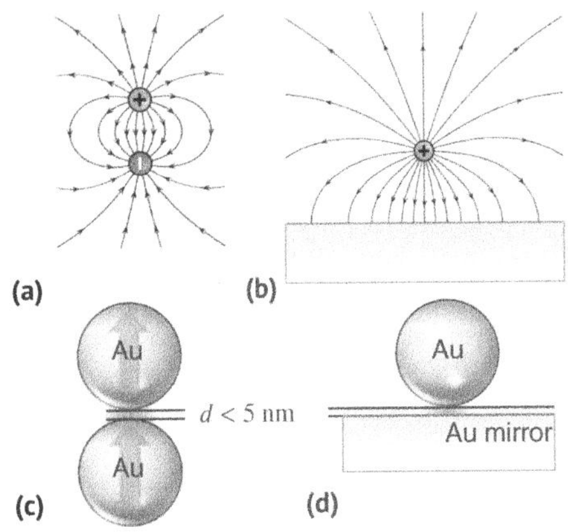

Figure 5.7 NPoM cavity. (a) Fringing fields between the two opposite charge is equivalent to (b) a charge placed above a polarizable surface. (c) Dipole–dipole coupling between the NP with gaps <5 nm is equivalent to (d) NP placed on metallic substrate with same dielectric gap.

by Aravind and Metlu, who predicted strong optical field enhancements and confinement in the gap. Nevertheless, studies on the complete picture of mode hybridization [33] and practical limits on the enhancements [34] in such gaps only emerged recently.

The optical excitation of the NPoM cavity is complex compared to a nanodimer system. The NPoM is best excited with plane waves at high angle (Figure 5.8a). When the electric field is polarized perpendicular (TM) to the metal film, the coupled mode (l_1) exhibits a strong low energy resonance (Figure 5.8b), controlled by the NP size (2R), gap distance (d), and refractive index of the gap contents (n).

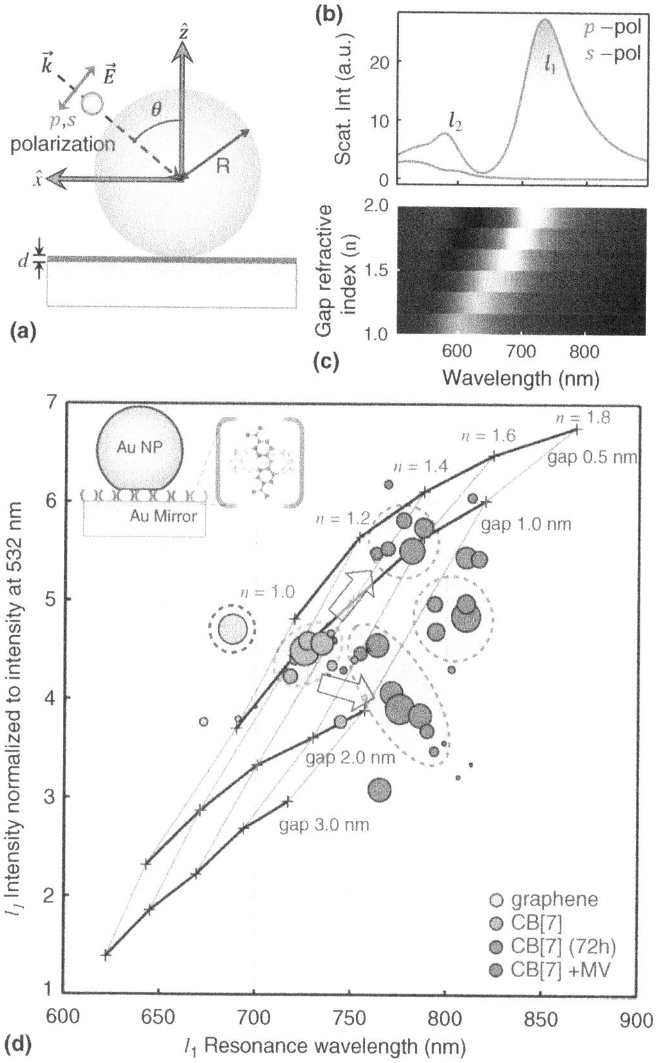

Figure 5.8 Molecular sensing. (a) Schematic of NPoM system, giving the polarization directions of incident light used for FDTD simulation. The light illumination is at an angle of 55° to the normal perpendicular to the metal film. (b) Scattering spectrum for two orthogonal polarizations (s- and p- as shown in [a]) for 80-nm particle with 1-nm gap and 1.4 refractive index. (c) Scattering spectrum obtained for various refractive indices at 4-nm gap. (d) Mapping the resonance position and intensity of the l_1 mode for various gap sizes and refractive index. The experimentally obtained points are for different spacer molecules. *Source*: Reproduced from [35].

5.6.2 Sensing Molecules in the Gap

Typically, in experiments, molecules are self-assembled on the metal surface before the deposition of NPs. This results in molecules being located directly within the field volume of the l_1 mode. As the fields are strongly confined in the gap, small changes in molecular properties will be seen in the modulation of the coupled mode. For example, if a $2R = 80$ nm particle with a gap size of 4 nm is considered, a change in the refractive index in the gap from $n = 1$ to 2 shifts the coupled mode from 600 to 710 nm and greatly increases its intensity (Figure 5.8c).

The l_1 resonance wavelength and its intensity are very sensitive to the type of molecule assembled in the gap, especially for $d < 3$ nm (Figure 5.8d). Simulations and experimental results on graphene (gap of 0.4 nm) and self-assembled molecules of cucurbit-7-uril (CB[7]) (gap of 0.9 nm) are in good agreement with each other [35]. Furthermore, loading CB[7] with guest molecules, such as methyl-viologen (MV), shows clear shifts in this spectral map, proving it is possible to monitor such atomic scale changes [34, 35].

5.6.3 Effect of Nanoparticle Size and Shape

The size of a NP can be used to tune the resonance of l_1 to a specific wavelength of interest (Figure 5.9). Benzene di-thiol (BPT) is one molecule that self-assembles on Au surfaces with a high degree of uniform packing density and orientation, resulting in gap size of 1.1 nm in the NPoM geometry. By systematically increasing the NP size from 40 to 100 nm the resonance l_1 mode can be monotonically tuned from 650 to 850 nm, which is in good agreement with theoretical predications. Optical fields are tightly confined in the gap in a similar way to the NP dimer system [26], and the width of this confinement is $\Delta_c \sim \sqrt{R}$ (Figure 5.9b). In practice, NPs are never spherical and colloidally synthesized NPs have a degree of crystalline faceting. The faceted NPs modify the optical resonance and confinement.

To account for these facets, the NP is simulated as a cropped sphere at the base contact with various circular facet sizes, w (Figure 5.10a). As the facet size increases (Figure 5.10b, c), the optical field spreads out within the cavity [37, 38], causing the capacitance to increase and the coupled mode (l_1) to decrease in energy. A secondary, potentially unwanted, effect of this is a decrease in the field enhancement (Figure 5.10d, e). Once the facets are large enough, they begin to support new

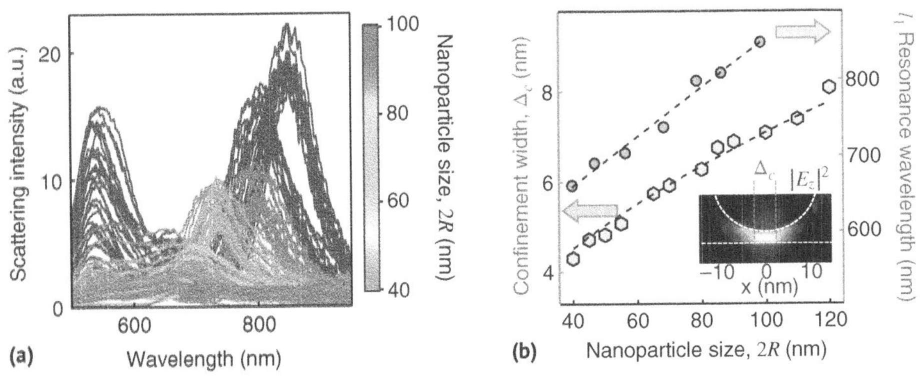

Figure 5.9 NP size. (a) Scattering spectrum from >1000 NPoMs for various batches of different NP size varying from 40 to 100 nm. *Source:* [36]/American Chemical Society/CC BY 4.0. (b) FDTD results calculating the resonance wavelength and field confinement width (Δ_c) for coupled mode of different NP size in NPoM geometry. The Δ_c is fitted to $\sqrt{R}$ (dashed black lines).

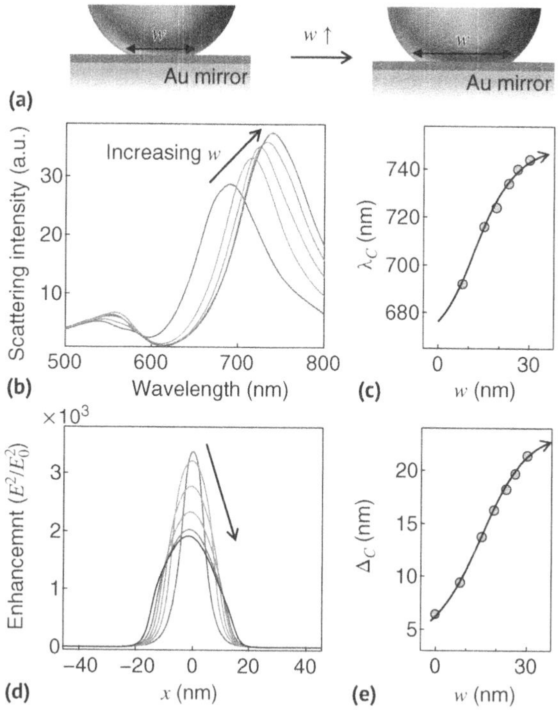

Figure 5.10 NP facets. (a) Schematic showing increasing NP facet widths (w). (b) Simulated scattering spectra for NPoMs with different facet widths $w = 8$–30 nm ($2R = 80$ nm, $n = 2.1$, and $d = 4.5$ nm). (c) Variation in the resonance wavelength, (d) line profiles for electric-field enhancement along the gap, and (e) FWHM confinement widths for increasing facet width.

sets of plasmonic modes, which intermix with l_1 modes. These effects become further complicated when the facet is not circular (as formed when using "ultraspherical" NPs), but are triangular or square, with the mode hierarchy and coupling being modified [39]. Further modifications are found when coating the NPoM with different refractive index materials [40].

Having introduced plasmonic cavities, we now return to the discussion of coupling of molecules to confined optical modes and how optical properties are modified due to this coupling.

5.7 Light–Matter Coupling

A model system to visualize the dynamics of light–matter interactions is the damped coupled harmonic oscillator (here a coupled pendulum), where light (plasmons) and matter (molecule-resonance) are considered as classical oscillators with resonance frequencies of ω_c and ω_e, where c and e correspond to cavity and emitter, respectively (Figure 5.11a). Treating the cavity and emitter as classical harmonic oscillators is a good approximation in the weak excitation limit. The damping associated with the plasmon and emitter is incorporated into the friction of each pendulum represented by $\gamma_{c,e}$. The coupling strength g between the plasmon and the emitter is modeled by a spring connecting the two pendulums.

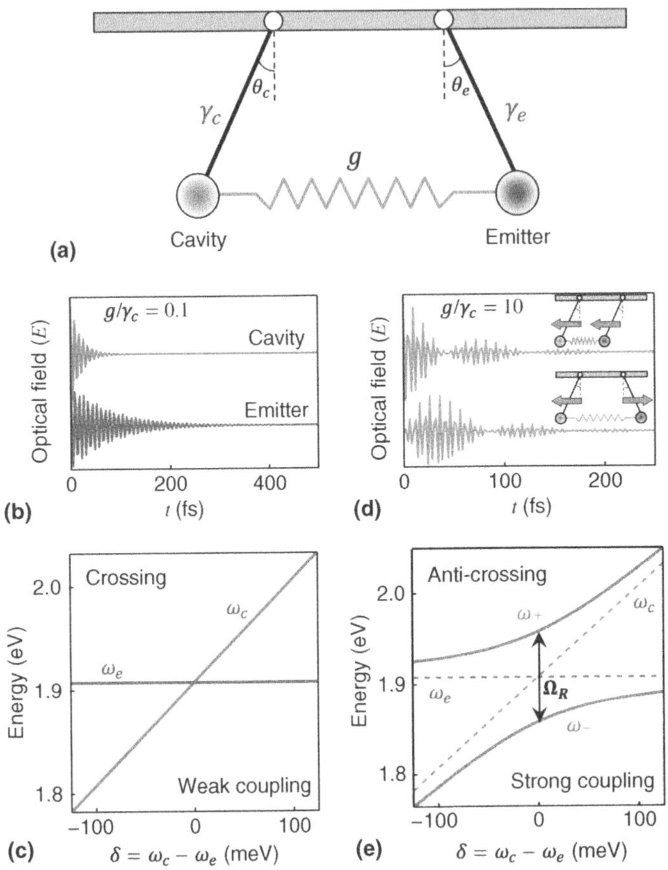

Figure 5.11 Damped coupled harmonic oscillator model for light–matter coupling. (a) Schematic of cavity and emitter as two damped oscillators connected by a spring. (b) Oscillations of the emitter and cavity in the weak-coupling regime. (c) Tuning the resonance of the cavity across the emitter resonance at $g/\gamma_c = 0.1$. (d) Oscillations of new coupled modes (ω_+, ω_-) in the strong-coupling regime with the phase of pendulum oscillations shown in inset. (e) Anticrossing of new polariton modes (ω_+, ω_-) on tuning the cavity resonance across the exciton energy.

This coupled system evolves according to a set of coupled differential equations characterized by the time-varying angles $\theta_c(t)$ and $\theta_e(t)$.

$$\frac{d^2\theta_e}{dt^2} + 2\gamma_e \frac{d\theta_e}{dt} + \omega_e^2\theta_e + g^2(\theta_e - \theta_c) = 0 \tag{5.14}$$

$$\frac{d^2\theta_c}{dt^2} + 2\gamma_c \frac{d\theta_c}{dt} + \omega_c^2\theta_c + g^2(\theta_c - \theta_e) = 0 \tag{5.15}$$

The energy transfer between the two oscillators is determined by the coupling constant g. The resulting nontrivial solution to the above-coupled equation is,

$$\Omega_\pm = \frac{1}{2}(\Omega_c + \Omega_e) \pm \frac{1}{2}\sqrt{(\Omega_e - \Omega_c)^2 + 4g^2} \tag{5.16}$$

where $\Omega_i = \omega_i - i\Gamma_i/2$ denotes the complex frequency of a resonance with frequency ω_i and linewidth Γ_i. This can be simplified to find the frequency of the two hybrid states $\omega_\pm$ as,

$$\omega_\pm = \frac{1}{2}(\omega_e + \omega_c) \pm \frac{1}{2} Re\left\{\sqrt{4g^2 + \delta^2 - (\gamma_e - \gamma_c)^2/4 - 2i\delta(\gamma_e - \gamma_c)/2}\right\} \tag{5.17}$$

where $\delta = (\omega_c - \omega_e)$ with $\omega_\pm$ being the energies of the new hybrid modes, which are determined by the coupling strength g. This purely classical picture covers the complete dynamics of light–matter coupling in linear-weak excitation limits, where a harmonic ladder of energies can be approximated to a two-level system. In fact, a full quantum mechanical solution obtained using the Jaynes–Cummings (JC) Hamiltonian provides eigenenergies similar to the classical solution.

The value of g is determined by the parametric strength of exciton dipole (μ) and the cavity mode volume (V_m),

$$\propto \mu\sqrt{V_m} \tag{5.18}$$

Depending on the coupling strength and damping associated with the system, two different regimes of light–matter interactions can be defined. (1) Weak coupling: In this limit $\frac{g}{\gamma_{c,e}} \ll 1$, the two oscillators behave independently and $\theta_{c,e}(t)$ converges to zero exponentially (Figure 5.11b). Here, the eigenfrequencies of the system are not inherently modified. Therefore, tuning the oscillation frequency of cavity (ω_c) across the exciton frequency (ω_e) causes an energy cross-over since the oscillators cannot effectively exchange energy (Figure 5.11c). (2) Strong coupling: Here $\frac{g}{\gamma_{c,e}} \gg 1$, and the cavity and emitter are able to exchange energy over several oscillation periods; therefore, resulting in the formation of new eigenmodes ($\omega_\pm$) (Figure 5.11d). $\omega_\pm$ corresponds to oscillations of the cavity and emitter in-phase (ω_-) and in antiphase (ω_+). These new eigenmodes depend entirely on the characteristic coupling constants of the system as their energies are given by $\omega_\pm = 1/2(\omega_c + \omega_e) \pm 1/2\sqrt{g^2 + \delta^2}$. Therefore, anticrossing behavior is observed when the cavity resonance is tuned across the exciton resonance ($\delta = \omega_c - \omega_e$) (Figure 5.11e).

The dispersion of $\omega_\pm$ tends to the uncoupled cavity and exciton energies at large detuning. For the scenario where $\delta = 0$, the energy splitting between $\omega_\pm$ is referred to as the vacuum Rabi splitting (Ω_R). The observation of $\Omega_R \neq 0$ is direct evidence that the system is operating in the strong-coupling regime and measuring Ω_R allows us to quantify the interaction strength. However, measurement of Ω_R is complicated by artifacts in experiments such as that reflection, transmission, and absorption spectra of the coupled system can give somewhat different values (Ω_R^r, Ω_R^t, and Ω_R^a, respectively). In particular, earlier works on dielectric microcavities by Savona et al. demonstrated that $\Omega_R^a \leq \Omega_R^t \leq \Omega_R^r$. This anomalous situation is referred to as the intermediate coupling regime, which occurs for lower quality factor cavities $Q < 100$. Similar observations are also reported in the coupling of molecular aggregates on plasmonic films. These effects are directly linked to the finite damping of the system dominated by absorption of excitons or SPPs. However, for localized plasmon systems, where coupling is generally probed via scattering measurements, achieving $g \gg \gamma_c + \gamma_e$ would represent the strong-coupling regime. The implications of weak and strong coupling are discussed in the following sections.

5.7.1 Weak-coupling Regime and Purcell Effect

In the weak-coupling regime ($g \ll \gamma_c, \gamma_e$), the energies are not directly affected by the coupling in the system; however, the presence of light-fields around the emitter profoundly affects the spontaneous decay rates. This effect was initially proposed for nuclear transitions in atoms by Purcell in

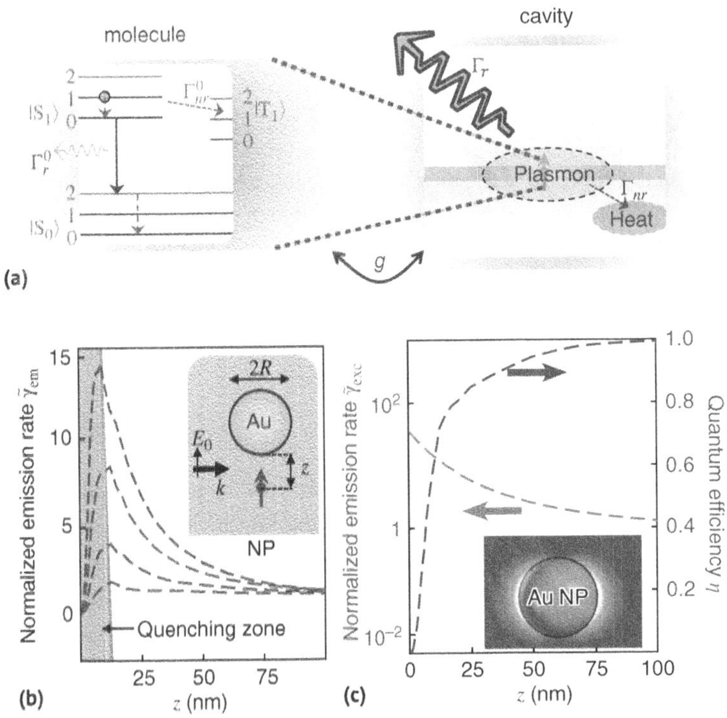

Figure 5.12 Decay pathways. (a) Sketch of the different decay pathways of the excited molecule coupling to a plasmonic cavity. (b) Calculated radiative emission of point dipole coupled to plasmonic nanosphere as a function of emitter-metal distance. *Source*: Reference [41]/American Chemical Society. (c) Calculated excitation rate and quantum efficiency of the emitter; near-field enhancement around NP shown in the inset. *Source*: Reproduced from [41]/American Chemical Society.

1946 and hence the ratio of modified (Γ_m) and free-space (Γ_0) decay rates is known as the Purcell factor:

$$F_P = \frac{\Gamma_m}{\Gamma_0} = \frac{3Q(\lambda/n)^3}{4\pi^2 V_m} \tag{5.19}$$

Here, λ/n is the wavelength associated with the transition frequency of an emitter in a medium of refractive index n. The cavity quality factor and mode volume are defined by Q and V_m. The emission and cavity resonance frequencies are assumed to be perfectly matched. In plasmonic cavities, where V_m is very small, this leads to large spontaneous decay rates. However, it is incorrect to assume that the enhanced decay always leads to an increase in emissions in the far field. This is especially true for an individual plasmonic nanostructure, such as a spherical Au NP, where the enhanced decay can be lost to absorption within the metal. This absorption is termed the nonradiative decay channel (Γ_{nr}) (note that this is different from the nonradiative decay channels inside the molecule itself [Γ_{nr}^0]) (Figure 5.12a). Therefore, the total radiative emission intensity extracted from a molecule near the metal (fluorescence enhancement) is determined by the competition between radiative (Γ_r) and nonradiative coupling rates (Γ_{nr}).

The radiative quantum efficiency of the system is defined as,

$$\eta = \frac{\Gamma_r + \Gamma_r^0}{\Gamma_r^0 + \Gamma_r + \Gamma_{nr}^0 + \Gamma_{nr}} \tag{5.20}$$

It is important to note that the coupling of a molecule to these different channels depends on the distance between the molecule and the NP. For an isolated NP, molecules closer to the NP experience a maximum excitation rate ($\gamma_{exc} \propto E^2$), but suffer from lower quantum yield. An optimal distance is between the Au NP and a molecule at which the emission intensity is maximized (~10 nm) (Figure 5.12b). At shorter distances, the emission is strongly suppressed by the coupling with higher order, mostly nonradiative modes of the cavity ($l_{2,3}, \ldots$). This regime in which the emission is suppressed is generally termed as the "quenching" zone. Another nonradiative pathway that is not represented here (Figure 5.12a) is a charge transfer process between the molecule and plasmon. A molecule in an excited state can undergo redox chemistry by transferring an electron to the metal (resonant electron transfer). The reverse process is also possible, in which excited plasmons relax by emission of high-energy electrons (hot-e), which are captured by molecules absorbed on the surface. However, the energy dynamics of these processes are still a matter of strong debate [42].

The quenching effect is suppressed to near-zero values for NPoM plasmonic cavities in which excitation rates are much larger than the quenching rates (Figure 5.12c). Plasmonic nanodimers and NPoM cavities supporting high field enhancement ($E/E_0 > 100$) in nanogaps are ideal for obtaining large light emission from molecules. In an NPoM cavity, the emission from the molecule in the gap is strongly enhanced even when the distance between the metal and the molecule is <1 nm (Figure 5.13a). At larger distances, the coupling between the NP and the film is weaker; therefore, larger NPs are required for any effects to be observed. However, at small gaps ($d < 5$ nm) smaller NPs ($2R = 40$ nm) give comparable enhancement. The competition between the radiative and nonradiative rates is crucial for NPoM systems with 1-nm gaps, ensuring Γ_r is larger than Γ_{nr}. Quantum efficiencies >0.5 can be reached in this geometry when the size of the NP is between 60 and 80 nm (Figure 5.13b). This compares to single NPs where emission is strongly quenched. Furthermore, when an emitter in an NPoM couples to the l_1 mode, radiative enhancements $>10^5$ can be achieved (Figure 5.13c). Higher order modes l_2 and l_3 also acquire a strong radiative component for small gaps due to their hybridization with dipolar modes, hence further suppressing the quenching zone.

From the enhancement rates, it is possible to extract the mode volume of the system, since the Purcell factor is a measure of the total decay rate, where $\Gamma_m = (\Gamma_r + \Gamma_{nr})$. In NPoM systems, $\Gamma_m/\Gamma_0 > 10^6$ at 650 nm, resulting in effective mode volumes <100 nm^3, and sub-ps emission rates.

It is important to note that emission intensity enhancement measured in experiments [43–46] (ratio of intensities with and without a cavity) is not equal to the Purcell factor in plasmonic cavities, because the intensity of incoming light is concentrated to the location of the emitters, increasing the amount of light they absorb. Therefore, the increased emission is mainly from this absorption enhancement. To circumvent this issue, the enhancement in fluorescence intensity needs to be measured at high laser intensities at which the absorption saturates and the emission intensity is dominated only by the radiative decay rates [47]. However, many molecules will undergo rapid photobleaching before saturation. On the other hand, the total number of photons emitted by the molecule before it bleaches is independent of the rate at which the molecule is excited, so the enhancement in total photons emitted is proportional to the radiative enhancement factor [48].

Early experiments [49] measured enhancements up to 1300 in the brightness of single-molecule fluorescence for gold bowtie nanoantennas with gaps <20 nm (Figure 5.14a, b). Electromagnetic simulations from the same study revealed that the increase in brightness was mainly due to enhanced absorption. However, recent studies [50] demonstrated enhancement in radiative decay rates by a factor of 1000 in narrow gaps (5 nm) formed between silver nanocubes and a gold film

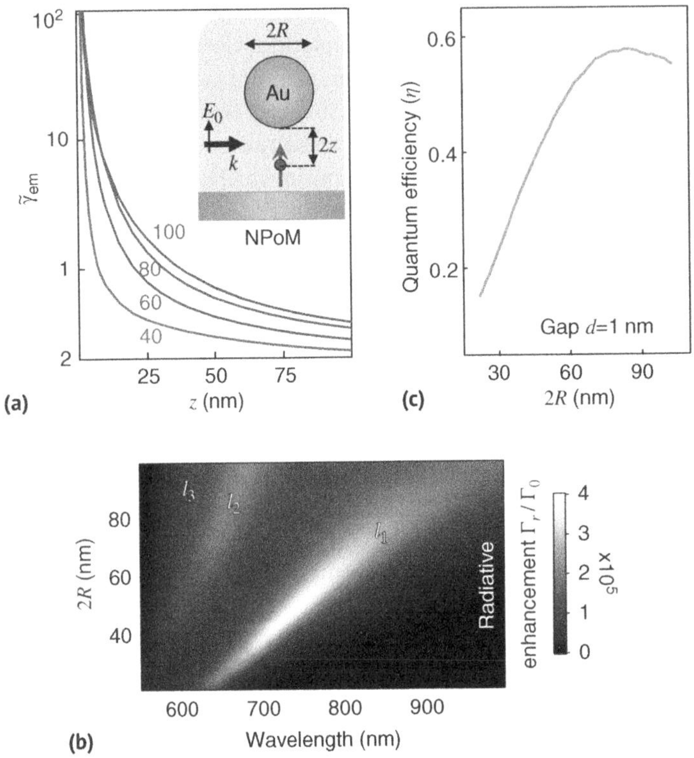

Figure 5.13 Emission from NPoM cavity. (a) Calculated emission from the system for various gap sizes while positioning a point emitter at the center of the cavity. *Source*: Reference [41]/American Chemical Society. (b) Radiative enhancement of an emitter placed in 1-nm gap NPoM cavity for various NP sizes, and (c) the quantum efficiency extracted from l_1 mode emission. *Source*: (a) and (b) reproduced from [41].

(Figure 5.14c, d). These measurements involved ensembles of emitters with unoptimized orientation of dipoles. Later experiments [52, 53] by the same group showed similar enhancements when coupling to a single QD in the same geometry, resulting in an ultrafast single-photon source at room temperature. For quantum technologies, it is important to have nanoscale antennas that can efficiently absorb light from single emitters and subsequently emit photons in a well-controlled manner. Akimov et al. [51] demonstrated the coupling of single-photon emission from a QD to the SPP mode of a silver nanowire (Figure 5.14e–g). The light emission from the other end of the nanowire was antibunched, confirming the quantized nature of SPPs.

Before discussing the strong-coupling regime, where energies of plasmon and molecule are intermixed to form new resonances of the system, it is important to consider that the spectral linewidths of the molecules and cavity are perturbed in the weak-coupling regime, as the emission by a molecule is not uniformly enhanced/suppressed. The linewidths of the emitter increase with increasing coupling strength (g) (Figure 5.15a, b), as the emission is faster in both radiative and nonradiative channels. This increase in linewidth, $\gamma_e = \gamma_0 + \gamma_m$ is proportional to the Purcell enhanced rate. γ_m continues to increase proportional to g^2 until the system reaches the strong-coupling regime, where new hybrid eigenstates emerge ($\omega_\pm$) (Figure 5.15c, d). This is a phase transition with the system switching at critical emitter damping, from the perturbative to the strong-coupling regime.

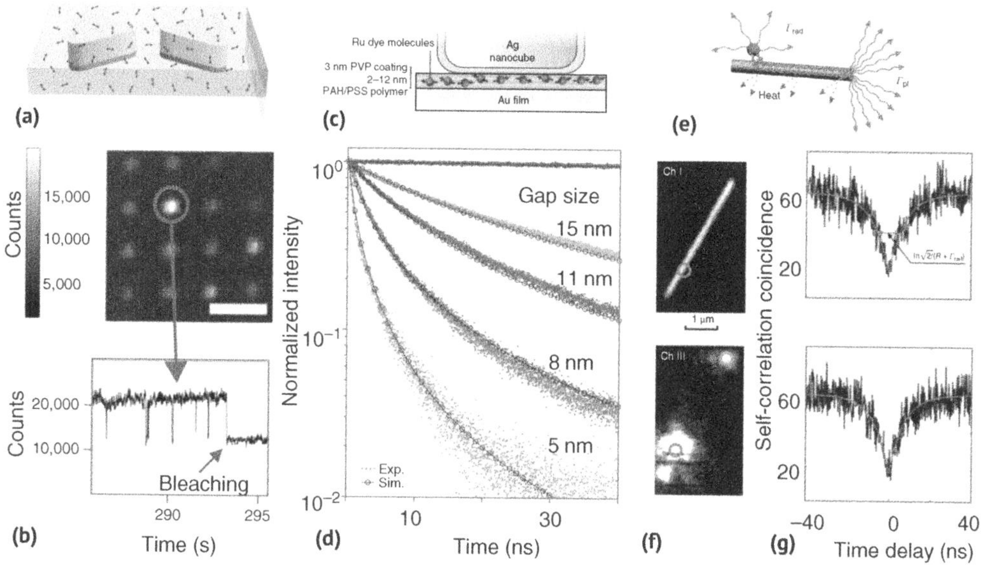

Figure 5.14 Enhanced light emission from plasmonic cavities. (a) Schematic of a bowtie gold nanoantenna coated with fluorescent molecules (black arrows). *Source*: Reference [43]/Springer Nature. (b) Confocal scan of 16 bowties coated with fluorescent molecules with an average of 3 molecules per gap. The fluorescence time trace of one such gap shown below represents blinking dynamics and eventual photobleaching of one molecule that has been enhanced by a factor of 1340. *Source*: Reference [49]/Springer Nature. (c) Cross-section of an experimental structure consisting of Ru dye intercalated into a polymer film and situated between a gold film and a silver nanocube. Arrows indicate typical directions of the transition dipole moments. *Source*: Reference [43]/Springer Nature. (d) Measured and simulated time-resolved ruthenium metal complex dye (Ru dye) emission for four gap thicknesses with normalized intensities. *Source*: Reproduced from [50]/Springer Nature. (e) Radiative coupling of QDs to silver nanowires. A coupled QD can emit photons into the guided surface plasmon modes of the nanowire. (f) (top) Optical image of a plasmonic nanowire with position of QD circled. (bottom) Fluorescence image when the excitation laser focused on the QD. The top-right circle indicates the light emission from the farthest end of the nanowire, used for photon cross-correlation measurements. (g) Second-order correlation function $g^{(2)}$ of single QD quantum dot fluorescence (top) without and (bottom) with coupling to nanowire. *Source*: (e–g) Reproduced from [51]/Springer Nature.

5.7.2 Strong Coupling

Strong coupling occurs in a system when $\omega_{\pm}$ results in real values at zero detuning with,

$$g^2 > (\gamma_c - \gamma_e)^2/16 \tag{5.21}$$

The time dynamics show periodic oscillations (Rabi oscillations), which are masked with finite damping of the system $(\gamma_e + \gamma_c)/2$ (Figure 5.15d). However, this clearly fails when $\gamma_c = \gamma_e$, since in this case, Eq. (5.21) always gives real solutions even though the linewidths may be larger than the splitting. To observe the splitting, modes clearly require at least,

$$\Omega_R > \overline{\Gamma} = (\gamma_c + \gamma_e)/2 \tag{5.22}$$

giving the condition $2g > \overline{\Gamma}$. These different scenarios are mapped in Figure 5.15e. There are two types of damping process that destroy the coherence of the Rabi oscillations: (1) Spontaneous decay to lower electronic states: this process occurs stochastically, breaking the coherence of the wavefunction. The rate of this damping process is set by the lifetime of the polariton states (both radiative and nonradiative parts). (2) Photon decoherence: the dephasing of photons within the cavity

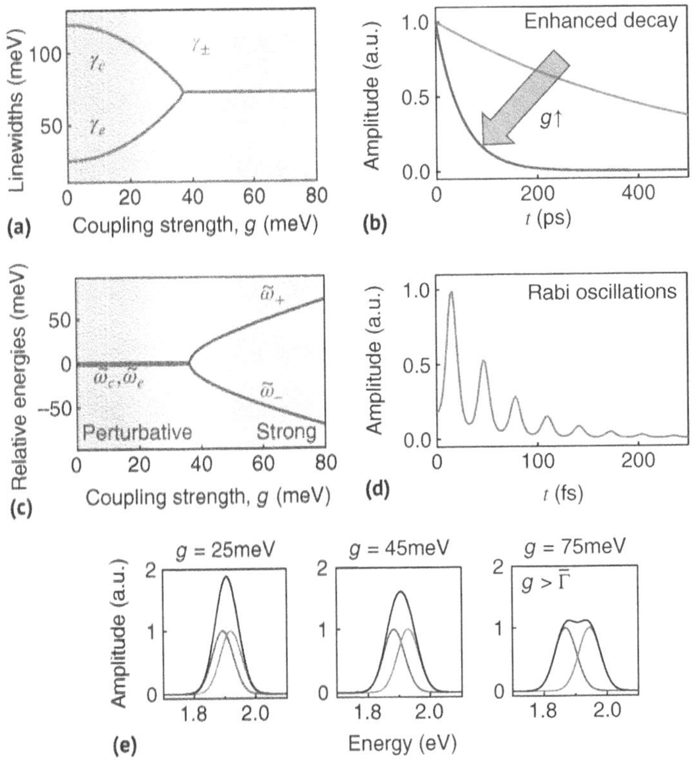

Figure 5.15 Purcell enhancement and Rabi oscillations. (a) Variation of linewidths and (b) relative energies of cavity, exciton, and polariton energies with change in coupling strength. Time dynamics of an emitter coupled to cavity in (c) weak- and (d) strong-coupling regime. In weak coupling, decay rate is enhanced, whereas in strong-coupling Rabi oscillations are observed. (e) Calculated polariton modes and the total summed response as a function of coupling strength, between the limits of Eqs. (5.21 and 5.22).

mode is complicated to understand for molecular systems with strong vibrational states, further influenced by the interaction with the environment that perturbs the phase of the wavefunction and destroys the Rabi oscillation dynamics.

The energy ladder obtained from solution of the JC Hamiltonian for strong light–matter coupling depicts the an harmonicity of the structure (Figure 5.16). This is also known as a JC ladder, giving rise to an intriguing optical effect termed photon blockade [54–56]. Photon blockade arises when the first excitation blocks absorption of a second photon of the same energy, enabling nonlinear response with a single photon. However, the spectrum of the system changes when more than one emitter is involved in the interaction. A system of N emitters coupled to the same electromagnetic mode acts like a giant quantum oscillator with dipole strength $\mu\sqrt{N}$. Hence, the strength of coupling correspondingly scales with the number of emitters, and in this case higher excitations remainharmonic and do not induce the effect of photon blockade within the limit of $N_{\text{ph}} \ll N$.

Earlier work on strong coupling in plasmonic systems focused on the coupling of molecular aggregates with SPP modes on metal films [57–64]. In many experiments, strong coupling was observed between a silver film and exciton modes of J-aggregates or aggregates of dye molecules. The strength of the coupling was revealed by measuring anticrossing in reflectivity measurements. Furthermore, the light emission from the coupled system traced the position of the lower polariton

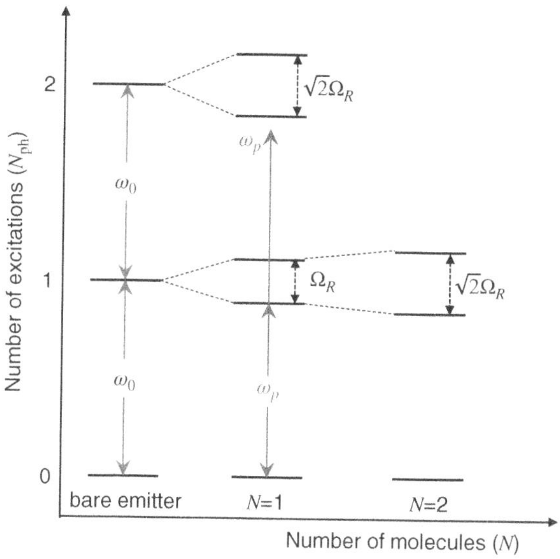

Figure 5.16 Nonlinearity of polariton states. Schematic of the hybridized energy levels: each excitation (N_{ph}) in the harmonic ladder has two states associated with it. The energy splitting of each state is given by $\Omega_R\sqrt{N_{ph}}$, with N being the number of emitters coupled in the system. This anharmonic splitting is shown in the diagram and explains how ω_p will only excite the first manifold and nothing above, creating a photon blockade.

branch with no evidence of emission from the upper polariton branch. Studies attributed this phenomenon to the presence of a large number of uncoupled excitons in the system, energy relaxation dynamics of different polariton states, the presence of N–1 dark states, or even the wavelength of excitation. This absence of upper polariton branch emission remains a not fully solved issue and is a subject of ongoing debate [65, 66].

Intriguing features start to appear when the number of emitters is drastically decreased to the quantum limit of the single emitter. To progress in this direction, localized plasmonic modes are utilized. In recent experiments with plasmonic dimers, nanorods, cubes, core-shell NPs, nanostars, and nanoprisms, the number of excitons is drastically reduced to <1000 excitons [67–70] (see Figure 5.17a–f for experiments on nanodimers and nanoprisms). There is also significant evidence of uncoupled excitons in such data. This is expected because, although J-aggregated molecules are coated onto nanostructures, the optical fields are strongly confined in the gaps or on the surface far from many of the molecules, and are sometimes shielded by oxide layers on the silver substrates, or else the excitons are not well aligned with the optical fields.

Achieving strong coupling at the single-emitter level requires meticulous alignment of the emitters in a highly confined plasmonic mode volume. Studies have utilized STM tips with subnanometer spatial control to couple confined plasmon modes near the tip with molecules deposited on the surface (Figure 5.17g–i). When the tip is in close proximity, Fano lineshapes are observed in tunneling electron-induced luminescence spectra [70]. However, the coupling strengths were still limited to values lower than the cavity damping rates due to the mismatch between the molecular dipole and cavity fields. In another work, Santhosh et al. [73] utilized capillary forces to assemble one to few QDs in the gaps of silver bowtie cavity (Figure 5.17j–l). They observed a transparency dip in the scattering spectra, registering coupling strengths of 120 meV, close to the strong-coupling regime.

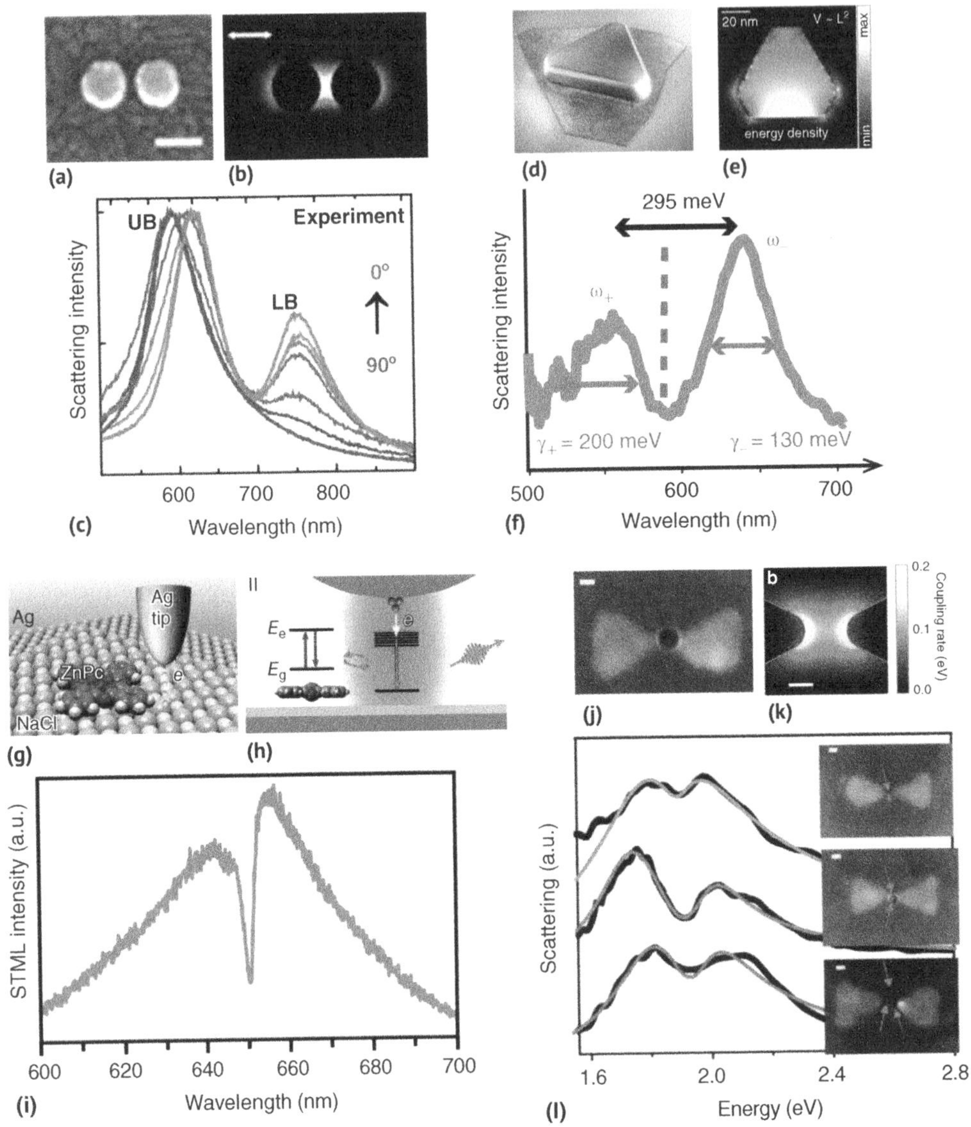

Figure 5.17 Strong coupling in plasmonic cavities. (a–c) Au plasmonic nanodisc dimer coupled with molecular *J*-aggregates. (a) SEM image of a plasmonic nanodisc dimer with a diameter of 85 nm and a gap of 15 nm. (b) Near-field enhancement map ($|E|^2 / |E_0|^2$) for (a), with maximum enhancement of 200 in the gap. (c) Polarization-dependent scattering spectra obtained from the nanodisk dimer system on coupling with exciton transition wavelength for *J*-aggregates at 693 nm. 0° represents detection polarization angle parallel to the dimer axis. *Source*: (a–c) Reproduced from [70]/American Chemical Society. (d) Schematic view of *J*-aggregate sheet coupled with silver nanoprism. (e) Energy density $|E|^2$ distribution around a silver nanoprism shows that V_m is approximated by the geometrical volume of the nanoprism. (f) Scattering spectrum of a single nanoprism strongly coupled to *J*-aggregates, resulting in a pronounced scattering dip. *Source*: (d–f) Reproduced from [71]/American Chemical Society. (g) Schematic of experimental STM used to achieve single-molecule Fano resonance. (h) Ag tip in close proximity to the molecule and (i) corresponding STM luminescence spectra representing a typical Fano lineshape. *Source*: (g–i) Reproduced from [72]/American Chemical Society. (j) SEM image of plasmonic bowties. (k) Scattering spectra of bowties with (from top to bottom) one, two, and three QDs in the gap. All spectra show a transparency dip due to Rabi splitting. (l) Spatial distribution of the coupling strength in the gap between the bowties. *Source*: (j–l) Reproduced from [73].

5.7.3 Single-molecule Strong Coupling

Creating strongly coupled mixed states from visible light and individual emitters is severely compromised by the hundred-fold difference in their spatial localization. High-quality cavities are used to boost interaction times and enhance coupling strengths as discussed in above-mentioned sections. However, in larger cavities, the longer round trip for photons to return to the same emitter decreases the coupling, which scales as $g \propto 1/\sqrt{V}$ with the effective cavity volume V. This coupling has to exceed both the cavity loss rate κ and the emitter scattering rate γ, in order for energy to cycle back and forth between matter and light components, requiring $2g > \gamma, \kappa$ [43]. For cryogenic emitters [44, 45] (laser-cooled atoms, vacancies in diamond, or semiconductor QDs), the suppressed emitter scattering allows large cavities (with high-quality factor $Q \propto \kappa^{-1}$) to reach strong coupling. However, severe technical challenges restrict energy, bandwidth, size, and complexity of devices. Progress toward room temperature is limited by the unavoidable increase in emitter scattering, and the difficulty of reducing the volume of dielectric-based microcavities at wavelength λ and refractive index n below $V_\lambda = (\lambda/n)^3$. At room temperature T, typical scattering rates for embedded dipoles, $\gamma \sim k_B T$, imply using $Q < 100$ which requires cavities $V_m < 10^{-5} V_\lambda$ to maximize the coupling. Improved confinement uses localized surface plasmons, combining oscillations of free electrons in metals with electromagnetic waves [46].

Several key aspects are critical to positioning a quantum emitter inside these small gaps. One is to prevent molecular aggregation, which commonly occurs. Another is to ensure that the transition dipole $\boldsymbol{\mu}_\mathrm{m}$ is perfectly aligned with the gap plasmon (along z). We use the common dye molecule MB with molecular transition at 665 nm, to which the plasmons are tuned. To avoid aggregation of the dye molecules and to assemble them in the proper orientation, recent studies utilize the host–guest chemistry of macrocyclic cucurbit[n]uril molecules [47]. These are pumpkin-shaped molecules with different hollow hydrophobic internal volume determined by the number of units in the ring n, in which guest molecules can sit (Figure 5.18a) [48]. Cucurbit[7]uril (CB[7]) is water-soluble and can accommodate only one MB molecule inside. Encapsulation of MB inside CB[7] is confirmed from absorption spectroscopy since MB dimers disappear on mixing low MB concentrations with CB[7] (1 : 10 molar ratio). Control experiments with smaller CB[5] molecules (into which MB cannot fit) do not remove this shoulder peak, ruling out parasitic binding. Placing single-MB molecules in CB[7] thus, avoids any aggregation. Carbonyl portals at either end of the 0.9-nm-high CB[n] molecules bind them with their rims flat onto the Au surface. When a monolayer of CB[7] is first deposited on the gold mirror and suitably filled with MB molecules, Au NPs bind on top to form the desired filled nanocavity (Figure 5.18a) with the MB molecule aligned vertically in the gap. Dark-field scattering spectra from individual NPoMs show the effect of aligning the emitter in different orientations (Figure 5.18b). With $\boldsymbol{\mu}_\mathrm{m}$ parallel to the mirror, the resonant scattering plasmonic peak is identical to NPoMs without any emitters. Instead, with $\boldsymbol{\mu}_\mathrm{m}$ perpendicular to the mirror, the spectra show two split peaks (ω_+ and ω_-) from the strong interaction between emitters and plasmon. Small fluctuations in peak wavelength are associated with ± 5 nm variations in NP size. When this NPoM is partially filled with MB inside the CB[7], two peaks at 610 and 750 nm are seen either side of the absorption peak of MB at ω_0 (Figure 5.18c), corresponding to the formation of hybrid plasmon–exciton ("plexciton") branches, $\omega_\pm = \omega_0 \pm g/2$. This yields a Rabi frequency $g = 380$ meV.

To probe single-molecule strong coupling, the mean number of MB molecules ($\overline{n}$) in the nanogap is systematically decreased by reducing the ratio of MB : CB[7] used[7]. With $\overline{n} = 2.5$ distinct systematic jumps matching the expected increase of g_n from $n = 1 - 3$ molecules was observed

(Figure 5.18d). The range in each g_n for the single-molecule regime arises because single molecules are located at different lateral positions within the gap plasmon, thus coupling with different strengths. Despite such advances, the yield and consistency of strong coupling are far from the realm needed to build realistic devices.

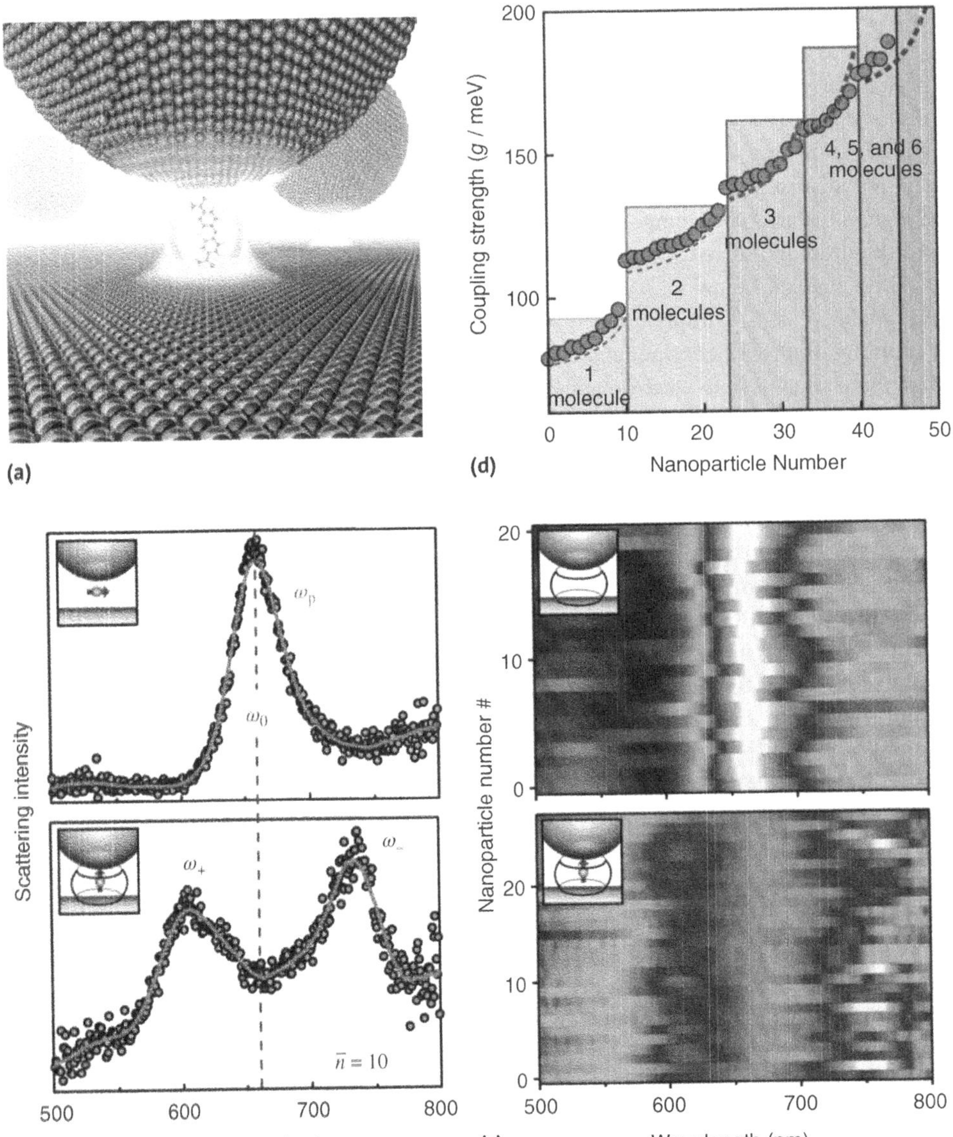

Figure 5.18 Strong coupling of individual NPoMs. (a) NPoM with a single dye molecule in the gap. (b) Scattering spectra from isolated NPoM coupled with dye μ_m oriented (top) perpendicular and (bottom) parallel to $|E_z|$ of gap plasmon (dashed line indicates dye absorption wavelength). (c) Comparison of scattering spectra from different NPoMs (schematics inset), with gaps filled by CB[7] monolayer, which are (top) empty, or (bottom) encapsulating MB dye molecules. (d) Coupling strength extracted from different NPoMs in sample of $\bar{n} = 2.5$. Theoretical coupling strength in perfect model (shaded bars) and random placement model (dashed lines).

The observation of strong coupling in inelastic light scattering, such as photoluminescence (PL) and Raman scattering, would be helpful to probe nonlinearities of the system. Despite this advantage, strong coupling of plasmonic systems is evidenced by two peaks in their scattering spectra. Energy splitting in PL provides less ambiguous evidence of strong coupling but has only been observed in very few studies with QDs [74–76]. Even these results are ambiguous, as PL peaks can arise from dark-plasmonic multimodes. Changes in lifetime of surface trap states or dark excitons may also give multiple peaks that are difficult to exclude based only on PL measurements. Only recently has consistent splitting been observed for QD NPoMs in both scattering and PL [77], giving strong optimism for future developments.

5.8 Conclusion

In this chapter, we discussed the coupling of a single molecule to localized nanocavity systems constructed from metal NPs. We highlight the challenges associated with fabrication of nanocavities with small gaps $d < 5$ nm as well as their demand for sensing. The self-assembled NPoM geometry overcomes many challenges and achieves reliable gaps of <2 nm using robust molecular spacers. From the context of light–matter interactions, such geometry successfully supports optical mode volumes of <100 nm^3, significantly accelerating Purcell-induced light emission and strong light–molecule coupling at room temperature.

References

1 Kaluzny, Y., Goy, P., Gross, M., Raimond, J. M., Haroche, S. Observation of self-induced rabi oscillations in two-level atoms excited inside a resonant cavity: the ringing regime of superradiance. *Phys. Rev. Lett.* **1983,** *51*(13), 1175–1178.

2 Raizen, M. G., Thompson, R. J., Brecha, R. J., Kimble, H. J., Carmichael, H. J. Normal-mode splitting and linewidth averaging for two-state atoms in an optical cavity. *Phys. Rev. Lett.* **1989,** *63*(3), 240–243.

3 Thompson, R. J., Rempe, G., Kimble, H. J. Observation of normal-mode splitting for an atom in an optical cavity. *Phys. Rev. Lett.* **1992,** *68*(8), 1132–1135.

4 Haroche, S. Nobel Lecture: controlling photons in a box and exploring the quantum to classical boundary. *Rev. Mod. Phys.* **2013,** *85*(3), 1083–1102.

5 Harris, D. C., Bertolucci, M. D. *Symmetry and Spectroscopy: An Introduction to Vibrational and Electronic Spectroscopy* (Dover Books on Chemistry) [Internet]. Dover Publications, **1989.** https://www.amazon.com/Symmetry-Spectroscopy-Introduction-Vibrational-Electronic/dp/048666144X?SubscriptionId=0JYN1NVW651KCA56C102&tag=techkie-20&linkCode=xm2&camp=2025&creative=165953&creativeASIN=048666144X.

6 Lakowicz, J. R. *Principles of Fluorescence Spectroscopy* [Internet]. Springer, **2011.** https://www.amazon.com/Principles-Fluorescence-Spectroscopy-Joseph-Lakowicz/dp/0387312781?SubscriptionId=0JYN1NVW651KCA56C102&tag=techkie-20&linkCode=xm2&camp=2025&creative=165953&creativeASIN=0387312781.

7 Zumofen, G., Mojarad, N. M., Sandoghdar, V., Agio, M. Perfect reflection of light by an oscillating dipole. *Phys. Rev. Lett.* **2008,** *101*(18), 180404.

8 Hilborn, R. C. Einstein coefficients, cross sections, f values, dipole moments, and all that. *Am. J. Phys.* **1982,** *50*(11), 982–986.

9 Lee, K. G., Chen, X. W., Eghlidi, H., Kukura, P., Lettow, R., Renn, A. et al. A planar dielectric antenna for directional single-photon emission and near-unity collection efficiency. *Nat. Photonics.* **2011,** *5*(3), 166–169.

10 Taminiau, T. H., Stefani, F. D., Segerink, F. B., van Hulst, N. F. Optical antennas direct single-molecule emission. *Nat. Photonics.* **2008,** *2*(4), 234–237.

11 Lodahl, P., Floris van Driel, A., Nikolaev, I. S., Irman, A., Overgaag, K., Vanmaekelbergh, D. et al. Controlling the dynamics of spontaneous emission from quantum dots by photonic crystals. *Nature* **2004,** 5;*430*(7000), 654–657.

12 Englund, D., Fattal, D., Waks, E., Solomon, G., Zhang, B., Nakaoka, T. et al. Controlling the spontaneous emission rate of single quantum dots in a two-dimensional photonic crystal. *Phys. Rev. Lett.* **2005,** *95*(1), 013904.

13 Pelton, M., Santori, C., Vučković, J., Zhang, B., Solomon, G. S., Plant, J. et al. Efficient source of single photons: a single quantum dot in a micropost microcavity. *Phys. Rev. Lett.* **2002,** 13;*89*(23), 233602.

14 Pelton, M. Modified spontaneous emission in nanophotonic structures. *Nat. Photonics.* **2015,** *9*(7), 427–435.

15 Schuller, J. A., Barnard, E. S., Cai, W., Jun, Y. C., White, J. S, Brongersma, M. L. Plasmonics for extreme light concentration and manipulation. *Nat. Mater.* **2010,** *9*(3), 193–204.

16 Novotny, L., van Hulst, N. Antennas for light. *Nat. Photonics.* **2011,** 83–90.

17 Maier, S.A. *Plasmonics: Fundamentals and Applications* [Internet]. Springer, **2007.** https://www.amazon.com/Plasmonics-Fundamentals-Stefan-Alexander-Maier/dp/0387331506?SubscriptionId=0JYN1NVW651KCA56C102&tag=techkie-20&linkCode=xm2&camp=2025&creative=165953&creativeASIN=0387331506.

18 Johnson, P. B., Christy, R. W. Optical constants of the noble metals. *Phys. Rev. B.* **1972,** *6*(12), 4370–4379.

19 Palik, E.D. *Handbook of Optical Constants of Solids* [Internet]. Academic Press, **1998.** https://www.amazon.com/Handbook-Optical-Constants-Solids-Edward-ebook/dp/B00SYJCYN0?SubscriptionId=0JYN1NVW651KCA56C102&tag=techkie-20&linkCode=xm2&camp=2025&creative=165953&creativeASIN=B00SYJCYN0.

20 Huber, A., Ocelic, N., Kazantsev, D., Hillenbrand, R. Near-field imaging of mid-infrared surface phonon polariton propagation. *Appl. Phys. Lett.* **2005,** *87*(8), 081103.

21 Gentile, M. J., Núñez-Sánchez, S., Barnes, W. L. Optical field-enhancement and subwavelength field-confinement using excitonic nanostructures. *Nano. Lett.* **2014,** *14*(5), 2339–2344.

22 Barnes, W. L. Surface plasmon–polariton length scales: a route to sub-wavelength optics. *J. Opt. Pure. Appl. Opt.* **2006,** *8*(4), S87.

23 Enoch, S., Bonod, N. (eds.). *Plasmonics: From Basics to Advanced Topics* (Vol. 167). Springer, **2012**.

24 Duan, H., Fernández-Domínguez, A. I, Bosman, M., Maier, S. A., Yang, J. K. W. Nanoplasmonics: classical down to the Nanometer Scale. *Nano. Lett.* **2012,** *12*(3), 1683–1689.

25 Ward, D. R., Corley, D. A., Tour, J. M., Natelson, D. Vibrational and electronic heating in nanoscale junctions. *Nat. Nanotechnol.* **2011,** 33–38.

26 Savage, K. J., Hawkeye, M. M., Esteban, R., Borisov, A. G., Aizpurua, J., Baumberg, J. J. Revealing the quantum regime in tunnelling plasmonics. *Nature* **2012,** *491*(7425), 574–577.

27 Cheng, L., Song, J., Yin, J., Duan, H. Self-assembled plasmonic dimers of amphiphilic gold nanocrystals. *J. Phys. Chem. Lett.* **2011,** *2*(17), 2258–62.

28 Chen, G., Wang, Y., Tan, L. H., Yang, M., Tan, L. S., Chen, Y. et al. High-purity separation of gold nanoparticle dimers and trimers. *J. Am. Chem. Soc.* **2009,** *131*(12), 4218–4219.

29 Kumar, J., Thomas, K. G. Surface-enhanced raman spectroscopy: Investigations at the nanorod edges and dimer junctions. *J. Phys. Chem. Lett.* **2011,** *2*(6), 610–615.

30 Lee, J. H., Nam, J. M., Jeon, K. S., Lim, D. K., Kim, H., Kwon, S. et al. Tuning and maximizing the single-molecule surface-enhanced Raman scattering from DNA-tethered nanodumbbells. *ACS Nano*. **2012,** *6*(11), 9574–9584.

31 Aravind, P. K., Metiu, H. The effects of the interaction between resonances in the electromagnetic response of a sphere-plane structure: applications to surface enhanced spectroscopy. *Surf. Sci.* **1983,** *124*(2), 506–528.

32 Aravind, P. K., Metiu, H. Use of a perfectly conducting sphere to excite the plasmon of a flat surface. 1. Calculation of the local field with applications to surface-enhanced spectroscopy. *J. Phys. Chem*. **1982,** *86*(26), 5076–5084.

33 Nordlander, P., Prodan, E. Plasmon hybridization in nanoparticles near metallic surfaces. *Nano. Lett*. **2004,** *4*(11), 2209–2213.

34 Ciracì, C., Hill, R. T., Mock, J. J., Urzhumov, Y., Fernández-Domínguez, A. I., Maier, S. A. et al. Probing the ultimate limits of plasmonic enhancement. *Science* **2012,** *337*(6098), 1072–1074.

35 de Nijs, B., Bowman, R. W., Herrmann, L. O., Benz, F., Barrow, S. J., Mertens, J. et al. Unfolding the contents of sub-nm plasmonic gaps using normalising plasmon resonance spectroscopy. *Faraday Discuss* **2015,** *178*(0), 185–193.

36 Benz, F., Chikkaraddy, R., Salmon, A., Ohadi, H., de Nijs, B., Mertens, J. et al. SERS of individual nanoparticles on a mirror: size does matter, but so does shape. *J. Phys. Chem. Lett.* **2016,***7*(12), 2264–2269.

37 Sigle, D. O., Mertens, J., Herrmann, L. O., Bowman, R. W., Ithurria, S., Dubertret, B. et al. monitoring morphological changes in 2D monolayer semiconductors using atom-thick plasmonic nanocavities. *ACS Nano*. **2015,** Jan 27;*9*(1), 825–830.

38 Tserkezis, C., Esteban, R., Sigle, D. O., Mertens, J., Herrmann, L. O., Baumberg, J. J. et al. Hybridization of plasmonic antenna and cavity modes: extreme optics of nanoparticle-on-mirror nanogaps. *Phys. Rev. A*. **2015,** *92*(5), 053811.

39 Elliott, E., Bedingfield, K., Huang, J., Hu, S., de Nijs, B., Demetriadou, A. et al. Fingerprinting the hidden facets of plasmonic nanocavities. *ACS Photonics* **2022,** *9*(8), 2643–2651.

40 Chikkaraddy, R., Huang, J., Kos, D., Elliott, E., Kamp, M., Guo, C. et al. Boosting optical nanocavity coupling by retardation matching to dark modes. *ACS Photonics* **2023,** *10*(2), 493–499.

41 Kongsuwan, N., Demetriadou, A., Chikkaraddy, R., Benz, F., Turek, V. A., Keyser, U. F. et al. Suppressed quenching and strong coupling of purcell-enhanced single-molecule emission in plasmonic nanocavities. *ArXiv161202611 Phys* [Internet]. **2016,** [cited 2017 Mar 16]. http://arxiv.org/abs/1612.02611.

42 Baumberg, J. J. Hot electron science in plasmonics and catalysis: what we argue about. *Faraday Discuss* **2019,** *214*(0), 501–511.

43 Törmä, P., Barnes, W. L. Strong coupling between surface plasmon polaritons and emitters: a review. *Rep. Prog. Phys*. **2015,** *78*(1), 013901.

44 Yoshie, T., Scherer, A., Hendrickson, J., Khitrova, G., Gibbs, H. M., Rupper, G. et al. Vacuum Rabi splitting with a single quantum dot in a photonic crystal nanocavity. *Nature* **2004,** *432*(7014), 200–203.

45 Thompson, J. D., Tiecke, T. G., de Leon, N. P., Feist, J., Akimov, A. V., Gullans, M. et al. Coupling a single trapped atom to a nanoscale optical cavity. *Science* **2013,** *340*(6137), 1202–1205.

46 Novotny, L., Hecht, B. *Principles of Nano-Optics*. Cambridge University Press, **2012**.

47 Chikkaraddy, R., de Nijs, B., Benz, F., Barrow, S. J., Scherman, O. A., Rosta, E. et al. Single-molecule strong coupling at room temperature in plasmonic nanocavities. *Nature* **2016,** *535*(7610), 127–130.

48 Kasera, S., Herrmann, L. O., Barrio, J. D., Baumberg, J. J., Scherman, O. A. Quantitative multiplexing with nano-self-assemblies in SERS. *Sci. Rep.* [Internet]. **2014,** *4*. http://www.nature.com/srep/2014/141030/srep06785/full/srep06785.html.

49 Kinkhabwala, A., Yu, Z., Fan, S., Avlasevich, Y., Müllen, K., Moerner W. E. Large single-molecule fluorescence enhancements produced by a bowtie nanoantenna. *Nat. Photonics* **2009,** *3*(11), 654–657.

50 Akselrod, G. M., Argyropoulos, C., Hoang, T. B., Ciracì, C., Fang, C., Huang, J. et al. Probing the mechanisms of large Purcell enhancement in plasmonic nanoantennas. *Nat. Photonics* **2014,** *8*(11), 835–840.

51 Akimov, A. V., Mukherjee, A., Yu, C. L., Chang, D. E., Zibrov, A. S., Hemmer, P. R. et al. Generation of single optical plasmons in metallic nanowires coupled to quantum dots. *Nature* **2007,** *450*(7168), 402–406.

52 Hoang, T. B., Akselrod, G. M., Mikkelsen, M. H. Ultrafast room-temperature single photon emission from quantum dots coupled to plasmonic nanocavities. *Nano. Lett.* **2016,** *16*(1), 270–275.

53 Hoang, T. B., Akselrod, G. M., Argyropoulos, C., Huang, J., Smith, D. R., Mikkelsen M. H. Ultrafast spontaneous emission source using plasmonic nanoantennas. *Nat. Commun.* [Internet]. **2015,** Jul 27 [accessed 2015 August 3]; *6*. http://www.nature.com/ncomms/2015/150727/ncomms8788/full/ncomms8788.html.

54 Birnbaum, K. M., Boca, A., Miller, R., Boozer, A. D., Northup, T. E., Kimble, H. J. Photon blockade in an optical cavity with one trapped atom. *Nature* **2005,** *436*(7047), 87–90.

55 Fink, J. M., Göppl, M., Baur, M., Bianchetti, R., Leek, P. J., Blais, A. et al. Climbing the Jaynes–Cummings ladder and observing its nonlinearity in a cavity QED system. *Nature* **2008,** *454*(7202), 315–318.

56 Kasprzak, J., Reitzenstein, S., Muljarov, E. A., Kistner, C., Schneider, C., Strauss, M. et al. Up on the Jaynes–Cummings ladder of a quantum-dot/microcavity system. *Nat. Mater.* **2010,** *9*(4), 304–308.

57 Sugawara, Y., Kelf, T. A., Baumberg, J. J., Abdelsalam, M. E., Bartlett, P. N. Strong coupling between localized plasmons and organic excitons in metal nanovoids. *Phys. Rev. Lett.* **2006,** *97*(26), 266808.

58 Ambjörnsson, T., Mukhopadhyay, G., Apell, S. P., Käll M. Resonant coupling between localized plasmons and anisotropic molecular coatings in ellipsoidal metal nanoparticles. *Phys. Rev. B.* **2006,** *73*(8), 085412.

59 Hakala, T. K., Toppari, J. J., Kuzyk, A., Pettersson, M., Tikkanen, H., Kunttu, H. et al. Vacuum rabi splitting and strong-coupling dynamics for surface-plasmon polaritons and rhodamine 6G molecules. *Phys. Rev. Lett.* **2009,** *103*(5), 053602.

60 Nagasawa, F., Takase, M., Murakoshi K. Raman enhancement via polariton states produced by strong coupling between a localized surface plasmon and dye excitons at metal nanogaps. *J. Phys. Chem. Lett.* **2013,** *5*(1), 14–19.

61 Valmorra, F., Bröll, M., Schwaiger, S., Welzel, N., Heitmann, D., Mendach S. Strong coupling between surface plasmon polariton and laser dye rhodamine 800. *Appl. Phys. Lett.* **2011,** *99*(5), 051110.

62 Baieva, S. V., Hakala, T. K., Toppari, J. J. Strong coupling between surface plasmon polaritons and Sulforhodamine 101 dye. *Nanoscale. Res. Lett.* **2012,** *7*(1), 191.

63 Koponen, M. A., Hohenester, U., Hakala, T. K., Toppari, J. J. Absence of mutual polariton scattering for strongly coupled surface plasmon polaritons and dye molecules with a large Stokes shift. *Phys. Rev. B.* **2013,** *88*(8), 085425.

64 Baieva, S., Ihalainen, J. A., Toppari, J. J., Strong coupling between surface plasmon polaritons and β-carotene in nanolayered system. *J. Chem. Phys.* **2013,** *138*(4), 044707.

65 Baumberg, J., Noginov, M., Aizpurua, J., Lin, K., Ebbesen, T., Kornyshev, A. A. et al. Quantum plasmonics, gain and spasers: general discussion. *Faraday Discuss* **2015,** *178*(0), 325–334.

66 George, J., Wang, S., Chervy, T., Canaguier-Durand, A., Schaeffer, G., Lehn, J. M. et al. Ultra-strong coupling of molecular materials: spectroscopy and dynamics. *Faraday Discuss* **2015,** *178*(0), 281–294.

67 Zengin, G., Johansson, G., Johansson, P., Antosiewicz, T. J., Käll, M., Shegai T. Approaching the strong coupling limit in single plasmonic nanorods interacting with J-aggregates. *Sci. Rep.* [Internet]. **2013,** *3*.

68 Fofang, N. T., Grady, N. K., Fan, Z., Govorov, A. O., Halas, N. J. Plexciton dynamics: exciton–plasmon coupling in a J-Aggregate–Au nanoshell complex provides a mechanism for nonlinearity. *Nano. Lett.* **2011,** *11*(4), 1556–1560.

69 Murray, W. A., Suckling J. R., Barnes, W. L. Overlayers on silver nanotriangles: field confinement and spectral position of localized surface plasmon resonances. *Nano. Lett.* **2006,** *6*(8), 1772–1777.

70 Schlather, A. E., Large, N., Urban, A. S., Nordlander, P., Halas, N. J. Near-field mediated plexcitonic coupling and giant rabi splitting in individual metallic dimers. *Nano. Lett.* **2013,** *13*(7), 3281–3286.

71 Zengin, G., Wersäll, M., Nilsson, S., Antosiewicz, T. J., Käll, M., Shegai T. Realizing strong light-matter interactions between single-nanoparticle plasmons and molecular excitons at ambient conditions. *Phys. Rev. Lett.* **2015,** *114*(15), 157401.

72 Zhang, Y., Meng, Q. S., Zhang, L., Luo, Y., Yu, Y. J., Yang, B. et al. Sub-nanometre control of the coherent interaction between a single molecule and a plasmonic nanocavity. *Nat. Commun.* **2017.**

73 Santhosh, K., Bitton, O., Chuntonov, L., Haran G. Vacuum Rabi splitting in a plasmonic cavity at the single quantum emitter limit. *Nat. Commun.* **2016,** *7*(1), ncomms11823.

74 Park, K. D., May, M. A., Leng, H., Wang, J., Kropp, J. A., Gougousi, T. et al. Tip-enhanced strong coupling spectroscopy, imaging, and control of a single quantum emitter. *Sci. Adv.* **2019,** *5*(7), eaav5931.

75 Leng, H., Szychowski, B., Daniel, M. C., Pelton M. Strong coupling and induced transparency at room temperature with single quantum dots and gap plasmons. *Nat. Commun.* **2018,** *9*(1), 4012.

76 Groß, H., Hamm, J. M., Tufarelli, T., Hess, O., Hecht B. Near-field strong coupling of single quantum dots. *Sci. Adv.* **2018,** *4*(3). eaar4906.

77 Hu, S., Huang, J., Arul, R., et al. Robust consistent single quantum dot strong coupling in plasmonic nanocavities. Nat Commun. **2024,** *15*, 6835. https://doi.org/10.1038/s41467-024-51170-7.

Part II

Spectroscopy and Dynamics

6

Nonlinear Spectroscopy Under Vibrational Strong Coupling

Adam D. Dunkelberger[1], *Cynthia G. Pyles*[1] *and Jeffrey C. Owrutsky*[1,2]

[1] *Chemistry Division, US Naval Research Laboratory, Washington, DC, USA*
[2] *Precise Systems, Inc., Lexington Park, MD, USA*

6.1 Introduction

There is growing interest in vibration-cavity polaritons because strong coupling to optical resonances, best demonstrated for microcavities, has been shown to modify chemistry in ways long dreamed of by scientists [1–4]. Alchemy initially aimed to turn lead into gold, and its modern analog emerged with the invention of lasers. Many hoped shining the right color laser – such as an infrared source with a frequency matching a vibration – would selectively break molecular bonds at will and enable previously inaccessible reaction paths [4, 5]. Alas, it was realized that molecules typically redistribute the energy faster than the rate of the laser-pumped reaction. Consequently, laser-driven selective chemistry has not panned out apart from carefully chosen instances relying on vibrational excitation to enhance reaction rates or carefully crafted coherent control of small gaseous molecules [3, 6, 7].

Vibrational strong coupling (VSC) produces molecular vibration polaritons (MVP), which typically consist of vibration-cavity polaritons employing Fabry–Pérot cavities [8, 9]. MVPs are but one type of the vast number of quasiparticles known as polaritons [10]. Plasmon and phonon polaritons are often explored in the arena of nanophotonics and are capable of producing enhanced optical interactions [11–14] such as surface-enhanced Raman scattering (SERS) but also found use for sensing with infrared absorption (surface-enhanced infrared absorption, SEIRA) [15–18]. In addition to recent applications to energy harvesting, biosensing, and phototherapy [19–23], plasmon and phonon polaritons can also provide optical resonances for strong coupling to molecular excitations [24–26]. Pertinent to our primary interest in this chapter, these materials and structures often have unique and complex nonlinear optical responses.

In contrast to plasmon and phonon polaritons, which typically rely on thin films or nanostructures of materials with appropriate optical constants, cavity polaritons are produced when a material in a Fabry–Pérot has an absorptive transition coincident with the transmissive mode of the cavity. When the coupling rate exceeds all damping processes, the system is said to be in the strong

Polariton Chemistry: Molecules in Cavities, First Edition. Edited by Joel Yuen-Zhou, Noel C. Giebink and Raphael F. Ribeiro.

coupling regime, characterized by the presence of two new hybrid light–matter states called polaritons with energies above (upper polariton, UP) and below (lower polariton, LP) that of the uncoupled material excitation [27–29]. The Rabi splitting between the polaritons depends on the square root of absorption per unit length of the material transition [8, 9, 28, 30–32].

The recent interest in MVP to alter reactions represents a confluence of quantum physics and chemical physics to achieve modified energy transfer and reaction dynamics. Early polariton studies involved strong coupling to electronic transitions (ESC) in cold atoms, and inorganic and organic polaritons [33–37]. The concepts extend naturally to VSC, albeit with two key differences: (i) the lack of photoluminescence from vibrational states and (ii) the much-reduced coupling strengths, such that the Rabi splitting is close to or below kT at room temperature. These differences complicate the interpretation of several exciting results in the field, and so there remains much confusion and debate concerning how VSC modifies chemical and physical phenomena. Leading hypotheses, addressed in detail elsewhere in this volume, include modified vibrational dynamics even at equilibrium. There have been extensive studies on nonstatistical reaction dynamics as a precursor to mode selectivity [38, 39]. These concepts are more easily realized in the rarefied conditions of the gas phase, where small-molecule unimolecular processes, such as intramolecular vibrational relaxation (IVR) and dissociation, are famously understood via Rice-Ramsperger-Kassel-Marcus theory [40–43]. These concepts have led us and others to begin to investigate the nonlinear spectroscopy of systems under VSC, seeking to understand the flow of vibrational energy in such systems to help explain the groundbreaking studies of VSC-modified chemistry. Along the way, the field has discovered a number of novel optical phenomena that can be generated by MVP.

MVPs follow the rules for strong coupling well established in the quantum electrodynamics community with exciton– polaritons [28, 29, 32]. Polariton modes give rise to new desirable properties and behaviors, including long-range energy transfer [44–47] and modified optical and electrical properties [48], at least in part due to the collective, delocalized nature of the polaritons. While there has been some reluctance to hail all the reported studies as unmitigated successes due to reports of irreproducibility [49, 50] and retracted or modified interpretations [51–53], a consensus is emerging that the modified rates and branching ratios are real [1, 2, 54]. It is no exaggeration to say that no one yet understands how the modified chemistry actually works, despite considerable effort [55–58]. One of the largest points of contention is the so-called *N* or dark-state problem. For systems undergoing VSC, most configurations of molecular absorbers lack the symmetry to contribute to the polariton modes and are, instead, dark or reservoir states that are nearly indistinguishable from the molecules without a cavity. It remains a critical challenge to stretch the QED theory (Jaynes Cummings [59] and Tavis Cummings [60]) to ensembles of molecular absorbers. Many experimental observations can be modeled with classical methods, fueling the debate of whether experiments are subject to quantum and coherent effects [61]. Nonlinear spectroscopy gives us the opportunity to begin to address these questions.

Finally, we briefly point out efforts to understand the nonlinear spectroscopy of systems under ESC. The luminescence of the polariton states in most ESC systems allows for more direct observation of the fate of the polaritons [62–66]. We would argue, however, that there is a broad spectrum of interpretation of these results [66–68]. Some early work in the field assigned features in the complex time-resolved spectra of exciton–polaritons to polariton-excited states. Recent work has shown that many spectral signatures can be described by changes in the optical properties of the medium, without invoking any polariton response at all. Accurate, thorough treatment of optical phenomena is critical for understanding the nonlinear spectroscopy of MVP.

Before detailing the nonlinear spectroscopy of systems under VSC, we begin with a brief discussion of the spectroscopic techniques typically employed to interrogate these systems. Molecular modes coupled to optical modes for VSC typically undergo vibrational population relaxation on the order of picoseconds. Therefore, measuring relaxation dynamics under VSC necessitates experimental methods using ultrafast infrared pulses [69–71]. The earliest vibrational-echo studies utilized free electron lasers [72, 73]. The development of Ti:Sapphire lasers and optical parametric amplifiers have made ultrafast infrared pulse generation increasingly accessible. Transient absorption and two-dimensional infrared (2D IR) spectroscopies prove particularly useful due to both the wealth of information provided [74, 75] and their ease of use aided by ongoing experimental advances [76, 77].

A typical transient absorption experiment tracks changes in sample absorption induced by an excitation (pump) pulse, achieving time resolution through careful control of the delay between the arrival of the excitation pulse and a much weaker probe pulse. The anharmonicity of vibrational modes energetically separates the 0–1 and 1–2 vibrational transitions, and this separation makes pump–probe experiments tractable for vibrational spectroscopy. The excitation pulse promotes a fraction of molecules in the sample from the vibrational ground state to the first vibrationally excited state, resulting in characteristic changes to the infrared spectrum. At the 0–1 frequency, fewer ground-state molecules are present to absorb probe light, leading to a ground-state bleach (GSB), and the emission from the excited molecules can be stimulated by the probe pulse. At the 1–2 frequency, molecules in the first excited state are present to absorb probe light. By the convention we adopt for this chapter, transient absorption responses are reported as $\Delta A = -\log_{10} \frac{T_{\text{pump on}}}{T_{\text{pump off}}}$, where T is the transmission of the probe light through the sample. Under this convention, the GSB and stimulated emission (SE) are negative whereas excited-state absorption (ESA) is positive. The summation of these terms yields a derivative-like lineshape. We will argue later in this chapter that understanding how VSC modifies this characteristic response is critical to understanding the nonlinear spectroscopy of coupled systems.

2D IR measurements employ more complex pulse sequences to extract new information about coherences and populations in vibrational states on ultrafast timescales [78, 79]. Instead of the single excitation pulse used for pump–probe measurements, 2D IR requires two excitation pulses separated by another delay time. These two pulses prepare coherences and populations which then interact with the probe pulse to generate a nonlinear signal. In practice, data are collected in the form of transient spectra as a function of the delay between excitation pulses. Altering the delay between the excitation and probe pulses yields time resolution. Fourier transforming the data with respect to the intra-excitation delay gives a new frequency axis that correlates to the excitation frequency. In the simplest description of 2D IR, a spectrum can be read as a map of the transient absorbance at a given pump–probe delay time as a function of both pump and probe frequencies. Despite the experimental differences, both transient absorption and 2D IR signatures can be described by third-order molecular response functions due to having the same number of light–matter interactions.

More sophisticated analysis of 2D IR data can reveal much richer information than pump–probe spectroscopy. Double-sided Feynman diagrams represent the different light–matter interactions available for a molecular system [74]. Transient absorption is sensitive to excitation pathways that involve populations in both the ground and excited states, a small subset of the Feynman pathways studied in 2D IR. Other pathways that involve short-lived coherences between vibrational states are much more readily interrogated by 2D IR, in large part to recent advances in technology for manipulating the infrared pulse sequences used to measure the nonlinear spectra. As we describe below, these pathways are especially important for the earlytime dynamics of strongly coupled systems, and 2D IR is extremely well suited to untangling the dynamics of coherence transfer present under VSC.

6.2 Experimental Considerations

We now turn to a brief description of several experimental details that become especially important when designing cavity-coupled systems for nonlinear experiments, beginning with the properties of the mirrors that form the cavities. In most experiments interrogating ground-state reactivity under VSC, the mirrors are made by evaporating thin layers of metal onto appropriate substrates. The 10–20 nm thickness used provides high enough reflectivity for the desired finesse ($F = \mathrm{FSR}/\Delta\nu = \pi\sqrt{R}/(1-R)$) while having reasonable transmission for modified chemistry and static spectroscopic studies. This method is popular also because of its relatively low cost and difficulty, broad spectral coverage, and because it can be accomplished with modest fabrication facilities. In almost all cases, however, the peak transmission at the polariton frequencies is less than a percent because of losses in the metal reflectors. For the first measurements of nonlinear spectroscopy under VSC, the sensitivity would have been limited by the infrared pulse probe transmission if metal mirrors were used [80]. To get more transmission, low-loss dielectric mirrors were obtained for operation near 5 μm, which provided ~10% transmission through the cavity modes. As discussed in Simpkins et al. [81], another difference between dielectric and metal reflectors is that the field of the optical mode can penetrate the dielectric surface to a significant extent. This spatial penetration effectively reduces the concentration of absorbers experienced by the optical mode compared to a metallic cavity with the same ensemble of absorbers and gives rise to a smaller Rabi splitting for the dielectric cavity. With experimental techniques now fairly well established, future experiments might seek to use metallic reflectors to better connect to ground-state reactivity experiments or to achieve even larger Rabi splittings. We offer one note of caution: the initial urge may be to use more intense probe pulses to overcome the low transmission of metal cavities. For the most popular VSC system, $W(CO)_6$, saturable absorption can rear its head in fluence regimes readily attainable with modern laser instruments [82]. Researchers working down this path should be sure to account for whether these phenomena occur and, if so, how the probe pulses themselves modify the transmission spectrum of the coupled system.

Another common experimental concern pertaining to nonlinear studies with liquid samples is the flatness, more specifically the parallelism, of the cavity devices. This is described in further detail in the chapter by Michon and Simpkins. To date, most experiments have relied on commercial polyetrafluoroethylene spacers to control the length of the cavities (which typically ranges from 12 to 50 μm). The use of dielectric mirrors gives a practical view of parallelism. For well-constructed cavities, interference patterns almost invariably appear, easily observed by the naked eye. One can roughly assume that the separation between these fringes corresponds to a half-wavelength of the visible light giving rise to the pattern, taken as 500 nm for the sake of simplicity. If, then, one observes five fringes across the cavity-coupled sample, one can conclude that there is approximately ($5 \times 250\,\mathrm{nm} = 1.25\,\mu\mathrm{m}$) range of lengths spread across the sample. Practically, defects in parallelism of this order of magnitude can dramatically degrade Fourier transform infrared spectroscopy (FTIR) spectra; the IR beam can be large compared to the distance over which the cavity length can vary significantly. The resultant spectrum therefore reports on a range of lengths, broadening transmission features. Typical ultrafast measurements, however, employ focused beams with diameters on the order of a few hundred micrometers. Over this distance, the mirrors are parallel enough for the polariton linewidths to not be broadened by parallelism. Under such conditions, we use parallelism as an advantage, allowing us to identify locations on the sample with the desired length to give the desired cavity wavelength without changing the angle of the sample relative to any of the incoming beams.

We now turn to the question of incidence angle. It is straightforward enough that changing the incidence angle of the interrogating light, be it an FTIR beam or a probe pulse, changes the cavity frequency and thus allows one to measure the dispersion of the cavity. For all nonlinear experiments reported so far, groups have taken great pain to ensure that the probe pulse intersects the sample at the same angle as normal as any excitation pulses. Keeping the angles the same ensures, in turn, that the excitation pulses and probe pulse experience the same polariton spectrum. The Kubarych group has implemented a sophisticated pulse-shaping technique to measure fully colinear 2D IR spectra and avoid this issue entirely [83, 84]. The much broader set of cases where these angles are *not* held constant has seen little attention in the literature, perhaps because the simpler case has been challenging enough to understand. Anecdotally, we have found that exciting at a different angle from the probe leads to smaller signals with distorted spectra. We expect that further work down this path will be fruitful. Even if most of the nonlinear response of the system arises from the reservoir, changing the excitation angle with respect to the probe angle might offer a new path toward interrogating extreme degrees of spectral diffusion in inhomogeneous systems.

6.3 Understanding the Nonlinear Response of MVP

We now turn to a review of the literature surrounding the nonlinear spectroscopy of MVPs and a discussion of the physics that underpins the nonlinear responses. We and others have recently reviewed much of this work in detail [27, 61, 85–88]; here, we aim to describe in more detail those aspects of the nonlinear spectroscopy that are best understood and to highlight areas where confusion remains.

6.4 Early Delays

From the first report of the pump–probe spectroscopy of vibration-cavity polaritons in 2016, oscillatory signals were observed at time delays shorter than the dephasing time of the polariton modes, as estimated from the measured linewidths [80]. It was quickly established that the frequency of this oscillatory signal matched the Rabi splitting of the coupled system and, thus, the authors inferred that the signal arose from Rabi oscillations, as had been previously reported under ESC [89].

After this initial report, Xiang et al. used 2D IR in a series of publications to better interrogate these oscillatory signals [90]. By Fourier filtering a series of 2D IR collected at varying waiting times, they were able to differentiate "AC" oscillatory signals from slowly evolving "DC" signals and isolate the spectrum of the oscillatory response [91]. We will argue in the next section that the DC response can be adequately understood as arising from the nonequilibrium distribution of reservoir modes induced by the pump pulse or pulses. The AC response, however, is likely the strongest evidence for polariton-specific transitions to be observed for MVPs. Using pulse sequences tailored to selectively excite at specific time-ordered combinations of UP and LP frequencies, Xiang et al. clearly demonstrated that the oscillatory signal arises from the generation of coherences between the UP and LP. Pulse sequences that would generate UP or LP populations led to the slowly evolving response.

The spectrum of the oscillatory response qualitatively resembles the steady-state transmission (or absorption) spectrum of the sample in that it is single-signed and its intensity as a function of frequency matches that of the steady-state spectrum. The overall intensity of the signal oscillates around zero at the Rabi frequency. The details of the origin of this response remain somewhat unclear. Xiang et al. [91] hypothesized that a "polariton bleach" could explain aspects of the response, suggesting that the response from the UP–LP coherences matches the response expected if the optical mode were coupled to a much broader molecular line. Xiang et al. also showed that the magnitude of the AC response is highly sensitive to parameters of the system, including concentration (and thus, Rabi splitting) and cavity length. Understanding the mechanism of this response in greater detail is a critical next step toward realizing optical and, perhaps, quantum devices based on the optical response of excited vibration-cavity polaritons.

The Rabi oscillations are accompanied by the first steps of the evolution of the reservoir modes. We and others have reported that the dynamics of this evolution can appear to differ dramatically from the dynamics measured in free space. The most important difference is the appearance of a delayed onset for the ESA from the first excited state of the reservoir. This effect was found in the first pump–probe measurements and has since been variously assigned to polariton excited states or cavity-enabled populations of other intermediate states with reservoir or uncoupled character [80, 88, 92]. One potentially important class of intermediate states is the higher-lying excited states of the reservoir modes. Both experimental and theoretical works have found that accidental overlaps between the LP frequencies and ω_{12} can lead to high degrees of population of the second vibrational excited state of the reservoir [91, 93, 94]. It is the decay of these higher-lying states that can lead to a delayed population of the first excited state. This is an important consequence of strong coupling: even if the nonlinear response of the system at later times can be described by reservoir excitation, as we will argue below, strong coupling does change the initial distributions of excited states that can be accessed by the pump pulse(s). More recent work from Hirschmann et al. [95] identified another candidate for an intermediate state in $W(CO)_6$: a Raman-active mode lying close in energy to the IR-active mode that is coupled to the cavity. This assignment was supported by evaluating the degree of intermediate-state population across a series of solvents with differing phonon densities of states (DOS). Higher phonon DOS should lead to more efficient scattering of the LP to the Raman-active mode, and indeed solvents with higher phonon DOS showed much larger fractions of the intermediate population. Altogether, this work suggests that molecule-specific details of the vibrational spectrum are critical to understanding energy flow and that new pathways might be opened by coupling to a cavity. This concept will be important for extending nonlinear spectroscopy under VSC to a broader range of molecules beyond metal carbonyls.

6.5 Later Delays

We now turn to later time delays, here taken to mean time delays longer than any dephasing or potential spectral diffusion timescales of the system. In practice, for the few systems reported on thus far [80, 96, 97], we specifically refer to time delays longer than about 10 ps. At these "late" delays, the nonlinear spectroscopy reports primarily on population relaxation dynamics of vibrationally excited molecules. In other words, a spectrum measured at these late delays is the spectrum of the system with a nonequilibrium distribution of populations in vibrational degrees of freedom of the system. In the course of this section, we will argue that only vibrational states of the reservoir are excited at these timescales, with scant evidence for polaritonic contributions to either the nonlinear response or the dynamic evolution of that response.

There exist two equivalent approaches to calculating the late-time nonlinear response: quantum mechanical input–output theory and transfer-matrix calculations [80, 90, 93]. Each requires recognizing the mechanical and electrical anharmonicities at play in the vibrational spectroscopy of molecules [93]. Mechanical anharmonicity arises from the Morse-like potential energy surface that dictates nuclear motion along the normal mode corresponding to a vibrational band. The dissociative nature of the high-displacement side of the potential forces the separation between vibrational energy levels to decrease as the quantum number increases. Higher-lying vibrational transitions occur at lower frequencies than the fundamental transition. Electrical anharmonicity modulates the intensity of higher-lying transitions such that the higher-lying transitions are different than would be expected for a harmonic oscillator (for a harmonic oscillator, the transition dipole moment scales with the quantum number of the final state such that the first ESA intensity is twice that of the ground state absorption). Both classes of anharmonicity affect the observations from nonlinear infrared spectroscopy. Anharmonicity leads to a fundamental, critical point that drives our interpretations: nonequilibrium vibrational distributions lead to red-shifted, somewhat weaker excited-state transitions and ground-state bleaching at the fundamental frequency. By understanding how these changes to the spectrum manifest in a cavity, we will show that one can easily reproduce the late-time nonlinear spectroscopy of coupled systems.

In the discussion that follows, we will lean heavily on the analytical expression from the transfer matrix method for transmission through a cavity, which is described in detail in previous descriptions of etalons and cavities as well as Chapter 13 of this volume [28, 98].

$$T_{cav} = \frac{T^2 e^{-\alpha L}}{1 + R^2 e^{-2\alpha L} - 2Re^{-\alpha L} \cos\left(4\pi n L \bar{\nu} + 2\varphi\right)} \tag{6.1}$$

Very briefly, the transmission spectrum of a filled cavity can be expressed in terms of the optical constants of the mirrors and medium within the cavity. If the cavity is empty or has a medium with a constant index of refraction (and no absorptions in the region of interest), the solution reduces to the standard three-layer transfer matrix method result for an etalon, with transmission fringes that depend on mirror reflectivity and cavity spacing. To model weak to strong coupling, the optical constants of the medium are modified to include one or more oscillators, as either a Lorentzian oscillator (for homogeneously broadened vibrational bands) or a Gaussian distribution of a finite number of Lorentzian oscillators (for inhomogeneously broadened bands) [8]. When all vibrational coherences have decayed and only population remains, we can calculate nonlinear spectra by computing transmission spectra that correspond to the pump-on and pump-off conditions of the experiment. The pump-off condition is simply the condition where all population is in the ground vibrational state. For the pump-on condition, we reduce the amplitude factor of the $v = 0$–1 transition (hereafter abbreviated ω_{01}) and introduce commensurate levels of amplitude to higher-lying transitions (primarily ω_{12} and ω_{23}). We then algebraically manipulate these transmission spectra to match the quantities reported in a given experiment, e.g., we can easily calculate $-\log(T_{pump\ on}/T_{pump\ off})$.

Calculating transient spectra shows that the anharmonicities inherent to vibrational modes dictate the nonlinear response of the system. Historically, we have divided the effects of the anharmonicity into two classes. The first arises because the concentration of ground-state oscillators is reduced. Reduction in concentration leads to a decrease in the coupling strength, and thus the Rabi splitting, between ω_{01} and the cavity mode. Without contributions from the higher-lying transitions, the pump-on polaritons contract compared to the pump-off, resulting in derivative-like lineshapes in the transient spectra. This effect is illustrated in Ribeiro et al. when only electrical anharmonicity is considered in a quantum-mechanical formulation (Figure 6.1). In practice, the predicted derivative-like lineshape is more readily observed when probing the UP region of the spectrum because the LP region is often dominated by ESA transitions of the system, whose influence we discuss next.

A set of coincidences combine to dictate the nonlinear spectroscopy of the LP region under VSC. With the notable exception of water, whose nonlinear spectroscopy under VSC has yet to be explored, achieving VSC typically requires obtaining Rabi splittings that overcome the 5–50 cm^{-1} linewidth of condensed-phase vibrational bands. The oscillator strength of vibrational transitions and solubility or density of species that support those transitions leads to Rabi splittings up to a few

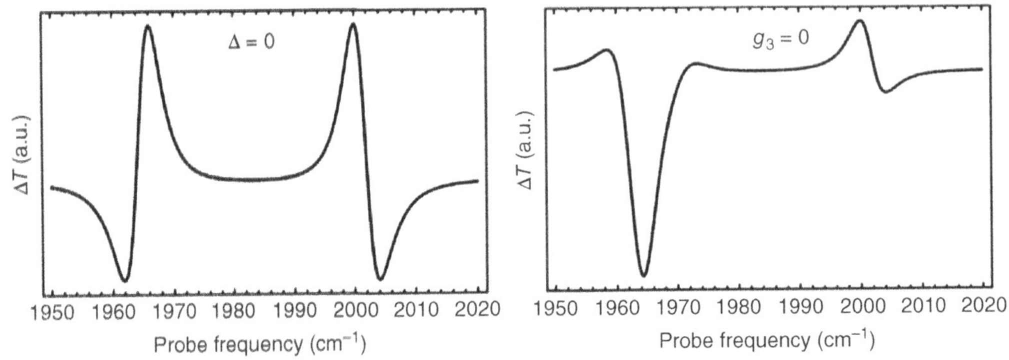

Figure 6.1 Calculated pump–probe spectra for hypothetical cases where (left) the nuclear anharmonicity, D, is 0 and the electrical anharmonicity, g_3, is nonzero or (right) g_3 is zero and D is nonzero. *Source*: [93] / American Chemical Society.

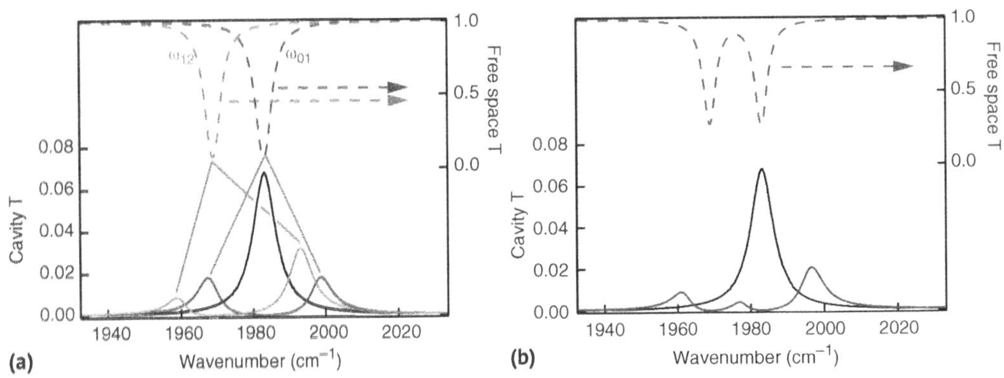

Figure 6.2 Calculated transmission spectra of Fabry–Pérot cavities with internal medium simulating neat hexane (single-peaked trace in darkest shade, a and b), oscillators at the ω_{01} frequency of $W(CO)_6$ in hexane (a), oscillators at the ω_{12} frequency of $W(CO)_6$ in hexane (a), and a mixture of oscillators at ω_{01} and ω_{12} (b). In each case, the dashed lines are the calculated transmission spectra of the medium in free space, i.e., with transparent windows instead of mirrors, with the oscillator intensity reduced by a factor of 6 for ease of comparison. *Source*: [61] / American Chemical Society.

hundred cm^{-1} but typically less than 100 cm^{-1}, yielding a separation between LP and ω_{01} equal to half the Rabi splitting when the cavity and vibrational modes are degenerate. The anharmonic shift of ω_{12} compared to ω_{01} is generally of the same order of magnitude. Critically, the anharmonic shift is not so large that one can ignore interactions between ω_{12} and the cavity mode. As shown for $W(CO)_6$ by Simpkins et al., the ESA transition can readily couple to the optical mode given sufficient amplitude (i.e., population in the excited state) (Figure 6.2a). Even small fractions of the population promoted to $v = 1$ produce large responses in the LP region. These observations are consistent with theoretical predictions of large nonlinear optical enhancements for strong coupling in infrared cavities [94, 99]. Conceptually, we think of the ω_{12} transition either weakly coupling to the cavity mode or experiencing enhanced absorption (or both) through its coincidence with the LP mode at low excitation fractions. As the excitation fraction increases, the system can enter a regime where both the ω_{01} and ω_{12} transitions are strongly coupled to the cavity mode (Figure 6.2b). These conceptual distinctions, together with the polariton contraction induced by ground-state bleaching, can be useful mnemonics but fail to capture the entire picture. We suggest that it is most correct and complete to holistically consider the optical response of the cavity and the medium inside (modified to reflect the nonequilibrium population distributions appropriate for the experiment) to interpret nonlinear spectra.

The analytical approach to calculating transient spectra can be used to gain important insight into the dynamics of systems under VSC. It is relatively straightforward to fit pump–probe spectra to formulations of the calculated spectra such that only the ratio of excited-state to ground-state population is a free parameter. We have had reasonable success using this method to fit data for both $W(CO)_6$ and $Fe(CN)_5(NO)^{2-}$ [80, 82, 97, 100]. In either case, fitting the transient data allows us to plot the fractional population of the excited state as a function of pump–probe delay (Figure 6.3). Disappointingly, in both cases, the temporal evolution of the excited-state population exactly matched the evolution measured in free space. In other words, we observe no evidence that the presence of the polariton modes modifies the relaxation of excited-state population in the reservoir.

Critically, fitting experimental data with this approach reveals one of the peculiarities of nonlinear measurements under VSC. Although the excited-state fraction as extracted from fitting to the analytical model evolves identically to free-space molecules, kinetic traces extracted at a particular probe frequency can vary dramatically. As demonstrated in Dunkelberger et al. [100] fitting these kinetic traces can introduce systematic errors on the order of 30% of the population lifetime, owing to the fact that the spectrum evolves as a function of the population such that the intensity at a given frequency no longer correlates linearly to the excitation fraction (Figure 6.4). Using the analytical expression for transmission and evaluating how the Rabi splitting evolves with time accounts for this spectral evolution and allows for more accurate fitting and reliable lifetimes.

The relative simplicity of the analytical approach and its success in reproducing late-time transient data make us confident in assigning the late-time response to a combination of polariton contraction and ESA from population in the reservoir. Polariton contraction is also observed in transient studies of ESC [66, 67]. The approach captures the relevant features of the experimental spectra without invoking polaritonic excited states, i.e., higher-lying transitions in the polaritonic ladder. That being said, there remains debate in the literature about the completeness of this

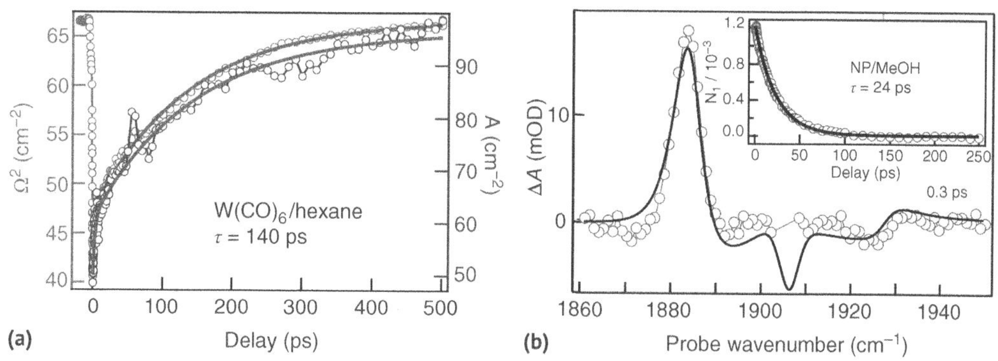

Figure 6.3 Approaches to extracting reservoir lifetimes from cavity-coupled spectra. (a) The square of the Rabi splitting as a function of time delay (left axis) and the fitted amplitude factor for ω_{01} (right axis) obtained for $W(CO)_6$ in hexane. Solid lines are fits, with resulting lifetimes indistinguishable from the free-space lifetime of 140 ps. *Source*: Adapted from [100]. (b) Representative pump–probe spectrum (open circles) with fit to spectrum (solid line) obtained for cavity-coupled nitroprusside in methanol. Inset: fraction of oscillator amplitude in ω_{12} extracted from spectral fitting as a function of time delay (circles). The solid line is a fit to the decay, with resulting lifetime indistinguishable from the free-space lifetime of 24 ps. *Source*: Adapted from [97].

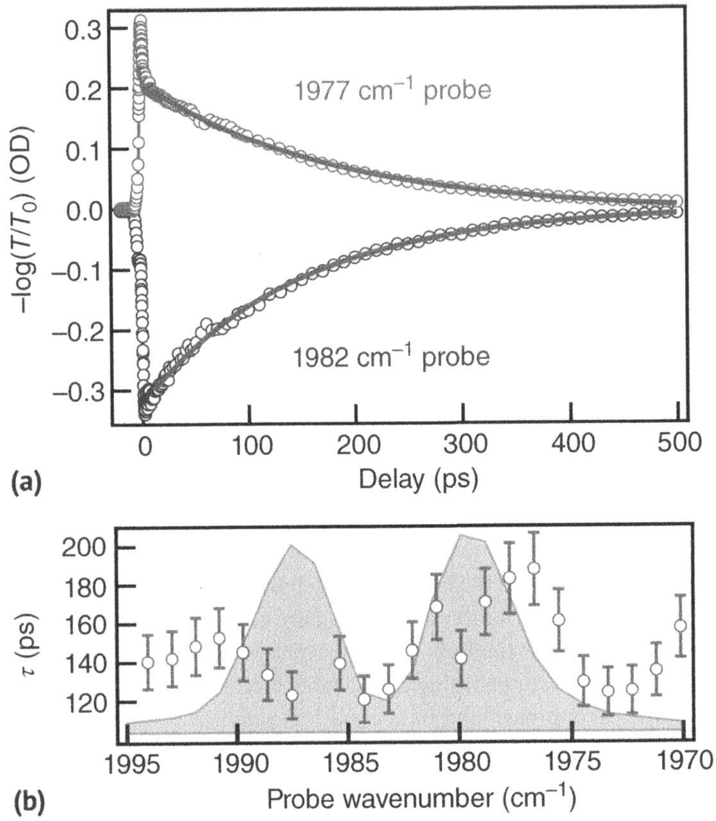

Figure 6.4 (a) Representative kinetic traces measured for a cavity-coupled solution of $W(CO)_6$ in hexane. (b) Exponential decay constants extracted from fits to kinetic traces plotted as a function of the frequency at which the kinetic trace was measured. The shaded area represents the transmission spectrum of the sample measured to obtain the data and fits. *Source*: Adapted from [100].

description and whether the polariton population (and thus, responses) might persist beyond the dephasing of the system [101–103]. The details of this debate become more pertinent for intermediate timescales that we discuss in the next section. Here, we only point readers to this debate and reiterate that the analytical expression for transmission appears to quantitatively reproduce late-time responses measured in transmission geometry without requiring the subtraction methods invoked by other works. While it is gratifying to be able to reproduce the experimental results with this simple classical model, it is somewhat disappointing that we can do so without including contributions from excited polaritons, since it is the excited polariton behavior we have been attempting to identify and understand.

Finally, we point out that the same concepts can be used in quantum mechanical treatments of the nonlinear response. Using input–output theory, Ribeiro et al. [93] computed transmission spectra and pump–probe transient spectra based entirely on the Hamiltonian for strong coupling rather than classical expressions for cavity transmission (Figure 6.5). Here too, it was critical to simultaneously include the effects of coupling between the optical mode, the ω_{01} transition, and the ω_{12} transition.

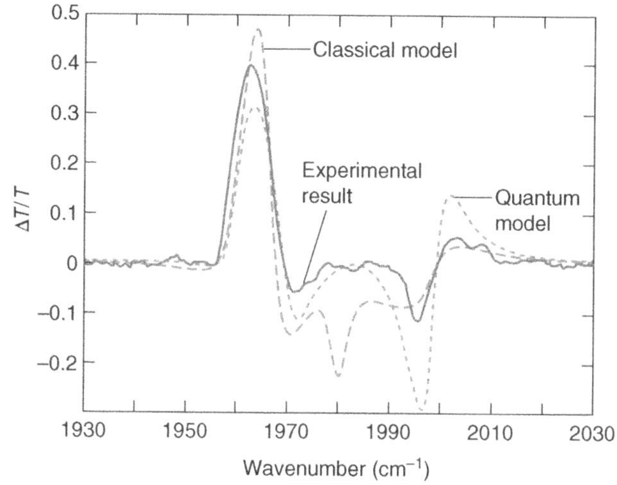

Figure 6.5 Representative experimentally measured pump–probe spectrum for $W(CO)_6$ in hexane. Both the "classical" model corresponding to the analytical approach described in this chapter and a "quantum" model, described in detail in Refs. [90] and [93], achieve reasonable agreement with the measured spectrum without including polariton excited-state transitions. *Source*: [90] / National Academy of Sciences.

6.6 Intermediate Delays

The interpretation of nonlinear spectroscopy under VSC is less straightforward at intermediate times, here taken to mean after dephasing is complete but before the spectral evolution is well described by the reservoir dynamics described above. In 2021, Grafton et al. [97] reported the 2D IR spectroscopy of $Fe(CN)_5(NO)^{2-}$ (nitroprusside, NP) with the nitrosyl stretching band coupled to a cavity. The authors (including two authors of this chapter) found that substantial differences between the measured spectra and spectra predicted from reservoir excitation persisted for more than 10 ps after excitation, much longer than the 2 ps dephasing time of the polariton modes. Finding reasonable, if not perfect, agreement with a multi-level quantum Rabi model of an anharmonic polariton ladder [104], Grafton et al. [97] hypothesized that these differences could arise from excited-state transitions in an anharmonic polariton ladder.

In the same year, Duan et al. [101] employed the same multi-level quantum Rabi model and an approach by which they attempted to subtract free-space nonlinear signals from vibration-cavity coupled signals to determine that polariton excited states might persist even longer, beyond the detection limit of their instrument. The debate has flared in the literature over whether this subtraction technique is appropriate, especially since it leads to a markedly different conclusion from that drawn from using the transfer-matrix formalism for transmission described in detail in the previous chapter. Herein we point out an important insight from Duan et al.

Duan et al. [101] realized and highlighted that 2D IR difference spectra reported in the Supporting Information of Grafton et al. [97] are remarkably similar to 2D IR difference spectra that can be generated by subtracting free-space 2D IR spectra for NP. In free space, the origin of the difference spectra is spectral inhomogeneity: at early waiting times, the inhomogeneously broadened band retains memory of its excitation, and the probed response is highly correlated to the excitation frequency. At late waiting times, this correlation vanishes as the oscillators sample different solvation environments (a well-known phenomenon called spectral diffusion). The prevailing wisdom in the field suggests that strong coupling removes the effects of inhomogeneous broadening through a mechanism qualitatively similar to motional narrowing in other systems [105]. To address this apparent contradiction, Duan et al. [101] posited that inhomogeneous contributions from spurious background signals contaminate data collected under strong coupling. Our group's alternative hypothesis is that the coupling for NP never reaches strengths sufficient to induce this narrowing and that some inhomogeneity in the reservoir is detectable in the coupled system. At present, our efforts are focused on developing a model to describe this persistent inhomogeneity and calculate cavity-coupled 2D IR spectra for inhomogeneous bands. While much work remains to be done to understand the details of spectral diffusion under VSC, it stands to reason that inhomogeneity might better describe the features described in Grafton et al. [97], rather than polariton excited states.

If the spectral diffusion hypothesis holds true, the interpretation diverges further from that of Duan et al. [101]. We would then surmise that after the earlytime coherent oscillations, the dynamics under VSC are completely described by appropriate treatments of the dynamics in free space, i.e., we can reproduce the experimental spectra by devising ways to correctly account for the reservoir frequency–frequency correlations and population evolution in the transmission of the sample. We would argue that there is scant evidence for long-lived polariton lifetimes. An important question arises here: why does inhomogeneity appear in the nonlinear spectroscopy under VSC? It has long been established, perhaps most famously by Houdre et al. [105] that polariton modes are narrowed to a homogeneous linewidth that is the average of the linewidth of the optical mode and the homogeneous linewidth of the molecular absorption through a mechanism reminiscent of motional narrowing. Again, more details remain to be resolved, but we surmise that the coupling in the case of NP is simply not strong enough to achieve this limit. There is substantial overlap between the polariton transmission and the wings of the nitrosyl absorption band, perhaps sufficient to allow direct excitation of the reservoir, which would then obey the same physics as the band in free space.

We end this section by pointing out recent work on a different class of MVPs comprising molecular vibrational modes coupled to optical resonances on metal nanostructures. Here, both the physics of coupling and the nonlinear spectroscopy are quite different from those in a Fabry–Pérot cavity. The mode volumes are much smaller in the nanoscale coupling case, such that the effective coupling strength per molecule is considerably higher. As we have recently reviewed [61], Cohn et al. [106] demonstrated that MVPs formed from nanoresonators exhibit nonlinear spectroscopic features that have yet to be observed in cavity-coupled systems, namely the apparent transfer of population from the reservoir to the LP. Linking this phenomenon to the macroscopic cavities used for the majority of other spectroscopic studies and for modifying chemical reactions is another important next step for the field.

6.7 Optical and Photophysical Opportunities

Having reviewed the nonlinear spectroscopy of systems under VSC, we now turn our attention to new opportunities presented by VSC. A great deal of attention has been paid to exciting work that suggests VSC can drastically modify chemical reactivity without excitation. Here, we review work that highlights how VSC leads to new optical behavior and modifies energy transfer and reactivity in coupled systems.

Most of the existing literature reporting on the nonlinear spectroscopy of $W(CO)_6$ and NP under VSC concerns solutions with Rabi splittings large enough that there is small overlap between the polariton modes and the molecular vibrational band of interest. Under these conditions, the experimental spectra can be matched well with calculated spectra that include only a few percent or less of the reservoir population promoted to the first vibrational excited state. Perhaps counterintuitively, samples for which the coupling is weaker (i.e., the Rabi splitting is smaller and the polaritons better overlap the reservoir absorption) can be excited to a much greater degree.

In 2018, Dunkelberger et al. [100] set out to use low-concentration $W(CO)_6$ to avoid overlap between the ω_{12} transition and the LP, potentially making interpretation of the pump–probe spectroscopy more straightforward. They found that, for 1 mM solutions, modest pump fluences was enough to completely collapse the Rabi splitting measured by the probe pulse. In other words, enough population was transferred to the excited state to force the system into the weak coupling regime. Until the excited-state population relaxed to refill the ground state, the probe pulse reported a transmission spectrum with only one peak at the uncoupled cavity frequency. Even under high-excitation conditions, the dynamics of the reservoir were shown not to differ from the free-space population dynamics. This low-concentration study represents important evidence for the concept of polariton or Rabi-splitting contraction, clearly showing that the polaritons can contract completely. That the contraction was shown to recover with the free-space population lifetime of the molecular excited state is strong evidence that polariton excited states are not responsible for the optical signature.

Under intense illumination, the collapse of the Rabi splitting can be observed even without a time-delayed probe pulse. In 2019, Dunkelberger et al. [82] revisited low-concentration $W(CO)_6$ and showed that the transmission of intense infrared pulses through the coupled samples is strongly dependent on the incident fluence. This phenomenon, observed for excitonic strong coupling in reflection and via photoluminescence [89, 107–110], is closely tied to the concept of saturable absorption: if the rate at which an ensemble absorbs photons is faster than the rate at which the excited levels relax, the absorption of the sample decreases as the incident fluence increases. In direct analogy, the observed Rabi splitting decreased as a function of the pulse fluence in Dunkelberger et al. [82] The saturation fluence, defined as the fluence at which the absorption of the sample is reduced by half, is typically an intrinsic absorber property directly linked to absorption cross-section. Under VSC, however, it was shown that the saturation fluence can be tuned by changing sample parameters like cavity length and concentration. The most obvious consequence of the polariton contraction is the dramatic increase in transmission at the uncoupled cavity frequency. However, this increase is accompanied by a decrease in transmission at the low-fluence polariton frequencies. This decrease in transmission is the basis of a patent awarded in 2023 (Simpkins et al. [103]) for optical limiting functionality.

Among the most promising reports of new phenomena enabled by MVPs are demonstrations of energy and coherence transfer between spatially separated species. For dilute solutions, vibrational energy transfer between solute molecules is unlikely because of the very low probability of any two molecules colliding, with any rare collision also having a low chance of transferring energy to a high-energy vibrational level compared to rotational or translational energy. Xiang et al. showed that simultaneously coupling two species with distinct vibrational spectra to the same optical mode can enable efficient vibrational energy transfer from one species to the other [111]. This important demonstration has also been reviewed recently [61, 85, 87]; here, we briefly point out two important implications. First, the delocalization itself is intriguing, linking the phenomenon of VSC to ESC, where delocalization has been demonstrated more extensively [44, 47, 112, 113]. Because MVPs cannot be directly measured through PL, and their excited-state signatures are elusive at best, direct evidence of their involvement, and thus even their presence, in excited-state processes, is critical. Second, the delocalization is a tantalizing clue for the mechanism of VSC-modified chemistry.

This evidence lends credence to arguments that invoke even the modest delocalization that might be expected for reservoir states to help explain effects on reactivity.

Delocalization also drives reports of coherence between spatially separated optical modes coupled to molecular vibrations. In analogy to the energy transfer experiments above, the Xiong group constructed optical devices with spatially separated cavities with transmission peaks separated by a few dozen cm^{-1}. Introducing a molecular absorber with a strong transition lying between the two transmission peaks leads to simultaneous coupling between the disparate cavity modes. Xiang et al. unequivocally showed that the MVPs spatially delocalize over tens of micrometers [114]. Subsequent work showed that coherences between specific combinations of polariton modes can be driven in these multi-cavity conditions [115]. Here, Xiang et al. showed that the energy separation between the polariton modes is a key parameter in determining the fate of the coherences. With 10 cm^{-1} or less separation, coherences were shown to survive and delocalize across the structured dual-cavity. With greater separation between the polariton modes, coherence could not be maintained [115].

To overcome this limitation and push delocalization to more widely separated polariton modes, Yang et al. recently reported an ingenious design that combines several classes of optical modes [116]. By carving a trough of the appropriate width into one reflector of a Fabry–Pérot cavity, Yang et al. devised a structure that supports traditional Fabry–Pérot modes as well as photonic particle-in-a-box modes confined to the trough. The dispersion of these confined modes is strikingly different from the dispersion of unconfined modes: it is flat for a given energy level until the in-plane momentum is sufficient to reach the next particle-in-a-box-like mode. With confined modes, coherences were found to survive with mode separations up to 40 cm^{-1}. Although the confined modes are not spatially separated, the two confined modes are orthogonal to one another. It is coupling to the molecular vibration that allows the modes to interact and these coherences to be driven. As the authors point out, this new paradigm for tailoring optical modes could lead to important new opportunities for quantum devices based on MVPs [87, 116, 117].

Finally, we review a recent work demonstrating, for the first time, modified photoisomerization under VSC [96]. Chen et al. examined the behavior of $Fe(CO)_5$ (iron pentacarbonyl, IPC) coupled to a Fabry–Pérot cavity. IPC is well understood to possess two strong vibrational modes near 2000 cm^{-1}. Energy can exchange between these modes via IVR or a competing pseudorotation channel that involves a rearrangement of the carbonyl ligands around the iron atom. The competition between these mechanisms has been extensively investigated via 2D IR [118]. Chen et al. found that, when polariton modes are excited, the energy exchange between the two IR-active modes is accelerated compared to when reservoir modes are excited. Furthermore, they used kinetic models to show that the polariton-excited pathway prefers IVR to pseudorotation, an inversion of the situation in free space [96]. While the physical origins of this change in mechanism remain somewhat in question, the report shows that strong coupling can modify even fundamental, unimolecular reactions. Targeting simple reactions like photoisomerization is likely to be an incisive path toward linking theoretical efforts with the more mysterious and controversial demonstrations of VSC-modified chemistry.

6.8 Concluding Remarks

Strong coupling adds rich complexity to the nonlinear spectroscopy of MVPs. We have shown here that much of the optical response can be described by considering changes in the effective optical constants of the intracavity medium. In our view, there is little evidence of a polaritonic population that survives past the dephasing time of the system. Polaritonic coherences, on the other hand, are readily observed and offer exciting new opportunities for manipulating light in the infrared. Nonlinear spectroscopy teaches us that the vibrational relaxation of the vast reservoir of dark modes is, apparently, unmodified from the relaxation in free space. Combined with the overwhelming dominance of the dark modes in the density of states of MVPs, these results seem to suggest that coupled systems in the dark might predominantly exchange energy at similar rates to free-space systems. Directly exciting the polariton modes, however, has now been shown to impact the reaction dynamics of simple, fundamental processes. We suggest that the most incisive future work in this field will follow one of these two paths: (i) continuing to develop a framework for understanding the relaxation of coupled systems for better connection to theoretical descriptions of modified chemistry in the dark and (ii) identifying simple photoreactions where reactive events can be systematically studied with or without strong coupling. Very recent reports along the second track clearly show that VSC does not affect hydrogen abstraction from alkanes by photogenerated halide radicals [119]. More work is required to determine to what extent direct excitation at polariton frequencies is required for reliable modification of photo-induced processes.

Acknowledgments

A.D.D. and J.C.O. acknowledge internal funding at the US Naval Research Laboratory for supporting this work. C.G.P. acknowledges the American Society for Engineering Education for administering her postdoctoral fellowship. The authors acknowledge Dr. Blake Simpkins, Dr. Igor Vurgaftman, and Prof. Wei Xiong for useful discussions of the content of the chapter and years of friendship and fruitful collaboration in this field.

References

1 Ahn, W., Triana, J. F., Recabal, F., Herrera, F., Simpkins, B. S. Modification of ground-state chemical reactivity via light-matter coherence in infrared cavities. *Science* **2023**, *380*(6650), 1165–1168.

2 Thomas, A., Lethuillier-Karl, L., Nagarajan, K., Vergauwe, R. M. A., George, J., Chervy, T. et al. Tilting a ground-state reactivity landscape by vibrational strong coupling. *Science* **2019**, *363*(6427), 615–619.

3 Shapiro, M., Brumer, P. *Principles of the Quantum Control of Molecular Processes*. Wiley-VCH: Weinheim, 2003; p. 250.

4 Crim, F. F. Vibrational state control of bimolecular reactions: discovering and directing the chemistry. *Acc. Chem. Res.* **1999**, *32*(10), 877–884.

5 Kohler, B., Krause, J. L., Raksi, F., Wilson, K. R., Yakovlev, V. V., Whitnell, R. M. et al. Controlling the future of matter. *Acc. Chem. Res.* **1995**, *28*(3), 133–140.

6 Heyne, K., Kuhn, O. Infrared laser excitation controlled reaction acceleration in the electronic ground state. *J. Am. Chem. Soc.* **2019**, *141*(30), 11730–11738.

7 Shin, J. Y., Shaloski, M. A., Crim, F. F., Case, A. S. First evidence of vibrationally driven bimolecular reactions in solution: reactions of Br atoms with dimethylsulfoxide and methanol. *J. Phys. Chem. B* **2017**, *121*(11), 2486–2494.

8 Long, J. P., Simpkins, B. S. Coherent coupling between a molecular vibration and Fabry–Perot optical cavity to give hybridized states in the strong coupling limit. *ACS Photonics* **2015**, *2*(1), 130–136.

9 Shalabney, A., George, J., Hutchison, J., Pupillo, G., Genet, C., Ebbesen, T. W. Coherent coupling of molecular resonators with a microcavity mode. *Nat. Commun.* **2015**, *6*, 5981.

10 Basov, D. N., Asenjo-Garcia, A., Schuck, P. J., Zhu, X. Y., Rubio, A. Polariton panorama. *Nanophotonics* **2021**, *10*(1), 549–577.

11 Caldwell, J. D., Lindsay, L., Giannini, V., Vurgaftman, I., Reinecke, T. L., Maier, S. A. et al. Low-loss, infrared and terahertz nanophotonics using surface phonon polaritons. *Nanophotonics* **2015**, *4*(1), 44–68.

12 Gramotnev, D. K., Bozhevolnyi, S. I. Plasmonics beyond the diffraction limit. *Nat. Photonics* **2010**, *4*(2), 83–91.

13 Schuller, J. A., Barnard, E. S., Cai, W. S., Jun, Y. C., White, J. S., Brongersma, M. L. Plasmonics for extreme light concentration and manipulation. *Nat. Mat.* **2010**, *9*(3), 193–204.

14 West, P. R., Ishii, S., Naik, G. V., Emani, N. K., Shalaev, V. M., Boltasseva, A. Searching for better plasmonic materials. *Laser Photonics Rev.* **2010**, *4*(6), 795–808.

15 Barnes, W. L., Dereux, A., Ebbesen, T. W. Surface plasmon subwavelength optics. *Nature* **2003**, *424*(6950), 824–830.

16 Neubrech, F., Huck, C., Weber, K., Pucci, A., Giessen, H. Surface-enhanced infrared spectroscopy using resonant nanoantennas. *Chem. Rev.* **2017**, *117*(7), 5110–5145.

17 Osawa, M. Surface-enhanced infrared absorption. In: *Near-field Optics and Surface Plasmon Polaritons*, Kawata, S., Ed. **2001**, Vol. 81, pp. 163–187. Springer Berlin, Heidelberg.

18 Willets, K. A., Van Duyne, R. P. Localized surface plasmon resonance spectroscopy and sensing. *Annu. Rev. Phys. Chem.* **2007**, *58*, 267–297.

19 Anker, J. N., Hall, W. P., Lyandres, O., Shah, N. C., Zhao, J., Van Duyne, R. P. Biosensing with plasmonic nanosensors. *Nat. Mat.* **2008**, *7*(6), 442–453.

20 Guo, C. F., Sun, T. Y., Cao, F., Liu, Q., Ren, Z. F. Metallic nanostructures for light trapping in energy-harvesting devices. *Light Sci. Appl.* **2014**, *3*, e161.

21 Huang, X. H., Jain, P. K., El-Sayed, I. H., El-Sayed, M. A. Plasmonic photothermal therapy (PPTT) using gold nanoparticles. *Lasers Med. Sci.* **2008**, *23*(3), 217–228.

22 Osterloh, F. E. Inorganic nanostructures for photoelectrochemical and photocatalytic water splitting. *Chem. Soc. Rev.* **2013**, *42*(6), 2294–2320.

23 Yu, S. J., Wilson, A. J., Kumari, G., Zhang, X. Q., Jain, P. K. Opportunities and challenges of solar-energy-driven carbon dioxide to fuel conversion with plasmonic catalysts. *ACS Energy Lett.* **2017**, *2*(9), 2058–2070.

24 Autore, M., Li, P. N., Dolado, I., Alfaro-Mozaz, F. J., Esteban, R., Atxabal, A. et al. Boron nitride nanoresonators for phonon-enhanced molecular vibrational spectroscopy at the strong coupling limit. *Light Sci. Appl.* **2018**, *7*, 17172.

25 Munkhbat, B., Wersall, M., Baranov, D. G., Antosiewicz, T. J., Shegai, T. Suppression of photo-oxidation of organic chromophores by strong coupling to plasmonic nanoantennas. *Sci. Adv.* **2018**, *4*(7), eaaas9552.

26 Chikkaraddy, R., de Nijs, B., Benz, F., Barrow, S. J., Scherman, O. A., Rosta, E. et al. Single-molecule strong coupling at room temperature in plasmonic nanocavities. *Nature* **2016**, *535*(7610), 127–130.

27 Herrera, F., Owrutsky, J. Molecular polaritons for controlling chemistry with quantum optics. *J. Chem. Phys.* **2020**, *152*(10), 100902.

28 Khitrova, G., Gibbs, H. M., Jahnke, F., Kira, M., Koch, S. W. Nonlinear optics of normal-mode-coupling semiconductor microcavities. *Rev. Mod. Phys.* **1999**, *71*(5), 1591–1639.

29 Torma, P., Barnes, W. L. Strong coupling between surface plasmon polaritons and emitters: a review. *Rep. Prog. Phys.* **2015**, *78*(1), 013901.

30 Dintinger, J., Klein, S., Bustos, F., Barnes, W. L., Ebbesen, T. W. Strong coupling between surface plasmon-polaritons and organic molecules in subwavelength hole arrays. *Phys. Rev. B* **2005**, *71*(3), 035424.

31 Lidzey, D. G., Bradley, D. D. C., Skolnick, M. S., Virgili, T., Walker, S., Whittaker, D. M. Strong exciton-photon coupling in an organic semiconductor microcavity. *Nature* **1998**, *395*(6697), 53–55.

32 Tischler, J. R., Bradley, M. S., Zhang, Q., Atay, T., Nurmikko, A., Bulovic, V. Solid state cavity QED: strong coupling in organic thin films. *Org. Electron.* **2007**, *8*(2–3), 94–113.

33 Kaluzny, Y., Goy, P., Gross, M., Raimond, J. M., Haroche, S. Observation of self-induced Rabi oscillations in 2-level atoms excited inside a resonant cavity – the ringing regime of super-radiance. *Phys. Rev. Lett.* **1983**, *51*(13), 1175–1178.

34 Raizen, M. G., Thompson, R. J., Brecha, R. J., Kimble, H. J., Carmichael, H. J. Normal-mode splitting and linewidth averaging for 2-state atoms in an optical cavity. *Phys. Rev. Lett.* **1989**, *63*(3), 240–243.

35 Agranovich, V., Benisty, H., Weisbuch, C. Organic and inorganic quantum wells in a microcavity: Frenkel-Wannier-Mott excitons hybridization and energy transformation. *Solid State Commun.* **1997**, *102*(8), 631–636.

36 Lidzey, D. G., Bradley, D. D. C., Armitage, A., Walker, S., Skolnick, M. S. Photon-mediated hybridization of Frenkel excitons in organic semiconductor microcavities. *Science* **2000**, *288*(5471), 1620–1623.

37 Weisbuch, C., Nishioka, M., Ishikawa, A., Arakawa, Y. Observation of the coupled exciton-photon mode splitting in a semiconductor quantum microcavity. *Phys. Rev. Lett.* **1992**, *69*(23), 3314–3317.

38 Gruebele, M., Wolynes, P. G. Vibrational energy flow and chemical reactions. *Acc. Chem. Res.* **2004**, *37*(4), 261–267.

39 Lourderaj, U., Hase, W. L. Theoretical and computational studies of non-RRKM unimolecular dynamics. *J. Phys. Chem. A* **2009**, *113*(11), 2236–2253.

40 Truhlar, D. G., Garrett, B. C., Klippenstein, S. J. Current status of transition-state theory. *J. Phys. Chem.* **1996**, *100*(31), 12771–12800.

41 Semmes, D. H., Baskin, J. S., Zewail, A. H. Direct observation of a mode-selective (non-RRKM) van der Waals reaction by picosecond photofragment spectroscopy. *J. Am. Chem. Soc.* **1987**, *109*(13), 4104–4106.

42 Leitner, D. M., Gruebele, M. Quantum model of restricted vibrational energy flow on the way to the transition state in unimolecular reactions. *Mol. Phys.* **2008**, *106*(2–4), 433–442.

43 Carpenter, B. K. Nonstatistical dynamics in thermal reactions of polyatomic molecules. *Annu. Rev. Phys. Chem.* **2005**, *56*, 57–89.

44 Zhong, X. L., Chervy, T., Zhang, L., Thomas, A., George, J., Genet, C. et al. Energy transfer between spatially separated entangled molecules. *Angew. Chem. Int. Ed.* **2017**, *56*(31), 9034–9038.

45 Georgiou, K., Michetti, P., Gai, L. Z., Cavazzini, M., Shen, Z., Lidzey, D. G. Control over energy transfer between fluorescent bodipy dyes in a strongly coupled microcavity. *ACS Photonics* **2018**, *5*(1), 258–266.

46 Galego, J., Garcia-Vidal, F. J., Feist, J. Cavity-induced modifications of molecular structure in the strong-coupling regime. *Phys. Rev. X* **2015**, *5*(4), 041022.

47 Coles, D. M., Somaschi, N., Michetti, P., Clark, C., Lagoudakis, P. G., Savvidis, P. G. et al. Polariton-mediated energy transfer between organic dyes in a strongly coupled optical microcavity. *Nat. Mat.* **2014**, *13*(7), 712–719.

48 Genet, C., Faist, J., Ebbesen, T. W. Inducing new material properties with hybrid light-matter states. *Phys. Today* **2021**, *74*(5), 42–48.

49 Imperatore, M. V., Asbury, J. B., Giebink, N. C. Reproducibility of cavity-enhanced chemical reaction rates in the vibrational strong coupling regime. *J. Chem. Phys.* **2021**, *154*(19), 191103.

50 Wiesehan, G. D., Xiong, W. Negligible rate enhancement from reported cooperative vibrational strong coupling catalysis. *J. Chem. Phys.* **2021**, *155*(24), 241103.

51 Ahn, W., Simpkins, B. S. Raman scattering under strong vibration-cavity coupling. *J. Phys. Chem. C* **2021**, *125*(1), 830–835.

52 Seidel, M., Chervy, T., Thomas, A., Akiki, E., Vergauwe, R. M. A., Shalabney, A. et al. Vibro-polaritonic IR emission in the strong coupling regime (Vol 5, Pg 217, 2018). *ACS Photonics* **2019**, *6*(7), 1823–1825.

53 Shalabney, A., George, J., Hiura, H., Hutchison, J. A., Genet, C., Hellwig, P. et al. Enhanced Raman scattering from vibro-polariton hybrid states. *Angew. Chem. Int. Ed.* **2015**, *54*(27), 7971–7975.

54 Thomas, A., George, J., Shalabney, A., Dryzhakov, M., Varma, S. J., Moran, J. et al. Ground-state chemical reactivity under vibrational coupling to the vacuum electromagnetic field. *Angew. Chem. Int. Ed.* **2016**, *55*(38), 11462–11466.

55 Campos-Gonzalez-Angulo, J. A., Poh, Y. R., Du, M., Yuen-Zhou, J. Swinging between shine and shadow: theoretical advances on thermally activated vibropolaritonic chemistry. *J. Chem. Phys.* **2023**, *158*(23), 230901.

56 Mandal, A., Taylor, M. A. D., Weight, B. M., Koessler, E. R., Li, X. Y., Huo, P. F. Theoretical advances in polariton chemistry and molecular cavity quantum electrodynamics. *Chem. Rev.* **2023**, *123*, 9786–9879.

57 Schäfer, C., Flick, J., Ronca, E., Narang, P., Rubio, A. Shining light on the microscopic resonant mechanism responsible for cavity-mediated chemical reactivity. *Nat. Commun.* **2022**, *13*(1), 7817.

58 Wang, D. S., Yelin, S. F. A roadmap toward the theory of vibrational polariton chemistry. *ACS Photonics* **2021**, *8*(10), 2818–2826.

59 Jaynes, E. T., Cummings, F. W. Comparison of quantum and semiclassical radiation theories with application to beam maser. *Proc. IEEE* **1963**, *51*(1), 89.

60 Tavis, M., Cummings, F. W. Exact solution for an N-molecule-radiation-field Hamiltonian. *Phys. Rev.* **1968**, *170*(2), 379.

61 Simpkins, B. S., Dunkelberger, A. D., Vurgaftman, I. Control, modulation, and analytical descriptions of vibrational strong coupling. *Chem. Rev.* **2023**, *123*(8), 5020–5048.

62 Agranovich, V. M., La Rocca, G. C., Bassani, F. Biexciton luminescence in semiconductor microcavities. *Phys. Status Solidi A* **1997**, *164*(1), 39–44.

63 Ballarini, D., De Giorgi, M., Gambino, S., Lerario, G., Mazzeo, M., Genco, A. et al. Polariton-induced enhanced emission from an organic dye under the strong coupling regime. *Adv. Opt. Mater.* **2014**, *2*(11), 1076–1081.

64 Melnikau, D., Esteban, R., Savateeva, D., Iglesias, A. S., Grzelczak, M., Schmidt, M. K. et al. Rabi splitting in photoluminescence spectra of hybrid systems of gold nanorods and J-aggregates. *J. Phys. Chem. Lett.* **2016**, *7*(2), 354–362.

65 Wersall, M., Munkhbat, B., Baranov, D. G., Herrera, F., Cao, J. S., Antosiewicz, T. J. et al. Correlative dark-field and photoluminescence spectroscopy of individual plasmon-molecule hybrid nanostructures in a strong coupling regime. *ACS Photonics* **2019**, *6*(10), 2570–2576.

66 Khazanov, T., Gunasekaran, S., George, A., Lomlu, R., Mukherjee, S., Musser, A. J. Embrace the darkness: an experimental perspective on organic exciton–polaritons. *Chem. Phys. Rev.* **2023**, 4(4), 041305.

67 Renken, S., Pandya, R., Georgiou, K., Jayaprakash, R., Gai, L. Z., Shen, Z. et al. Untargeted effects in organic exciton-polariton transient spectroscopy: a cautionary tale. *J. Chem. Phys.* **2021**, *155*(15), 154701.

68 Groenhof, G., Climent, C., Feist, J., Morozov, D., Toppari, J. J. Tracking polariton relaxation with multiscale molecular dynamics simulations. *J. Phys. Chem. Lett.* **2019**, *10*(18), 5476–5483.

69 Arrivo, S. M., Dougherty, T. P., Grubbs, W. T., Heilweil, E. J. Ultrafast infrared-spectroscopy of vibrational co-stretch up-pumping and relaxation dynamics of $W(CO)_6$. *Chem. Phys. Lett.* **1995**, *235*(3–4), 247–254.

70 Heilweil, E. J., Cavanagh, R. R., Stephenson, J. C. Population relaxation of CO ($v = 1$) vibrations in solution phase metal carbonyl complexes. *Chem. Phys. Lett.* **1987**, *134*(2), 181–188.

71 Owrutsky, J. C., Raftery, D., Hochstrasser, R. M. Vibrational-relaxation dynamics in solutions. *Annu. Rev. Phys. Chem.* **1994**, *45*, 519–555.

72 Zimdars, D., Tokmakoff, A., Chen, S., Greenfield, S. R., Fayer, M. D., Smith, T. I. et al. Picosecond infrared vibrational photon echoes in a liquid and glass using a free electron laser. *Phys. Rev. Lett.* **1993**, *70*(18), 2718–2721.

73 Rella, C. W., Kwok, A., Rector, K., Hill, J. R., Schwettman, H. A., Dlott, D. D. et al. Vibrational echo studies of protein dynamics. *Phys. Rev. Lett.* **1996**, *77*(8), 1648–1651.

74 Hamm, P., Zanni, M. *Concepts and Methods of 2D Infrared Spectroscopy*. Cambridge University Press: Cambridge, 2011.

75 Shim, S.-H., Gupta, R., Ling, Y. L., Strasfeld, D. B., Raleigh, D. P., Zanni, M. T. Two-dimensional IR spectroscopy and isotope labeling defines the pathway of amyloid formation with residue-specific resolution. *Proc. Natl. Acad. Sci. USA* **2009**, *106*(16), 6614–6619.

76 Fuller, F. D., Wilcox, D. E., Ogilvie, J. P. Pulse shaping based two-dimensional electronic spectroscopy in a background free geometry. *Opt. Express* **2014**, *22*(1), 1018–1027.

77 Al-Mualem, Z. A., Chen, X., Shirley, J. C., Xu, C., Baiz, C. R. Boxcars 2D IR spectroscopy with pulse shaping. *Opt. Express* **2023**, *31*(2), 2700–2709.

78 Ganim, Z., Chung, H. S., Smith, A. W., Deflores, L. P., Jones, K. C., Tokmakoff, A. Amide I two-dimensional infrared spectroscopy of proteins. *Acc. Chem. Res.* **2008**, *41*(3), 432–441.

79 Zanni, M. T., Hochstrasser, R. M. Two-dimensional infrared spectroscopy: a promising new method for the time resolution of structures. *Curr. Opin. Struct. Biol.* **2001**, *11*(5), 516–522.

80 Dunkelberger, A. D., Spann, B. T., Fears, K. P., Simpkins, B. S., Owrutsky, J. C. Modified relaxation dynamics and coherent energy exchange in coupled vibration-cavity polaritons. *Nat. Commun.* **2016**, *7*, 13504.

81 Simpkins, B. S., Fears, K. P., Dressick, W. J., Spann, B. T., Dunkelberger, A. D., Owrutsky, J. C. Spanning strong to weak normal mode coupling between vibrational and Fabry–Perot cavity modes through tuning of vibrational absorption strength. *ACS Photonics* **2015**, *2*(10), 1460–1467.

82 Dunkelberger, A. D., Grafton, A. B., Vurgaftman, I., Soykal, O. O., Reinecke, T. L., Davidson, R. B. et al. Saturable absorption in solution-phase and cavity-coupled tungsten hexacarbonyl. *ACS Photonics* **2019**, *6*(11), 2719–2725.

83 Baiz, C. R., Schach, D., Tokmakoff, A. Ultrafast 2D IR microscopy. *Opt. Express* **2014**, *22*(15), 18724–18735.

84 Duan, R., Kubarych, K. J. Dual-mode vibrational polariton dynamics probed with collinear 2D-IR. *The International Conference on Ultrafast Phenomena (UP) 2022.* **2022**. https://doi.org/10.1364/UP.2022.Tu1A.5.

85 Dunkelberger, A. D., Simpkins, B. S., Vurgaftman, I., Owrutsky, J. C. Vibration-cavity polariton chemistry and dynamics. *Annu. Rev. Phys. Chem.* **2022**, *73*, 429–451.

86 Simpkins, B. S., Dunkelberger, A. D., Owrutsky, J. C. Mode-specific chemistry through vibrational strong coupling (or a wish come true). *J. Phys. Chem. C* **2021**, *125*(35), 19081–19087.

87 Xiong, W. Molecular vibrational polariton dynamics: what can polaritons do? *Acc. Chem. Res.* **2023**, *56*(7), 776–786.

88 Xiang, B., Xiong, W. Molecular vibrational polariton: its dynamics and potentials in novel chemistry and quantum technology. *J. Chem. Phys.* **2021**, *155*(5), 050901.

89 Vasa, P., Pomraenke, R., Cirmi, G., De Re, E., Wang, W., Schwieger, S. et al. Ultrafast manipulation of strong coupling in metal-molecular aggregate hybrid nanostructures. *ACS Nano* **2010**, *4*(12), 7559–7565.

90 Xiang, B., Ribeiro, R. F., Dunkelberger, A. D., Wang, J. X., Li, Y. M., Simpkins, B. S. et al. Two-dimensional infrared spectroscopy of vibrational polaritons. *Proc. Natl. Acad. Sci. USA* **2018**, *115*(19), 4845–4850.

91 Xiang, B., Ribeiro, R. F., Li, Y. M., Dunkelberger, A. D., Simpkins, B. B., Yuen-Zhou, J. et al. Manipulating optical nonlinearities of molecular polaritons by delocalization. *Sci. Adv.* **2019**, *5*(9), eaax5196.

92 Xiang, B., Ribeiro, R. F., Chen, L. Y., Wang, J. X., Du, M., Yuen, O. E. Z. et al. State-selective polariton to dark state relaxation dynamics. *J. Phys. Chem. A* **2019**, *123*(28), 5918–5927.

93 Ribeiro, R. F., Dunkelberger, A. D., Xiang, B., Xiong, W., Simpkins, B. S., Owrutsky, J. C. et al. Theory for nonlinear spectroscopy of vibrational polaritons. *J. Phys. Chem. Lett.* **2018**, *9*(13), 3766–3771.

94 Ribeiro, R. F., Campos-Gonzalez-Angulo, J. A., Giebink, N. C., Xiong, W., Yuen-Zhou, J. Enhanced optical nonlinearities under collective strong light-matter coupling. *Phys. Rev. A* **2021**, *103*(6), 063111.

95 Hirschmann, O., Bhakta, H. H., Xiong, W. The role of IR inactive mode in $W(CO)_6$ polariton relaxation process. *Nanophotonics* **2023**, *13*, 2029–2034.

96 Chen, T. T., Du, M. T., Yang, Z. M., Yuen-Zhou, J., Xiong, W. Cavity-enabled enhancement of ultrafast intramolecular vibrational redistribution over pseudorotation. *Science* **2022**, *378*(6621), 790–793.

97 Grafton, A. B., Dunkelberger, A. D., Simpkins, B. S., Triana, J. F., Hernández, F. J., Herrera, F. et al. Excited-state vibration-polariton transitions and dynamics in nitroprusside. *Nat. Commun.* **2021**, *12*(1), 214.

98 Zhu, Y. F., Gauthier, D. J., Morin, S. E., Wu, Q. L., Carmichael, H. J., Mossberg, T. W. Vacuum Rabi splitting as a feature of linear-dispersion theory – analysis and experimental-observations. *Phys. Rev. Lett.* **1990**, *64*(21), 2499–2502.

99 Li, T. E., Nitzan, A., Subotnik, J. E. Cavity molecular dynamics simulations of vibrational polariton-enhanced molecular nonlinear absorption. *J. Chem. Phys.* **2021**, *154*(9), 094124.

100 Dunkelberger, A. D., Davidson, R. B., Ahn, W., Simpkins, B. S., Owrutsky, J. C. Ultrafast transmission modulation and recovery via vibrational strong coupling. *J. Phys. Chem. A* **2018**, *122*(4), 965–971.

101 Duan, R., Mastron, J. N., Song, Y., Kubarych, K. J. Isolating polaritonic 2D-IR transmission spectra. *J. Phys. Chem. Lett.* **2021**, *12*(46), 11406–11414.

102 Duan, R., Mastron, J. N., Song, Y., Kubarych, K. J. Reply to "Comment on: 'Isolating vibrational polariton 2D-IR transmission spectra'". *J. Phys Chem. Lett.* **2023**, *14*(4), 1046–1051.

103 Simpkins, B. S., Yang, Z. M., Dunkelberger, A. D., Vurgaftman, I., Owrutsky, J. C., Xiong, W. Comment on "Isolating polaritonic 2D-IR transmission spectra". *J. Phys. Chem. Lett.* **2023**, *14*(4), 983–988.

104 Hernández, F. J., Herrera, F. Multi-level quantum Rabi model for anharmonic vibrational polaritons. *J. Chem. Phys.* **2019**, *151*(14), 144116.

105 Houdre, R., Stanley, R. P., Ilegems, M. Vacuum-field Rabi splitting in the presence of inhomogeneous broadening: resolution of a homogeneous linewidth in an inhomogeneously broadened system. *Phys. Rev. A* **1996**, *53*(4), 2711–2715.

106 Cohn, B., Das, K., Basu, A., Chuntonov, L. Infrared open cavities for strong vibrational coupling. *J. Phys Chem. Lett.* **2021**, *12*(29), 7060–7066.

107 Butté, R., Delalleau, G., Tartakovskii, A. I., Skolnick, M. S., Astratov, V. N., Baumberg, J. J. et al. Transition from strong to weak coupling and the onset of lasing in semiconductor microcavities. *Phys. Rev. B* **2002**, *65*(20), 205310.

108 Houdre, R., Gibernon, J. L., Pellandini, P., Stanley, R. P., Oesterle, U., Weisbuch, C. et al. Saturation of the strong-coupling regime in a semiconductor microcavity – free-carrier bleaching of cavity polaritons. *Phys. Rev. B* **1995**, *52*(11), 7810–7813.

109 Lagoudakis, P. G., Martin, M. D., Baumberg, J. J., Malpuech, G., Kavokin, A. Coexistence of low threshold lasing and strong coupling in microcavities. *J. Appl. Phys.* **2004**, *95*(5), 2487–2489.

110 Norris, T. B., Rhee, J. K., Citrin, D. S., Nishioka, M., Arakawa, Y. Coherent and incoherent dynamics of excitons in semiconductor microcavities. *Nuovo Cimento Della Soc. Ital. Fis. D* **1995**, *17*(11–12), 1295–1303.

111 Xiang, B., Ribeiro, R. F., Du, M., Chen, L., Yang, Z., Wang, J. et al. Intermolecular vibrational energy transfer enabled by microcavity strong light–matter coupling. *Science* **2020**, *368*(6491), 665–667.

112 Reitz, M., Mineo, F., Genes, C. Energy transfer and correlations in cavity-embedded donor-acceptor configurations. *Sci. Rep.* **2018**, *8*, 9050.

113 Zhong, X. L., Chervy, T., Wang, S. J., George, J., Thomas, A., Hutchison, J. A. et al. Non-radiative energy transfer mediated by hybrid light-matter states. *Angew. Chem. Int. Ed.* **2016**, *55*(21), 6202–6206.

114 Xiang, B., Wang, J., Yang, Z., Xiong, W. Nonlinear infrared polaritonic interaction between cavities mediated by molecular vibrations at ultrafast time scale. *Sci. Adv.* **2021**, *7*(19), eabf6397.

115 Xiang, B., Yang, Z., You, Y.-Z., Xiong, W. Ultrafast coherence delocalization in real space simulated by polaritons. *Adv. Opt. Mater.* **2022**, *10*(5), 2102237.

116 Yang, Z., Bhakta, H. H., Xiong, W. Enabling multiple intercavity polariton coherences by adding quantum confinement to cavity molecular polaritons. *Proc. Natl. Acad. Sci. USA* **2023**, *120*(1), e2206062120.

117 Yang, Z., Xiong, W. Molecular vibrational polaritons towards quantum technologies. *Adv. Quantum Technol.* **2022**, *5*(8), 2100163.

118 Cahoon, J. F., Sawyer, K. R., Schlegel, J. P., Harris, C. B. Determining transition-state geometries in liquids using 2D-IR. *Science* **2008**, *319*(5871), 1820–1823.

119 Fidler, A. P., Chen, L., McKillop, A. M., Weichman, M. L. Ultrafast dynamics of CN radical reactions with chloroform solvent under vibrational strong coupling. *J. Chem. Phys.* **2023**, *159*(16), 164302.

7

Quantum Dynamics, Optical Signals, and Spectroscopy of Molecular Polaritons

Zhedong Zhang[1,2]

[1] *Department of Physics, City University of Hong Kong, Kowloon, Hong Kong SAR*
[2] *City University of Hong Kong, Shenzhen Research Institute, Shenzhen, Guangdong, China*

7.1 Introduction

Strong coupling of material excitations to light produces a set of hybrid quantum states known as polaritons. The polaritons have emerged as an attractive class of mesoscopic materials, thanks to their incredible quantum coherence properties [1–9]. The confined geometry due to the cavities leads to the quantized radiation fields, important for the formation of polaritons. Therefore, the polaritons may create new paradigm for the microscopic dynamics and interactions in complex materials [10–14].

Recent advancements in optics and material synthesis have provided new opportunities for molecules interacting with photons, resulting in a sort of new quasiparticles: the molecular polaritons [6, 7, 15–21]. Notable efforts were devoted to such a hybridization of matter and light, so as to explore unprecedented schemes of light-emission sources and control of material properties [22–31]. The light-molecule interactions in confined geometry have been shown to alter the molecular spectroscopy significantly [32–34]. The collective coupling of a number of molecules to photons is of fundamental importance, leading to various dramatic effects, such as high-temperature condensations [35–38], superradiance [39–42], and boost of chemical reactivity [1–3, 43–46]. Organic molecules present a particularly favorable candidate to achieve these goals even at room temperature due to their large dipole moments and stability. In contrast to atoms, molecular polaritons are more complicated. This arises from rich degrees of freedom of organic molecules, which result in multiscale interactions and entanglement between electrons, nuclears and solvents. The couplings to microcavities lead to the formation of various polaritons covering a wide spectrum of wavelengths resonant with different channels of relaxation [24–30, 47–51]. A new challenge for optical signals and spectroscopy is therefore indicated. It turns out that the polariton dynamics interplaying with its optical response is responsible for the unusual correlation and fluctuation properties of molecules from single molecules to the meso and macroscopic scale [52–54]. These call for extensive efforts in theory so as to gain an unified understanding of different but correlated aspects, for instance, dark states with a high mode density [30, 55–59], nonlinearities and strongly-correlated excitons [60–63].

Polariton Chemistry: Molecules in Cavities, First Edition. Edited by Joel Yuen-Zhou, Noel C. Giebink and Raphael F. Ribeiro. Published 2026 by John Wiley & Sons, Inc.

In this chapter, an overview of microscopic theory for molecular polaritons and their spectroscopy will be presented, in the basis of quantum electrodynamics (QED). Using the density matrix and Heisenberg–Langevin equation, a quantum-mechanical theory and approach will be developed for the multidimensional resonant spectroscopy of molecules interacting with cavity photons [26, 64]. Elaborate formalism and calculations will be provided with a foci of vibrational and exciton polaritons, along with the cutting-edge research performed recently in polariton spectroscopy [6, 16, 28, 65]. The results show real-time monitoring of cooperative dynamics of cavity polaritons against local motions of molecules arising from the polaron and disorder effects [66, 67]. Moreover, a multidimensional projection of optical signals and spectroscopy clearly reveals the long-range coherence of polaritons in a crosstalk with the solvent motion and vibronic coupling that may lead to localization effects [64].

The chapter will be organized as follows: Section 7.2 gives a QED framework for molecules interacting with cavities; Section 7.3 presents the theory for pump-probe signal for molecular polaritons; Section 7.4 develops a gateway-window theory for the 2D infrared spectroscopy for vibrational polaritons, revealing collective dynamics of polaritons against local disorder from solvent motion; Section 7.5 develops a microscopic theory for the 2D electronic spectroscopy of molecular exciton polaritons using Heisenberg–Langevin approach, showing a capability of accessing many-particle effects with a large number of molecules as well as delocalization–localization transition.

7.2 Quantum Electrodynamics of Molecular Polaritons

The electron and nuclear motions in molecules underline the scenarios of the interactions with the electromagnetic waves. This results in the response of molecules to the external fields, primarily in the visible and infrared regimes where the electronic structure and dynamics of molecules are imprinted. There are multiple levels of description for the coupling of molecules to radiation fields, ranging from single-electron scale to quasiparticle picture. In this chapter, we will adopt the quasiparticle description to develop the microscopic theory for polariton dynamics and optical spectroscopy. This is because of the visible and infrared wavelengths of our interest, wherein the valence excitations of molecules dominate. We will start with the full quantum light–matter interaction-both molecules and radiation are quantized, for a sake of generality. The convention $\hbar = k_B = 1$ will be used for the unit. Based on quantum electrodynamics (QED), the coupling of molecules to field is of the form [68]

$$H = \sum \frac{1}{2M}\left(\mathbf{p} - \frac{q}{c}\mathbf{A}\right)^2 + qU + \frac{1}{8\pi}\int (\mathbf{E}^2 + \mathbf{B}^2)\,\mathrm{d}\mathbf{r} \tag{7.1}$$

where $\mathbf{p}$, M, and q is the respective canonical momentum, mass and electric charge of the particles. c is the speed of light in vacuum. $\mathbf{A}$ and U are the vector and scalar potentials of the electromagnetic waves, respectively, so that

$$\mathbf{B} = \nabla \times \mathbf{A}, \quad \mathbf{E} = -\nabla U - \frac{1}{c}\frac{\partial \mathbf{A}}{\partial t} \tag{7.2}$$

In the nonrelativistic limit, one can apply the Coulomb gauge, i.e., $\nabla \cdot \mathbf{A} = 0$ to ensure the transverse fields and $U = U(\mathbf{r})$ that is time-independent. We will adopt this hereafter.

The intensity of the EM fields is usually not so strong that

$$\frac{2c}{q}|\mathbf{p} \cdot \mathbf{A}| \gg \mathbf{A}^2 \tag{7.3}$$

apart from the strong-field effects. The $\mathbf{A}^2$ term can thus be dropped. Using the second quantization, one recasts Eq. (7.1) into $H = \int \mathscr{H} d\mathbf{r}$ and

$$\mathscr{H} \approx \Psi^\dagger \left(-\frac{1}{2M}\nabla^2 + U\right)\Psi + \frac{1}{8\pi}\left(\mathbf{E}^2 + \mathbf{B}^2\right) - \frac{1}{c}\Psi^\dagger \mathbf{J}\Psi \cdot \mathbf{A} \tag{7.4}$$

is the Hamiltonian density, where $\mathbf{J} = q\mathbf{p}/M$ stands for the current operator. Ψ denotes the field operator for the matter field, e.g., the quasiparticles in molecules.

In the context of quantum physics in which the $\mathbf{A}^2$ term does not contribute significantly; it would be more convenient to work with the **E** and **B** fields rather than the **A** field. To this end, one applies a canonical transformation

$$\Psi \rightarrow \exp(-iq\mathbf{r} \cdot \mathbf{A})\Psi \tag{7.5}$$

to Eq. (7.1), and invokes the dipole approximation, which yields

$$\mathscr{H}_d \approx \sum_i \Psi^\dagger \left(-\frac{1}{2M_i}\nabla_i^2 + U_i\right)\Psi + \frac{1}{8\pi}\left(\mathbf{E}^2 + \mathbf{B}^2\right) - \sum_i \Psi^\dagger \mathbf{P}_i \Psi \cdot \mathbf{E} \tag{7.6}$$

after a lengthy but straightforward algebra, where $\mathbf{P}_i = q_i\mathbf{r}_i$ is the electric dipole operator. The third term in Eq. (7.6) gives the molecule–radiation interaction. The index i in Eq. (7.6) runs over all molecules and all the degrees of freedom in each individual molecule, e.g., electrons and nuclears. Equation (7.6) does not contain the direct many particle interactions such as Coulomb and dipole–dipole forces. It is not a big deal to add these terms, so that we will skip to avoid redundancy.

In molecules, the electron-field and radiation operators have the Fourier expansion, i.e.,

$$\Psi(\mathbf{r}, \mathbf{R}) = \sum_{i,\sigma} c_{i,\sigma}\varphi_{i,\sigma}(\mathbf{r}, \mathbf{R}) \tag{7.7a}$$

$$\mathbf{E}(\mathbf{r}, t) = \sum_{\mathbf{k}} \sqrt{\frac{2\pi\nu_{\mathbf{k}}}{V}}\mathbf{e}_{\mathbf{k}}\left[a_{\mathbf{k}}e^{i(\mathbf{k}\cdot\mathbf{r}-\nu_{\mathbf{k}}t)} + a_{\mathbf{k}}^\dagger e^{-i(\mathbf{k}\cdot\mathbf{r}-\nu_{\mathbf{k}}t)}\right] \tag{7.7b}$$

with $a_{\mathbf{k}}$ annihilating photons at mode **k** and $\left[a_{\mathbf{k}}, a_{\mathbf{k}'}^\dagger\right] = \delta_{\mathbf{k}\mathbf{k}'}.c_\sigma$ annihilates the quasielectrons in molecules so that $\left\{c_{i,\sigma}, c_{j,\lambda}^\dagger\right\} = \delta_{ij}\delta_{\sigma\lambda}.\varphi_{i,\sigma}(\mathbf{r}, \mathbf{R})$ denotes the orbital wave function of a quasi-electron at level (i, σ) whereby **R** denotes the nuclear coordinates. $\mathbf{e}_{\mathbf{k}}$ is the unit vector of the polarization of electric field and V is the mode volume. Assuming

$$\left(-\frac{1}{2M_j}\nabla_j^2 + U_j\right)\varphi_{j,\sigma}(\mathbf{r}, \mathbf{R}) = \omega_{j,\sigma}(\mathbf{R})\varphi_{j,\sigma}(\mathbf{r}, \mathbf{R}) \tag{7.8}$$

which can be solved by the numerical methods of the single-body Schrödinger's equation and inserting Eqs. (7.7a) and (7.7b) into Eq. (7.6), one finds the full Hamiltonian

$$H = \sum_{i,j,\sigma} t_{i,j}^\sigma(\mathbf{P}, \mathbf{R})c_{i,\sigma}^\dagger c_{j,\sigma} + \sum_{\mathbf{k}} \nu_{\mathbf{k}}a_{\mathbf{k}}^\dagger a_{\mathbf{k}} - \sum_{i,\mathbf{k}}\sum_{\sigma,\sigma'} f_{\sigma\sigma'}^i \mathscr{E}_{\mathbf{k}}c_{i,\sigma}^\dagger c_{i,\sigma'}\left(a_{\mathbf{k}}e^{-i\nu_{\mathbf{k}}t} + a_{\mathbf{k}}^\dagger e^{i\nu_{\mathbf{k}}t}\right) \# \tag{7.9}$$

where $\mathscr{E}_{\mathbf{k}} = \sqrt{2\pi\nu_{\mathbf{k}}/V}$ and $f_{\sigma\sigma'}^i = \langle\varphi_{i,\sigma}|\,\mathbf{P}_i \cdot \mathbf{e}_{\mathbf{k}}\,|\varphi_{i,\sigma'}\rangle$ is the matrix element of the dipole moment. $\mathbf{P} = -i\nabla_{\mathbf{R}}$ is the momentum operator of nuclears. Notice that $t_{i,i}^\sigma(\mathbf{P}, \mathbf{R}) = \omega_{i,\sigma}(\mathbf{R})$ is the term in the Born–Oppenheimer regime whereas $t_{i,j}^\sigma(\mathbf{P}, \mathbf{R}); i \neq j$ involving the nuclear velocity gives the nonadiabatic processes.

Equation (7.9) lays the foundation for exploring the relaxation and the optical responses of molecular polaritons, when subject to external fields. Using lasers to kick the systems, the coupling term reads

$$H_{\text{int}}(t) = -\sum_{i} \Psi^{\dagger}\mathbf{P}_{i}\Psi \cdot \mathbf{E}(\mathbf{r},t) = -\sum_{i}\sum_{\sigma,\sigma'} c_{i,\sigma}^{\dagger}c_{i,\sigma'}\mathbf{f}_{\sigma\sigma'}^{i} \cdot \mathbf{E}(\mathbf{r},t) \tag{7.10}$$

which involves the dipolar interaction only, for the sake of simplicity. The complete form of the Hamiltonian is $H_{\text{m}} = H + H_{\text{int}}(t)$. The time-evolving dynamics of the polariton systems are governed by the Liouville equation

$$\dot{\rho} = -i\,[H_{\text{m}}, \rho] \tag{7.11}$$

which gives the free propagator without external fields

$$\rho\,(t-t_0) = U(t-t_0)\,\rho\,(t_0)\,U^{\dagger}\,(t-t_0),\ U(t) = \mathcal{T}e^{-i\int^{t} H\mathrm{d}t'} \tag{7.12}$$

where $\mathcal{T}$ is the time-ordered operator. Subject to the external fields, one obtains the formal solution

$$\rho(t) = \mathcal{T}e^{-i\int^{t}\mathrm{d}t'[V_{\text{int}}(t'),*]}\rho\,(t_0),\ V_{\text{int}}(t) = U^{\dagger}(t)H_{\text{int}}(t)U(t) \tag{7.13}$$

and $V_{\text{int}}(t)$ is the matter-field coupling in the interactive picture. Normally, Eq. (7.13) is hard to be solved for broadband fields. We thus seek the solution by assuming weak laser fields in Eq. (7.10), which allow the perturbative expansion with respect to $V_{\text{int}}(t)$. Notice that we treat the coupling to external fields that is classical as perturbation only, whereas not the quantum radiation fields in Eq. (7.9). This works for the most of cavity polariton systems, in that the confined geometry may lead to strong-coupling regime that makes the perturbative method breakdown.

The measured quantity in optics-related experiments is the polarization, i.e., $\mathbf{P}(t) = \text{Tr}[\boldsymbol{\mu}\rho(t)] = \text{Tr}\left[\boldsymbol{\mu}\mathcal{T}e^{-i\int^{t}\mathrm{d}t'[V_{\text{int}}(t'),*]}\rho\,(t_0)\right] = \mathbf{P}^{(0)}(t) + \mathbf{P}^{(1)}(t) + \mathbf{P}^{(2)}(t) + \cdots$ where $\boldsymbol{\mu}$ denotes the dipole operator of molecules. Making use of the Dyson series for Eq. (7.13), one finds

$$\mathbf{P}^{(n)}(t) = (-i)^{n}\int^{t}[\mathrm{d}\tau]\,\langle\boldsymbol{\mu}\,[V_{\text{int}}\,(\tau_n), [V_{\text{int}}\,(\tau_{n-1}), \cdots, [V_{\text{int}}\,(\tau_1), \rho\,(t_0))]\cdots]]\rangle \tag{7.14}$$

with the notation $\int^{t}[\,\mathrm{d}\tau] \equiv \int^{t}\mathrm{d}\tau_n\int^{\tau_n}\mathrm{d}\tau_{n-1}\cdots\int^{\tau_2}\mathrm{d}\tau_1$. The leading order gives rise to the linear response immediately, i.e., $\mathbf{P}^{(1)}(t) = \int_{-\infty}^{t}\chi^{(1)}(t-\tau)\mathbf{E}(\tau)\mathrm{d}\tau$ where

$$\chi^{(1)}(t) = -i\langle[\boldsymbol{\mu}(t), \boldsymbol{\mu}\,(t_0)]\rangle,\ \boldsymbol{\mu} = \sum_{i}\sum_{\sigma,\sigma'} c_{i,\sigma}^{\dagger}c_{i,\sigma'}\mathbf{f}_{\sigma\sigma'}^{i} \tag{7.15}$$

Therefore, the absorption spectrum of the cavity-polariton systems is given by

$$S_{\text{A}}(\omega) = \Re\int_{0}^{\infty} e^{i\omega t}\,\langle[\boldsymbol{\mu}(t), \boldsymbol{\mu}(0)]\rangle\,\mathrm{d}t \tag{7.16}$$

Although the full Hamiltonian $H_{\text{m}} = H + H_{\text{int}}(t)$ where H is given by Eq. (7.9) dictates the microscopic dynamics of cavity-coupled molecules, numerous degrees of freedom may result in a Hilbert space of extremely high dimensional [5, 69]. This is often a great challenge in theory and computation, at the expense of a blurry picture of molecular structure and dynamics. An alternative way is to use effective Hamiltonian involving fewer degrees of freedom (but could be very redundant) as ramification from Eq. (7.9), in which the quasiparticles serve as the building blocks. We will adopt this strategy in the next sections.

7.3 Pump-probe Spectra for Molecular Polaritons

As the quantum dynamics of molecular polaritons draw much attention recently and are of fundamental importance for controlling the chemical reactivity, the study has to go beyond the scope of the linear absorption spectrum for a complete understanding of the underlying mechanism. To this end, let us consider the pump-probe spectra which has been studied extensively in molecular polaritons. Because the polariton dynamics is read out by the probe pulse, we consider the quantum-mechanical interaction between the N molecules and the probe field

$$V_{\text{int}}(t) = -\mathbf{P}\cdot\mathbf{E}\left(t-\tau_{\text{pr}}\right) = -\sqrt{N}\left(\boldsymbol{\mu}\cdot\mathbf{e}_{\text{pr}}\right)\mathscr{E}\left(t-\tau_{\text{pr}}\right)e^{-i\nu_{\text{pr}}(t-\tau_{\text{pr}})} - \text{h.c.} \tag{7.17}$$

following Eq. (7.10), whereas the pump field comes into the effect in the polariton density matrix. $\boldsymbol{\mu} = N^{-1/2}\sum_{i=1}^{N}\boldsymbol{\mu}_i e^{i\mathbf{k}\cdot\mathbf{r}_i}$ is the collective dipole operator. $\mathbf{e}_{\text{pr}}$ is the unit polarization vector of the probe field and $\mathscr{E}(t-\tau)$ is the pulse envelop centered at time τ. Suppose the system is pumped into the state ρ_0, Eq. (7.14) results in the First-order correction to the polarization and the transmission of the probe field reads

$$S_{\text{pp}}\left(\omega,\tau_{\text{pr}}\right) \propto \Im[\mathbf{E}(\omega)\cdot\mathbf{P}(\omega)] \tag{7.18}$$

On substituting Eq. (7.14) into (7.18) and carrying out the Fourier transform, one finds the pump-probe signal

$$S_{\text{pp}}\left(\omega,\tau_{\text{pr}}\right) \propto N\Re\int_{-\infty}^{\infty}\text{d}t\int_{-\infty}^{t}\text{d}\tau e^{i\omega(t-\tau_{\text{pr}})}R(t,\tau)E_{\text{pr}}^{*}(\omega)E_{\text{pr}}\left(\tau-\tau_{\text{pr}}\right) \tag{7.19}$$

where the response function is

$$R(t,\tau) = \text{Tr}\{\mu(t-\tau)[\mu,\rho(\tau)]\},\ \mu \equiv \boldsymbol{\mu}\cdot\mathbf{e}_{\text{pr}} \tag{7.20}$$

Exchanging the integration order in Eq. (7.19) yields

$$S_{\text{pp}}\left(\omega,\tau_{\text{pr}}\right) \propto N\Re\int_{-\infty}^{\infty}\text{d}\tau\text{Tr}\{\mu(\omega)[\mu,\rho(\tau)]\}e^{i\omega(\tau-\tau_{\text{pr}})}E_{\text{pr}}\left(\tau-\tau_{\text{pr}}\right)E_{\text{pr}}^{*}(\omega) \tag{7.21a}$$

$$\mu(\omega) = \int_{-\infty}^{\infty}\mu(t)e^{i\omega t}\text{d}t = \int_{-\infty}^{\infty}U^{\dagger}(t)\left(\boldsymbol{\mu}\cdot\mathbf{e}_{\text{pr}}\right)U(t)e^{i\omega t}\text{d}t \tag{7.21b}$$

and Eq. (7.21b) involves the polariton resonances, i.e. spectral lines and their shapes. Eq. (7.21a) evidences how the pump-probe signal does capture the real-time dynamics of cavity polaritons described by $\rho(\tau)$. To this clearly, we take the matrix element $\langle\psi_m|\,\mu(t)\,|\psi_n\rangle = \langle\psi_m|\,U^{\dagger}(t)\left(\boldsymbol{\mu}\cdot\mathbf{e}_{\text{pr}}\right)U(t)\,|\psi_n\rangle\,(m\neq n)$ in the basis $\{|\psi_n\rangle\}$. It gives

$$\mu_{mn}(\omega) = \sum_{m',n'}\mu_{m'n'}\int_{-\infty}^{\infty}U_{m'm}^{*}(t)U_{n'n}(t)e^{i\omega t}\text{d}t,\ \mu_{mn} = \boldsymbol{\mu}_{mn}\cdot\mathbf{e}_{\text{pr}} \tag{7.22}$$

in which the time integral is the Fourier component of the coherence. $\mu_{mn}(\omega)$ thus provides the resonances of molecular polaritons and their spectral line shapes. The trace part in Eq. (7.21a) can be written into

$$\mathrm{Tr}\{\mu(\omega)[\mu,\rho(\tau)]\} = \sum_{m>n} \mu_{mn}(\omega)\mu_{nm}\left[\rho_{mm}(\tau) - \rho_{nn}(\tau)\right] + (\text{Quantum beats}) \tag{7.23}$$

which indeed resolves the coupled dynamics of polariton and dark states. The quantum beating terms come from the coherence $\rho_{n'n}(\tau)$ of polaritons, which usually shows a faster decay than the populations.

Invoking the impulsive approximation such that the pulse duration is much shorter than the timescales of the excited-state dephasing and solvent relaxation [70–73], Eq. (7.21a) reads a simple form

$$S_{\mathrm{pp}}\left(\omega,\tau_{\mathrm{pr}}\right) \approx \text{const.} \times N\Re\mathrm{Tr}\left\{\mu(\omega)\left[\mu,\rho\left(\tau_{\mathrm{pr}}\right)\right]\right\}\left|E_{\mathrm{pr}}(\omega)\right|^2 \tag{7.24}$$

7.4 Multidimensional Infrared Spectroscopy for Vibrational Polaritons: Density-matrix Theory

Multidimensional spectroscopy provides a powerful tool for studying the polariton dynamics. It resolves the material dynamics by introducing several controllable time and frequency scales to reveal the information about polariton correlation functions, which are not possible through the 1D techniques. Coherent 2D spectra for complex molecules has led to the advancements of polaritons, demonstrating the strong-coupling effects which considerably modify the photoluminescence properties [1, 2, 6]. In this section, we will discuss the 2D resonant spectroscopy of polaritons, with a foci of the vibrational polaritons (VPs) in molecules [26]. The results will possess a multiscale dynamics of the VPs, revealing the collective nature in a trade-off with the nonlinearity and disorder that may localize the waves. The coupling of the VPs to the dark states will also be clarified.

7.4.1 Gateway-window Formalism

To begin with, let us consider a group of N molecules in an infrared cavity. Each molecule has a bond vibration interacting with the cavity; b_i and $a_{\mathbf{k}}$ denote the bosonic annihilation operators for the vibrations and the cavity modes, respectively. The dipolar interaction between the system and pulse series is of the form

$$V(t) = -\sum_{n=1}^{3}\sum_{j=1}^{N}\left(\boldsymbol{\mu}_j^* \cdot \mathbf{e}_n\right) b_j^\dagger(t) E\left(t-\overline{\tau}_n\right) e^{\mathrm{i}\mathbf{k}_n\cdot\mathbf{r}_j} - \text{h.c.} \tag{7.25}$$

in the interactive picture, where $\boldsymbol{\mu}_j$ is the infrared dipole moment of the jth molecule and $\mathbf{e}_n$ is the unit polarization vector of the nth laser pulse. $E\left(t-\overline{\tau}_n\right) = \mathscr{E}(t-\overline{\tau}_n)\, e^{-\mathrm{i}\nu_n(t-\overline{\tau}_n)}$ denotes the positive-frequency component of the electric field and $\mathscr{E}\left(t-\overline{\tau}_n\right)$ gives the time-domain pulse shape. $\overline{\tau}_n$ is the arrival time of the pulse, yielding the time delay between two adjacent pulses $T_n = \overline{\tau}_{n+1} - \overline{\tau}_n$. Using Eq. (7.14) the third-order polarization of the VPs reads

$$P_3(t) \propto \iiint \mathrm{d}\tau_3\,\mathrm{d}\tau_2\,\mathrm{d}\tau_1 R^{(3)}\left(t-\tau_3,\tau_3-\tau_2,\tau_2-\tau_1\right) E\left(\tau_3-\overline{\tau}_3\right) E\left(\tau_2-\overline{\tau}_2\right) E^*\left(\tau_1-\overline{\tau}_1\right) \tag{7.26}$$

where the third-order response function is

$$R^{(3)}(t_3,t_2,t_1) = R_1(t_3,t_2,t_1) + R_2(t_3,t_2,t_1) - R_3(t_3,t_2,t_1) \tag{7.27}$$

following the phase-cycling condition $\omega = -\nu_1 + \nu_2 + \nu_3$ and

$$R_1(t_3,t_2,t_1) = |\mu|^4 N^2 \langle \Psi_0| \left[\mathbf{b}(\tau_1)\cdot \mathbf{e}_1^*\right]\left[\mathbf{b}^\dagger(\tau_3)\cdot \mathbf{e}_3\right]\left[\mathbf{b}(t)\cdot \mathbf{e}^*\right]\left[\mathbf{b}^\dagger(\tau_2)\cdot \mathbf{e}_2^*\right]|\Psi_0\rangle \tag{7.28a}$$

$$R_2(t_3,t_2,t_1) = |\mu|^4 N^2 \langle \Psi_0| \left[\mathbf{b}(\tau_1)\cdot \mathbf{e}_1^*\right]\left[\mathbf{b}^\dagger(\tau_2)\cdot \mathbf{e}_2^*\right]\left[\mathbf{b}(t)\cdot \mathbf{e}^*\right]\left[\mathbf{b}^\dagger(\tau_3)\cdot \mathbf{e}_3\right]|\Psi_0\rangle \tag{7.28b}$$

$$R_3(t_3,t_2,t_1) = |\mu|^4 N^2 \langle \Psi_0| \left[\mathbf{b}(\tau_1)\cdot \mathbf{e}_1^*\right]\left[\mathbf{b}(t)\cdot \mathbf{e}^*\right]\left[\mathbf{b}^\dagger(\tau_3)\cdot \mathbf{e}_3\right]\left[\mathbf{b}^\dagger(\tau_2)\cdot \mathbf{e}_2^*\right]|\Psi_0\rangle \tag{7.28c}$$

and $|\Psi_0\rangle$ is the VP state prior to the pulse action. The collective vector operator is defined as

$$\mathbf{b} = \frac{1}{\mu\sqrt{N}} \sum_{n=1}^{N} \boldsymbol{\mu}_n b_n,\ |\mu|^2 = \sum_{n=1}^{N} \boldsymbol{\mu}_n^* \cdot \boldsymbol{\mu}_n \tag{7.29}$$

Notice that the dipole approximation was assumed above, i.e., wavelength ≫ sample size, which falls into the parameter regime of most experiments. And we are focusing on the photo-echo signal $\omega = -\nu_1 + \nu_2 + \nu_3$ through the phase cycling, whereby the other components can be obtained in a similar manner. The heterodyne-detected signal is measured normally in the laboratories, so that one has $\Im\left[P_3(\omega)E_{\text{lo}}^*(\omega)\right]$ with $E_{\text{lo}}(\omega)$ being the local-oscillator field. Taking Fourier transform over the first-time delay $T_1 = \overline{\tau}_2 - \overline{\tau}_1$, the 2D coherent signal is found to be

$$S_{\text{2D}}(\Omega_3, T_2, \Omega_1) \propto \Im \iint_0^\infty dt\, dT_1 e^{i(\Omega_3 t + \Omega_1 T_1)} P_3(t) E_{\text{lo}}^*(\Omega_3) \tag{7.30}$$

Before model calculations, it is essential to present some general results for the 2D signal by incorporating the reduced description of VP dynamics, namely, the reduced density matrix. Using Eqs. (7.28a–7.28c) we calculate

$$\begin{aligned}&\int_{-\infty}^{\infty} dt e^{i\Omega_3(t-\overline{\tau}_3)} \text{Tr}\left[b_i(t) b_k^\dagger(\tau_2)\rho_0(q) b_l(\tau_1) b_j^\dagger(\tau_3)\right]\\ &= e^{i\Omega_3(\tau_3-\overline{\tau}_3)} \text{Tr}\left[b_i(\Omega_3)\cdot \mathcal{G}(q,\tau_3-\tau_2)\left(b_k^\dagger \rho_0(q) b_l(\tau_1-\tau_2)\right)\cdot b_j^\dagger\right]\end{aligned} \tag{7.31}$$

$$\begin{aligned}&\int_{-\infty}^{\infty} dt e^{i\Omega_3(t-\overline{\tau}_3)} \text{Tr}\left[b_i(t) b_j^\dagger(\tau_3)\rho_0(q) b_l(\tau_1) b_k^\dagger(\tau_2)\right]\\ &= e^{i\Omega_3(\tau_3-\overline{\tau}_3)} \text{Tr}\left[b_i(\Omega_3) b_j^\dagger \cdot \mathcal{G}(q,\tau_3-\tau_2)\left(\rho_0(q) b_l^\dagger(\tau_1-\tau_2) b_k^\dagger\right)\right]\end{aligned} \tag{7.32}$$

$$\begin{aligned}&\int_{-\infty}^{\infty} dt e^{i\Omega_3(t-\overline{\tau}_3)} \text{Tr}\left[b_i(t) b_j^\dagger(\tau_3) b_k^\dagger(\tau_2)\rho_0(q) b_l(\tau_1)\right]\\ &= e^{i\Omega_3(\tau_3-\overline{\tau}_3)} \text{Tr}\left[b_i(\Omega_3) b_j^\dagger \cdot \mathcal{G}(q,\tau_3-\tau_2)\left(b_k^\dagger \rho_0(q) b_l(\tau_1-\tau_2)\right)\right]\end{aligned} \tag{7.33}$$

where $\rho_0(q)$ is the reduced density matrix of VPs before the pulse actions (q denotes the bath coordinates) and the Fourier component is

$$b_i(\Omega) = \mathcal{F}\left[b_i(t)\right] \Leftrightarrow \mathbf{b}(\Omega) = \mathcal{F}[\mathbf{b}(t)] \tag{7.34}$$

which gives rise to the spectral lines and shape. $\mathscr{G}(q,t)$ is the propagator of the VPs in the absence of pulses, which is subject to the VP dynamics as detailed later, e.g., Eq. (7.51) for the reduced density matrix of VPs.

The third-order polarization can be partitioned into $P_3(t) = P_3^{(1)}(t) + P_3^{(2)}(t) - P_3^{(3)}(t)$ associated with the three response functions. Inserting Eqs. (7.28a–7.28c) into Eq. (7.26) and using Eqs. (7.31–7.33) in which defining new variables $t_2 = \tau_3 - \tau_2, t_1 = \tau_2 - \tau_1$, one has $P_3(\Omega_3) = \mathscr{F}[P_3(t)]$ so that

$$P_3^{(1)}(\Omega_3) = \int_{-\infty}^{\infty} d\tau_3 \int_0^{\infty} dt_2 \int_0^{\infty} dt_1 e^{i\Omega_3\tau_3} E(\tau_3) E(\tau_3 + T_2 - t_2) E^*(\tau_3 + T_2 + T_1 - t_2 - t_1)$$
$$\times \mathrm{Tr}\left[(\mathbf{b}(\Omega_3)\cdot\mathbf{e}^*)\cdot\mathscr{G}(q,t_2)\left(\mathbf{e}_2\cdot\mathbf{b}^\dagger\rho_0(q)\mathbf{b}(-t_1)\cdot\mathbf{e}_1^*\right)\cdot\left(\mathbf{b}^\dagger\cdot\mathbf{e}_3\right)\right] \tag{7.35}$$

$$P_3^{(2)}(\Omega_3) = \int_{-\infty}^{\infty} d\tau_3 \int_0^{\infty} dt_2 \int_0^{\infty} dt_1 e^{i\Omega_3\tau_3} E(\tau_3) E(\tau_3 + T_2 - t_2) E^*(\tau_3 + T_2 + T_1 - t_2 - t_1)$$
$$\times \mathrm{Tr}\left[(\mathbf{b}(\Omega_3)\cdot\mathbf{e}^*)\left(\mathbf{b}^\dagger\cdot\mathbf{e}_3\right)\cdot\mathscr{G}(q,t_2)\left(\rho_0(q)\mathbf{b}(-t_1)\cdot\mathbf{e}_1^*\mathbf{e}_2\cdot\mathbf{b}^\dagger\right)\right] \tag{7.36}$$

$$P_3^{(3)}(\Omega_3) = \int_{-\infty}^{\infty} d\tau_3 \int_0^{\infty} dt_2 \int_0^{\infty} dt_1 e^{i\Omega_3\tau_3} E(\tau_3) E(\tau_3 + T_2 - t_2) E^*(\tau_3 + T_2 + T_1 - t_2 - t_1)$$
$$\times \mathrm{Tr}\left[(\mathbf{b}(\Omega_3)\cdot\mathbf{e}^*)\left(\mathbf{b}^\dagger\cdot\mathbf{e}_3\right)\cdot\mathscr{G}(q,t_2)\left(\mathbf{e}_2\cdot\mathbf{b}^\dagger\rho_0(q)\mathbf{b}(-t_1)\cdot\mathbf{e}_1^*\right)\right] \tag{7.37}$$

When the second pump and the probe pulses are well separated, namely, $T_2 \gg$ pulse duration, the 2D spectroscopic experiment can be viewed as a three-step process: preparation, propagation and detection. The two pump pulses create an initial doorway state of the VPs, which then propagates, and is finally probed at the time delay T_2 with a window state. For a precise definition of the three-step process, we recast Eqs. (7.35–7.37) into

$$P_3(\Omega_3) = \int_{-\infty}^{\infty} d\tau_3 \int_{-\infty}^{\tau_3+T_2} dt' \int_0^{\infty} dt_1 e^{i\Omega_3\tau_3} E(\tau_3) E(t') E^*(t' + T_1 - t_1)$$
$$\times\{\mathrm{Tr}\left[\mathbf{b}^\dagger\cdot\mathbf{e}_3, \mathbf{b}(\Omega_3)\cdot\mathbf{e}^*\right]\cdot\mathscr{G}(q,\tau_3)\mathscr{G}(q,T_2)\mathscr{G}(q,-t')\left(\mathbf{e}_2\cdot\mathbf{b}^\dagger\rho_0(q)\mathbf{b}(-t_1)\cdot\mathbf{e}_1^*\right)$$
$$+ \mathrm{Tr}(\mathbf{b}(\Omega_3)\cdot\mathbf{e}^*)\left(\mathbf{b}^\dagger\cdot\mathbf{e}_3\right)\cdot\mathscr{G}(q,\tau_3)\mathscr{G}(q,T_2)\mathscr{G}(q,-t')\left(\rho_0(q)\mathbf{b}(-t_1)\cdot\mathbf{e}_1^*\mathbf{e}_2\cdot\mathbf{b}^\dagger\right) \tag{7.38}$$

where the identity $\mathscr{G}(q,t_2) = \mathscr{G}(q,\tau_3)\mathscr{G}(q,T_2)\mathscr{G}(q,-t')$ has been used. Given the delay T_2 larger than the pulse duration, we may extend the upper limit $\tau_3 + T_2$ of integral over t' towards ∞, up to a proper approximation. This allows the definition of the gateway operators

$$D^{(e)}(q,T_1) = \int_{-\infty}^{\infty} dt' \int_{-t'}^{\infty} dt'' \left(\mathbf{e}_2\cdot\mathbf{b}^\dagger(t')\right)\rho_0(q)\left(\mathbf{b}(-t'')\cdot\mathbf{e}_1^*\right)E(t')E^*(T_1 - t'') \tag{7.39a}$$

$$D^{(g)}(q,T_1) = \int_{-\infty}^{\infty} dt' \int_{-t'}^{\infty} dt'' \rho_0(q)\left(\mathbf{b}(-t'')\cdot\mathbf{e}_1^*\right)\left(\mathbf{e}_2\cdot\mathbf{b}^\dagger(t')\right)E(t')E^*(T_1 - t'') \tag{7.39b}$$

$$\mathcal{D}^{(j)}(q,\Omega_1)=\int_0^{\infty} e^{i\Omega_1 T_1} D^{(j)}(q,T_1)\,\mathrm{d}T_1;\; j=e,g \tag{7.39c}$$

and the window operators

$$\mathcal{W}(\Omega_3)=\int_{-\infty}^{\infty} \mathrm{d}\tau_3 E(\tau_3)\left[\mathbf{b}^\dagger(\tau_3)\cdot\mathbf{e}_3,\mathbf{b}(\Omega_3)\cdot\mathbf{e}^*\right] \tag{7.40a}$$

$$\mathcal{V}(\Omega_3)=\int_{-\infty}^{\infty} \mathrm{d}\tau_3 E(\tau_3)\left(\mathbf{b}(\Omega_3)\cdot\mathbf{e}^*\right)\left(\mathbf{b}^\dagger(\tau_3)\cdot\mathbf{e}_3\right) \tag{7.40b}$$

These recast Eq. (7.38) into

$$P_3(\Omega_3)\propto N^2\mathrm{Tr}\left[\mathcal{W}(\Omega_3)\cdot\mathcal{G}(q,T_2)D^{(e)}(q,T_1)+\mathcal{V}(\Omega_3)\cdot\mathcal{G}(q,T_2)D^{(g)}(q,T_1)\right] \tag{7.41}$$

On substituting Eq. (7.41) into Eq. (7.30) the 2DIR signal for the VPs can be obtained

$$\begin{aligned} S_{2\mathrm{D}}(\Omega_3,T_2,\Omega_1)\propto N^2\Im\mathrm{Tr}\big[&\mathcal{W}(\Omega_3)\cdot\mathcal{G}(q,T_2)\mathcal{D}^{(e)}(q,\Omega_1)\\ &+\mathcal{V}(\Omega_3)\cdot\mathcal{G}(q,T_2)\mathcal{D}^{(g)}(q,\Omega_1)\big]\end{aligned} \tag{7.42}$$

Equations (7.39a–7.42) provide the most general formulation for the 2D infrared coherent signal for the VPs. To elaborate the gateway and window operations, we will further invoke the *impulsive approximation* such that the pulse duration is much shorter than the timescales of molecular dephasing and solvent motions [70, 74]. This enables us to carry out the time integrals so that

$$\mathcal{D}^{(e)}(q,\Omega_1)\approx\left(\mathbf{e}_2\cdot\mathbf{b}^\dagger\right)\rho_0(q)\left(\mathbf{b}(-\Omega_1)\cdot\mathbf{e}_1^*\right)\mathcal{M} \tag{7.43a}$$

$$\mathcal{D}^{(g)}(q,\Omega_1)\approx\rho_0(q)\left(\mathbf{b}(-\Omega_1)\cdot\mathbf{e}_1^*\right)\left(\mathbf{e}_2\cdot\mathbf{b}^\dagger\right)\mathcal{M} \tag{7.43b}$$

$$\mathcal{W}(\Omega_3)\approx\left[\mathbf{b}^\dagger\cdot\mathbf{e}_3,\mathbf{b}(\Omega_3)\cdot\mathbf{e}^*\right]\mathcal{E}(\nu_3) \tag{7.43c}$$

$$\mathcal{V}(\Omega_3)\approx\left(\mathbf{b}(\Omega_3)\cdot\mathbf{e}^*\right)\left(\mathbf{b}^\dagger\cdot\mathbf{e}_3\right)\mathcal{E}(\nu_3) \tag{7.43d}$$

where $\mathcal{M}=\int_{-\infty}^{\infty}\mathrm{d}t'\int_{-t'}^{\infty}\mathrm{d}t''E(t')E^*(T_1-t'')\approx\mathcal{E}(\nu_2)\mathcal{E}^*(\nu_1)$. Along with Eq. (7.42) and Eqs. (7.43a–d), the VPs are created via a spectral gateway $\mathcal{D}^{(e)}(q,\Omega_1)(\mathcal{D}^{(g)}(q,\Omega_1))$ that gives spectral lines at Ω_1, then undergo a free propagation at $\mathcal{G}(q,T_2)$, and is finally probed via a spectral window $\mathcal{W}(\Omega_3)(\mathcal{V}(\Omega_3))$ that gives spectral lines at Ω_3.

7.4.2 Cooperativity Versus Localization

Equation (7.42) contains the cross terms $(i,j,k,l=1,2,\ldots,N)$

$$\mathcal{D}_{kl}^{(\cdots)}(q,\Omega_1),\ \mathcal{W}_{ji}(\Omega_3),\ \text{and}\ \mathcal{V}_{ji}(\Omega_3);\ j\neq i,k\neq l \tag{7.44}$$

arising from cooperative interactions in the VPs, i.e., the collective operators $\mathbf{b},\mathbf{b}^\dagger$ [Eq. (7.29)] which create the many-particle excitations cooperatively. These terms would not show significant effects in the molecular samples in condensed phase [74], due to ultraweak correlation between molecules that yields

$$\rho_0(q) = \otimes \prod_{n}^{N} \rho_0^{(n)}(q_n) + \otimes \prod_{n}^{N} \prod_{m \neq n}^{N} \rho_0^{(nm)}(q_n, q_m) + \cdots \tag{7.45}$$

with the first term dominating. Such a product form for the molecular density matrix, however, is no longer valid when the systems possess long-range coherence or correlations, which for instance applies to the collectivity in molecular polaritons. In other words, Eq. (7.45) is of explicit significance to all orders. The cavity-induced collective effects will be detailed later in the study of VP dynamics of liquid-phase molecules.

It is worth noting that the collective operators $\mathbf{b}, \mathbf{b}^\dagger$ cannot form the integrals of motion, in the presence of the disorder and the nonlinearity which may cause localization of the waves. This has been indicated from the structure of the eigenstates of Eq. (7.46), i.e., the Hamiltonian H not to be written in terms of $\mathbf{b}, \mathbf{b}^\dagger$ and $a_\mathbf{k}, a_\mathbf{k}^\dagger$ only. In other words, from the perspective of group theory, neither $\mathbf{b}, \mathbf{b}^\dagger$ and $a_\mathbf{k}, a_\mathbf{k}^\dagger$ nor their combinations can be defined properly as the ladder operators for the system, due to the violation of the permutation symmetry [75–77]. In the optical signal in Eq. (7.42), the nonradiative transition channels – the ones induced by the disorder in the QSLE and quantified by $\mathscr{G}(q, T_2)$ – may therefore enable the connections to and the response of the states other than the ones through the action of the collective operators.

In what follows we will discuss the 2DIR spectroscopic properties of the VPs in realistic molecules, e.g., $W(CO)_6$, in order to gain deeper understanding of the formalism above in an intimate connection to recent experiments.

7.4.3 Stochastic Model for Vibrational Polaritons

To calculate the 2D signal above, one has to know the time-dependent $b_m(t)$, which is governed by the equation of motion (EOM) for the VPs. A generic Hamiltonian for the VPs emerges from Eq. (7.9), i.e.,

$$H = \sum_{i=1}^{N} \left[\omega_i(q_i) b_i^\dagger b_i + \Delta_i b_i^\dagger b_i^\dagger b_i b_i + \sum_{\mathbf{k}} g_{i,\mathbf{k}} \left(b_i^\dagger a_\mathbf{k} + b_i a_\mathbf{k}^\dagger \right) \right] + \sum_{\mathbf{k}} \nu_\mathbf{k} a_\mathbf{k}^\dagger a_\mathbf{k} \tag{7.46}$$

where $\omega_i(q_i) = \varpi_i + \delta\omega_i q_i$ and $g_{i,\mathbf{k}} = (\mu_i \cdot \mathbf{e}_\mathbf{k}) \sqrt{2\pi\nu_\mathbf{k}/V}$. ϖ_i and Δ_i are the vibrational frequency and the anharmonicity in the ith molecule, respectively. $q_i \in [-1, 1]$ is a random coordinate describing the motions of low-frequency vibrations and solvent locally surrounding individual molecule. There are two ways to solve the dynamics with Eq. (7.46): (i) Ensemble average and (ii) quantum stochastic Liouville equation (QSLE). For (i), the time-evolution operator reads

$$U_q(t) = e^{-i \int^{t} H(t') dt'} \tag{7.47}$$

which allows one to calculate the observable

$$\langle M \rangle (t) = \mathrm{Tr}_{0,q} \left[M U_q(t) \rho_0(q) U_q^\dagger(t) \right] \tag{7.48}$$

$\mathrm{Tr}_{0,q}$ includes the quantum-mechanical and ensemble averages, and $\rho_0(q)$ denotes the initial density matrix of the system. Calculating Eq. (7.48) is straightforward via numerical methods like stochastic trajectory and Monte Carlo. This may require a heavy load of computational resources for the VPs, because the low-frequency vibrations and solvents contain a large number of states that are densely distributing. And the many-body effects will also make the calculation of Eq. (7.48) breakdown, when many molecules are involved in the samples in cavities.

When using ultrashort pulses, or when the eigenstate spectrum is very dense, the full-time-evolving dynamics in Eq. (7.48) become impractical because of a large number of states. It is therefore advantageous to employ the QSLE, which provides a reduced description for the VPs [26, 78]. Suppose the coordinates follow the Markovian process, the probability obeys the equation

$$\dot{P}(q) = \hat{W}(q)P(q) \tag{7.49}$$

and $\hat{W}(q)$ can be either a differential operator or a matrix. Truncating to the second order and tracing out the bath states, one finds the QSLE for the density matrix of VP + bath system

$$\dot{\rho}(q,t) = -i[H(q), \rho(q,t)] + \hat{W}(q)\rho(q,t) \tag{7.50}$$

given the Hamiltonian $H(q)$ by Eq. (7.46). The formal solution to Eq. (7.50) can be obtained, and it defines the VP propagator $\mathscr{G}(q,t)$ in Liouville space ($H_-(q) \equiv [H(q), *]$)

$$\rho(q,t) = \mathscr{G}(q,t)\rho_0(q) = e^{\mathscr{L}(q)t}\rho_0(q), \;\mathscr{L}(q) = -iH_-(q) + \hat{W}(q) \tag{7.51}$$

There are two typical models commonly used for the stochastic process of the coordinates q's [79]:

1. The multiple-state jump model for spectral line shape and chemical exchanges;
2. The Brownian oscillator model for spectral diffusion.

Case 1 It is assumed that the bath hops between M discrete states, causing time-dependent spectral jumps between stable configurations of the VP complexes. The coordinates q are allowed to take M_i values, so that the q-dependent term in Eq. (7.46) reads the form of

$$\omega_i(q_i) = \text{const.} + \delta\omega_i q_i = \sum_{s=1}^{M_i} \delta\omega_i^{(s)}|s\rangle\langle s|;\; i = 1, 2, \dots, N \tag{7.52}$$

where $|s\rangle\langle s|$ is the projection operator in q space. The action of the W operator in Eq. (7.51) thus yields

$$\begin{aligned}\hat{W}(q)\rho = \sum_{i=1}^{N}\sum_{m>n}\Big[&\gamma_{mn}^{(i),+}\left(\tau_{mn}^{(i),+}\rho\tau_{mn}^{(i),-} - \tau_{mn}^{(i),-}\tau_{mn}^{(i),+}\rho\right)\\ &+\gamma_{mn}^{(i),-}\left(\tau_{mn}^{(i),-}\rho\tau_{mn}^{(i),+} - \tau_{mn}^{(i),+}\tau_{mn}^{(i),-}\rho\right)\Big] + \text{h.c.}\end{aligned} \tag{7.53}$$

where $\tau_{sr}^{(i),+} = |s\rangle\langle r|$ and $\tau_{sr}^{(i),-} = |r\rangle\langle s|$ denote the respective raising and lowering operators for the jumps between M_i states. $\gamma_{sr}^{(i),+}$ and $\gamma_{sr}^{(i),-}$ are the rates of the up- and down-transitions, respectively. At finite temperature, it allows the Boltzmann law, i.e., $\gamma_{sr}^{(i),+}/\gamma_{sr}^{(i),-} = e^{-\left[\delta\omega_i^{(s)} - \delta\omega_i^{(r)}\right]/T}$.

Case 2 The solvent motion surrounding individual molecule is described by an over-damped Brownian oscillator associated with the dimensionless continuous stochastic variables: coordinate Q and momentum P. This was used to describe the spectral diffusion processes in chemical physics, as seen from recent work on modeling the signatures of conformational fluctuations in trialanine in water, incorporating the Brownian oscillator coordinate model of fluctuations [80]. The q-dependent term in Eq. (7.46) is

$$\omega_i(q_i) = \text{const.} \times \delta\omega_i q_i, \quad H_s = \sum_{i=1}^{N} \frac{v_i}{2}\left(p_i^2 + q_i^2\right) \tag{7.54a}$$

whereby the local motion of the solvent is assumed to be a group of harmonic oscillators. For the reduced description of solvent motion, it may be advantageous to use the Wigner phase space representation, which is particularly suitable for semiclassical approximations [74]. The Wigner representation of an operator A is obtained by the Fourier transform of the coordinate representation

$$A(\mathbf{p},\mathbf{q}) = \int_{-\infty}^{\infty} e^{-i\mathbf{p}\cdot\mathbf{s}} \langle \mathbf{q}+\mathbf{s}/2|A|\mathbf{q}-\mathbf{s}/2\rangle \, d\mathbf{s} \tag{7.55}$$

which yields the average of the operator

$$\langle A\rangle = \iint \rho(\mathbf{p},\mathbf{q})A(\mathbf{p},\mathbf{q})d\mathbf{p}\, d\mathbf{q} \tag{7.56}$$

In phase space, the solvent motion follows a semiclassical picture, governed by the Langevin equation, i.e.,

$$\dot{q}_i = \nu_i p_i, \quad \dot{p}_i = -\nu_i q_i - \gamma_i p_i + f_i(t) \tag{7.57}$$

where $f_i(t)$ denotes a white-noise random force, i.e., $\langle f_i(t) f_j(t)\rangle = 2D_i\delta_{ij}\delta(t-t')$. Therefore, by projecting into the phase space the W operator in Eq. (7.51) has the form

$$\hat{W}(p,q)\rho(p,q) = \sum_{i=1}^{N}\left[\frac{\partial}{\partial p_i}(\gamma_i p_i + \nu_i q_i) + \frac{\partial}{\partial q_i}(-\nu_i p_i) + D_i\frac{\partial^2}{\partial p_i^2}\right]\rho(p,q) \tag{7.58}$$

In most circumstances, the solvent moves much slower than the intramolecular vibrations. The kinetic energy term can be neglected properly, i.e., $p_i \approx 0$ which falls into the overdamped regime. Using Eq. (7.57), one obtains $\dot{q}_i = -\gamma_i q_i + f_i(t)$ with $\frac{\nu_i^2}{\gamma_i} \to \gamma_i$, $\frac{\nu_i}{\gamma_i} f_i(t) \to f_i(t)$. Equation (7.58) then reduces to

$$\hat{W}(q)\rho(q) = \sum_{i=1}^{N}\left(\gamma_i\frac{\partial}{\partial q_i}q_i + D_i\frac{\partial^2}{\partial q_i^2}\right)\rho(q) \tag{7.59}$$

which gives the so-called Ornstein–Uhlenbeck process.

7.4.4 Simulations of 2DIR Spectra for VPs

To elaborate the approach of using the QSLE, we consider a group of identical molecules in a single-mode infrared cavity [26]. The intramolecular vibrational modes are set to be resonant with the cavity photons whereby the solvent induces frequency fluctuations at each individual molecule. The Hamiltonian is of the form

$$H_v = \sum_{i=1}^{N} [(\delta + \delta\omega_i\eta_i)]\,|v_i\rangle\langle v_i| + g\left(|v_i\rangle\langle G|a + |G\rangle\langle v_i|a^\dagger\right) \tag{7.60}$$

in the rotating frame of the photons, where a is the annihilation operator for cavity photons. $|v_i\rangle$ is the first vibrational excited state of the ith molecule (vibrational frequency is ω) and $|G$ denotes the vibrational ground state of this N-molecule system. In Eq. (7.60) the anharmonicity is assumed to be large, i.e., $\Delta_i \to \infty$ causing the higher excitations of each molecule far detuned from the cavity photons and laser pulses. Here $\delta = \omega - \nu$ denotes the vibration-cavity detuning and ν is the photon frequency. The operator $\eta_i = \frac{1}{2}(1-\tau_i^z)$ and $\tau_i^z, \tau_i^\pm$represent the Pauli matrices acting on solvent coordinates at the ith molecule, i.e.,

$$\tau_i^+\,|q_i| = \sqrt{1-q_i}\,|1-q_i|,\ \tau_i^-\,|q_i| = \sqrt{q_i}\,|1-q_i|,\ \tau_i^z\,|q_i| = (1-2q_i)\,|q_i| \tag{7.61}$$

The terms in Eq. (7.60) proportional to η_i describe the dynamical disorder, yielding a random shift for the vibrational frequency such that $\omega \to \omega + \delta\omega_i(q)$ associated with a distribution $P(q,t)$ of the solvent coordinates q_l s. Here, we assume a single coordinate q_i for the ith molecule and a two-state description for the solvent motion, i.e., $q_i = 0, 1$ [26, 79]. The remaining degrees of freedom act as thermal baths, leading to the random jump between these states. The full dynamics of the density matrix ρ of the VPs involving the disorder is governed by the QSLE from Eq. (7.51), i.e.,

$$\dot{\rho} = -i[H_v, \rho] + \sum_{i=1}^{N} \frac{\gamma_i}{2} \left[\bar{n}_{\nu_i}\left(\tau_i^+ \rho \tau_i^- - \tau_i^- \tau_i^+ \rho\right) + \left(\bar{n}_{\nu_i} + 1\right)\left(\tau_i^- \rho \tau_i^+ - \tau_i^+ \tau_i^- \rho\right)\right] + \frac{v}{2Q}\left(a\rho a^\dagger - a^\dagger a \rho\right) + \text{h.c.} \tag{7.62}$$

where ν_i is the typical frequency of the solvent motion in the vicinity of the ith molecule and $\gamma_i = \pi \sum_s \lambda_{i,s}^2 \delta\left(\nu_i - \nu_s^{(i)}\right)$ with $\lambda_{i,s}$ being the solvent-bath coupling. Equation (7.62) dictates the superoperator W of damping and thus $\mathscr{L} = -iH_{v,-} + W$. This leads to the formal solution to the QSLE $\rho(q,t) = e^{\mathscr{L}t}\rho_0(q)$. The initial state $\rho_0(q)$ is at thermal equilibrium $\left(|G_q\rangle = |G\rangle \otimes |q\right.$

$$\rho_0(q) = \sum_q P(q)|G_q|\langle G_q\rangle, \; P(q) = 2^{-N} \prod_{s=1}^{N} \left[1 + (-1)^{\delta_{q_s,1}} \tanh\left(\frac{\nu_s}{2T}\right)\right] \tag{7.63}$$

Let q denote the configuration $\{q_1, q_2, \ldots, q_N\}$ of the solvent motion, $|v_n, q\rangle = |v_n\rangle \otimes |q$ and we use v_n to denote the singly excite state of the joint vibration-photon system where $v_n; n = 1, 2, \ldots, N$ are for the single excitation of vibrations at the nth molecule and v_{N+1} is for the single photon state of the infrared cavity (Figure 7.1a). The Hamiltonian given by Eq. (7.60) is block diagonal with respect to the solvent configuration $\{q_1, q_2, \ldots, q_N\}$. All molecules, in resonance with cavity photons (0–0 transition in Figure 7.1b), result in the two polariton states: upper vibrational polariton (UVP) and lower vibrational polariton (LVP). This is obtained by diagonalizing the H_v in $\{0_1, 0_2, \ldots, 0_N\}$ block. All other modes are dark. Diagonalizing the H_v in other configurations with different q_l s produces the local dark states (LDSs) which may erode the collective nature of the VPs. Equation (7.62) describes the disorder-induced couplings between the VPs, dark states and LDSs, as a result

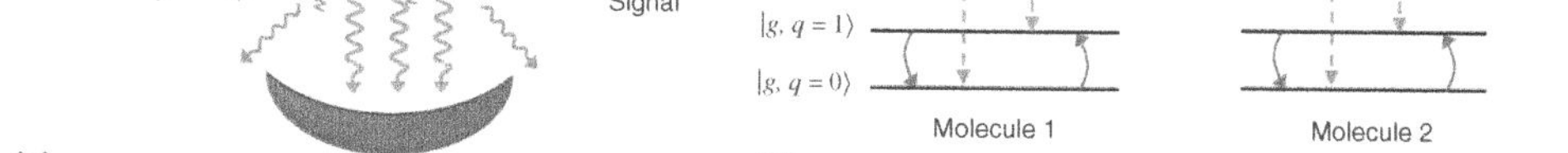

(a) **(b)**

Figure 7.1 Schematic of 2D infrared spectroscopy for molecular vibrational polaritons. (a) Three time-ordered picosecond pulses incident off the cavity axis, with three controllable time delays T_1, T_2, T_3. (b) Level diagrams of molecules including disorder. Transition $|g, q = 0 \leftrightarrow |e, q = 0$ (red dashed arrow) is resonant with cavity photons whereas transition $|g, q = 1 \leftrightarrow |e, q = 1$ (green dashed arrow) is off-resonant with a detuning $\delta\omega$ given by the vibration–solvent interaction $\delta\omega_i\eta_i |v_i\rangle\langle v_i|$ in Eq. (7.60); disorder is caused by random transitions denoted by purple curved arrows. *Source*: [26]/American Chemical Society.

of the random jumps. One would expect cross-talks between these states associated with certain timescales. This can be readily monitored by the 2DIR spectroscopic signal.

By replacing the operator b_i to $\sigma_i^- = |G\rangle\langle v_i|$, $b_i^\dagger$ to $\sigma_i^+ = |v_i\rangle\langle G|$ in Eqs. (7.39–7.42), and defining the collective scalar operators

$$\gamma^\pm = \frac{1}{\sqrt{N}}\sum_{n=1}^{N}\sigma_n^\pm \tag{7.64}$$

the gateway and window operators have the form

$$\mathscr{D}^{(e)}(q,\Omega_1) \propto \int_0^\infty dT_1 \int_{-\infty}^\infty dt' \int_{-t'}^\infty dt'' e^{i\Omega_1 T_1}\gamma^+(t')\rho_0(q)\gamma^-(-t'')E(t')E^*(T_1-t''),$$

$$\mathscr{D}^{(g)}(q,\Omega_1) \propto \int_0^\infty dT_1 \int_{-\infty}^\infty dt' \int_{-t'}^\infty dt'' e^{i\Omega_1 T_1}\rho_0(q)\gamma^-(-t'')\gamma^+(t')E(t')E^*(T_1-t''),$$

$$\mathscr{W}(\Omega_3) \propto \int_{-\infty}^\infty d\tau_3 E(\tau_3)[\gamma^+(\tau_3),\gamma^-(\Omega_3)],$$

$$\mathscr{V}(\Omega_3) \propto \int_{-\infty}^\infty d\tau_3 E(\tau_3)\gamma^-(\Omega_3)\gamma^+(\tau_3),$$

up to prefactors that depend on the polarization of pulses. On substituting into Eq. (7.42) the 2DIR signal can be found therein, with the propagator $\mathscr{G}(q,T_2) = e^{\mathscr{L}T_2}$ given by Eq. (7.62).

In general, calculating the signal in Eq. (7.42) is hard, due to the integrals over pulse shapes. To simplify the calculations, we invoke the impulsive approximation which may enable the multifold convolutions over the pulse envelopes. After some tedious algebra we are able to obtain the 2DIR signal. All the discussions are detailed in Ref. [26].

To reveal the effect of disorder against the polariton-induced delocalization, we first assume cryogenic solvent so that the temperature $T \approx 0$. The finite-temperature effect will be considered later. The top row in Figure 7.2 (adopted from Ref. [26]) shows the tomography of the 2DIR signal $S_{2D}(\Omega_3, T_2, \Omega_1)$ with various T_2 delay. We note that the line-broadening along the diagonal is larger than that along the anti-diagonal. This is reasonable because of the inhomogeneous broadening attributed to the solvent. As prolonging T_2, the cross peaks above the diagonal show up and their intensities keep increasing. Cross peak at $\Omega_1 - \nu = 3.6\ \text{cm}^{-1}$ ($\Omega_1 - \nu = -3.6\ \text{cm}^{-1}$), $\Omega_3 - \nu = 18\ \text{cm}^{-1}$ manifests the excitation transfer from UVP (LVP) to LDSs that weakly coupled to cavity photons. This is manifested by the fixed probe frequency Ω_3 without much shift as T_2 increases. Besides, the cross peak positioned at $\Omega_1 - \nu = 3.6\ \text{cm}^{-1}, \Omega_3 - \nu = -3.6\ \text{cm}^{-1}$ indicates the excitation transfer from UVP to LVP, while the cross peak at $\Omega_1 - \nu = -3.6\ \text{cm}^{-1}, \Omega_3 - \nu = 3.6\ \text{cm}^{-1}$ indicates the excitation transfer from LVP to UVP. Lower peak intensity at $\Omega_1 - \nu = \pm 3.6\ \text{cm}^{-1}, \Omega_3 - \nu = \mp 3.6\ \text{cm}^{-1}$ than the one at $\Omega_1 - \nu = \pm 3.6\ \text{cm}^{-1}, \Omega_3 - \nu = \mp 18\ \text{cm}^{-1}$ implies the system of interest is more in favor of populating LDSs than populating polaritonic states. We can further observe from these peak variations that the polariton–polariton transfer occurring in $\sim$ 100 ps is slower than the $\sim$ 30 ps polariton-LDSs transfer. Moreover, the top row of Figure 7.2 illustrates that the rate of UP $\rightarrow$ LDSs is higher than that of LP $\rightarrow$ LDSs. This is consistent with the time evolution of both populations and intermolecular coherence obtained before.

In the bottom row of Figure 7.2, we plot the 2DIR signal $S_{2D}(\Omega_3, T_2, \Omega_1)$ for different T_2 delays, by including thermal excitations of the solvent whereas the molecules are in the vibrational ground state prior to the pulse actions. Compared to the top row, we now observe cross peaks both above

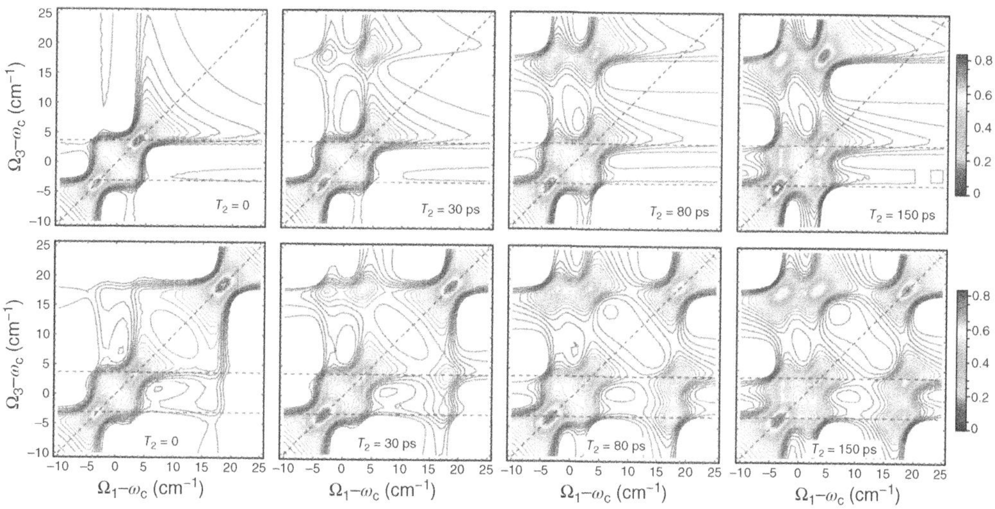

Figure 7.2 2DIR signal variation with T_2 delay, given by Eq. (7.42), for three $W(CO)_6$ placed in an infrared microcavity with a linear fashion along the cavity axis. Prior to pulse action, (top) the solvent motion is not considered and (bottom) the solvent is at thermal equilibrium under room temperature. Ω_1 and Ω_3 correspond to excitation and emission frequencies, respectively. Molecular parameters are $\omega = 1983$ cm^{-1}, $\nu = 1983$ cm^{-1}, $\delta\omega_i = 18$ cm^{-1}, $g = 2.1$ cm^{-1}, $\nu_i = 62$ cm^{-1}, $\gamma_i = 0.18$ cm^{-1}, $\nu/Q = 0.04$ cm^{-1} and $= 300$ K; Pulse shape is Gaussian with the parameters $\sigma_1 = \sigma_2 = \sigma_3 = \sigma_{lo} = 50$ cm^{-1}, $\omega_1 = \omega_2 = 1983$ cm^{-1} and $\omega_3 = \omega_{lo} = 1993$ cm^{-1}. *Source*: [26]/American Chemical Society.

and below the diagonal line. In addition to the information given by the top row of Figure 7.2, an extra peak positioned at $\Omega_1 - \nu = \Omega_3 - \nu \approx 18$ cm^{-1} can be observed even at $T_2 = 0$. This indicates that the thermal motion of the solvent may make the LDSs visible in absorption spectra, manifesting the modifications of the polaritonic structure of molecular vibrations. The excitation transfer from LDSs to UP (LP) is further evident by the cross peaks at $\Omega_1 - \nu \approx 18$ cm^{-1}, $\Omega_3 - \nu \approx 3.6$ cm^{-1} ($\Omega_3 - \nu \approx -3.6$ cm^{-1}) below the diagonal, when the LDSs are populated. The population transfer between LDSs and polariton states is faster than that between LP and UP, as evident from the polariton dynamics.

So far, the analysis based on Figure 7.2 clearly elucidates how the cooperativity in the joint vibration-photon system is eroded by the local fluctuations (i.e., disorder), associated with the presence of extra cross peaks other than those characterizing the polariton modes. This allows us to explain and better understand the recent experiments [28], where the coupling between bright polariton and LDSs was demonstrated. However, the positions of LDSs relative to polariton states in 2D IR spectra predicted by our work shows the deviation from these experiments. This is due to the fewer molecules considered in our simulations (for $= 3, g\sqrt{N} \approx 3.6$ cm^{-1} $\ll \delta\omega_i$) than many experiments which had $\sim$ 4000 $W(CO)_6$ in the infrared cavity. To support this, one can further calculate the distribution of dipole moments for an ensemble with 4137 $W(CO)_6$, as depicted in figure 4 in Ref. [26] wherein several molecules out of 4137 being large-detuned off cavity frequency under the influence of solvent-induced disorder. The extra peaks sandwiched in between the two polariton branches may be attributed to the LDSs since the associated frequency shift of a few cm^{-1} S is much smaller than the Rabi splitting $= 2g\sqrt{N} \approx 40$ cm^{-1} between the two polariton branches. These, combined with Figure 7.2, thereby help in understanding the experimental results. Exploring the properties with a large number of molecules will be an important topic which will be the goal of the subsequent chapter.

7.5 Multidimensional Electronic Spectroscopy for Exciton Polaritons: Heisenberg–Langevin Theory

The collective interaction between many molecules and photons is of fundamental importance, leading to interesting phenomena, e.g., superradiance and cooperative dynamics of polaritons [26, 39, 64, 76, 81]. In contrast to atoms, whereby superradiance and cavity polaritons are well understood, molecular exciton polaritons are more complex. This arises from the complicated couplings between electronic and nuclear degrees of freedom, which possess new challenges for optical spectroscopy. Recently, exact diagonalization of molecular Hamiltonian was used to calculate the optical responses, by taking only a few vibrational excitations into account [5, 82–84]. Here, we will focus on the 2D spectroscopy for exciton polaritons of molecules where the optical cavities are in use and the electronic excitations are created. The polaritonic relaxation pathways involving the population and coherence dynamics, however, still remain elusive. Elaborate experiments on nonlinear optical properties have demonstrated unusual correlations and many-body dynamics of molecular polaritons. These call for an extensive understanding of dark states with a high mode density [30, 55–59], nonlinearities, and multiexciton correlations [60–63].

Previous spectroscopic studies of cavity polaritons were mostly based on wave function methods involving nuclear dynamics [85–88], Redfield theory, and quantum chemistry simulations of low excitations [43, 69, 78, 89, 90]. Absorption and emission associated with multiple phonons and optically dark states depend on a strong polariton–polaron interaction, which raises a fundamental issue in exciton polaritons and, however, complicates the simulation of multidimensional spectroscopy.

7.5.1 Langevin Model for Exciton Polaritons

To begin with, let us consider N identical molecules in an optical cavity [64]. Each has two energy surfaces corresponding to electronically ground and excited states, i.e., $|g_j\rangle, |e_j\rangle\,(j = 1, 2, \ldots, N)$, respectively. There is a minimum offset between the two surfaces as shown in Figure 7.3b, causing the exciton coupling with the intramolecular vibrations. The induced electric dipole formed between the electron and hole leads to the interaction between excitons and cavity photons. The Hamiltonian is of the

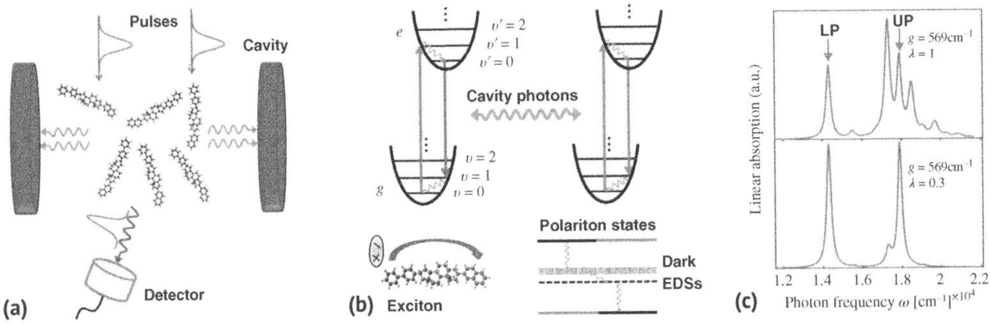

Figure 7.3 Schematic illustration of time-resolved spectroscopy for molecular polaritons. (a) Emission signal is collected along a certain direction, once the molecules are excited by laser pulses. (b) Exciton–photon interaction in molecules in the presence of vibronic coupling attached to individual molecule. This results in the dark states and local dark states (LDSs) weakly interacting with cavity, apart from the upper and lower polariton modes; Rich timescales and channels of excited-state relaxation are thus expected. (c) Linear absorption of molecular polaritons with 10 organic molecules in an optical cavity. The parameters are $\omega_D = 16113\ \text{cm}^{-1}, \delta = \delta_c = 0, \Gamma = 20\ \text{cm}^{-1}, \gamma = 1\ \text{cm}^{-1}, \gamma_c = 0.9\ \text{cm}^{-1}, \omega_v = 1200\ \text{cm}^{-1}$, typically from cyanine dyes [91]. *Source*: [64]/American Physical Society.

Holstein-Tavis-Cummings form

$$H = \sum_{n=1}^{N} \left[\delta\sigma_n^+\sigma_n^- + \omega_v b_n^\dagger b_n - \lambda\omega_v\sigma_n^+\sigma_n^- \left(b_n + b_n^\dagger\right) + g\left(\sigma_n^+ a + \sigma_n^- a^\dagger\right)\right] + \delta_c a^\dagger a \tag{7.65}$$

in the rotating frame of photons, where δ denotes the detuning between excitons and external laser pulse. $\sigma_n^+ = |e_n\rangle\langle g_n|$ and $\sigma_n^- = |g_n\rangle\langle e_n|$ are the respective raising and lowering operators for the excitons in the nth molecule, satisfying the $su(2)$ Lie algebra

$$[\sigma_n^-, \sigma_m^+] = \sigma_n^z\delta_{nm}, \quad [\sigma_n^\pm, \sigma_m^z] = \pm 2\sigma_n^\pm\delta_{nm} \tag{7.66}$$

b_n denotes the bosonic annihilation operator of the vibrational mode with a high frequency ω_v, in the nth molecule, and a annihilates cavity photons. Each molecule has one high-frequency vibrational mode. This is straightforward to be generalized to many high-frequency vibrational modes, which is closer to the reality but is more redundant. In addition to the strong coupling to the single-longitudinal cavity mode, the molecules are subject to three temporally separated laser pulses carrying electric fields $E_j\left(t - T_j\right)e^{-i\nu_j(t-T_j)}; j = 1, 2, 3$. The interaction reads

$$V(t) = -\sum_{n=1}^{N}\sum_{j=1}^{3} \left[\sigma_n^+\Omega_j\left(t - T_j\right)e^{-i(\nu_j-\nu_3)t}e^{i\nu_j T_j} + \text{h.c.}\right] \tag{7.67}$$

with the Rabi frequency $\Omega_j\left(t - T_j\right) = \mu_{\text{eg}}E_j\left(t - T_j\right)$ for the jth pulse field and μ_{eg} as the molecular electric dipole moment. We have adopted the Condon approximation that μ_{eg} is independent of the nuclear coordinates [70, 74]. The full Hamiltonian thus reads $H(t) = H + V(t)$ and the most general form of the quantum Langevin equation (QLE) is thus $\left(\mathbf{v} = [\sigma_1^-, \sigma_2^-, \cdots, \sigma_N^-, a]^{\mathrm{T}}\right)$

$$\dot{\mathbf{v}} = -i[H(t), \mathbf{v}] - \Upsilon\mathbf{v} + \mathbf{f}(t), \ \mathbf{f}(t): \text{ noise operator.} \tag{7.68}$$

On substituting Eqs. (7.65) and (7.67), one recasts the QLE into

$$\dot{\mathbf{v}} = -\mathrm{K}\mathbf{v} + \mathbf{f}(t) - i\sum_{j=1}^{3}\Omega_j\left(t - T_j\right)e^{-i(\nu_j-\nu_3)t}e^{i\nu_j T_j}\mathbf{w} \tag{7.69a}$$

$$\dot{b}_n = -\left(i\omega_v + \Gamma\right)b_n + i\lambda\omega_v\sigma_n^+\sigma_n^- + \sqrt{2\Gamma}b_n^{\text{in}}(t) \tag{7.69b}$$

with $\mathbf{f}(t) = \left[\sqrt{2\gamma}\sigma_1^{-,\text{in}}(t), \sqrt{2\gamma}\sigma_2^{-,\text{in}}(t), \cdots, \sqrt{2\gamma}\sigma_N^{-,\text{in}}(t), \sqrt{2\gamma_c}a^{\text{in}}(t)\right]^{\mathrm{T}}$ and

$$\mathrm{K} = \begin{pmatrix} i\delta_1 + \gamma & 0 & \cdots & 0 & ig\sigma_1^z \\ 0 & i\delta_2 + \gamma & \cdots & 0 & ig\sigma_2^z \\ \vdots & \vdots & & \vdots & \vdots \\ 0 & 0 & \cdots & i\delta_N + \gamma & ig\sigma_N^z \\ ig & ig & \cdots & ig & i\delta_c + \gamma_c \end{pmatrix}, \quad \mathbf{w} = \begin{pmatrix} 2n_1 - 1 \\ 2n_2 - 1 \\ \vdots \\ 2n_N - 1 \\ 0 \end{pmatrix} \tag{7.70}$$

where $\delta_n = \delta - \lambda\omega_v\left(b_n + b_n^\dagger\right); n = 1, 2, \ldots, N$ and the noise operators obey the fluctuation dissipation relation

$$\left\langle b_j^{\text{in},\dagger}(t)b_{j'}^{\text{in}}(t')\right\rangle = \bar{n}_{\text{th}}\delta_{jj'}\delta(t - t'), \quad \left\langle b_j^{\text{in}}(t)b_{j'}^{\text{in},\dagger}(t')\right\rangle = \left(\bar{n}_{\text{th}} + 1\right)\delta_{jj'}\delta(t - t') \tag{7.71a}$$

$$\left\langle a^{\text{in},\dagger}(t)a^{\text{in}}(t')\right\rangle = 0,\ \left\langle a^{\text{in}}(t)a^{\text{in},\dagger}(t')\right\rangle = \delta(t-t') \tag{7.71b}$$

$$\left\langle \sigma_j^{\text{in},+}(t)\sigma_{j'}^{\text{in},-}(t')\right\rangle = 0,\ \left\langle \sigma_j^{\text{in},-}(t)\sigma_{j'}^{\text{in},+}(t')\right\rangle = \left\langle \sigma_j^- \sigma_j^+\right\rangle \delta_{jj'}\delta(t-t') \tag{7.71c}$$

We incorporate the polaron transform via the displacement operator

$$D = \otimes \prod_{n=1}^{N} D_n,\ D_n = e^{-\lambda\left(b_n - b_n^\dagger\right)} \tag{7.72}$$

into the QLEs. This allows to define the dressed lowering operator $\tilde{\sigma}_n^- = \sigma_n^- D_n^\dagger$, which gives

$$\dot{D}_n = \lambda D_n\left[i\omega_v\left(b_n + b_n^\dagger - 2\lambda\sigma_n^+\sigma_n^-\right) - \Gamma\left(b_n^\dagger - b_n\right) + \sqrt{2\Gamma}\left(b_n^{\text{in},\dagger}(t) - b_n^{\text{in}}(t)\right) + \lambda\left(i\omega_v + \Gamma\right)\right] \tag{7.73}$$

The dressed lowering operator is to involve the vibronic coupling to all orders, as it is normally moderate or strong. Using Eqs. (7.69a) and (7.73) and taking the time derivative of $\tilde{\sigma}_n$, some algebra gives rise to

$$\begin{aligned}\dot{\tilde{\sigma}}_n^- = & -\left(i\tilde{\delta} + \gamma + \lambda^2\Gamma\right)\tilde{\sigma}_n^- - ig\sigma_n^z a D_n^\dagger + i\sum_{j=1}^{3} \sigma_n^z D_n^\dagger \Omega_j\left(t - T_j\right) e^{-i(v_j - v_3)t} e^{iv_j T_j} \\ & + \lambda\Gamma\left(b_n^\dagger - b_n\right)\tilde{\sigma}_n^- - \lambda\sqrt{2\Gamma}\left[b_n^{\text{in},\dagger}(t) - b_n^{\text{in}}(t)\right]\tilde{\sigma}_n^- + \sqrt{2\gamma}\tilde{\sigma}_n^{-,\text{in}}(t)\end{aligned} \tag{7.74}$$

and $\tilde{\sigma}_n^{-,\text{in}}(t) = \sigma_n^{-,\text{in}} D_n^\dagger$. It turns out that the real-time dynamics of the population $\tilde{\sigma}_n^+\tilde{\sigma}_n^-$ eliminates the contribution from the term $\lambda\sqrt{2\Gamma}\left[b_n^{\text{in},\dagger}(t) - b_n^{\text{in}}(t)\right]$ such that

$$\dot{n}_l \to -\lambda\sqrt{2\Gamma}\left[b_l^{\text{in},\dagger} - b_l^{\text{in}} + \text{h.c.}\right] n_l = 0 \tag{7.75}$$

The term $\lambda\sqrt{2\Gamma}\left[b_n^{\text{in},\dagger}(t) - b_n^{\text{in}}(t)\right]$ can thus be properly dropped, for a pleasant approximation. One observes that the nuclear velocity $\propto i\left(b_n^\dagger - b_n\right)$ is much slower than the electron and hole motions. We may further drop the term correspondingly and neglect the back influence from exciton to nuclear motion along the line of the stochastic Liouville equation [26, 78], where the latter recasts Eq. (7.69b) into $\dot{b}_n \approx -\left(i\omega_v + \Gamma\right)b_n + \sqrt{2\Gamma}b_n^{\text{in}}(t)$. We then solve for the vibration dynamics

$$b_n(t) \approx e^{-(i\omega_v+\Gamma)t}b_n(0) + \sqrt{2\Gamma}\int_0^t e^{-(i\omega_v+\Gamma)(t-t')}b_n^{\text{in}}(t')\,dt' \tag{7.76}$$

which posses

$$\left[b_n(t), b_m^\dagger(t')\right] = \begin{cases} e^{-(i\omega_v+\Gamma)(t-t')}\delta_{nm}, & \text{for } t \geqslant t' \\ e^{(i\omega_v-\Gamma)(t-t')}\delta_{nm}, & \text{for } t < t' \end{cases} \tag{7.77}$$

Equation (7.77) holds the canonical commutation relation when $t = t'$. From Eq. (7.74) using the polaron-dressed operators, we can recast the QLEs Eq. (7.69a) into ($\delta := \delta - \lambda^2\omega_v, \gamma := \gamma + \lambda^2\Gamma$)

$$\begin{aligned}\dot{\sigma}_l^- = & -(i\delta + \gamma)\tilde{\sigma}_l^- - ig\sigma_l^z a D_l^\dagger - i\sum_{j=1}^{3}\left(2n_l - 1\right)D_l^\dagger\Omega_j\left(t - T_j\right)e^{-i(v_j - v_3)t}e^{iv_jT_j} \\ & + \sqrt{2\gamma}\tilde{\sigma}_l^{-,\text{in}}(t),\end{aligned} \tag{7.78a}$$

$$\dot{a} = -(i\delta_c + \gamma_c)a - ig\sum_{n=1}^{N} \tilde{\sigma}_n^- D_n + \sqrt{2\gamma_c}a^{-,\text{in}}(t) \tag{7.78b}$$

which can be reformed into a compact form

$$\dot{\text{V}} = -\text{MV} + \text{V}^{\text{in}}(t) - i\sum_{j=1}^{3} \Omega_j\left(t - T_j\right) e^{i\nu_j T_j} e^{-i(\nu_j - \nu_3)t}\,\text{W} \tag{7.79}$$

where $\text{V} = \left[\tilde{\sigma}_1^-, \tilde{\sigma}_2^-, \ldots, \tilde{\sigma}_N^-, a\right]^{\text{T}}, \text{V}^{\text{in}}(t) = \left[\sqrt{2\gamma}\tilde{\sigma}_1^{-,\text{in}}(t), \sqrt{2\gamma}\tilde{\sigma}_2^{-,\text{in}}(t), \ldots, \sqrt{2\gamma}\tilde{\sigma}_N^{-,\text{in}}(t), \sqrt{2\gamma_c}a^{\text{in}}(t)\right]^{\text{T}}$ and other matrices are

$$\text{M} = \begin{pmatrix} i\delta+\gamma & 0 & \cdots & 0 & ig\sigma_1^z D_1^\dagger \\ 0 & i\delta+\gamma & \cdots & 0 & ig\sigma_2^z D_2^\dagger \\ \vdots & \vdots & & \vdots & \vdots \\ 0 & 0 & \cdots\, i\delta+\gamma & & ig\sigma_N^z D_N^\dagger \\ igD_1 & igD_2 & \cdots & igD_N & i\delta_c+\gamma_c \end{pmatrix}, \quad \text{W} = \begin{pmatrix} (2n_1-1)D_1^\dagger \\ (2n_2-1)D_2^\dagger \\ \vdots \\ (2n_N-1)D_N^\dagger \\ 0 \end{pmatrix} \tag{7.80}$$

Eq. (7.79) allows for the formal solution through the interaction picture, i.e.,

$$\begin{aligned} \text{V}(t) - \text{V}_0 = &-i\sum_{j=1}^{3}\int_0^t d\tau\Omega_j\left(\tau - T_j\right) e^{i\nu_j T_j} G(t-\tau)\text{W}(\tau)e^{-i(\nu_j-\nu_3)\tau} \\ &+ \int_0^t d\tau G(t-\tau)\text{V}^{\text{in}}(\tau) \end{aligned} \tag{7.81}$$

with the free propagator $G(t) = \mathscr{T}\exp\left[-\int_0^t \text{M}(\tau)d\tau\right]$. Notice that Eq. (7.81) has to be solved for final solution self-consistently, due to the involvement of the population terms n_j in $\text{W}(\tau)$. In what follows, we will work in the regime of weak coupling to the laser pulses, along with the parameter regimes feasible in recent experiments [16, 92].

Equation (7.81) represents the dynamics of molecular polaritons. The perturbation theory of the molecule–field interaction $V(t)$ will be used, and we will calculate the two-dimensional optical signals from of photon emission off the cavity axis, as depicted in Figure 7.3a. These signals are governed by multipoint Green's functions of dipole operators, which are determined by the exact solution Eq. (7.81) to the QLEs [64].

7.5.2 Correlation Functions of Vibrations

Before calculating the optical signal via Eq. (7.81), the vibration correlation functions as an essential component have to be studied. Since the third-order signal is of the most interest, the calculations will proceed with a foci of the two-point and four-point correlation functions [64]. The following identities may be useful

$$\begin{aligned} e^{\lambda\left[b_n^\dagger(t) - b_n(t)\right]} = e^{-\lambda^2/2}e^{\lambda b_n^\dagger(t)}e^{-\lambda b_n(t)} = e^{\lambda^2/2}e^{-\lambda b_n(t)}e^{\lambda b_n^\dagger(t)} \\ e^{-\lambda b_n(t)}|0 = |0 \end{aligned} \tag{7.82}$$

For the two-point correlator, Eqs. (7.77) and (7.82) enable

$$
\begin{aligned}
e^{\lambda^2}\left\langle 0|D_i^\dagger(t)D_j(t')\,|0\right\rangle &= \left\langle 0|e^{\lambda b_i(t)}e^{\lambda b_j^\dagger(t')}|0\right\rangle \\
&= \left\langle 0|e^{\lambda b_j^\dagger(t')}e^{\lambda b_i(t)}e^{\lambda^2\left[b_i(t),b_j^\dagger(t')\right]}|0\right\rangle \\
&= \theta\,(t-t')\exp\left[\delta_{ij}\lambda^2 e^{-(i\omega_v+\Gamma)(t-t')}\right] \\
&\quad +\theta\,(t'-t)\exp\left[\delta_{ij}\lambda^2 e^{(i\omega_v-\Gamma)(t'-t)}\right]
\end{aligned}
\tag{7.83}
$$

For the four-point correlator, Eqs. (7.77) and (7.82) enable

$$
\begin{aligned}
&e^{2\lambda^2}\left\langle 0|D_{j'}(t)D_j^\dagger\,(t')\,D_l^\dagger\,(t'')\,D_i\,(t''')\,|0\right\rangle \\
&= \left\langle 0|e^{-\lambda b_{j'}(t)}e^{-\lambda b_j^\dagger(t')}e^{\lambda b_j(t')}e^{-\lambda b_l^\dagger(t'')}e^{\lambda b_l(t'')}e^{\lambda b_i^\dagger(t''')}|0\right\rangle \\
&= e^{\lambda^2\left[b_{j'}(t),b_j^\dagger(t')\right]}e^{\lambda^2\left[b_l(t''),b_j^\dagger(t''')\right]}e^{-\lambda^2\left[b_j(t'),b_l^\dagger(t'')\right]}e^{\lambda^2\left[b_{j'}(t),b_l^\dagger(t'')\right]}e^{\lambda^2\left[b_j(t'),b_i^\dagger(t''')\right]}e^{-\lambda^2\left[b_{j'}(t'),b_i^\dagger(t''')\right]}
\end{aligned}
\tag{7.84}
$$

which is general. For the segment $t \leqslant t' \leqslant t'' \leqslant t'''$, it yields

$$
\begin{aligned}
&e^{2\lambda^2}\left\langle 0|D_{j'}(t)D_j^\dagger\,(t')\,D_l^\dagger\,(t'')\,D_i\,(t''')\,|0\right\rangle\theta\,(t'''-t'')\,\theta\,(t''-t')\,\theta\,(t'-t) \\
&= \exp\left[\lambda^2\delta_{j'j}e^{(i\omega_v-\Gamma)(t'-t)}\right]\exp\left[\lambda^2\delta_{il}e^{(i\omega_v-\Gamma)(t'''-t'')}\right]\exp\left[-\lambda^2\delta_{jl}e^{(i\omega_v-\Gamma)(t''-t')}\right] \\
&\quad\times\exp\left[\lambda^2\delta_{j'l}e^{(i\omega_v-\Gamma)(t''-t)}\right]\exp\left[\lambda^2\delta_{ij}e^{(i\omega_v-\Gamma)(t'''-t')}\right]\exp\left[-\lambda^2\delta_{ij'}e^{(i\omega_v-\Gamma)(t'''-t)}\right]
\end{aligned}
\tag{7.85}
$$

For the other 23 segments, the calculations proceed as usual by using Eq. (7.77). The results will be presented elsewhere.

7.5.3 Absorption Spectrum

We first present the absorption spectrum of molecular polaritons, which has led to intensive measurements in recent experiments. Assuming $\omega_v \gg T$ that applies for organic molecules at room temperature, the vibrational correlation functions can be evaluated with a vacuum state. Using the linear response theory and noting the 1 in W responsible for the absorption, the far-field polarization reads

$$
P_A(t) = -i\sum_{i=1}^{N}\sum_{l=1}^{N}\int_{-\infty}^{t} d\tau E(\tau - T)e^{-iv(\tau-T)}\left\langle 0|G_{il}(t-\tau)D_i(t)D_l^\dagger(\tau)|0\right\rangle \tag{7.86}
$$

The Fourier component of $P_A(t)$ in Eq. (7.86) yields the first-order response function $\chi^{(1)}(\omega)$, which gives the absorption spectra, i.e., $S_A(\omega) = \Im\left[\chi^{(1)}(\omega)\right]$. Some algebra leads to

$$
S_A(\omega) = \sum_{i,l=1}^{N}\sum_{m=0}^{\infty} e^{-\lambda^2}\frac{\lambda^{2m}}{m!}\delta_{il}^m\Re\left[\mathscr{G}_{il}\left(\omega - \xi_m^*\right)\right],\ \xi_m \equiv m\left(\omega_v + i\Gamma\right) \tag{7.87}
$$

where $\mathscr{G}(\Omega) = \int_0^\infty G(t)e^{i\Omega t}\,dt$ is the Fourier component of the propagator $G(t)$ and $S_m^\lambda = e^{-\lambda^2}\lambda^{2m}/m\,!$ is the Franck–Condon factor [64, 93]. $S_A(\omega)$ resolves the lower polariton (LP) and upper polariton (UP) at $\omega_{\mathrm{LP/UP}}$, the localized dark states (LDSs) at $\omega_D + m\omega_v$ decoupled from the cavity. The dark states at ω_D are not visible due to the destructive interference. To see these closely, we assume $\gamma = \gamma_c$ and $\delta = \delta_c = 0$. The peak intensities are found

$$\frac{S_A(\omega_D + m\omega_v)}{S_A(\omega_{\mathrm{LP/UP}})} = \frac{\lambda^{2m}}{m!}\left(1 - \frac{1}{N}\right)\frac{2\gamma}{m\Gamma} \approx \frac{\lambda^{2m}}{m!}\frac{2\gamma}{m\Gamma} \text{ when } N \gg 1 \tag{7.88a}$$

$$\frac{S_A(\omega_{\mathrm{LP/UP}} + m\omega_v)}{S_A(\omega_{\mathrm{LP/UP}})} = \frac{\lambda^{2m}}{m!N}\frac{\gamma}{\gamma + m\Gamma} \approx 0 \text{ when } N \gg 1 \tag{7.88b}$$

The LDSs may be of comparable intensities with polariton modes. Such spectral-line properties will be shown to be generally true in the time-resolved spectroscopic signals.

7.5.4 Time-resolved Emission of Polaritons

There are a variety of schemes for the time-resolved spectroscopy. For a general understanding, we present the time-resolved emission spectrum of cavity polaritons which builds up the first floor for the multidimensional resonant spectroscopy. Subject to a probe pulse, Eq. (7.81) solves for the far-field dipolar radiation governed by the macroscopic polarization of molecules $P(t) \propto \sum_{i=1}^{N} \langle \sigma_{i;1}^-(t) \rangle$. The Fourier component of $P(t)$ gives rise to the emission signal

$$\begin{aligned} P(\omega, T) \propto\ & 2i \sum_{i=1}^{N} \sum_{l=1}^{N} \iint_{-\infty}^{\infty} dt d\tau \theta(t - \tau) e^{i\omega t} E(\tau - T) \\ & \times e^{-i\upsilon(\tau - T)} \left\langle \mathrm{G}_{il}(t - \tau) n_l(\tau) D_l^\dagger(\tau) D_i(t) \right\rangle \end{aligned} \tag{7.89}$$

We note that, from the dressed populations $n_l(\tau)D_l^\dagger(\tau)$, the cavity polaritons of molecules undergo a dynamics against the local fluctuations arising from the polaron effect. The polaron-induced localization as a result of the LDSs will compete with the polariton cooperativity [26, 59, 64]. These can be visualized from the emission signal, which is a real-time monitoring of polariton dynamics through pulse shaping and grating. More advanced information will be elaborated by the multidimensional projections of the emission signal, which will therefore result in the multidimensional spectroscopy of the molecules in cavities.

7.5.5 Two-dimensional Polariton Spectroscopy

To have multidimensional projections of the signals, sequential laser pulses have to interact with the molecular polaritons [64]. We are seeking the solution for Eq. (7.81), in series of the molecule–field interactions, i.e., $V = V_0 + V_1 + V_2 + V_3 + \cdots$. The zeroth-order term V_0 is a result of free propagation in the absence of laser pulses, while the First-order term denotes the absorption under the first pulse, creating the coherence between the ground and electronically excited states. Namely, it reads

$$V_0(t) = G(t)V(0) + i \int_0^t G(t - t') V^{\mathrm{in}}(t') \, dt' \tag{7.90a}$$

$$V_1(t) = -i \sum_{s=1}^{3} \int_0^t dt' \Omega_s(t' - T_s) e^{i\nu_s T_s} G(t - t') \overline{\mathrm{W}}(t') e^{-i(\nu_s - \nu_3)t'} \tag{7.90b}$$

and $\overline{\mathrm{W}}(t') = \left[D_1^\dagger(t'), D_2^\dagger(t'), \cdots, D_N^\dagger(t'), 0\right]^{\mathrm{T}}$, which yield the electronic polarization to the First order

$$\tilde{\sigma}_{l;1}^{-}(t) = -i\sum_{s=1}^{3}\sum_{j=1}^{N}\int_{0}^{t} dt'\Omega_s(t'-T_s)e^{i\nu_s T_s}G_{lj}(t-t')D_j^{\dagger}(t')e^{-i(\nu_s-\nu_3)t'} \tag{7.91}$$

Knowing that the excitations are created by the first two pulses, one has $n_{l;2}(t) = \sigma_{l;1}^{+}(t)\sigma_{l;1}^{-}(t)$ from Eq. (7.91), yielding

$$\begin{aligned} n_{l;2}(t) = \sum_{j=1}^{N}\sum_{j'=1}^{N}\iint_{0}^{t} & dt''dt'\Omega_1^{*}(t'-T_1)\Omega_2(t''-T_2)e^{-i(\nu_1 T_1-\nu_2 T_2)}e^{-i(\nu_3-\nu_1)t'}e^{i(\nu_3-\nu_2)t''} \\ & \times G_{lj'}^{*}(t-t')G_{lj}(t-t'')D_{j'}(t')D_j^{\dagger}(t'') + [\text{Other combinations}] \end{aligned} \tag{7.92}$$

It is obvious that the second-order correction $V_2(t)$ must involve the noise operators in the form of $\tilde{\sigma}_i^{\pm,\text{in}}(t)$ and $\tilde{\sigma}_i^{+,\text{in}}(t)\tilde{\sigma}_j^{-,\text{in}}(t')\tilde{\sigma}_k^{-,\text{in}}(t'')+$ their Hermitian conjugates. To the third-order correction, the noise-operator-dependent terms remain the combinations $\tilde{\sigma}_i^{+,\text{in}}(t)\tilde{\sigma}_j^{-,\text{in}}(t')\tilde{\sigma}_k^{+,\text{in}}(t'')\tilde{\sigma}_l^{-,\text{in}}(t''')$ as well as $\tilde{\sigma}_i^{+,\text{in}}(t)\tilde{\sigma}_j^{+,\text{in}}(t')\tilde{\sigma}_k^{-,\text{in}}(t'')\tilde{\sigma}_l^{-,\text{in}}(t''')$. These will not contribute to the third-order signals eventually, by noting

$$\left\langle\tilde{\sigma}_i^{+,\text{in}}(t)\tilde{\sigma}_j^{-,\text{in}}(t')\tilde{\sigma}_k^{+,\text{in}}(t'')\tilde{\sigma}_l^{-,\text{in}}(t''')\right\rangle = 0 \tag{7.93a}$$

$$\left\langle\tilde{\sigma}_i^{+,\text{in}}(t)\tilde{\sigma}_j^{+,\text{in}}(t')\tilde{\sigma}_k^{-,\text{in}}(t'')\tilde{\sigma}_l^{-,\text{in}}(t''')\right\rangle = 0 \tag{7.93b}$$

given $\left\langle\tilde{\sigma}_i^{+,\text{in}}(t)\tilde{\sigma}_j^{-,\text{in}}(t')\right\rangle = 0$. This is reasonable because creating the excitons via environmental fluctuations is of extremely low chance, at room temperature. Keeping the significant components for the signals, we have $V_2(t) = 0$ therein, by noting the fact $V(0) = 0$ prior to the pulse action. We therefore proceed to the third-order correction via Eq. (7.79), obtaining

$$V_3(t) = 2i\int_{0}^{t} d\tau\Omega_3(\tau-T_3)e^{i\nu_3 T_3}G(t-\tau)\mathrm{W}_y(\tau) + [\text{Other combinations}] \tag{7.94}$$

where $\mathrm{W}_y(\tau) = \left[n_{1;2}(\tau)D_1^{\dagger}(\tau), n_{2;2}(\tau)D_2^{\dagger}(\tau), \cdots, n_{N;2}(\tau)D_N^{\dagger}(\tau), 0\right]^{\mathrm{T}}$. As the three pulses are time-ordered, the transition pathways are selective. This enables us to simplify the third-order polarization, resulting from the term cancellation. Along with the phase cycling $\nu = -\nu_1 + \nu_2 + \nu_3$, the far-field dipolar radiation reads from Eq. (7.94), i.e.,

$$\sigma_{i;3}^{-}(t) = 2i\sum_{l=1}^{N}\int_{0}^{t} d\tau\Omega_3(\tau-T_3)e^{-i\nu_3(t-T_3)}G_{il}(t-\tau)n_{l;2}(\tau)D_l^{\dagger}(\tau)D_i(t) \tag{7.95}$$

So far, we have derived in a mathematical rigor the third-order polarization of molecules, keeping all the essential terms. Thus, Eq. (7.95) must contain the transient absorption component, provided that the eigenvalues of the matrix M normally involve the frequencies corresponding to the higher transition ladders, i.e., $e \to f$. This is an important aspect for a self-consistent formalism of the spectroscopic signals, as the ultrashort pulses having broad bandwidth would induce considerably the transitions to the doubly excited manifold. For instance, a large number of identical molecules that may enable the approximation $g\sigma_l^z \approx g$ when including up to the doubly excited manifold (applicable for few-photon cavities), the cascading transitions $g \to e, e \to f$ will have the same frequencies. This can be seen from the eigenvalues of the M, given no direct many-particle couplings between molecules.

Inserting Eq. (7.91) into Eq. (7.95) one obtains the macroscopic far-field polarization $P(t) \propto \sum_{i=1}^{N} \langle \sigma^{-}_{i;3}(t) \rangle$ which yields

$$P(\omega) \propto 2i \sum_{i,l=1}^{N} \sum_{j,j'=1}^{N} \iiint \int_{0}^{\infty} dt d\tau dt'' dt' e^{i\omega t} \mathscr{E}_3(\tau - T_3)\mathscr{E}_2(t'' - T_2)\mathscr{E}_1^*(t' - T_1)$$
$$\times \left\langle 0 | G_{il}(t-\tau) G^{\dagger}_{lj'}(\tau - t') G_{lj}(\tau - t'') D_{j'}(t') D^{\dagger}_{j}(t'') D^{\dagger}_{l}(\tau) D_i(t) | 0 \right\rangle \tag{7.96}$$

where $\mathscr{E}_s(t) = e^{-i\nu_s t}\Omega_s(t); s = 1,2,3$ and the four-point correlation function of vibrations $\left\langle 0 | D_{j'}(t') D^{\dagger}_{j}(t'') D^{\dagger}_{l}(\tau) D_i(t) | 0 \right\rangle$ has to be evaluated explicitly. The 2D signal is usually detected via a reference laser beam (local oscillator field) interfering with the emission. This leads to the heterodyne-detected signal that can be obtained by inserting Eq. (7.96) into Eq. (7.30).

In general, calculating the signal with Eq. (7.96) is hard due to the integrals over pulse shapes. The procedures can be simplified by invoking the impulsive approximation such that the pulse is shorter than the dephasing and solvent timescales [70, 71]. This works for many systems of condensed-phase molecules whose dephasing time is typically $\sim$ 100 fs longer than the laser pulses. The solvent relaxation usually takes place within more than a few picoseconds, noting from the reorganization energy $\lesssim$ 30 cm^{-1} [72, 73]. In the optical regime, the ultrafast molecular spectroscopy normally acquires the laser pulses with a duration of $\sim 6 - 30$ fs. We further consider the few-photon cavity that draws much attention in recent experiments and notice the vibronic coupling predominately accounted for by the polarons. The most significant terms may remain, allowing the approximation $g\sigma^z_l D^{\dagger}_l \approx g D^{\dagger}_l \approx g$ in the M matrix in Eq. (7.80). The higher-order corrections will be presented elsewhere. We obtain an analytical solution to the 2D polariton signal (2DPS), i.e.,

$$S(\Omega_3, T, \Omega_1) \propto ie^{i\phi} \sum_{i,l=1}^{N} \sum_{j,j'=1}^{N} \sum_{p=1}^{N+1} \sum_{\{m\}=0}^{\infty} S^{\lambda}_{\{m\}} \beta^{iljj'}_{\{m\}} (-1)^{m_3+m_6} \mathscr{G}_{il}\left(\Omega_3 + \xi_{m_2+m_5+m_6}\right)$$
$$\times \mathscr{G}^*_{pj'}\left(-\Omega_1 - \xi_{m_1+m_4+m_6}\right) G^*_{lp}(T) G_{lj}(T) e^{i\xi_{m_3+m_4+m_5+m_6}T} \tag{7.97}$$

subsequently from Eq. (7.96), where $S^{\lambda}_{\{m\}} = \prod_{s=1}^{6} S^{\lambda}_{m}$ and $\beta^{iljj'}_{\{m\}} = \delta^{m_1}_{j'j}\delta^{m_2}_{il}\delta^{m_3}_{jl}\delta^{m_4}_{j'l}\delta^{m_5}_{ij}\delta^{m_6}_{ij'}$. ϕ encodes a global phase from the four laser pulses. More details of the derivation of the signals via QLEs are given in Ref. [64].

Figure 7.4 shows the snapshots of the 2DPS signal at various time delays T, to simulate the molecular polariton dynamics [Eq. (7.96)]. The two rows illustrate the signal for $N = 1$ and 10 cyanine dyes in cavity. The LDSs are resolved for $N = 10$ but not for $N = 1$, evident by the absence of the peaks at $\Omega_{1,3} = \omega_D \pm n\omega_v (n = 0, 1, 2, \ldots)$ in the lower row [i.e., Figures 7.4(a-d), Figures 7.4(e-h)]. This is what we expect from the level structure of the Hamiltonian in Eq. (7.65).

When the delay T varies, Figures 7.4(a-d) show the real-time dynamics of the polaritons in a crosstalk with the LDSs, including the population and coherence components. One notes the absorption resonances at Ω_1 = 14300, 17300, and 17900 cm^{-1} in an agreement with Figure 7.3c. The time variation of the cross peaks with $\Omega_1 - \Omega_3 \neq$ integer $\times\omega_v$ for different T s monitors the dynamics of quantum coherences; this reveals oscillations, e.g., at the frequency of 3000 cm^{-1} for the peak at $(\Omega_1 = 17900\ \text{cm}^{-1}, \Omega_3 = 14900\ \text{cm}^{-1})$ which evidences the superposition $\alpha |\omega_D - \omega_v\rangle + \beta |\omega_{UP}\rangle$. With longer time delays, i.e., $T > 300$fs (upper row of Figure 7.4), the cross peaks imposing $\Omega_1 - \Omega_3 \neq$ integer $\times\omega_v$ predominately resolve the energy transfer between UP, LP, and LDSs. For instance, the energy transfer pathways, i.e., UP $\rightarrow$ LDS$\omega_D - \omega_v$ [at $\Omega_1 = 17900\ \text{cm}^{-1}, \Omega_3 = 14900\ \text{cm}^{-1}$], UP $\rightarrow$ LP $\left[\text{at } \Omega_1 = 17900\ \text{cm}^{-1}, \Omega_3 = 14300\ \text{cm}^{-1}\right]$ and LDS $\omega_D + \omega_v \rightarrow$ LP $\left[\text{at } \Omega_1 = 17300\ \text{cm}^{-1}, \Omega_3 = 14300\ \text{cm}^{-1}\right]$, can be visualized. It should be

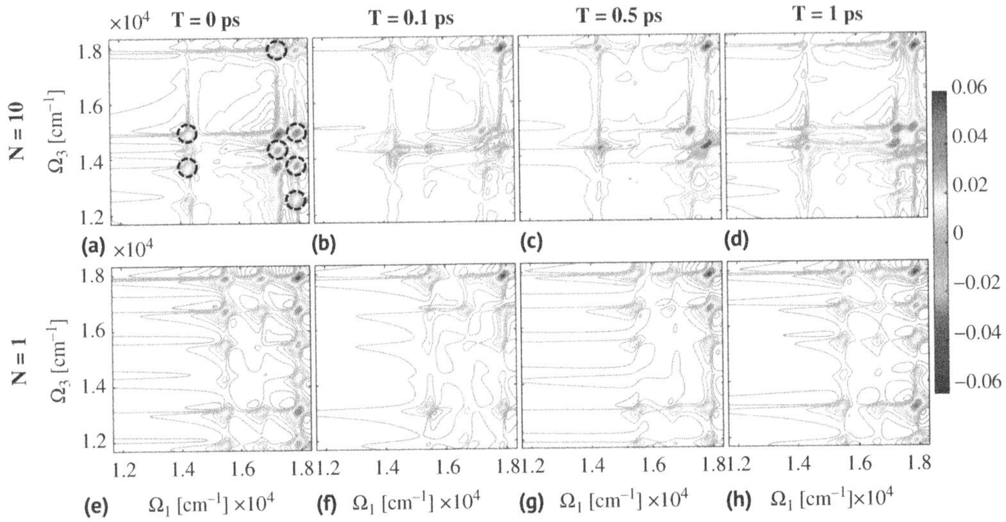

Figure 7.4 2D time- and frequency-resolved signal for molecular polaritons where (up) $N = 10$ organic molecules and (down) $N = 1$ organic molecule. T is varied, denoting the delay between the second and the probe laser pulses. Horizontal and vertical axis are for absorption and emission frequencies, respectively. $\lambda = 1$ and $g\sqrt{N}/\omega_v = 1.5$. Other parameters are the same as Figure 7.3. *Source*: [64]/American Physical Society.

further noted that the pathways UP $\rightarrow$ LDS$\omega_D - \omega_v$ and LDS $\omega_D + \omega_v$ $\rightarrow$ LP dominate, during a longer timescale $T >$ 500 fs as depicted in Figures 7.4(c,d). In addition to these cross peaks below the diagonal, a few cross peaks above the diagonal can be observed from Figure 7.4 (upper row), revealing the up-hill energy transfer pathways, i.e., LP $\rightarrow$ LDSs, LDSs $\rightarrow$ UP and LP $\rightarrow$ UP which are much weaker. These multiple pathways and timescales have been verified in recent 2D spectroscopic experiments conducted for vibrational and exciton polaritons of organic molecules [7, 16, 28, 44, 92].

7.5.6 Connection to Polariton Pump-probe Spectra

Lastly, it should be noted the intrinsic connection of the 2DPS to the pump-probe spectra of molecules in cavities [64]. This can be readily obtained by letting $T_1 = T_2, \Omega_1 = \Omega_2$ and $\nu_1 = \nu_2$ in Eq. (7.96), so as to account for the nonrephasing term in the calculation of $n_{l;2}(t)$ from Eq. (7.91). The signal reads $S_{\rm pp}(\omega, T) = \Im\left[E_3^*(\omega)P(\omega)\right]$ with $P(\omega) = P_{\rm r}(\omega) + P_{\rm nr}(\omega)$ where the rephasing and nonrephasing components are

$$P_{\rm r}(\omega) = 2i \sum_{i,l=1}^{N} \sum_{j,j'=1}^{N} \iiint\int_0^\infty dtd\tau dt''dt' e^{i\omega t}\mathscr{E}_3(\tau - T_3)\,\mathscr{E}_1(t'' - T_1)\,\mathscr{E}_1^*(t' - T_1)$$
$$\times \left\langle 0|G_{il}(t-\tau)G^\dagger_{lj'}(\tau - t')\,G_{lj}(\tau - t'')\,D_{j'}(t')\,D_j^\dagger(t'')\,D_l^\dagger(\tau)D_i(t)|0\right\rangle \tag{7.98}$$

$$P_{\rm nr}(\omega) = 2i \sum_{i,l=1}^{N} \sum_{j,j'=1}^{N} \iiint\int_0^\infty dtd\tau dt''dt' e^{i\omega t}\mathscr{E}_3(\tau - T_3)\,\mathscr{E}_1^*(t'' - T_1)\,\mathscr{E}_1(t' - T_1)$$
$$\times \left\langle 0|G_{il}(t-\tau)G_{lj'}(\tau - t')\,G^\dagger_{lj}(\tau - t'')\,D_j(t'')\,D_{j'}^\dagger(t')\,D_l^\dagger(\tau)D_i(t)|0\right\rangle \tag{7.99}$$

A compact form can be obtained by the impulsive approximation, i.e.,

$$S_{\rm pp}(\omega,T)=\sum_{i,l=1}^{N}\sum_{j,j'=1}^{N}\sum_{\{m\}=0}^{\infty}S^{\lambda}_{\{m\}}\delta_{il}^{m_1}\left(\delta_{j'l}-\delta_{jl}\right)^{m_2}\left(\delta_{ij}-\delta_{ij'}\right)^{m_3}$$
$$\times\,\Re\left[\mathcal{G}_{il}\left(\omega+\xi_{m_1+m_3}\right)G^{*}_{lj'}(T)G_{lj}(T)e^{i\xi_{m_2+m_3}T}\right] \tag{7.100}$$

The pump-probe signal may reveal the molecular polariton dynamics, through simulating Eq. (7.100). This is straightforward but not quite necessary, as the pump-probe spectra has simpler form from which some generic features can be observed. We will work for the sake of generality, to manifest the time-resolved properties of the signal. The ultrashort pulses having broadband smear out the mode selectivity in the absorption process, whereas the time grating makes the emission spectrally resolved [See from $\mathcal{G}_{il}\left(\omega+\xi_{m_1+m_3}\right)$]. Similar as the 2DPS, the LP and UP modes separated by $2g\sqrt{N}$ as well as the LDSs at $\omega=\omega_D-\omega_v,\omega_D-2\omega_v,\ldots$ can be resolved in $S_{\rm pp}(\omega,T)$. At $T=0$, the peak intensity distribution over the modes follows the absorption spectrum in Eq. (7.87), apart from the anti-Stokes shifts of (integer) $\times\,\omega_v$. This has been well understood by recalling the absorption and fluorescence spectrums in molecular spectroscopy [74]. As varying the time delay, the spectral-line intensity $S_{\rm pp}\left(\omega_{\rm UP/LP},T\right)$ shows a phase difference from the $S_{\rm pp}\left(\omega_D-m\omega_v,T\right)$, associated with different damping rates that are responsible for the incoherent channels of relaxations. Because of the density of states $\sim N$ for LDSs, higher than the one for polaritons, the LDSs show the localized nature. The polariton states featuring the extended waves are therefore trading off with the LDS-induced localization, as seen from the multiple channels and timescales of the dynamics given by Eq. (7.100).

In summary, we presented microscopic theory of multidimensional spectroscopy for molecular polaritons, using the quantum Langevin equation. The exciton-photon-phonon interaction that is tripartite and complicated can be adopted naturally, yielding the polaron–polariton coupling that used to be challenge in previous works. The fast evolving dynamics of polaritons and dark states subject to various timescales can be readily visualized in the 2DPS. The polaron effect thus appears as the peaks at the energy decoupled from cavity photons, which resolves the LDSs against the polariton dynamics. The quantized Brownian oscillator model allows us to incorporate different modes with a multitude of timescales. These fall into a different category from the cavity QED for atoms, where no crosstalk between superradiant and subradiant states can be observed [77, 94–100]. Our model used a simplified description for aggregated molecules, i.e., a group of two-level systems coupled to intramolecular vibrations undergoing the Brownian oscillation. To account for more complicated dynamics like the photochemical kinetics and geometric phases, extensive efforts need to be devoted to the anharmonicity and multilevel systems. This may, however, need heavier loads of numerical methods.

Acknowledgments

Z.Z. gratefully acknowledges the support of the Early Career Scheme from Hong Kong Research Grants Council (No. 21302721), the National Science Foundation of China (No. 12104380), and the National Science Foundation of China/RGC Collaborative Research Scheme (No. CRS-CUHK401/22).

Z.Z. would also like to thank Girish S. Agarwal (Texas A&M University), Marlan O. Scully (Texas A&M University), Shaul Mukamel (University of California Irvine), Markus Kowalewski (Stockholm University), Dangyuan Lei (City University of Hong Kong), Frank Schlawin (University of

Hamburg & Max Planck Institute for the Structure and Dynamics of Matter) and Chunfeng Zhang (Nanjing University) for their useful discussions.

References

1 Thomas, A., George, J., Shalabney, A., Dryzhakov, M., Varma, S. J., Moran, J. et al. *Angew. Chem., Int. Ed.* **2016**, *55*, 11462 (2016).
2 Hutchison, J., Schwartz, T., Genet, C., Devaux, E., Ebbesen, T. W. *Angew. Chem.* **2012**, *124*, 1624.
3 Herrera, F., Spano, F. C. *Phys. Rev. Lett.* **2016**, *116*, 238301.
4 Li, X., Mandal, A., Huo, P. *Nat. Commun.* **2021**, *12*, 1315.
5 Spano, F. C. *J. Chem. Phys.* **2015**, *142*, 184707.
6 Coles, D. M., Somaschi, N., Michetti, P., Clark, C., Lagoudakis, P. G., Savvidis, P. G. et al. *Nat. Mater.* **2014**, *13*, 712.
7 Dunkelberger, A. D., Spann, B. T., Fears, K. P., Simpkins, B. S., Owrutsky, J. C. *Nat. Commun.* **2016**, *7*, 13504.
8 Martinez-Martinez, L., Ribeiro, R., Campos-Gonzalez- Angulo, J., Yuen-Zhou, J. *ACS Photonics* **2018**, *5*, 167.
9 Galego, J., Garcia-Vidal, F. J., Feist, J. *Nat. Commun.* **2016**, *7*, 13841.
10 Byrnes, T., Kim, N. Y., Yamamoto, Y. *Nat. Phys.* **2014**, *10*, 803–813.
11 Kavokin, A., Liew, T. C. H., Schneider, C., Lagoudakis, P. G., Klembt, S., Hoefling, S. *Nat. Rev. Phys.* **2022**, *4*, 435–451.
12 Kowalewski, M., Bennett, K., Mukamel, S. *J. Phys. Chem. Lett.* **2016**, *7*, 2050.
13 Ribeiro, R. F., Dunkelberger, A. D., Xiang, B., Xiong, W., Simpkins, B. S., Owrutsky, J. C. et al. *J. Phys. Chem. Lett.* **2018**, *9*, 3766.
14 Wang, X., Zhang, Z. D., Wang, J. *Phys. Rev. A* **2021**, *103*, 013516.
15 Yang, Z., Bhakta, H. H., Xiong, W. *Proc. Natl. Acad. Sci. USA* **2022**, *120*, e2206062120.
16 Fassioli, F., Park, K. H., Bard, S. E., Scholes, G. D. *J. Phys. Chem. Lett.* **2021**, *12*, 11444.
17 Liu, B., Menon, V. M., Sfeir, M. Y. *ACS Photonics* **2020**, *7*, 2292.
18 Plumhof, J. D., Stöferle, T., Mai, L., Scherf, U., Mahrt, R. F. *Nat. Mater.* **2014**, *13*, 247.
19 Thomas, A., Lethuillier-Karl, L., Nagarajan, K., Vergauwe, R. M. A., George, J., Chervy, T. et al. *Science* **2019**, *363*, 615.
20 Georgiou, K., Michetti, P., Gai, L., Cavazzini, M., Shen, Z., Lidzey, D. G. *ACS Photonics* **2018**, *5*, 258.
21 Stranius, K., Hertzog, M., Börjesson, K. *Nat. Commun.* **2018**, *9*, 2273.
22 Polak, D., Jayaprakash, R., Lyons, T. P., Martinez-Martinez, L. A., Leventis, A., Fallon, K. J. et al. *Chem. Sci.* **2020**, *11*, 343.
23 Wilson, M. W. B., Rao, A., Clark, J., Sai Santosh Kumar, R., Brida, D., Cerullo, G. et al. *J. Am. Chem. Soc.* **2011**, *133*, 11830.
24 Flick, J., Ruggenthaler, M., Appel, H., Rubio, A. *Proc. Natl. Acad. Sci. USA* **2017**, *114*, 3026.
25 Savvidis, P. G., Connolly, L. G., Skolnick, M. S., Lidzey, D. G., Baumberg, J. J. *Phys. Rev. B* **2006**, *74*, 113312.
26 Zhang, Z. D., Wang, K., Yi, Z., Zubairy, M. S., Scully, M. O., Mukamel, S. *J. Phys. Chem. Lett.* **2019**, *10*, 4448.
27 Zhang, Z. D., Saurabh, P., Dorfman, K. E., Debnath, A., Mukamel, S. *J. Chem. Phys.* **2018**, *148*, 074302.

28 Xiang, B., Ribeiro, R. F., Dunkelberger, A. D., Xiong, W. *Proc. Natl. Acad. Sci. USA*. **2018**, *115*, 4845.

29 Martinez-Martinez, L. A., Du, M., Ribeiro, R. F., Kena-Cohen, S., Yuen-Zhou, J. *J. Phys. Chem. Lett.* **2018**, *9*, 1951.

30 Galego, J., Garcia-Vidal, F. J., Feist, J. *Phys. Rev. X* **2015**, *5*, 041022.

31 Sanchez-Barquilla, M., Fernandez-Dominguez, A. I., Feist, J., Garcia-Vidal, F. J. *ACS Photonics* **2022**, *9*, 1830.

32 Wang, D., Kelkar, H., Martin-Cano, D., Rattenbacher, D., Shkarin, A., Utikal, T. et al. *Nat. Phys.* **2019**, *15*, 483.

33 Hulkko, E., Pikker, S., Tiainen, V., Tichauer, R. H., Groenhof, G., Toppari, J. J. *J. Chem. Phys.* **2021**, *154*, 154303.

34 Saurabh, P., Mukamel, S. *J. Chem. Phys.* **2016**, *144*, 124115.

35 Kasprzak, J., Solnyshkov, D. D., Andre, R., Dang, L. S., Malpuech, G. *Phys. Rev. Lett.* **2008**, *101*, 146404.

36 Kasprzak, J., Richard, M., Kundermann, S., Baas, A., Jeambrun, P., Keeling, J. M. J. et al. *Nature* **2006**, *443*, 409.

37 Deng, H., Haug, H., Yamamoto, Y. *Rev. Mod. Phys.* **2010**, *82*, 1489.

38 Zhang, Z. D., Zhao, S., Lei, D. *Phys. Rev. B* **2022**, *106*, L220306.

39 Spano, F. C., Kuklinski, J. R., Mukamel, S. *Phys. Rev. Lett.* 1990, *65*, 211.

40 Hepp, K., Lieb, E. H. *Ann. Phys.* **1973**, *76*, 360.

41 Spano, F. C., Mukamel, S. *J. Chem. Phys.* **1989**, *91*, 683.

42 De Boer, S., Wiersma, D. A. *Chem. Phys. Lett.* **1990**, *165*, 45.

43 Phuc, N. T. *J. Chem. Phys.* **2021**, *155*, 014308.

44 Chen, T., Du, M., Yang, Z., Yuen-Zhou, J., Xiong, W. *Science* **2022**, *378*, 790.

45 Bennett, K., Kowalewski, M., Mukamel, S. *Faraday Discuss.* **2016**, *194*, 259.

46 Satapathy, S., Khatoniar, M., Parappuram, D. K., Liu, B., John, G., Feist, J. et al. *Sci. Adv.* **2021**, *7*, eabj0997.

47 Xiong, W. *Acc. Chem. Res.* **2023**, *56*, 776.

48 TÓrmä, P., Barnes, W. L. *Rep. Prog. Phys.* **2015**, *78*, 013901.

49 Nagarajan, K., Thomas, A., Ebbesen, T. W. *J. Am. Chem. Soc.* **2021**, *143*, 16877.

50 Li, T. E., Cui, B., Subotnik, J. E., Nitzan, A. *Annu. Rev. Phys. Chem.* **2022**, *73*, 43.

51 Gu, B., Nenov, A., Segatta, F., Garavelli, M., Mukamel, S. *Phys. Rev. Lett.* **2021**, *126*, 053201.

52 Herrera, F., Peropadre, B., Pachon, L. A., Saikin, S. K., Aspuru-Guzik, A. *J. Phys. Chem. Lett.* **2014**, *5*, 3708.

53 Xiang, B., Ribeiro, R. F., Li, Y., Dunkelberger, A. D., Simpkins, B. B., Yuen-Zhou, J. et al. *Sci. Adv.* **2019**, *5*, eaax5196.

54 Xiang, B., Wang, J., Yang, Z., Xiong, W. *Sci. Adv.* **2021**, 7, eabf6397.

55 Zhang, Z. D., Mukamel, S. *Chem. Phys. Lett.* **2017**, *683*, 653.

56 Zeb, M. A., Kirton, P. G., Keeling, J. *ACS Photonics* **2018**, *5*, 249.

57 Csehi, A., Vibók, Á., Halász, G. J., Kowalewski, M. *Phys. Rev. A* **2019**, *100*, 053421.

58 Du, M., Yuen-Zhou, J. *Phys. Rev. Lett.* **2022**, *128*, 096001.

59 Vendrell, O. *Phys. Rev. Lett.* **2018**, *121*, 253001.

60 Gu, J., Walther, V., Waldecker, L., Rhodes, D., Raja, A., Hone, J. C. et al. *Nat. Commun.* **2021**, *12*, 2269.

61 Mukamel, S. *J. Chem. Phys.* **2016**, *145*, 041102.

62 Zhong, J.-H., Vogelsang, J., Yi, J.-M., Wang, D., Wittenbecher, L., Mikaelsson, S. et al. *Nat. Commun.* **2020**, *11*, 1464.

63 Axt, V. M., Mukamel, S. *Rev. Mod. Phys.* **1998**, *70*, 145.

64 Zhang, Z. D., Nie, X., Lei, D., Mukamel, S. *Phys. Rev. Lett.* **2023**, *130*, 103001.

65 Cheng, C.-Y., Krainova, N., Brigeman, A. N., Khanna, A., Shedge, S., Isborn, C. et al. *Nat. Commun.* **2022**, *13*, 7937.

66 Engelhardt, G., Cao, J. *Phys. Rev. Lett.* 130, 213602 (2023).

67 Gera, T., Sebastian, K. L. *J. Chem. Phys.* **2022**, *156*, 194304.

68 Landau, L. D., Lifshitz, E. M. *Quantum Mechanics: Non-relativistic Theory*, 3rd ed. Butterworth-Heinemann: Oxford, 1981.

69 Herrera, F., Spano, F. C. *Phys. Rev. A* **2017**, *95*, 053867.

70 Yan, Y. J., Mukamel, S. *Phys. Rev. A* **1990**, *41*, 6485.

71 Yan, Y. J., Mukamel, S. *J. Chem. Phys.* **1991**, *94*, 179.

72 Cheng, Y.-C., Fleming, G. R. *Annu. Rev. Phys. Chem.* **2009**, *60*, 241.

73 Cao, J., Silbey, R. J. *J. Phys. Chem. A* **2009**, *113*, 13825.

74 Mukamel, S. *Principles of Nonlinear Optical Spectroscopy*. Oxford University Press: Oxford, 1999.

75 Dicke, R. H. *Phys. Rev.* **1954**, *93*, 99.

76 Tavis, M., Cummings, F. W. *Phys. Rev.* **1968**, *170*, 379.

77 Agarwal, G. S. *Quantum Optics*. Springer-Verlag: Berlin, 1974.

78 Tanimura, Y. *J. Phys. Soc. Jpn.* **2006**, *75*, 082001.

79 Abramavicius, D., Palmieri, B., Voronine, D. V., Šanda, F., Mukamel, S. *Chem. Rev.* **2009**, *109*, 2350.

80 la Cour Jansen, T., Zhuang, W., Mukamel, S. *J. Chem. Phys.* **2004**, *121*, 10577.

81 Garcia-Vidal, F. J., Feist, J., del Pino, J. *New J. Phys.* **2015**, *17*, 053040.

82 Herrera, F., Spano, F. C. *Phys. Rev. Lett.* **2017**, *118*, 223601.

83 Reitz, M., Sommer, C., Genes, C. *Phys. Rev. Lett.* **2019**, *122*, 203602.

84 Spano, F. C., Silva, C. *Annu. Rev. Phys. Chem.* **2014**, *65*, 477.

85 Lacombe, L., Hoffmann, N. M., Maitra, N. T. *Phys. Rev. Lett.* **2019**, *123*, 083201.

86 Kowalewski, M., Bennett, K., Mukamel, S. *J. Chem. Phys.* **2016**, *144*, 054309.

87 Flick, J., Appel, H., Ruggenthaler, M., Rubio, A. *J. Chem. Theory Comput.* **2017**, *13*, 1616.

88 Gudem, M., Kowalewski, M. *J. Phys. Chem. A* **2021**, *125*, 1142.

89 Breuer, H.-P., Petruccione, F. *The Theory of Open Quantum Systems*. Oxford University Press: Oxford, 2002.

90 Saez-Blazquez, R., Feist, J., Romero, E., Fernandez-Domínguez, A. I., García-Vidal, F. J. *J. Phys. Chem. Lett.* **2019**, *10*, 4252.

91 Guarin, C. A., Villabona-Monsalve, J. P., López-Arteaga, R., Peon, J. *J. Phys. Chem. B* **2013**, *117*, 7352.

92 DelPo, C. A., Kudisch, B., Park, K. H., Khan, S.-U.-Z., Fassioli, F., Fausti, D. et al. *J. Phys. Chem. Lett.* **2020**, 11, 2667.

93 Franck, J. *Trans. Faraday Soc.* **1925**, *21*, 536.

94 Wang, Z., Li, H., Feng, W., Song, X., Song, C., Liu, W. et al. *Phys. Rev. Lett.* **2020**, *124*, 013601.

95 Gross, M., Haroche, S. *Phys. Rep.* **1982**, *93*, 301.

96 Raimond, J. M., Brune, M., Haroche, S. *Rev. Mod. Phys.* **2001**, *73*, 565.

97 Agarwal, G. S. *Phys. Rev. Lett.* **1984**, *53*, 1732.

98 Pavolini, D., Crubellier, A., Pillet, P., Cabaret, L., Liberman, S. *Phys. Rev. Lett.* **1985**, *54*, 1917.

99 Evans, R. E., Bhaskar, M. K., Sukachev, D. D., Nguyen, C. T., Sipahigil, A., Burek, M. J. et al. *Science* **2018**, *362*, 662.

100 Guerin, W., Araujo, M. O., Kaiser, R. *Phys. Rev. Lett.* **2016**, *116*, 083601.

8

Molecular Dynamics Simulations of Exciton–Polaritons in Organic Microcavities

Gerrit Groenhof [1], Ruth H. Tichauer [2] and Ilia Sokolovskii [1]

[1] *Nanoscience Center and Department of Chemistry, University of Jyväskylä, Jyväskylä, Finland*
[2] *Departamento de Física Teórica de la Materia Condensada and Condensed Matter Physics Center (IFIMAC), Universidad Autónoma de Madrid, Madrid, Spain*

8.1 Introduction

Placing a photo-active organic material inside an optical microcavity or near a plasmonic nano-resonator can lead to measurable changes in the material's properties, and consequently modify its propensity for energy transfer [1–8], charge transport [9–12], lasing [13–15], or even its photochemistry [16–20]. By restricting the number of available optical modes and confining the electromagnetic (EM) field into small volumes, micro- and nano-resonators modify the light–matter interaction strength [21]. When this interaction exceeds the intrinsic losses of both the photoactive material and the optical resonator, excitons and confined light modes hybridize into new light–matter states called polaritons [22–25]. With low effective masses and high group velocities from their photonic component, and large interaction cross-sections from their matter component, polaritons constitute attractive quasi-particles for, in particular, Bose–Einstein condensation (BEC) and energy transport. However, because of the high complexity of organic materials, it remains unclear how the formation of these polaritonic states leads to the experimentally observed changes in material properties [26], in particular, because the large majority of states that form are *dark*, meaning that they lack a significant photonic contribution, and hence remain largely unaffected and therefore similar to the molecular excited states. While the prospect of using optical cavities to control photochemistry is very appealing, the lack of theoretical understanding has so far prevented leveraging this concept in a predictable, controllable, and systematic manner.

To bridge this gap, the theory community has fervently sought for ways to integrate quantum optics into quantum chemistry and develop new models for predicting the effects of strong light–matter coupling on molecular properties [27, 28]. However, while the majority of such developments were focused on modeling the coupling between a single molecule and a single photon mode with high accuracy [29–32], changes to material properties have so far only been observed when very many molecules *collectively* couple to the very many confined light modes of a Fabry–Pérot microcavity [16, 18, 33]. To go beyond such single-molecule and/or single-cavity mode descriptions and model the collective interaction between *many* molecules and *multiple* cavity modes,

Polariton Chemistry: Molecules in Cavities, First Edition. Edited by Joel Yuen-Zhou, Noel C. Giebink and Raphael F. Ribeiro. Published 2026 by John Wiley & Sons, Inc.

we proposed an alternative strategy, based on semiclassical molecular dynamics (MD) simulations with a multiscale quantum mechanics/molecular mechanics (QM/MM) Hamiltonian [34–36]. In this contribution, we discuss the underlying assumptions of this model and bring together our developments from recent years into a concise overview of the current state of the art of the model. We then highlight three recent applications in which we could use MD simulations to acquire atomistic insights into the (i) relaxation [37], (ii) propagation of organic exciton–polaritons [38–40], and (iii) ultra-fast photochemistry in the collective strong coupling regime [41]. In all cases, the insights acquired from the simulations would have been very difficult to obtain by experiments alone due to restrictions on spatial and time resolution in the latter.

8.2 Molecular Dynamics in the Collective Strong Coupling Regime

The established model of quantum optics, the Tavis–Cummings model [42, 43], provides conceptual insight into polariton formation but does not account for the chemical details of the molecules nor does it include the mode structure of the cavities, both of which are essential to fully capture and predict how strong light–matter coupling affects the physicochemical properties of materials. To overcome these limitations and simulate the dynamics when a large number, N, of molecules is strongly coupled to the n_{mode} modes of a Fabry–Pérot microcavity, as illustrated in Figure 8.1, we replace the two-level systems in the Tavis–Cummings model with molecules, and introduce multiple discrete modes to account for the cavity dispersion.

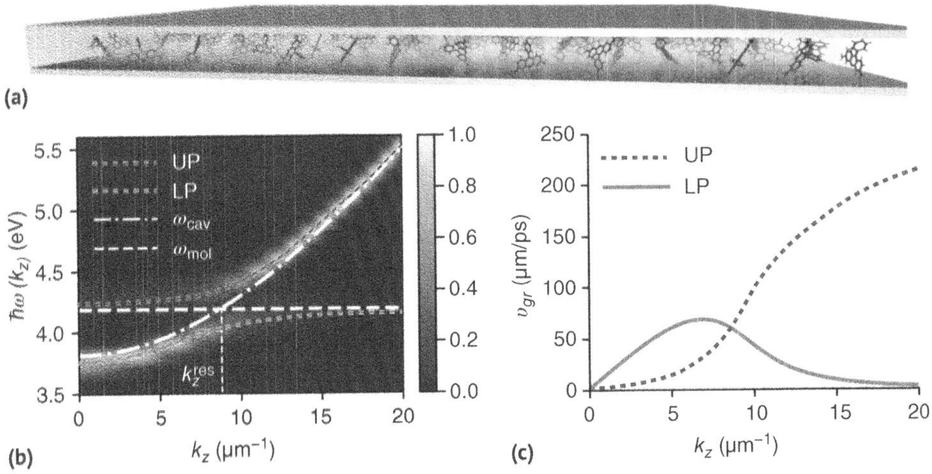

Figure 8.1 Schematic illustration of a Fabry–Pérot microcavity filled with Rhodamine molecules (a). Due to the strong light–matter coupling between the molecular excitations and the modes of the cavity, a lower polariton (LP) and upper polariton (UP) branch of bright light–matter states form (b) that are separated by the Rabi splitting, defined as the energy gap between the branches at the wave vector (k_z) at which the cavity dispersion intersects the absorption maximum of the molecules (4.18 eV at the CIS/3-21G//Amber03 level of QM/MM theory, dashed lines). Because the polaritonic branches are dispersive, the associated polaritons possess group velocity (c), defined as $v_{gr} = \partial\omega(k_z)/\partial k_z$.

8.2.1 Born–Oppenheimer Approximation in the Electronic Strong Coupling Regime

Our starting point for modeling strong coupling between *electronic* excitations of molecules and the confined light modes of an optical cavity is the Born–Oppenheimer approximation, which we apply to separate the nuclei on the one hand, from the electrons *plus* confined light modes on the other hand [44]. To describe the combined electronic and cavity mode degrees of freedom, we replace the traditional two-level systems in the Tavis-Cummings Hamiltonian by the adiabatic electronic states of the molecules, which depend parametrically on the nuclear coordinates:

$$\begin{aligned}\hat{H}^{\mathrm{TC}} &= \sum_j^N \sum_m h\nu_j^m(\mathbf{R}_j)\hat{\sigma}_{j,m}^+\hat{\sigma}_{j,m}^- + \sum_j^N V_{\mathrm{S}_0}^{\mathrm{mol}}(\mathbf{R}_j) + \sum_p^{n_{\mathrm{mode}}} \hbar\omega_{\mathrm{cav}}(\mathbf{k}_p)\hat{a}_p^\dagger\hat{a}_p \\ &+ \sum_j^N \sum_m \boldsymbol{\mu}_m^{\mathrm{TMD}}(\mathbf{R}_j)\cdot[\hat{\sigma}_{j,m}^+ + \hat{\sigma}_{j,m}^-][\hat{a}_p^\dagger\mathbf{f}_p^*(\mathbf{R}_c^j) + \hat{a}_p\mathbf{f}_p(\mathbf{R}_c^j)]\end{aligned} \tag{8.1}$$

Here, $\hat{\sigma}_{j,m}^+ = |\mathrm{S}_m^j\rangle\langle\mathrm{S}_0^j|$ is an operator that excites molecule j with nuclear coordinates $\mathbf{R}_j$, from the electronic ground state $|\mathrm{S}_0^j\rangle$ into m^{th} electronic excited state $|\mathrm{S}_m^j\rangle$ with excitation energy $h\nu_j^m$. Likewise, $\hat{\sigma}_{j,m}^- = |\mathrm{S}_0^j\rangle\langle\mathrm{S}_m^j|$ de-excites molecule j from electronic excited state $|\mathrm{S}_m^j\rangle$ into the electronic ground state $|\mathrm{S}_0^j\rangle$ with energy $V_{\mathrm{S}_0}^{\mathrm{mol}}(\mathbf{R}_j)$. Operators $\hat{a}_p^\dagger$ and $\hat{a}_p$ create and annihilate a photon of energy $\hbar\omega_{\mathrm{cav}}(\mathbf{k}_p)$ in cavity mode p with in-plane momentum $\mathbf{k}_p$.

Within the long-wavelength approximation, the light–matter interaction is calculated as the inner product between the transition dipole moment, $\boldsymbol{\mu}_m^{\mathrm{TDM}}(\mathbf{R}_j)$, associated with an excitation of molecule j into the mth excited electronic state on the one hand, and the vacuum field associated with an excitation of a cavity mode p and described by the mode function $\mathbf{f}_p(\mathbf{R}_c^j)$, with $\mathbf{R}_c^j$ the center of molecule j, on the other hand. The second term in Eq. (8.1) is the total energy of the combined molecule-cavity system in the ground state, with no excitations in neither the molecules nor in the cavity modes. Note that in the Hamiltonian (Eq. 8.1), we do not include *direct* Coulomb or excitonic interactions between molecules. Although adding such interactions is straightforward [45], we assume that the molecules are sufficiently diluted in the cavity medium so that these interactions can be neglected.

8.2.2 Quantum Mechanics/Molecular Mechanics

Computing electronic wave functions of molecules, including transition dipole moments ($\boldsymbol{\mu}_m^{\mathrm{TDM}}(\mathbf{R}_j)$) and potential energy surfaces ($V_{\mathrm{S}_m}^{\mathrm{mol}}(\mathbf{R}_j)$), is the domain of quantum chemistry and one of the first practical applications of QM. Thanks to relentless developments in this field, quantum chemistry has become an important and predictive technique in chemistry, physics, and material science [46]. However, despite tremendous advances in computer power, the materials that have been studied experimentally in the context of polaritonic chemistry are still too large to be described at any level of ab initio electronic structure theory or density functional theory (DFT) [47].

To go beyond such size limitations, MM force fields have been introduced, in which the potential energy is approximated by analytical functions that correspond to intuitive chemical concepts, such as bonds, angles, and torsions at short intermolecular distances, and pairwise Coulomb and Lennard-Jones potentials at larger distances [48]. Electrons are ignored at this level of theory, but their influence is expressed by the empirical parameters in these functions, which are valid for the ground state of a given covalent structure. Therefore, processes that involve electronic

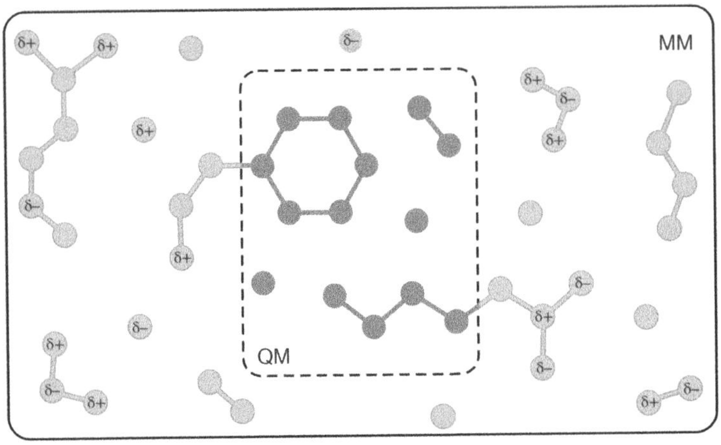

Figure 8.2 Illustration of the QM/MM concept [48]. A small region (inside the dashed box) in which a photochemical reaction occurs, and therefore cannot be described with a force field, is treated at a sufficiently high level of QM theory. The remainder of the system (outside the dashed box) is modeled at the MM level. Sticks represent chemical bonds, while $\delta-$ or $\delta+$ indicate that an atom has a partial negative or positive charge in the MM description.

rearrangements, such as photo-excitation into an electronically excited state [49] or chemical reactions in which bonds are formed and broken [50], cannot be described with an MM force field.

To overcome the system size limitations of a full QM description on the one hand, and the lack of reactivity of a full MM treatment on the other hand, methods have been developed that treat a small part of the system at the level of quantum chemistry (QM), while retaining the computationally more affordable force field (MM) for the larger part. This multiscale QM/MM strategy, introduced almost five decades ago by Warshel and Levitt [51], is schematically illustrated in Figure 8.2 [52]. The justification for dividing a system into regions that are described at different levels of theory is the local character of chemical or photochemical processes in condensed phases. A distinction can, therefore, often be made between a "reaction center" with atoms that are directly involved in the process and a "spectator" region, in which the atoms do not directly participate, but provide an electrostatic or steric background.

The multiscale QM/MM potential energy operator thus includes interactions between (i) atoms in the QM region, (ii) atoms in the MM region, and (iii) QM and MM atoms. The interactions within the QM and MM regions are described at the QM and MM levels, respectively. Modeling the interactions between the two subsystems is more challenging and can be done at various degrees of sophistication [48]. The most popular approaches are to either describe all interactions between the two subsystems at the force field level, or, alternatively, include all charged MM atoms into the electronic Hamiltonian as additional one-electron operators, and hence allow the MM environment to polarize the QM subsystem. In our simulations, we routinely use both embedding schemes, depending on the characteristics of the system we want to investigate.

8.2.3 Multiscale Tavis–Cummings Hamiltonian

Within the Born–Oppenheimer approximation, the nuclear degrees of freedom evolve on adiabatic potential energy surfaces that are the parametric nuclear-position-dependent eigenstates of the Tavis–Cummings Hamiltonian in Eq. (8.1). To obtain these eigenstates, we introduce a basis, in which each state is a tensor product between the molecular electronic states, computed at the QM/MM level described above (Section 8.2.2), and the Fock states of the cavity modes:

$$\begin{aligned}|\phi_{\{j,m_j\},\{p,n_p\}}\rangle &= \Pi_{\{j,m_j\}}\Pi_{\{p,n_p\}}\hat{\sigma}^+_{j,m_j}(\hat{a}^\dagger_p)^{n_p}|\mathrm{S}_0^1\mathrm{S}_0^2..\mathrm{S}_0^{N-1}\mathrm{S}_0^N\rangle \otimes |00..0\rangle \\ &= \Pi_{\{j,m_j\}}\Pi_{\{p,n_p\}}\hat{\sigma}^+_{j,m_j}(\hat{a}^\dagger_p)^{n_p}|\Pi_i^N\mathrm{S}_0^i\rangle \otimes |\Pi_p^{n_{\mathrm{mode}}}0_p\rangle \\ &= \Pi_{\{j,m_j\}}\Pi_{\{p,n_p\}}\hat{\sigma}^+_{j,m_j}(\hat{a}^\dagger_p)^{n_p}|\phi_0\rangle\end{aligned} \quad (8.2)$$

where state $|\phi_0\rangle$ is the ground state of the molecule-cavity system with no excitations in neither the molecules nor cavity modes:

$$|\phi_0\rangle = |\mathrm{S}_0^1\mathrm{S}_0^2..\mathrm{S}_0^{N-1}\mathrm{S}_0^N\rangle \otimes |00..0\rangle = |\Pi_i^N\mathrm{S}_0^i\rangle \otimes |\Pi_p^{n_{\mathrm{mode}}}0_p\rangle \quad (8.3)$$

The basis states in Eq. (8.2) represents $1, 2, 3, 4, \cdots, N$ excitations among the N molecules, with the constraint that each molecule, j, can at most have a single excitation into one of its multiple excited electronic states $|\mathrm{S}^j_m\rangle$. In contrast, because the cavity modes are bosonic, there is no restriction on the number of excitations, n_p, these modes can have. Therefore, the number of basis states in principle is infinite.

The light–matter interaction terms in the Hamiltonian of Eq. (8.1) either couple basis states with the same number of excitations (i.e., $\hat{\sigma}^+_{j,m}\hat{a}_p$ and $\hat{\sigma}^-_{j,m}\hat{a}^\dagger_p$ exchange a single excitation between one of the molecules and one of the cavity modes), or basis states that differ by two excitations, of which one is in a molecule and the other is in a cavity mode (i.e., $\hat{\sigma}^+_{j,m}\hat{a}^\dagger_p$ and $\hat{\sigma}^-_{j,m}\hat{a}_p$ create or annihilate an excitation in both a molecule and a cavity mode *simultaneously*). Because molecules only have a limited number of accessible (bound) electronic states, the total number of basis states required to find the adiabatic eigenstates of the Hamiltonian (Eq. 8.1) is finite in practice, but still computationally intractable for all but the smallest systems. Therefore, further approximations are needed.

Under the assumption that the light–matter interaction strength, as manifested by the Rabi splitting between the lower polariton (LP) and upper polariton (UP) branches (Figure 8.1b), is below 10% of the excitation energy of the molecules [53], and the molecular excitations are resonant with the cavity modes (i.e. $h\nu_j \approx \hbar\omega_{\mathrm{cav}}$), the so-called counter-rotating terms ($\hat{\sigma}^+_{j,m}\hat{a}^\dagger_p$ and $\hat{\sigma}^-_{j,m}\hat{a}_p$) are small and can thus be neglected. Within this rotating wave approximation (RWA), manifolds of basis states with different numbers of excitations become uncoupled. In addition, under weak driving conditions, usually employed in experiments on organic cavities, the Hamiltonian can furthermore be restricted to the single-excitation subspace. Thus, we only consider (i) basis states in which either one of the molecules is in the first electronic excited state (i.e. $|\mathrm{S}^j_1\rangle$), while all other molecules are in their ground state (i.e. $|\mathrm{S}_0^{i\neq j}\rangle$) and none of the cavity modes is excited:

$$|\phi_{j\leq N}\rangle = \hat{\sigma}^+_j|\phi_0\rangle \quad (8.4)$$

and (ii) basis states in which all molecules are in the electronic ground state, while one of the cavity modes is excited:

$$|\phi_{j>N}\rangle = \hat{a}^\dagger_{j-N}|\phi_0\rangle \quad (8.5)$$

Thus, under the RWA and within the single-excitation subspace, the multiscale Tavis–Cummings Hamiltonian (Eq. 8.1) rearranges into:

$$\begin{aligned}\hat{H}^{\mathrm{TC,RWA}} &= \sum_j^N h\nu_j(\mathbf{R}_j)\hat{\sigma}_j^+\hat{\sigma}_j^- + \sum_p^{n_{\mathrm{mode}}} \hbar\omega_{\mathrm{cav}}(\mathbf{k}_p)\hat{a}_p^\dagger\hat{a}_p \\ &+ \sum_j^N\sum_p^{n_{\mathrm{mode}}} \boldsymbol{\mu}^{\mathrm{TMD}}(\mathbf{R}_j)\cdot[\hat{\sigma}_j^+\hat{a}_p\mathbf{f}_p(\mathbf{R}_c^j) + \hat{\sigma}_j^-\hat{a}_p^\dagger\mathbf{f}_p^*(\mathbf{R}_c^j)] \\ &+ \sum_j^N V_{\mathrm{S}_0}^{\mathrm{mol}}(\mathbf{R}_j)\end{aligned} \tag{8.6}$$

8.2.4 Multimode Fabry–Pérot Cavities

In Fabry–Pérot microcavities, the material is sandwiched between two parallel mirrors (Figure 8.1a). To conserve energy and momentum, incoming light is confined to a standing wave in the direction perpendicular to the cavity mirrors, defined as x, and a plane-wave in the y, z-plane parallel to the mirrors, with in-plane momentum, or wave vector $\mathbf{k}_{\|}$. The standing wave arises from constructive interference of frequencies satisfying $\omega_q = q\pi c/(n_r L_x)$, where n_r is the refractive index of the material inside the cavity, c is the speed of light in vacuum, L_x is the cavity width, and q is an integer indicating the number of half wavelengths within L_x. As a consequence, the energy of a microcavity photon is quantized in q, but continuous in $\mathbf{k}_{\|}$ via the dispersion relation (dashed-dotted curve in Figure 8.1b):

$$\hbar\omega_{\mathrm{cav}}(\mathbf{k}) = \frac{\hbar c|\mathbf{k}|}{n_{\mathrm{r}}} = \frac{\hbar c}{n_{\mathrm{r}}}\sqrt{k_x^2 + k_y^2 + k_z^2} = \frac{\hbar c}{n_{\mathrm{r}}}\sqrt{\left(q\frac{\pi}{L_x}\right)^2 + |\mathbf{k}_{\|}|^2} \tag{8.7}$$

with k_x fixed at $\omega_q n_r/c$, and k_y and k_z the y and z components of the in-plane wave vector ($|\mathbf{k}_{\|}|^2 = k_x^2 + x_z^2$). The EM field within the microcavity at the position of molecule j in the middle of the cavity (i.e. $\mathbf{R}_c^j$) is described by

$$\mathbf{f}_p(\mathbf{R}_j) = \mathbf{u}_\lambda\sqrt{\hbar\omega_{\mathrm{cav}}(\mathbf{k})/2\epsilon_0 V_{\mathrm{cav}}}e^{-i\mathbf{k}_{\|}\cdot\mathbf{R}_c^j} \tag{8.8}$$

with $\mathbf{u}_\lambda$ a unit vector indicating the direction or polarization (s or p) of the cavity vacuum field at the position of molecule j; ϵ_0 the vacuum permittivity; and V_{cav} the mode volume.

Following Michetti and La Rocca [54], we assume a one-dimensional (1D) cavity in the z direction ($k_y = 0$) such that the vacuum field points in the y-direction ($\mathbf{u}_\lambda = \mathbf{u}_y$), as illustrated in Figure 8.3. If we consider the fundamental mode only (i.e. $q = 1$ in $\omega_q = q\pi c/n_r L_x$), $k_x = \pi/L_x$, the energy of the cavity photon depends solely on the z-component of the incident wave vector ($\hbar\omega_{\mathrm{cav}}(k_z)$), and thus, the cavity modes are fully characterized by k_z. Likewise, the positions of the molecules are now restricted to the cavity z-axis, and we use z_j rather than $\mathbf{R}_c^j$, to indicate their positions in what follows.

By imposing periodic boundary conditions in the z-direction of the 1D cavity, the continuous wave vector in the z-direction, k_z, is restricted to discrete values [54]: $k_{z,p} = 2\pi p/L_z$ with $p \in \mathbb{Z}$ and L_z the length of the 1D cavity. With a suitable cut-off for the dispersion (Eq. 8.7), typically sufficiently high above the molecular excitation energy, the number of modes n_{mode} is finite, and the Tavis–Cummings Hamiltonian in Eq. (8.6) becomes an $(N+n_{\mathrm{mode}})$ by $(N+n_{\mathrm{mode}})$ matrix containing four blocks [35]:

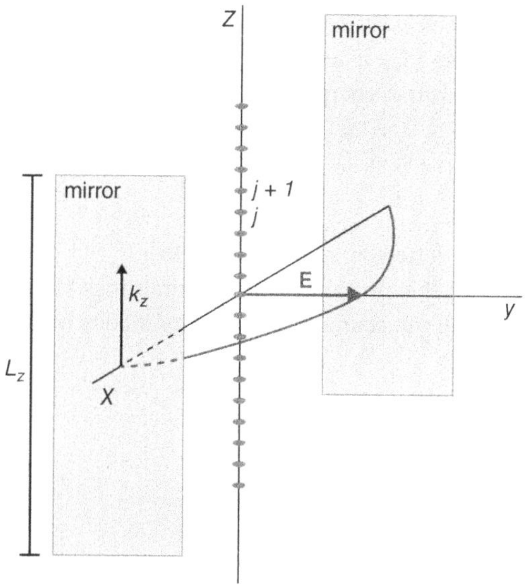

Figure 8.3 One-dimensional (1D) Fabry–Pérot microcavity model. Two reflecting mirrors located at $-\frac{1}{2}x$ and $\frac{1}{2}x$, confine light modes along this direction, while free propagation along the z direction is possible for plane waves with in-plane momentum k_z and energy $\hbar\omega_{\mathrm{cav}}(k_z)$. The vacuum field vector (E) points along the y-axis, reaching a maximum amplitude at $x = 0$ where the N molecules (ellipses) are placed, distributed along the z-axis at positions z_j with $1 \leq j \leq N$.

$$\mathbf{H}^{\mathrm{TC,RWA}} = \begin{pmatrix} \mathbf{H}^{\mathrm{mol}} & \mathbf{H}^{\mathrm{int}} \\ \mathbf{H}^{\mathrm{int}\dagger} & \mathbf{H}^{\mathrm{cav}} \end{pmatrix} \tag{8.9}$$

The upper left block, $\mathbf{H}^{\mathrm{mol}}$, is an $N \times N$ matrix containing the single-photon excitations of the molecules. Because direct excitonic interactions between molecules are neglected, this block is diagonal, with elements labeled by the molecule indices j:

$$H^{\mathrm{mol}}_{j,j} = \langle \phi_0 | \hat{\sigma}_j \hat{H}^{\mathrm{TC}} \hat{\sigma}_j^{+} | \phi_0 \rangle \tag{8.10}$$

for $1 \leq j \leq N$. Each matrix element of $\mathbf{H}^{\mathrm{mol}}$ thus represents the potential energy $V^{\mathrm{mol}}_{\mathrm{S}_1}(\mathbf{R}_j)$ of a molecule, j, in the electronic excited state $|\mathrm{S}_1^j(\mathbf{R}_j)\rangle$ while all other molecules, $i \neq j$, are in the electronic ground state $|\mathrm{S}_0^i(\mathbf{R}_i)\rangle$:

$$H^{\mathrm{mol}}_{j,j} = V^{\mathrm{mol}}_{\mathrm{S}_1}(\mathbf{R}_j) + \sum_{i \neq j}^{N} V^{\mathrm{mol}}_{\mathrm{S}_0}(\mathbf{R}_i) \tag{8.11}$$

The potential energy surfaces of the molecule in the ground, $V^{\mathrm{mol}}_{\mathrm{S}_0}(\mathbf{R}_j)$, and excited states, $V^{\mathrm{mol}}_{\mathrm{S}_1}(\mathbf{R}_j)$, are computed at a suitable level of QM/MM theory (Section 8.2.2).

The lower right block, $\mathbf{H}^{\mathrm{cav}}$, is an $n_{\mathrm{mode}} \times n_{\mathrm{mode}}$ matrix containing the single-photon excitations of the cavity modes, and is also diagonal:

$$H^{\mathrm{cav}}_{p,p} = \langle \phi_0 | \hat{a}_p \hat{H}^{\mathrm{TC}} \hat{a}_p^{\dagger} | \phi_0 \rangle \tag{8.12}$$

for $1 \leq p \leq n_{\text{mode}}$. Here, $\hat{a}_p^\dagger$ excites cavity mode p with wave vector $k_{z,p} = 2\pi p/L_z$. In these matrix elements, all molecules are in the electronic ground state (S_0). Their energies are, therefore, the sum of the cavity energy at $k_{z,p}$ and the molecular ground state energies:

$$H_{p,p}^{\text{cav}} = \hbar\omega_{\text{cav}}(k_{z,p}) + \sum_j^N V_{S_0}^{\text{mol}}(\mathbf{R}_j) \tag{8.13}$$

where $\omega_{\text{cav}}(k_{z,p})$ is the cavity dispersion (Eq. 8.7, dashed-dotted curve in Figure 8.1b).

The two $N \times n_{\text{mode}}$ off-diagonal blocks $\mathbf{H}^{\text{int}}$ and $\mathbf{H}^{\text{int}\dagger}$ in the multimode Tavis–Cummings Matrix (Eq. 8.9) model the light–matter interactions between the molecules and the cavity modes within the long-wavelength approximation:

$$\begin{aligned} H_{j,p}^{\text{int}} &= -\boldsymbol{\mu}^{\text{TDM}}(\mathbf{R}_j) \cdot \mathbf{u}_y \sqrt{\frac{\hbar\omega_{\text{cav}}(k_{z,p})}{2\epsilon_0 V_{\text{cav}}}} \langle \phi_0 | \hat{\sigma}_j \hat{\sigma}_j^+ \hat{a}_p e^{i2\pi p z_j/L_z} \hat{a}_p^\dagger | \phi_0 \rangle \\ &= -\boldsymbol{\mu}^{\text{TDM}}(\mathbf{R}_j) \cdot \mathbf{u}_y \sqrt{\frac{\hbar\omega_{\text{cav}}(k_{z,p})}{2\epsilon_0 V_{\text{cav}}}} e^{i2\pi p z_j/L_z} \end{aligned} \tag{8.14}$$

for $1 \leq j \leq N$ and $1 \leq p \leq n_{\text{mode}}$. Here, z_j is the position of the geometric center of molecule j on the z-axis of the cavity.

Because the computation of the electronic energies, $V_{S_0}^{\text{mol}}(\mathbf{R}_j)$ and $V_{S_1}^{\text{mol}}(\mathbf{R}_j)$, and transition dipole moments, $\boldsymbol{\mu}^{\text{TDM}}(\mathbf{R}_j)$, can be performed in parallel, cavities with thousands of molecules can be efficiently modeled on high-performance computing resources. Indeed, we could perform simulations of 1600 Rhodamine dyes, including their solvent environment (43200 QM and 17700800 MM atoms) on 38400 CPU cores [34].

8.2.5 Semiclassical Molecular Dynamics

Diagonalizing the matrix representation in the basis of the single-excitation product states (Eq. 8.9) of the multiscale Tavis–Cummings Hamiltonian (Eq. 8.6) yields the $N + n_{\text{mode}}$ adiabatic hybrid light–matter eigenstates, $|\psi^m\rangle$:

$$|\psi^m\rangle = \left(\sum_j^N \beta_j^m \hat{\sigma}_j^+ + \sum_p^{n_{\text{mode}}} \alpha_p^m \hat{a}_p^\dagger \right) |\phi_0\rangle \tag{8.15}$$

with eigenenergies E_m. The β_j^m and α_p^m expansion coefficients reflect the contribution of the molecular excitations, $|S_1^j(\mathbf{R}_j)\rangle$, and the cavity mode excitations, $|1_p\rangle$, to polariton $|\psi^m\rangle$. Due to their parametric dependence on the nuclear degrees of freedom, these eigenenergies form adiabatic potential energy surfaces $E_m(\mathbf{R})$, with $\mathbf{R}$ the $3 \times N_{\text{mol}} \times N_{\text{atoms}}$ coordinates of all atoms in the system.

Because for all but the smallest systems [55–60] a direct application of the time-dependent Schrödinger equation for the propagation of the combined photonic–electronic-nuclear degrees of freedom is computationally intractable, we model the nuclear degrees of freedom *classically* using MD instead [61]. Within this approximation, the classical trajectories are evolved by numerically integrating Newton's equations of motion under the influence of the forces due to the quantum degrees of freedom, for which the wave function, $|\Psi(t)\rangle$, is propagated along the trajectory.

Since the eigenstates, $|\psi^m\rangle$, of the Hermitian QM/MM Tavis–Cummings Hamiltonian (Eq. 8.15) form a complete set, we expand the wave function, $|\Psi(t)\rangle$, in these states

$$|\Psi(t)\rangle = \sum_m c_m(t)|\psi^m\rangle \tag{8.16}$$

The equations of motion for the time-dependent expansion coefficients, $c_m(t)$, are obtained by substituting the wave function into the time-dependent Schrödinger equation:

$$\dot{c}_m = -\frac{i}{\hbar}E_m c_m - \sum_n \sum_j \langle \psi^m | \nabla_n \psi^j \rangle \cdot \dot{\mathbf{r}}_n = -\frac{i}{\hbar}E_m c_m - \sum_j c_j \mathbf{d}_{mj} \cdot \dot{\mathbf{R}} \tag{8.17}$$

with $\dot{\mathbf{r}}_n$ the velocity vector of nucleus n, and $\langle \psi^m | \nabla_n \psi^j \rangle$ the nonadiabatic coupling vector between states m and j for a displacement of that nucleus. Likewise, $\dot{\mathbf{R}}$ and $\mathbf{d}_{mj}$ are the $3 \times N_{\text{mol}} \times N_{\text{atoms}}$ dimensional velocity and nonadiabatic coupling vectors of all atoms in the system.

The nonadiabatic coupling vectors, $\mathbf{d}_{mj}$, account for the fact that adiabatic states, in contrast to diabatic states, can change along the nuclear trajectory. Thus, in the semiclassical MD simulations, nuclear motions can induce population transfer between adiabatic states [61]. While the time-propagation of the QM degrees of freedom is relatively straightforward, it remains ambiguous how to couple that evolution to the classical trajectories, despite decades of intensive research [62]. In our simulations, we rely on two approaches: mean-field or Ehrenfest MD and surface hopping.

8.2.5.1 Ehrenfest Molecular Dynamics

In Ehrenfest MD, the trajectory evolves on the expectation value of the potential energy surface with respect to the total wave function (Eq. 8.16):

$$E^{\text{mf}}(\mathbf{R}) = \langle \Psi | \hat{H}^{\text{TC,RWA}} | \Psi \rangle = \sum_j \sum_m c_j^* c_m \langle \psi^j | \hat{H}^{\text{TC,RWA}} | \psi^m \rangle \tag{8.18}$$

Thus, in Ehrenfest MD, the matrix representation of the QM/MM Tavis–Cummings Hamiltonian $\hat{H}^{\text{TC,RWA}}$ (Eq. 8.9) is diagonalized at every time step of the simulation, using the basis functions $|\phi_j\rangle$ (Eqs. 8.4 and 8.5), obtained from QM/MM calculations of the electronic states of the molecules [61]. Then, the derivatives of the mean-field potential energy surface (Eq. 8.18) are computed from the molecular S_0 and S_1 gradients in combination with the expansion coefficients, $c_m(t)$, and used to integrate the positions of the classical nuclei over a discrete time interval, Δt [63]. In parallel, the wavefunction is propagated according to Eq. (8.17) using the unitary propagation scheme proposed by Grannuci et al. [64]. Because Ehrenfest dynamics performs well if there are many states close in energy [65], it should be ideally suited for simulating polaritonic systems composed of very many collectively coupled molecules, in which the light–matter eigenstates form a dense manifold of nearly parallel potential energy surfaces [66].

8.2.5.2 Surface Hopping

Tully proposed an alternative approach, in which a distinction is made between the *total* wave function, $|\Psi(t)\rangle$ (Eq. 8.16), and an *auxiliary* adiabatic wave function, $|\psi^m\rangle$ (Eq. 8.15) [67]. As in Ehrenfest MD, the total wave function defines all quantum observables, such as the populations of the adiabatic states, and is propagated in time along with the classical trajectory. The auxiliary adiabatic wave function defines the potential energy surface on which the classical trajectory evolves and is not explicitly propagated in time, but can stochastically change (hop) to other adiabatic states.

In order to maintain the correct quantum statistical distribution of states in an ensemble of surface hopping trajectories, the probabilities of switching between different adiabatic states must reflect the populations of those adiabatic states in the primary electronic wave function. Based on Eq. (8.17), the rate at which the populations of the adiabatic states ($|c_m|^2$) change along a nuclear trajectory is

$$\begin{aligned}\frac{\partial}{\partial t}|c_m|^2 &= c_m\dot{c}_m^* + c_m^*\dot{c}_m \\ &= -\sum_j 2\Re[c_j^* c_m \mathbf{d}_{mj} \cdot \dot{\mathbf{R}}] \\ &= -\sum_j b_{mj}\end{aligned} \tag{8.19}$$

Tully argued that the total number of switches should be minimal [67]. A minimum number of such switches is realized in N^{traj} simulations if, during a small time interval δt, $N^{\text{traj}} b_{mj}\delta t$ switches from $|\psi^j\rangle$ to $|\psi^m\rangle$ occur, and none from $|\psi^m\rangle$ to $|\psi^j\rangle$. Since there are initially $N^{\text{traj}}|c_j|^2$ trajectories in state $|\psi^j\rangle$, the probability that one of these will switch to $|\psi^m\rangle$ is

$$p_{j\to m} = \frac{N^{\text{traj}} b_{mj}\delta t}{N^{\text{traj}}|c_j|^2} = \frac{b_{mj}\delta t}{|c_j|^2} \tag{8.20}$$

and the probability of a switch from $|\psi_m\rangle$ to $|\psi_j\rangle$ should be zero. Therefore, the fewest switches algorithm computes the transition probabilities as

$$p_{j\to m} = \max\left[0, \frac{b_{mj}\Delta t}{|c_j|^2}\right] \tag{8.21}$$

where Δt is the integration time step in the simulation. The probabilities in Eq. (8.21) are compared to a random number ζ drawn from a uniform distribution between 0 and 1, and a decision to hop from an adiabatic state $|\psi^j\rangle$ to another adiabatic state $|\psi^m\rangle$ is made if $\sum_{q\neq j}^{m} p_{j\to q} < \zeta < \sum_{q\neq j}^{m+1} p_{j\to q}$. After the hop, the evolution of the classical trajectory continues on the adiabatic potential energy surface, $E_m(\mathbf{R})$, of state $|\psi^m\rangle$. Being based on the rates at which populations change, the fewest switches algorithm ensures that surface hops only occur if the trajectory is in regions where the nonadiabatic coupling vectors $\mathbf{d}_{jm}$ are nonzero. Outside those regions, the nonadiabatic coupling vectors are negligible and the adiabatic state populations no longer change.

In order to conserve energy, the velocities of the nuclei are adjusted when a state switch is made to an adiabatic state that has a different energy at the current nuclear configuration. Since nuclei can only induce transitions between states if their velocity vectors overlap with the nonadiabatic coupling vectors ($\dot{\mathbf{R}}\cdot\mathbf{d}_{mj}$), the components of the nuclear velocities are adjusted parallel to these coupling vectors [68], as if the nuclei were subjected to a force in the direction of the nonadiabatic coupling vector and the work done by that force were used to promote the system to another adiabatic state [69–71].

While Tully's fewest switches surface hopping algorithm [67] has been highly successful for modeling nonadiabatic dynamics in systems with few electronic states [62], the number of simulations required to achieve convergence when there are hundreds to thousands of near-degenerate states poses a huge computational challenge for modeling the dynamics of the large ensembles of strongly coupled molecules. Furthermore, because hops are stochastic, the trajectories are not deterministic, which not only violates the principles of classical dynamics but can also lead to inconsistencies between the potential energy surface on which the trajectory evolves and the total polaritonic wave function, $|\Psi(t)\rangle$ (Eq. 8.16). While for bare molecules such inconsistencies can be overcome with decoherence corrections [72–74], these corrections are ad hoc and lack a physical basis. Therefore, it is not a priori clear how to apply such corrections to polaritonic systems. Nevertheless, for simulations with fewer molecules, in particular, when these molecules can undergo photochemical reactions, surface hopping remains the semiclassical method of choice [75–77].

8.2.5.3 Cavity Decay

The mirrors of optical cavities used in experiments are not perfect in the sense that confined photons can only be reflected a limited number of times. Excitation, therefore, irreversibly decays by leaking out through the cavity mirrors, and the cavity mode excitations have a finite lifetime (τ_{cav}, characterized by the Q-factor, defined as $Q = \omega_{\text{cav}}\tau_{\text{cav}}$) or decay rate, $\gamma_{\text{cav}} = 1/\tau_{\text{cav}}$. To include the effect of such losses in our simulations, we implemented an ad hoc first-order decay of population from adiabatic states, $|\psi^m\rangle$, with a rate that is proportional to the total contribution of the cavity mode excitations to that state (i.e. $\Gamma_m = \gamma_{\text{cav}}\sum_p^{n_{\text{modes}}}|\alpha_p^m|^2$, in Eq. 8.15) [35]. Due to this decay, the norm of the wave function, $\sum_m |c_m(t)|^2$ (Eq. 8.16), is not conserved, but decreases to reflect the radiative losses. More recently, we have introduced an effective non-Hermitian Hamiltonian approach to account for cavity decay more consistently [36]. In this approach, the cavity losses are added as imaginary decay terms (i.e. $-\frac{i}{2}\sum_p^{n_{\text{modes}}}\hbar\gamma_p\hat{a}_p^\dagger\hat{a}_p$) to the Tavis–Cummings Hamiltonian (Eq. 8.6).

8.3 Applications

In contrast to inorganic semiconductors, the oscillator strength of organic materials is often high enough to reach the strong coupling regime at room temperature. However, because organic molecules tend to be more flexible, understanding the effect of molecular structure and dynamics on polariton formation is essential to leverage strong coupling in future applications or devices. Among the most promising applications of polaritons are (i) room temperature BEC, (ii) enhanced energy transport in organic media, and (iii) manipulating photochemistry. We have used our multiscale MD simulation approach to investigate the molecular processes underlying these phenomena in atomistic detail.

8.3.1 Polariton Relaxation

BECs are coherent macroscopic states that manifest when the De Broglie wavelength λ_{DB} is larger than the inter-particle spacing of the bosons forming the condensate [78]. Because the De Broglie wavelength is inversely proportional to the square root of the mass of the bosons (i.e. $\lambda_{\text{DB}} \propto 1/\sqrt{mk_{\text{B}}T}$, where k_{B} is the Boltzmann constant), the very low effective mass of exciton–polaritons should allow BEC at high temperatures [79]. Indeed, BEC has been achieved at room temperature in strongly coupled organic materials [80–82]. Because the condensate emerges from the macroscopic occupation of a single state, emission from polaritons forming that condenstate, should be coherent. Furthermore, since polariton emission does not require population inversion, polaritonic BEC can potentially provide a low-power alternative to traditional lasers relying on such population inversion for obtaining coherent emission [83].

Upon condensation, the lowest energy state of the polariton dispersion ($\mathbf{k}_{\|} = \mathbf{0}$ in Fabry–Pérot resonators Figure 8.4b) becomes macroscopically occupied via stimulated scattering of polaritons [84]. Because this stimulated scattering process requires the population of the $\mathbf{k}_{\|} = \mathbf{0}$ state to be above a certain threshold, polariton relaxation toward the bottom of the LP branch constitutes a key requirement for the formation of polariton condensates. However, due to losses in the system that limit the polariton lifetime, creating a sufficiently high population at the bottom of the LP branch for condensation requires pumping thresholds that are currently too high for electrical injection [85, 86], which will be essential for practical applications of polariton lasing.

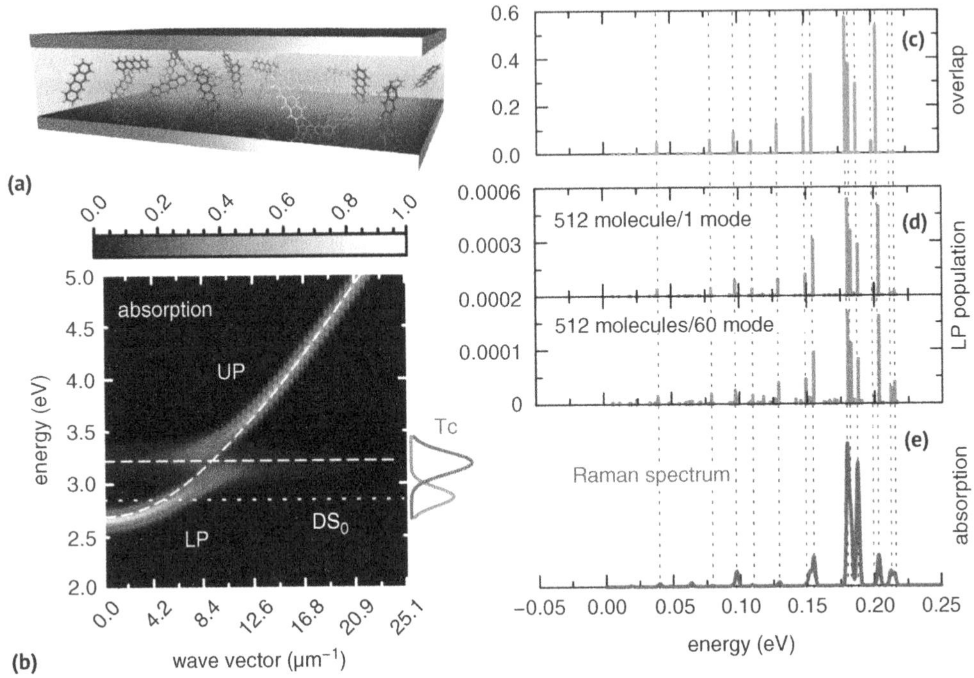

Figure 8.4 Polariton relaxation in single and multimode cavities containing 512 Tetracene molecules. Schematic illustration of the cavity system (a, not to scale) and the dispersion when 512 Tetracenes are strongly coupled to the 60 modes of a 1D Fabry–Pérot cavity (b). At the TDA-CAMB3LYP/6-31G(d) level of time-dependent density functional theory, the absorption maximum of Tetracene at the S_0 minimum energy geometry is 3.22 eV (long dashed line in b) and 2.85 eV in the S_1 minimum (short dashed line). In panel (c) we show the overlap of the vibrations and the nonadiabatic coupling vector for population exchange between the DS_0 state, in which excitation is localized on the Tetracene molecule in the S_1 geometry, and the lower polariton (LP), as a function of the vibrational energy. The extent of population transfer when these vibrations are selectively excited is visualized in panel (d) for single-mode (top) and multimode (bottom) cavities. Panel (e) shows the computed Raman spectrum of Tetracene.

Increasing the cavity Q-factor to reduce the losses during the relaxation process, would in principle be a straightforward manner to lower that threshold [82, 87], but is difficult to achieve in practice due to the constraints on mirror materials. Alternatively, the pumping threshold may also be reduced if the rate of population transfer into the LP minimum can be enhanced, but that requires a detailed understanding at the molecular level of the relaxation mechanisms. Experiments [14, 23, 88–100] and theory [35, 54, 66, 101–113] suggest that population transfer into the LP is dominated by two relaxation processes: radiative pumping (RP) and vibrationally assisted scattering (VAS).

RP requires overlap between the molecular emission spectrum and the LP branch [14, 93, 100, 114], and occurs at rates that match the fluorescence lifetime of the molecules. Therefore, RP has been interpreted as an exchange of a photon between an uncoupled molecule in the exciton reservoir and a polaritonic state via spontaneous emission and re-absorption. While exciton–polariton condensation has been achieved through RP by tuning the fundamental cavity mode to optimize the overlap between the LP minimum and the maximum of the Stokes-shifted emission [13, 14, 115], controlling the rate of RP is challenging, as the fluorescence lifetime of the molecules is determined by the strength of the vacuum field fluctuations.

VAS was first observed by Lidzey et al., who found that for *J*-aggregates, which have a negligible Stokes shift, relaxation from the dark states can occur on a much shorter timescale than the fluorescence lifetime [94, 95]. Such ultra-fast relaxation process was originally hypothesized by Agranovich et al., who postulated a scattering mechanism involving Raman-active vibrations of the molecules in the cavity [101, 102]. Indeed, in transient spectroscopy experiments, ultra-fast relaxation was observed when the energy gap between the dark states and the LP matches the frequency of such vibrations [95]. Because VAS provides a much faster relaxation channel than RP, enhancing this channel by engineering the molecules or the cavity could pave the road toward electrically driven polariton lasers.

Optimizing relaxation into the lowest energy state of the LP requires a detailed understanding of the underlying mechanisms, which is currently incomplete. To acquire atomistic insights into the relaxation processes, we performed Ehrenfest MD simulations of 512 Tetracene molecules in single and multimode cavities (Figure 8.4) and tracked the relaxation of the population from the dark state manifold into the LP. We calculated 84 short semiclassical MD trajectories that were initiated in the lowest energy dark state, DS_0 (Figure 8.4b), in which the excitation is localized on one Tetracene that is in the S_1, or Stokes-shifted, minimum geometry, while all other molecules remain in the S_0 geometry [37, 101].

In each simulation, we selectively activated one of the 84 vibrations of the molecule in the S_1 minimum energy geometry by providing initial atomic velocities in the direction of the displacements associated with that vibration. In Figure 8.4d we plot the extent of population transfer into the LP state after 1.0 fs as a function of the energy of the vibration along which the initial velocities were directed, for both the single and multimode cavities. Population is transferred from the lowest energy dark state, DS_0, into the LP only if vibrations with a displacement that overlaps with the nonadiabatic coupling vector ($\mathbf{d}_{mj}$, defined in Eq. 8.17), are activated.

The nonadiabatic coupling vector $\mathbf{d}_{DS_0,LP}$ contains displacements along the totally symmetric vibrations, which are all Raman active. Therefore, the same vibrations are also present in the calculated Raman spectrum (Figure 8.4e). Because the latter observation is in line with the theoretical predictions of Agranovich [101, 102] and experimental findings of Lidzey et al. [94, 95], we initially identified the relaxation process involving displacements along these vibrations as VAS [37]. However, the nonadiabatic coupling vector that determines the rate of population exchange between the DS_0 and LP states, scales as $1/N$, with N the number of Tetracene molecules in the cavity. Therefore, in the limit of an experimentally relevant number of molecules, namely 10^5 or higher [111, 112, 116], the relaxation rate would be significantly slower, on the order of nanoseconds, which is more comparable to the rates associated with RP. Because VAS was found to be much faster, even in the thermodynamic limit [95], further work is needed to disentangle these two relaxation processes and reveal the molecular mechanism underlying VAS.

8.3.2 Polariton Transport

Because of their large binding energies in organic molecules, excitons are localized and diffuse through the material via incoherent hops, until the exciton is trapped or lost due to radiative or nonradiative decay. Exciton diffusion is, therefore, limited to a few nanometers in typical organic materials [117]. Recent experiments suggest that coupling excitons to the confined light modes of an optical resonator to form polaritons can overcome this limitation owing to (i) the long-range coherence, (ii) the small effective masses, and (iii) the large group velocities ($v_g = \partial\omega(k_z)/\partial k_z$, Figure 8.1c) of these quasi-particles. While the coherence of polaritions within the entire mode volume of the cavity has been exploited to drive energy transfer perpendicular to the cavity mirrors

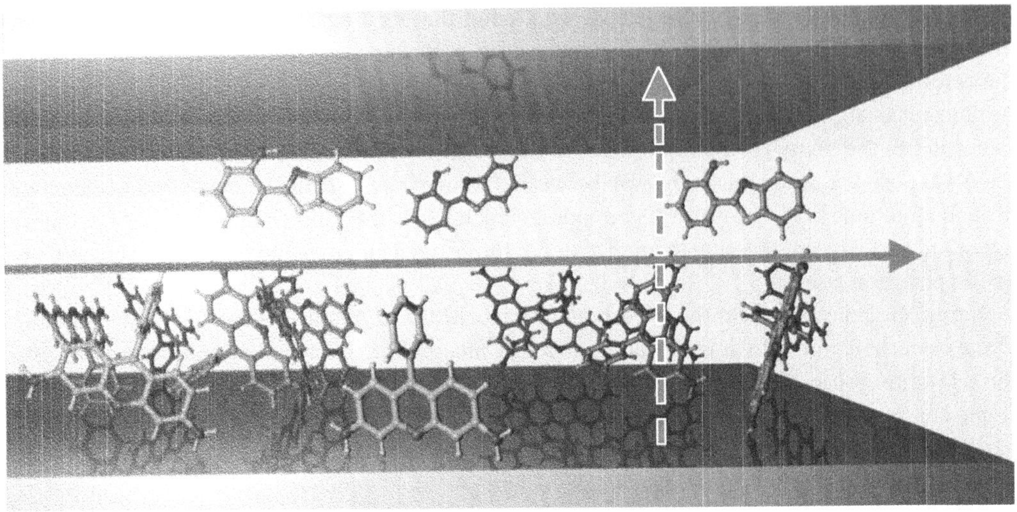

Figure 8.5 Schematic depiction of polariton transport in the out-of-plane (i.e. perpendicular to the mirrors, vertical arrow) and in-plane (i.e. parallel to the mirrors, horizontal arrow) direction of an optical cavity [128].

(vertical arrow in Figure 8.5) [1, 4, 118–122], the low effective mass in combination with the high group velocity has been used to achieve exciton transport parallel to the mirrors (horizontal arrow in Figure 8.5) that significantly exceeds the intrinsic exciton diffusion length of the material [2, 3, 123–127].

Although for the in-plane transport of exciton–polaritons, theory predicts ballistic motion at their group velocities [104, 106, 129, 130], as in inorganic microcavities [131, 132], experiments on organic cavity systems suggest a more complex picture: stationary microscopy measurements, which provide information on the distance over which polaritons propagate [2, 5, 6, 123], suggest both ballistic [2, 6] and diffusive [5] transport, while transient microscopy, which also yields insight into the time-evolution of the propagation [3, 7, 8, 124, 126, 127] suggests propagation can be both ballistic [7] and diffusive [3], or a combination of the two [8, 124]. In addition, the measured velocities in the time-resolved experiments are significantly lower than the group velocities of the polariton branches.

To acquire atomistic insights into the propagation of exciton–polaritons in organic materials, we performed multiscale Ehrenfest MD simulations of 1024 molecules (Rhodamine, Tetracene, and Methylene Blue) strongly coupled to a 1D Fabry–Pérot microcavity (Figure 8.1) [38–40]. Because in the experiments the strongly coupled materials were excited either resonantly into the bright polaritonic states [7, 124, 126], or off-resonantly into an uncoupled molecular electronic state [2, 3, 5, 6, 8], we performed simulations for both initial conditions. To also investigate the effect of the cavity Q-factor on the polariton propagation, we varied the cavity decay rates in our simulations (Section 8.2.5.3).

In our simulations, we observed that immediately after resonantly exciting a wavepacket of bright LP states in an ideal lossless cavity with a broad-band excitation pulse, the wavepacket, visualized in Figure 8.6a for the Rhodamine cavity, propagates ballistically with central group velocity, as indicated by the quadratic increase of the mean-square displacement (MSD) in Figure 8.6c. During this ballistic propagation, the population transfers from the initially excited bright polaritonic states into the dark states, as evidenced by the increase of the excitonic contribution to the wavepacket (black line in Figure 8.6b). As dark states have no group velocity, these population

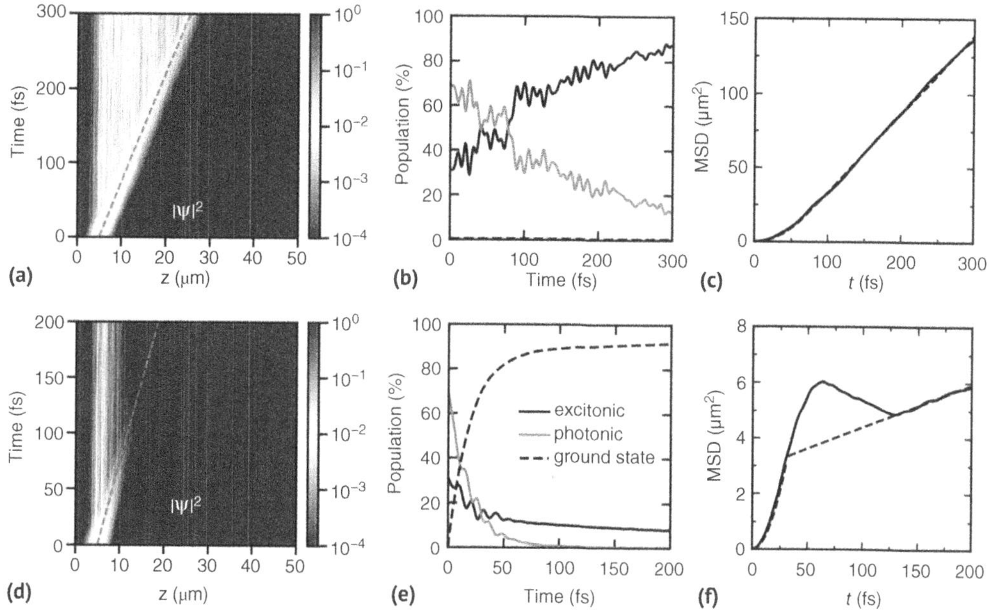

Figure 8.6 Propagation of exciton–polaritons after resonant excitation of a wavepacket of LP states in a perfect lossless cavity ($\gamma_{cav} = 0$ ps^{-1}, top) and lossy cavity ($\gamma_{cav} = 66.7$ ps^{-1}, bottom) containing 1024 Rhodamine molecules. Panels a and d: the total wavepacket as a function of position and time. Panels b and e: contributions of the molecular excitons (black), and cavity mode excitations (gray) to $|\Psi(z,t)|^2$ as a function of time in the lossy cavity. The population in the ground state, created by radiative decay through the imperfect mirrors, is plotted in dashed lines. Panels c and f: Mean squared displacement (MSD, i.e. $\langle (z(t) - z(0))^2 \rangle$). Dashed lines are, respectively, quadratic and linear fits to the MSD.

transfers interrupt the ballistic motion. However, because the population transfers, which are mediated by molecular vibrations that overlap with the nonadiabatic coupling vector (Section 8.3.1) [37], are reversible, the population also transfers from the dark state manifold back into the LP states. Such reversible exchange of population between the propagating polaritonic states and the stationary dark states leads to a transition around 100 fs from the initial ballistic phase into a diffusion regime. Because our model excludes direct exciton–exciton couplings between molecules, the diffusion process observed in our simulations originates from continuously interrupted transient propagation of bright states at their respective group velocities. These results, therefore, suggest a microscopic rationale for the observation of these two transport regimes in recent experiments [3, 6, 8, 124].

As shown in Figure 8.6d, e, including losses to account for the imperfect cavity mirrors used in experiments, restricts the propagation distance, but has no impact on the mechanism, which still involves a transition from a fast initial ballistic regime into a slower diffusion process. A signature of the losses is a contraction of the wavepacket, manifested by a reduction of the MSD in Figure 8.6f, which was also observed experimentally when UP states were resonantly excited with a broadband laser pulse [7]. The results of our simulations of transport initiated by resonant excitation into either the LP [38] or UP [39], suggest that such wavepacket contraction is due to irreversible radiative decay of population from states with the highest group velocities. Because the cavity lifetime controls both the duration and length of the initial ballistic propagation phase of exciton–polaritons, the Q-factor provides an effective means to tune energy transport in the strong coupling

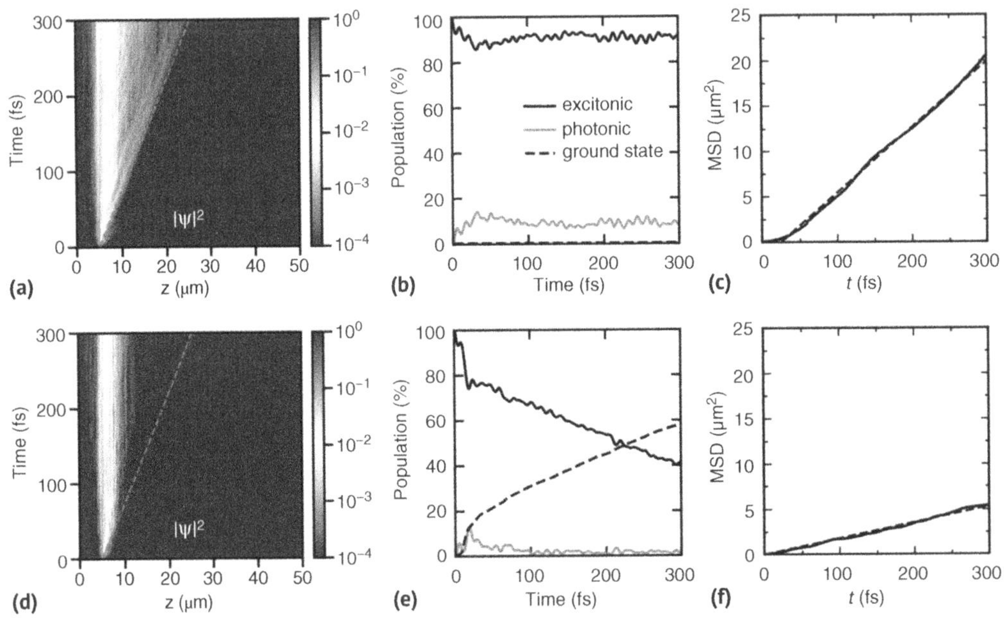

Figure 8.7 Propagation of exciton–polaritons after off-resonant excitation of a single molecule into the S_1 state in a perfect lossless cavity ($\gamma_{cav} = 0\ ps^{-1}$, top) and lossy cavity ($\gamma_{cav} = 66.7\ ps^{-1}$, bottom), containing 1024 Rhodamine molecules. Panels a and d: the total wavepacket as a function of position and time. Panels b and e: contributions of the molecular excitons (black), and cavity mode excitations (gray) to $|\Psi(z,t)|^2$ as a function of time in the lossy cavity. The population in the ground state, created by radiative decay through the imperfect mirrors, is plotted in dashed lines. Panels c and f: Mean squared displacement (MSD, i.e. $\langle (z(t) - z(0))^2 \rangle$). Dashed lines are linear fits to the MSD.

regime [7, 39]. Furthermore, in line with experiments, propagation was observed on timescales that significantly exceeded the cavity mode lifetime. The extension of the overall lifetime is due to the population "surviving" in dark states, which lack photonic contribution and are hence protected against decay [101, 102].

Under off-resonant excitation conditions, in which a single molecule is excited into a localized S_1 electronic state, the propagation, shown in Figure 8.7, proceeds as a diffusion process from the start of the simulation, as evidenced by the linear increase of the MSD of the wavepacket (Figure 8.7c). As in the second propagation stage after on-resonant excitation, the diffusion is due to a continuous and reversible exchange of population between the stationary dark states and propagating bright polaritonic states, which is driven by molecular vibrations that overlap with the nonadiabatic coupling vector ($\mathbf{d}_{mj}$, Eq. 8.17). Because propagation occurs via the polaritonic states, adding a radiative decay restricts the distance over which diffusion occurs, in qualitative agreement with experimental observations [3, 5, 8].

8.3.3 Polaritonic Photochemistry

Arguably the most intriguing application of strong light–matter coupling is the modification of photochemical reactions, as reported in several research articles in the past decade [16–20]. Even if few of these spectacular findings have been contested [116, 133], the tantalizing prospect of using cavities as catalysts continues to appeal to the imagination of chemists.

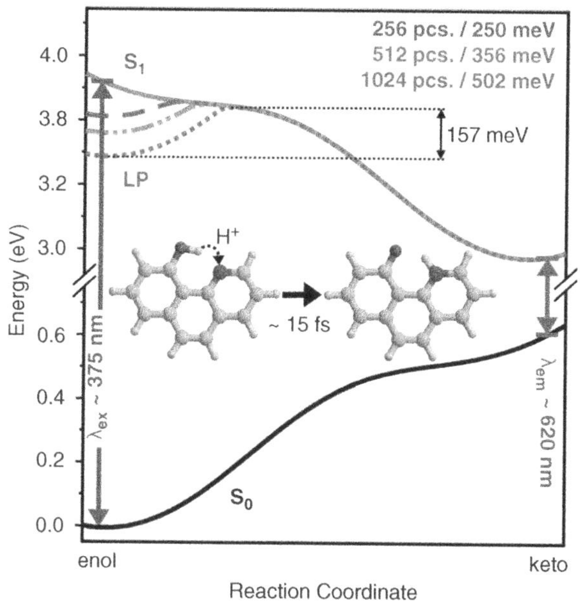

Figure 8.8 Potential energy profiles evaluated at the TDA-CAMB3LYP/6-31G(d) level of theory, for intra-molecular proton transfer in the electronic ground state (S_0, black), first singlet electronic excited state (S_1, solid), in the lower polariton (LP), when 256 (dashed), 512 (dashed-dotted), or 1024 (dotted) *identical* HBQ molecules are strongly coupled to a single confined light mode of an optical cavity that is resonant with the $S_0 \rightarrow S_1$ transition in HBQ. The Rabi splittings are listed in the upper right corner. The energy barrier for proton transfer in the LP state with the highest Rabi splitting is indicated by the double-headed arrow.

The paradigm often invoked to explain the experimental observations is based on a theoretical model of a one-dimensional reaction coordinate that predicts suppression of photochemical reactivity in the collective strong coupling regime due to the formation of a polaritonic potential energy surface with a local minimum in the Franck–Condon region [134]. In Figure 8.8, we illustrate this concept for the ultra-fast excited-state intra-molecular proton transfer in 10-hydroxybenzo[h]quinoline (HBQ). Outside the cavity, this process is barrierless in the S_1 state, and can hence occur on a very short timescale of 15 fs [135, 136]. Without disorder, coupling N *identical* HBQ molecules to a cavity that is resonant with the $S_0 \rightarrow S_1$ electronic transition, leads to the formation of the UP and LP, as well as $N - 1$ dark states. The Rabi splitting creates a local minimum on the potential energy surface of the LP state. As a consequence, the excited-state intra-molecular proton transfer in the lowest-energy excited state of the molecule-cavity system is no longer barrierless (Figure 8.8), and because the Rabi splitting scales as $\sqrt{N}$, the barrier increases with increasing the number of molecules [134].

While the assumptions of a single reaction coordinate and no disorder in this model permitted an exact QM treatment of the problem, they may have little to no relevance for the actual experiments, which have so far been performed with molecules that have many degrees of freedom, and at elevated temperatures. To test the hypothesis by Galego et al. that photochemistry is suppressed in the collective strong coupling regime [134], we performed semiclassical surface hopping MD simulations of 512 HBQ molecules solvated in cyclohexane and equilibrated at 300 K, coupled to a single-mode cavity that is tuned at the excitation energy of HBQ [41].

As shown in Figure 8.9a, the absorption spectrum of HBQ splits inside the cavity, suggesting that the molecule-cavity system is in the strong coupling regime with a Rabi splitting of 230 meV. In the

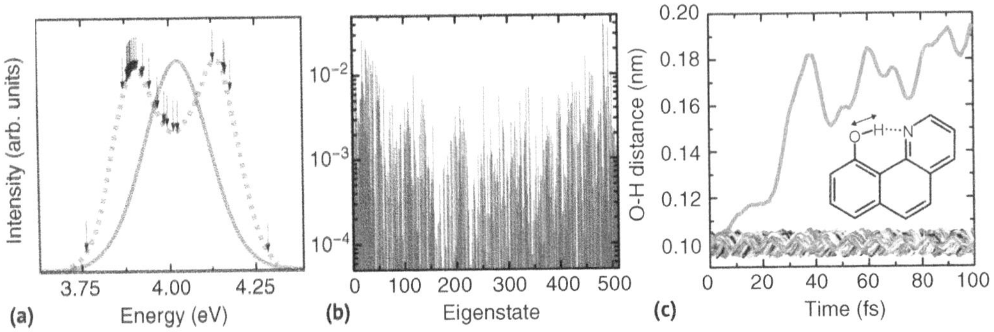

Figure 8.9 Panel a: Absorption spectra of HBQ in cyclohexane at 300 K outside (solid line) and inside (dashed line) of a single-mode microcavity. Arrows indicate the eigenstates in which MD simulations were started. Panel b: Contribution of the cavity mode ($|\alpha^m|^2$, Eq. 8.15) to the eigenstates of the coupled HBQ-cavity system. Panel c: Time evolution of the distance between the oxygen atom (O) and the proton (H) in each of the 512 strongly coupled HBQ molecules (all colors) after excitation into an optically bright polaritonic state (i.e. state 19, second arrow from the left in panel a). As schematically illustrated in the inset, the O–H distance is the reaction coordinate for excited-state intra-molecular proton transfer and a distance of ~ 0.10 nm corresponds to the situation where the proton is bound to the phenol oxygen, whereas a distance around 0.19 nm indicates that the reaction has occurred and the proton has transferred to the nitrogen (N). Here, the latter happened in molecule 400, for which the O–H distance is shown in thick gray line.

22 MD simulations, each of which was started in a different eigenstate of the molecule-cavity system (vertical arrows in Figure 8.9a), intra-molecular proton transfer occurs in one of the molecules, irrespective of the polaritonic eigenstate that was initially excited (Figure 8.9c), and with a rate that is similar to bare HBQ. Thus, even if the Rabi splitting implies strong coupling, the formation of polaritonic states does not lead to a suppression of the photochemistry, in contrast to the theoretical prediction [134].

A closer inspection of the eigenstates suggests that due to the disorder among the HBQ molecules at 300 K, the simple picture for 1D systems at 0 K (Figure 8.8), in which only the lower and upper polaritons are light–matter hybrid states with an altered potential energy surface, is no longer valid [19, 137]. Instead, the cavity mode is distributed over *all* eigenstates of the molecule-cavity system, as shown in Figure 8.9b. Because molecular excitations dominate these states, rather than the cavity mode, the potential energy surfaces are hardly perturbed with respect to the bare molecules.

In conclusion, the results of the MD simulations suggest that in the strong coupling regime, the bright polaritonic states merely act as a gateway to access the reactive dark states and that the efficiency of this gateway, or "optical filter," depends on the overlap between the polaritonic absorption spectrum and the molecular density of states [66]. Therefore, to suppress the reaction, such overlap needs to be avoided, which for HBQ would require entering the regime of ultra strong coupling (USC), in which the Rabi splitting exceeds the excitation energy by more than 10% [53], and the RWA may no longer be valid [29]. However, the USC regime is not only more difficult to reach experimentally due to the limited solubility of HBQ but is also much more challenging for our simulations, as the RWA on which our model is based may break down in this regime [29]. Nevertheless, new experiments, in particular time-resolved spectroscopy measurements, in which the number of molecules in the cavity is varied systematically, are urgently needed to verify the validity of theoretical models and advance our understanding of how collective strong light–matter coupling affects photochemistry.

8.4 Summary and Outlook

While MD simulations have come of age, in particular in structural biology [49, 138] and in material science [46], their use in polaritonic chemistry has remained rather limited. Yet, because the potential applications of strong coupling would typically be under ambient conditions, in which dynamical disorder plays a key role and the traditional two-level models of quantum optics no longer suffice, MD simulations are the method of choice to complement experiment and acquire insights into the mechanisms by which polariton formation influences the material properties at the molecular level.

However, current constraints on computational hardware and software preclude simulations of optical cavity systems with as many molecules as in reality. Therefore, to further narrow down the gap between theory and experiment, it will be essential to "meet in the middle" and develop strategies for performing experiments with as many (or as few) molecules as computer simulations can handle with the aim of facilitating a direct comparison. One way to achieve this goal is to use nanoplasmonics [139, 140]. The sub-wavelength confinement of light by plasmonic nanostructures provides vacuum field strengths that, in combination with the large transition dipole moment of typical organic molecules, suffice to reach the strong coupling regime with far fewer molecules than in Fabry–Pérot microcavities [139, 141, 142]. Indeed, strong coupling has been achieved very recently with as few as five dye molecules in a self-assembled plasmonic nano-cavity [143]. These developments, in combination with the emergence of computational approaches for quantizing the EM field in such cavities [144–146], are promising indications that theory and experiment will meet "in the middle" to provide a complete understanding of the possibilities and limitations of manipulating matter with confined light.

References

1 Coles, D. M., Somaschi, N., Michetti, P., Clark, C., Lagoudakis, P. G., Savvidis, P. G. et al. Polariton-mediated energy transfer between organic dyes in a strongly coupled optical microcavity. *Nat. Mater.* **2014**, *13*, 712–719.

2 Lerario, G., Ballarini, D., Fieramosca, A., Cannavale, A., Genco, A., Mangione, F. et al. High-speed flow of interacting organic polaritons. *Light Sci. Appl.* **2017**, *6*, e16212.

3 Rozenman, G. G., Akulov, K., Golombek, A., Schwartz, T. Long-range transport of organic exciton-polaritons revealed by ultrafast microscopy. *ACS Photonics* **2018**, *5*, 105–110.

4 Georgiou, K., Jayaprakash, R., Othonos, A., Lidzey, D. G. Ultralong-range polariton-assisted energy transfer in organic microcavities. *Angew. Chem. Int. Ed.* **2021**, *60*, 16661–16667.

5 Berghuis, M. A., Tichauer, R. H., de Jong, L., Sokolovskii, I., Bai, P., Ramezani, M. et al. Controlling exciton propagation in organic crystals through strong coupling to plasmonic nanoparticle arrays. *ACS Photonics* **2022**, *9*, 123.

6 Hou, S., Khatoniar, M., Ding, K., Qu, Y., Napolov, A., Menon, V. M. et al. Ultralong-range energy transport in a disordered organic semiconductor at room temperature via coherent exciton-polariton propagation. *Adv. Mater.* **2020**, *32*(28), 2002127.

7 Pandya, R., Ashoka, A., Georgiou, K., Sung, J., Jayaprakash, R., Renken, S. et al. Tuning the coherent propagation of organic exciton-polaritons through dark state delocalization. *Adv. Sci.* **2022**, *9*, 2105569.

8 Balasubrahmaniyam, M., Simkovich, A., Golombek, A., Ankonina, G., Schwartz, T. From enhanced diffusion to ultrafast ballistic motion of hybrid light–matter excitations. *Nat. Mater.* **2023**, *22*, 338–344.

9 Orgiu, E., George, J., Hutchison, J. A., Devaux, E., Dayen, J. F., Doudin, B. Conductivity in organic semiconductors hybridized with the vacuum field. *Nat. Mater.* **2015**, *14*, 1123–1129.

10 Krainova, N., Grede, A. J., Tsokkou, D., Banerji, N., Giebink, N. C. Polaron photoconductivity in the weak and strong light-matter coupling regime. *Phys. Rev. Lett.* **2020**, *124*, 177401.

11 Nagarajan, K., George, J., Thomas, A., Devaux, E., Chervy, T., Azzini, S. et al. Conductivity and photoconductivity of a p-type organic semiconductor under ultrastrong coupling. *ACS Nano* **2020**, *14*, 10219–10225.

12 Bhatt, P., Kaur, K., George, J. Enhanced charge transport in two-dimensional materials through light–matter strong coupling. *ACS Nano* **2021**, *15*, 13616–13622.

13 Kéna-Cohen, S., Forrest, S. R. Room-temperature polariton lasing in an organic single-crystal microcavity. *Nat. Photonics* **2010**, *4*, 371–375.

14 Akselrod, G. M., Young, E. R., Bradley, M. S., Bulović, V. Lasing through a strongly-coupled mode by intra-cavity pumping. *Opt. Express* **2013**, *21*, 12122–12128.

15 Hakala, T. K., Moilanen, A. J., Väkeväinen, A. I., Guo, R., Martikainen, J. P., Daskalakis, K. S. et al. Bose–Einstein condensation in a plasmonic lattice. *Nat. Phys.* **2018**, *14*(7), 739.

16 Hutchison, J. A., Schwartz, T., Genet, C., Devaux, E., Ebbesen, T. W. Modifying chemical landscapes by coupling to vacuum fields. *Angew. Chem. Int. Ed.* **2012**, *51*, 1592–1596.

17 Munkhbat, B., Wersäll, M., Baranov, D. G., Antosiewicz, T. J., Shegai, T. Suppression of photo-oxidation of organic chromophores by strong coupling to plasmonic nanoantennas. *Sci. Adv.* **2018**, *4*, eaas9552.

18 Stranius, K., Herzog, M., Börjesson, K. Selective manipulation of electronically excited states through strong light-matter interactions. *Nat. Commun.* **2018**, *9*, 2273.

19 Mony, J., Climent, C., Petersen, A. U., Moth-Poulsen, K., Feist, J., Börjesson, K. Photoisomerization efficiency of a solar thermal fuel in the strong coupling regime. *Adv. Funct. Mater.* **2021**, *31*, 2010737.

20 Yu, Y., Mallick, S., Wang, M., Börjesson, K. Barrier-free reverse-intersystem crossing in organic molecules by strong light-matter coupling. *Nat. Commun.* **2021**, *12*, 3255.

21 Vahala, K. J. Optical microcavities. *Nature* **2003**, *424*, 839–846.

22 Agranovich, V., Benisty, H., Weisbuch, C. Organic and inorganic quantum wells in a microcavity: Frenkel-wannier-mott excitons hybridization and energy transformation. *Solid State Commun.* **1997**, *102*, 631–636.

23 Lidzey, D. G., Bradley, D. D. C., Skolnick, M. S., Virgili, T., Walker, S., Whittaker, D. M. Strong exciton-photon coupling in an organic semiconductor microcavity. *Nature* **1998**, *395*, 53–55.

24 Törmä, P., Barnes, W. L. Strong coupling between surface plasmon polaritons and emitters: a review. *Rep. Prog. Phys.* **2015**, *78*, 013901.

25 Rider, M. S., Barnes, W. L. Something from nothing: linking molecules with virtual light. *Contemp. Phys.* **2022**, *62*(4), 217–232.

26 Mandal, A., Taylor, M. A., Weight, B. M., Koessler, E. R., Li, X., Huo, P. Theoretical advances in polariton chemistry and molecular cavity quantum electrodynamics. *Chem. Rev.* **2023**, *123*, 9786–9879.

27 Ruggenthaler, M., Sidler, D., Rubio, A. Understanding polaritonic chemistry from ab initio quantum electrodynamics. *Chem. Rev.* **2023**, *123*, 11191–11229.

28 Fregoni, J., Garcia-Vidal, F. J., Feist, J. Theoretical challenges in polaritonic chemistry. *ACS Photonics* **2022**, *9*(4), 1096–1107.

29 Flick, J., Ruggenthaler, M., Appel, H., Rubio, A. Cavity Born-Oppenheimer approximation for correlated electron-nuclear-photon systems. *J. Chem. Theory Comput.* **2017**, *13*, 1616–1625.

30 Flick, J., Ruggenthaler, M., Appel, H., Rubio, A. Atoms and molecules in cavities: From weak to strong coupling in qed chemistry. *Proc. Natl. Acad. Sci. USA* **2017**, *114*, 3026–3034.

31 Haugland, T. S., Ronca, E., KjØnstad, E. F., Rubio, A., Koch, H. Coupled cluster theory for molecular polaritons: changing ground and excited states. *Phys. Rev. X* **2020**, *10*, 041043.

32 Fábri, C., Halász, G. J., Cederbaum, L. S., Vibók, Á. Born–Oppenheimer approximation in optical cavities: from success to breakdown. *Chem. Sci.* **2021**, *12*, 1251–1258.

33 Hertzog, M., Wang, M., Mony, J., Börjesson, K. Strong light–matter interactions: a new direction within chemistry. *Chem. Soc. Rev.* **2019**, *48*, 937–961.

34 Luk, H.-L., Feist, J., Toppari, J. J., Groenhof, G. Multiscale molecular dynamics simulations of polaritonic chemistry. *J. Chem. Theory Comput.* **2017**, *13*, 4324–4335.

35 Tichauer, R. H., Feist, J., Groenhof, G. Multiscale simulations of molecular polaritons: the effect of multiple cavity modes on polariton relaxation. *J. Chem. Phys.* **2021**, *154*, 104112.

36 Sokolovskii, I., Groenhof, G. Non-hermitian molecular dynamics simulations of exciton–polaritons in lossy cavities. *J. Chem. Phys.* **2024**, *160*, 092501.

37 Tichauer, R. H., Morozov, D., Sokolovskii, I., Toppari, J. J., Groenhof, G. Identifying vibrations that control non-adiabatic relaxation of polaritons in strongly coupled molecule-cavity systems. *J. Phys. Chem. Lett.* **2022**, *13*, 6259–6267.

38 Sokolovskii, I., Tichauer, R. H., Morozov, D., Feist, J., Groenhof, G. Multi-scale molecular dynamics simulations of enhanced energy transfer in organic molecules under strong coupling. *Nat. Commun.* **2023**, *14*, 6613.

39 Tichauer, R. H., Sokolovskii, I., Groenhof, G. Tuning coherent propagation of organic exciton-polaritons through the cavity q-factor. *Adv. Sci.* **2023**, *10*, 2302650.

40 Sokolovskii, I., Groenhof, G. Photochemical initiation of polariton-mediated exciton propagation. *Nanophotonics* **2024**, *13*, 2687–2694.

41 Dutta, A., Tiainen, V., Sokolovskii, I., Duarte, L., Markešević, N., Morozov, D. et al. Thermal disorder prevents the suppression of ultra-fast photochemistry in the strong light-matter coupling regime. *Nat. Commun.* **2024**, *15*, 6600.

42 Jaynes, E. T., Cummings, F. W. Comparison of quantum and semiclassical radiation theories with application to the beam maser. *Proc. IEEE* **1963**, *51*, 89–109.

43 Tavis, M., Cummings, F. W. Approximate solutions for an n-molecule radiation-field Hamiltonian. *Phys. Rev.* **1969**, *188*, 692–695.

44 Galego, J., Garcia-Vidal, F. J., Feist, J. Cavity-induced modifications of molecular structure in the strong-coupling regime. *Phys. Rev. X* **2015**, *5*, 041022.

45 Sisto, A., Glowacki, D. R., Martinez, T. J. Ab initio nonadiabatic dynamics of multichromophore complexes: a scalable graphical-processing-unit-accelerated exciton framework. *Acc. Chem. Res.* **2014**, *47*, 2857–2866.

46 Michaelides, A., Martinez, T. J., Alavi, A., Kresse, G., Manby, F. R. Preface: Special topic section on advanced electronic structure methods for solids and surfaces. *J. Chem. Phys.* **2015**, *143*, 102601.

47 Hohenberg, P., Kohn, W. Imhomogeneous electron gas. *Phys. Rev.* **1964**, *136*, 864–871.

48 Groenhof, G. Introduction to QM/MM simulations. *Methods Mol. Biol.* **2013**, *924*, 12489–12491.

49 Groenhof, G., Modi, V., Morozov, D. Observe while it happens: catching photoactive proteins in the act with non-adiabatic molecular dynamics simulations. *Curr. Opin. Struct. Biol.* **2020**, *61*, 106–112.

50 Brunk, E., Rothlisberger, U. Mixed quantum mechanical/molecular mechanical molecular dynamics simulations of biological systems in ground and electronically excited states. *Chem. Rev.* **2015**, *115*, 6217–6263.

51 Warshel, A., Levitt, M. Theoretical studies of enzymatic reactions: dielectric, electrostatic and steric stabilization of carbonium ion in the reaction of lysozyme. *J. Mol. Biol.* **1976**, *103*, 227–249.

52 Groenhof, G. Solving chemical problems with a mixture of quantum-mechanical and molecular mechanics calculations: Nobel Prize in chemistry 2013. *Angew. Chem. Int. Ed.* **2013**, *52*, 12489–12491.

53 Forn-Díaz, P., Lamata, L., Rico, E., Kono, J., Solano, E. Ultrastrong coupling regimes of light-matter interaction. *Rev. Mod. Phys.* **2019**, *91*, 025005.

54 Michetti, P., La Rocca, G. C. Polariton states in disordered organic microcavities. *Phys. Rev. B.* **2005**, *71*, 115320.

55 Kowalewski, M., Bennett, K., Mukamel, S. Cavity femtochemistry: manipulating nonadiabatic dynamics at avoided crossings. *J. Phys. Chem. Lett.* **2016**, *7*, 2050–2054.

56 Davidsson, E., Kowalewski, M. Simulating photodissociation reactions in bad cavities with the lindblad equation. *J. Chem. Phys.* **2020**, *153*, 234304.

57 Vendrell, O. Coherent dynamics in cavity femtochemistry: application of the multi-configuration time-dependent hartree method. *Chem. Phys.* **2018**, *509*, 55–65.

58 Vendrell, O. Collective Jahn-Teller interactions through light-matter coupling in a cavity. *Phys. Rev. Lett.* **2018**, *121*, 253001.

59 Ulusoy, I. S., Gomez, J. A., Vendrell, O. Modifying the nonradiative decay dynamics through conical intersections via collective coupling to a cavity mode. *J. Phys. Chem. A* **2019**, *123*, 8832–8844.

60 Ulusoy, I. S., Vendrell, O. Dynamics and spectroscopy of molecular ensembles in a lossy microcavity. *J. Chem. Phys.* **2020**, *153*, 044108.

61 Boggio-Pasqua, M., Burmeister, C. F., Robb, M. A., Groenhof, G. Photochemical reactions in biological systems: probing the effect of the environment by means of hybrid quantum chemistry/molecular mechanics simulations. *Phys. Chem. Chem. Phys.* **2012**, *14*, 7912–7928.

62 Crespo-Otero, R., Barbatti, M. Recent advances and perspectives on nonadiabatic mixed quantum-classical dynamics. *Chem. Rev.* **2018**, *118*, 7026–7068.

63 Verlet, L. Computer "experiments" on classical fluids. I. Thermodynamical properties of Lennard-Jones molecules. *Phys. Rev.* **1967**, *159*, 98–103.

64 Granucci, G., Persico, M., Toniolo, A. Direct semiclassical simulation of photochemical processes with semiempirical wave functions. *J. Chem. Phys.* **2001**, *114*, 10608–10615.

65 Tully, J. Mixed quantum–classical dynamics. *Faraday Discuss.* **1998**, *110*, 407–419.

66 Groenhof, G., Climent, C., Feist, J., Morozov, D., Toppari, J. J. Tracking polariton relaxation with multiscale molecular dynamics simulations. *J. Chem. Phys. Lett.* **2019**, *10*, 5476–5483.

67 Tully, J. C. Molecular dynamics with electronic transitions. *J. Chem. Phys.* **1990**, *93*, 1061–1071.

68 Hammes-Schiffer, S., Tully, J. C. Proton transfer in solution: molecular dynamics with quantum transitions. *J. Chem. Phys.* **1994**, *101*, 4657–4667.

69 Pechukas, P. Time-dependent semiclassical scattering theory. I. Potential scattering. *Phys. Rev.* **1969**, *181*, 166–174.

70 Pechukas, P. Time-dependent semiclassical scattering theory. II. Atomic collisions. *Phys. Rev.* **1969**, *181*, 174–185.

71 Coker, D. F., Xiao, L. Methods for molecular dynamics with nonadiabatic transitions. *J. Chem. Phys.* **1995**, *102*, 496–510.

72 Zhu, C., Nangia, S., Jasper, A. W., Truhlar, D. G. Coherent switching with decay of mixing: an improved treatment of electronic coherence for non-Born–Oppenheimer trajectories. *J. Chem. Phys.* **2004**, *121*, 7658–7670.

73 Granucci, G., Persico, M. Critical appraisal of the fewest switches algorithm for surface hopping. *J. Chem. Phys.* **2007**, *126*, 134114.

74 Vindel-Zandbergen, P., Ibele, L. M., Ha, J. K., Min, S. K., Curchod, B. F., Maitra, N. T. Study of the decoherence correction derived from the exact factorization approach for nonadiabatic dynamics. *J. Chem. Theory Comput.* **2021**, *17*, 3852–3862.

75 Fregoni, J., Granucci, G., Coccia, E., Persico, M., Corni, S. Manipulating azobenzene photoisomerization through strong light–molecule coupling. *Nat. Commun.* **2018**, *9*, 4688.

76 Fregoni, J., Corni, S., Persico, M., Granucci, G. Photochemistry in the strong coupling regime: a trajectory surface hopping scheme. *J. Comp. Chem.* **2020**, *41*, 2033–2044.

77 Antoniou, P., Suchanek, F., Varner, J. F., Foley IV, J. J. Role of cavity losses on nonadiabatic couplings and dynamics in polaritonic chemistry. *J. Phys. Chem. Lett.* **2020**, *11*, 9063–9069.

78 Anderson, M. H., Ensher J. R, Matthews M. R, Wieman C. E, Cornell. E. A. Observation of Bose-Eeinstein condensation in a dilute atomic vapor. *Science* **1995**, *269*, 198–201.

79 Deng, H., Haug, H., Yamamoto, Y. Exciton-polariton Bose-Einstein condensation. *Rev. Mod. Phys.* **2010**, *82*, 1489.

80 Ramezani, M., Halpin, A., Fernández-Domínguez, A. I., Feist, J., Rodriguez, S. R. K., Garcia-Vidal, F. J. et al. Plasmon-exciton-polariton lasing. *Optica* **2016**, *4*, 31–37.

81 Dietrich, C. P., Steude, A., Tropf, L., Schubert, M., Kronenberg, N. M., Ostermann, K. et al. An exciton-polariton laser based on biologically produced fluorescent protein. *Sci. Adv.* **2016**, *2*, e1600666.

82 Ramezani, M., Halpin, A., Wang, S., Berghuis, M., Rivas, J. G. Ultrafast dynamics of nonequilibrium organic exciton-polariton condensates. *Nano Lett.* **2019**, *19*, 8590–8596.

83 Imamoglu, A., Ram, R. J., Pau, S., Yamamoto, Y. Nonequilibrium condensates and lasers without inversion: Exciton-polariton lasers. *Phys. Rev. A* **1996**, *53*, 4250.

84 Keeling, J., Kéna-Cohen, S. Bose-Einstein condensation of exciton-polaritons in organic microcavities. *Annu. Rev. Phys. Chem.* **2020**, *71*, 435–459.

85 Tischler, J. R., Bradley, M. S., Bulović, V., Song, J. H., Nurmikko, A. Strong coupling in a microcavity led. *Phys. Rev. Lett.* **2005**, *95*, 036401.

86 Schneider, C., Rahimi-Iman, A., Kim, N. Y., Fischer, J., Savenko, I. G., Amthor, M. et al. An electrically pumped polariton laser. *Nature* **2013**, *497*(7449), 348–352.

87 Castellanos, G. W., Ramezani, M., Murai, S., Gómez Rivas, J. Non-equilibrium Bose–Einstein condensation of exciton-polaritons in silicon metasurfaces. *Adv. Opt. Mater.* **2023**, *11*, 2202305.

88 Lidzey, D. G., Virgili, T., Bradley, D. D. C., Skolnick, M. S., Walker, S., Whittaker, D. M. Observation of strong exciton-photon coupling in semiconductor microcavities containing organic dyes and j-aggregates. *Opt. Mater.* **1999**, *12*, 243–247.

89 Lidzey, D. G., Bradley, D. D. C., Virgili, T., Armitage, A., Skolnick, M., Walker, S. Room temperature polariton emission from strongly coupled organic semiconductor microcavities. *Phys. Rev. Lett.* **1999**, *82*, 3316–3319.

90 Lidzey, D. G., Bradley, D. C., Armitage, A., Walker, S., Skolnick, M. S. Photon-mediated hybridization of Frenkel excitons in organic semiconductor microcavities. *Science* **2000**, *288*, 1620–1623.

91 Tartakovskii, A. I., Emam-Ismail, M., Lidzey, D. G., Skolnick, M. S., Bradley, D. D. C., Walker, S. et al. Raman scattering in strongly coupled organic semiconductor microcavities. *Phys. Rev. B* **2001**, *63*, 121302.

92 Lidzey, D. G., Fox, A. M., Rahn, M. D., Skolnick, M. S., Agranovich, V. M., Walker, S. Experimental study of light emission from strongly coupled organic semiconductor microcavities following nonresonant laser excitation. *Phys. Rev. B* **2002**, *65*, 195312-1–195312-10.

93 Lodden, G. H., Holmes, R. J. Electrical excitation of microcavity polaritons by radiative pumping from a weakly coupled organic semiconductor. *Phys. Rev. B* **2010**, *82*, 125317.

94 Coles, D. M., Michetti, P., Clark, C., Tsoi, W. C., Adawi, A. M., Kim, J. S. et al. Vibrationally assisted polariton-relaxation processes in strongly coupled organic-semiconductor microcavities. *Adv. Funct. Mater.* **2011**, *21*, 3691–3696.
95 Somaschi, N., Mouchliadis, L., Coles, D., Perakis, I. E., Lidzey, D. G., Lagoudakis, P. G. et al. Ultrafast polariton population build-up mediated by molecular phonons in organic microcavities. *Appl. Phys. Lett.* **2011**, *99*, 143303.
96 Coles, D. M., Grant, R., Lidzey, D. G., Clark, C., Lagoudakis, P. G. Imaging the polariton relaxation bottleneck in strongly coupled organic semiconductor microcavities. *Phys. Rev. B* **2013**, *88*, 121303.
97 Baieva, S., Hakamaa, O., Groenhof, G., Heikkilä, T. T., Toppari, J. J. Dynamics of strongly coupled modes between surface plasmon polaritons and photoactive molecules: the effect of the stokes shift. *ACS Photonics* **2017**, *4*, 28–37.
98 Mewes, L., Wang, M., Ingle, R. A., Börjesson, K., Chergui, M. Energy relaxation pathways between light-matter states revealed by coherent two-dimensional spectroscopy. *Commun. Phys.* **2020**, *3*, 157.
99 Hulkko, E., Pikker, S., Tiainen, V., Tichauer, R. H., Groenhof, G., Toppari, J. J. Effect of molecular stokes shift on polariton dynamics. *J. Chem. Phys.* **2021**, *154*, 154303.
100 Lüttgens, J. M., Berger, F. J., Zaumseil, J. Population of exciton-polaritons via luminescent sp^3 defects in single-walled carbon nanotubes. *ACS Photonics* **2021**, *8*, 182–193.
101 Agranovich, V. M., Litinskaia, M., Lidzey, D. G. Cavity polaritons in microcavities containing disordered organic semiconductors. *Phys. Rev. B* **2003**, *67*, 085311.
102 Litinskaya, M., Reineker, P., Agranovich, V. M. Fast polariton relaxation in strongly coupled organic microcavities. *J. Lumin.* **2004**, *110*, 364–372.
103 Litinskaya, M., Reineker, P. Balance between incoming and outgoing cavity polaritons in a disordered organic microcavity. *J. Lumin.* **2007**, *1*(22–123), 418–420.
104 Agranovich, V. M., Gartstein, Y. N. Nature and dynamics of low-energy exciton polaritons in semiconductor microcavities. *Phys. Rev. B* **2007**, *75*, 075302.
105 Litinskaya, M. Propagation and localization of polaritons in disordered organic microcavities. *Phys. Lett. A* **2008**, *372*, 3898–3903.
106 Michetti, P., La Rocca, G. C. Simulation of J-aggregate microcavity photoluminescence. *Phys. Rev. B* **2008**, *77*(19), 195301.
107 Mazza, L., Fontanesi, L., La Rocca, G. C. Organic-based microcavities with vibronic progressions: photoluminescence. *Phys. Rev. B* **2009**, *80*, 235314.
108 Michetti, P., La Rocca, G. C. Exciton-phonon scattering and photoexcitation dynamics in J-aggregate microcavities. *Phys. Rev. B* **2009**, *79*, 35325.
109 Michetti, P., La Rocca, G. C. Polariton-polariton scattering in organic microcavities at high excitation densities. *Phys. Rev. B* **2010**, *82*, 115327.
110 Mazza, L., Kéna-Cohen, S., Michetti, P., La Rocca, G. C. Microscopic theory of polariton lasing via vibronically assisted scattering. *Phys. Rev. B* **2013**, *88*, 075321.
111 del Pino, J., Feist, J., Garcia-Vidal, F. J. Quantum theory of collective strong coupling of molecular vibrations with a microcavity mode. *N. J. Phys.* **2015**, *17*, 053040.
112 Martínez-Martínez, L. A., Eizner, E., Kéna-Cohen, S., Yuen-Zhou, J. Triplet harvesting in the polaritonic regime: a variational polaron approach. *J. Chem. Phys.* **2019**, *151*, 054106.
113 Arnardottir, K. B., Moilanen, A. J., Strashko, A., Törmä, P., Keeling, J. Multimode organic polariton lasing. *Phys. Rev. Lett.* **2020**, *125*, 233603.
114 Grant, R. T., Michetti, P., Musser, A. J., Gregoire, P., Virgili, T., Vella, E. et al. Efficient radiative pumping of polaritons in a strongly coupled microcavity by a fluorescent molecular dye. *Adv. Opt. Mater.* **2016**, *4*, 1615–1623.

115 Moilanen, A. J., Daskalakis, K. S., Taskinen, J. M., Törmä, P. Spatial and temporal coherence in strongly coupled plasmonic Bose-Einstein condensates. *Phys. Rev. Lett.* **2021**, *127*, 255301.
116 Eizner, E., Martínez-Martínez, L. A., Yuen-Shou, J., Kéna-Cohen, S. Inverting singlet and triplet excited states using strong light-matter coupling. *Sci. Adv.* **2019**, *5*, eaax4482.
117 Mikhnenko, O. V., Blom, P. W., Nguyen, T. Q. Exciton diffusion in organic semiconductors. *Energy Environ. Sci.* **2015**, *8*, 1867–1888.
118 Zhong, X., Chervy, T., Wang, S., George, J., Thomas, A., Hutchison, J. A. et al. Non-radiative energy transfer mediated by hybrid light-matter states. *Angew. Chem. Int. Ed.* **2016**, *55*, 6202–6206.
119 Zhong, X., Chervy, T., Zhang, L., Thomas, A., George, J., Genet, C. et al. Energy transfer between spatially separated entangled molecules. *Angew. Chem. Int. Ed.* **2017**, *56*, 9034–9038.
120 Georgiou, K., Jayaprakash, R., Askitopoulos, A., Coles, D. M., Lagoudakis, P. G., Lidzey, D. G. Generation of anti-stokes fluorescence in a strongly coupled organic semiconductor microcavity. *ACS Photonics* **2018**, *5*, 4343–4351.
121 Son, M., Armstrong, Z. T., Allen, R. T., Dhavamani, A., Arnold, M. S., Zanni, M. T. Energy cascades in donor-acceptor exciton-polaritons observed by ultrafast two-dimensional white-light spectroscopy. *Nat. Commun.* **2022**, *13*, 7305.
122 Bhatt, P., Dutta, J., Kaur, K., George, J. Long-range energy transfer in strongly coupled donor–acceptor phototransistors. *Nano Lett.* **2023**, *23*(11), 5004–5011.
123 Zakharko, Y., Rother, M., Graf, A., Hähnlein, B., Brohmann, M., Pezoldt, J., Zaumseil, J. Radiative pumping and propagation of plexcitons in diffractive plasmonic crystals. *Nano Lett.* **2018**, *18*, 4927–4933.
124 Pandya, R., Chen, R. Y., Gu, Q., Sung, J., Schnedermann, C., Ojambati, O. S. et al. Microcavity-like exciton-polaritons can be the primary photoexcitation in bare organic semiconductors. *Nat. Commun.* **2021**, *12*, 6519.
125 Wurdack, M., Estrecho, E., Todd, S., Yun, T., Pieczarka, M., Earl, S. K. et al. Motional narrowing, ballistic transport, and trapping of room-temperature exciton polaritons in an atomically-thin semiconductor. *Nat. Commun.* **2021**, *12*, 5366.
126 Xu, D., Mandal, A., Baxter, J. M., Cheng, S. W., Lee, I., Su, H. et al. Ultrafast imaging of coherent polariton propagation and interactions. *Nat. Commun.* **2023**, *14*, 3881.
127 Jin, L., Sample, A. D., Sun, D., Gao, Y., Deng, S., Li, R. et al. Enhanced two-dimensional exciton propagation via strong light–matter coupling with surface lattice plasmons. *ACS Photonics* **2023**, *10*, 1983–1991.
128 Groenhof, G., Toppari, J. J. Coherent light harvesting through strong coupling to confined light. *J. Phys. Chem. Lett.* **2018**, *9*, 4848–4851.
129 Allard, T. F., Weick, G. Disorder-enhanced transport in a chain of lossy dipoles strongly coupled to cavity photons. *Phys. Rev. B* **2022**, *106*, 245424.
130 Engelhardt, G., Cao, J. Polariton localization and dispersion properties of disordered quantum emitters in multimode microcavities. *Phys. Rev. Lett.* **2023**, *130*, 213602.
131 Freixanet, T., Sermage, B., Tiberj, A., Planel, R. In-plane propagation of excitonic cavity polaritons. *Phys. Rev. B* **2000**, *61*, 7233.
132 Myers, D. M., Mukherjee, S., Beaumariage, J., Snoke, D. W., Steger, M., Pfeiffer, L. N. et al. Polariton-enhanced exciton transport. *Phys. Rev. B* **2018**, *98*, 235302.
133 Thomas, P. A., Tan, W. J., Kravets, V. G., Grigorenko, A. N., Barnes, W. L. Non-polaritonic effects in cavity-modified photochemistry. *Adv. Mater.* **2024**, *36*, 2309393.
134 Galego, J., Garcia-Vidal, F. J., Feist, J. Suppressing photochemical reactions with quantized light fields. *Nat. Commun.* **2016**, *7*, 13841.

135 Kim, C. H., Joo, T. Coherent excited state intramolecular proton transfer probed by time-resolved fluorescence. *Phys. Chem. Chem. Phys.* **2009**, *11*, 10266–10269.

136 Lee, J., Kim, C. H., Joo, T. Active role of proton in excited state intramolecular proton transfer reaction. *J. Phys. Chem. A* **2013**, *117*, 1400–1405.

137 Houdré, R., Stanley, R. P., Ilegems, M. Vacuum-field rabi splitting in the presence of inhomogeneous broadening: resolution of a homogeneous linewidth in an inhomogeneously broadened system. *Phys. Rev. A* **1996**, *53*, 2711–2715.

138 Hollingsworth, S. A., Dror, R. O. Molecular dynamics simulation for all. *Neuron* **2018**, *99*, 1129–1143.

139 Chikkaraddy, R., de Nijs, B., Benz, F., Barrow, S. J., Scherman, O. A., Rosta, E. et al. Single-molecule strong coupling at room temperature in plasmonic nanocavities. *Nature* **2016**, *535*, 127–130.

140 Hugall, J. T., Singh, A., van Hulst, N. F. Plasmonic cavity coupling. *ACS Photonics* **2018**, *5*, 43–53.

141 Wersäll, M., Cuadra, J., Antosiewicz, T. J., Balci, S., Shegai, T. Observation of mode splitting in photoluminescence of individual plasmonic nanoparticles strongly coupled to molecular excitons. *Nano Lett.* **2017**, *17*, 551–558.

142 Bitton, O., Haran, G. Plasmonic cavities and individual quantum emitters in the strong coupling limit. *Acc. Chem. Res.* **2022**, *55*, 1659–1668.

143 Heintz, J., Markesevic, N., Gayet, E. Y., Bonod, N., Bidault, S. Few-molecule strong coupling with dimers of plasmonic nanoparticles assembled on DNA. *ACS Nano* **2021**, *15*, 14732–14743.

144 Medina, I., García-Vidal, F. J., Fernández-Domínguez, A. I., Feist, J. Few-mode field quantization of arbitrary electromagnetic spectral densities. *Phys. Rev. Lett.* **2021**, *126*, 093601.

145 Fregoni, J., Haugland, T. S., Pipolo, S., Giovannini, T., Koch, H., Corni, S. Strong coupling between localized surface plasmons and molecules by coupled cluster theory. *Nano. Lett.* **2021**, *21*, 6664–6670.

146 Sánchez-Barquilla, M., García-Vidal, F. J., Fernández-Domínguez, A. I., Feist, J. Few-mode field quantization for multiple emitters. *Nanophotonics* **2022**, *11*, 4363–4374.

9

Disorder in Cavity-modified Transport and Chemistry

David Hagenmüller[1], Jérôme Dubail[1,2], Francesco Mattiotti[1], Guido Pupillo[1] and Johannes Schachenmayer[1]

[1] *CESQ, ISIS (UMR 7006), Université de Strasbourg and CNRS, Strasbourg, France*
[2] *LPCT (UMR 7019), Université de Lorraine and CNRS, Nancy, France*

9.1 Introduction

When a matter excitation couples to the electromagnetic field confined in a cavity, new hybrid polariton modes can be formed. This typically occurs when the light–matter coupling strength g exceeds the linewidths of the bare modes, a situation referred to as the strong coupling (SC) regime of cavity quantum-electrodynamics (cQED) [1, 2]. As long as the coupling strength remains a negligible fraction of the bare mode frequencies, one can distinguish two different situations: (i) the individual SC, where a single quantum emitter is coupled to one or several cavity modes and (ii) the *collective* SC regime, where N emitters are coupled to the photonic degrees of freedom. A minimal cQED model for a single cavity mode at resonance with the emitter transition, (i) is given by the Jaynes–Cummings model [3], which gives rise to a nonlinear polariton ladder, where the two polaritons in each subspace containing n excitations are separated by a splitting $2g\sqrt{n}$. Conversely, (ii) is described by the Tavis–Cummings (TC) model [4], whose single-excitation linear spectrum consists of 2 polaritons separated by a splitting $2g_c$ (with $g_c = g\sqrt{N}$ the collective coupling strength) and $N - 1$ dark states at the energy of the bare modes [5, 6].

Since the first experiments showing that SC of a molecular electronic transition to a cavity mode can give rise to a significant change in the photochemical reaction rate [7], polaritonic states in organic materials have been explored as a tool to control chemical reactions involving both excited [8–15] and ground electronic manifolds [16–18]. This has given rise to a new active field known as polaritonic chemistry [19–22]. A fundamental issue that has currently been discussed for theoretical explanations in polaritonic chemistry is what is known as "$1/N$ problem": while collective coupling strengths can be made very large (e.g. with g_c on the eV scale for electronic SC), individual single-molecule couplings $g \propto 1/N$ are consequently typically small since a huge number $N \gg 1$ of molecules are expected to be coupled per cavity mode in experiments. This raises an important question about how the small individual couplings lead to modified chemistry on the single molecule level?

Polariton Chemistry: Molecules in Cavities, First Edition. Edited by Joel Yuen-Zhou, Noel C. Giebink and Raphael F. Ribeiro.

A similar issue arises in the context of cavity-modified transport under SC at the level of single carriers that can be either excitons or charges. For excitons, the effect of polaritons on the energy transfer between donor and acceptor molecules has been well investigated in the context of organic optoelectronics [23–27], while polariton-mediated ultrafast transport in a chain of quantum emitters was highlighted as a way to overcome localization in disordered organic systems [28, 29].

The interplay between SC and disorder induced by, e.g., inhomogeneous broadening of the excitonic transitions in organic semiconductors may provide an answer to the $1/N$ issue. While it is known that motional narrowing occurring for strong enough light–matter coupling efficiently protects (collective) polaritonic states against disorder-induced broadening [30–32], disorder leads to an hybridization of cavity photons with the continuum of dark states as a result of the broken translational invariance [33]. In the linear single excitation regime, each one of the $N-1$ dark states acquires a finite photon weight $\sim 1/N$, which thus provides a total photon weight (of order one) comparable to the photon weight of the polaritons in the thermodynamic limit $N \to \infty$ [34–36].

In this context, the dark state contribution to exciton transport and localization properties under SC has been the topic of recent theoretical [34, 35, 37–43] and experimental investigations [36, 43]. In particular, it was shown that disorder-induced hybridization of the dark states with a cavity mode can lead to a particular type of localization that we denote as "semilocalization," which corresponds to a localization distributed over multiple sites regardless of the distance separating them. This semilocalization has been shown to be able to contribute to a diffusive-like exciton transport in the large-N limit [34, 35]. Furthermore, excitation transfer through semilocalized states plays a crucial role in modifying vibrational wave-packet dynamics as shown in [44], which hints to an important role of disorder and semilocalization in polaritonic chemistry. Also, in recent experimental observations of modified chemical reaction rates under vibrational SC [45] disorder has been argued to play a key role.

The main goal of this book chapter is to introduce recent theoretical advances that are based on a bottom-up minimal quantum optics model approach to cavity-coupled molecules. A full theoretical quantum treatment of realistic molecules with many degrees of electronic and motional degrees of freedom coupled to the full electromagnetic environment is impossible in the large N limit without strong approximations. To understand and isolate fundamental physical effects, it can instead be instructive to reduce the physics of the problem to models that include the relevant degrees of freedom in their simplest form. For considering only electro-photonic degrees of freedom, this leads us to consider disordered Tavis–Cummings (TC) Hamiltonians. When adding nuclear motional degrees of freedom, we can consider them in their simplest form in terms of a disordered Holstein–Tavis–Cummings (HTC) model [10, 44, 46]. Here, we introduce these models and demonstrate how they help us to understand the role of disorder. In particular, we will introduce the main concepts of semilocalization and its impact on light–matter-coupled molecules in disordered TC models in Section 9.2. There, we will explain how disorder modifies eigenstate properties and impacts long-range excitation transport. In Section 9.3, we then review how disorder-induced semilocalization modifies vibrational dynamics in disordered HTC models and can result in non-Gaussian vibrational wave-packets. We provide a conclusion and an outlook on the field in Section 9.4.

9.2 Semilocalization

In this section, we discuss how strong light–matter coupling extending over the entire volume of the system can compete with localization and lead to the phenomenon of *semilocalization*. Coherent transport in disordered systems can be strongly inhibited due to Anderson localization [47], which stems from constructive interferences between loop trajectories that do not contribute to transport in the thermodynamic limit. While an arbitrarily small disorder leads to the localization of eigenstates in a d-dimensional space for $d = 1$ and $d = 2$, a metal–insulator transition occurs for $d \geq 3$ [48, 49]. In Section 9.2.1 we will first introduce the main theoretical concepts around localization in an archetypal disordered TC model with nearest-neighbor hopping. We show that strong collective coupling can deeply affect the Anderson transition for $d = 3$ through dark-state semilocalization. Different physical quantities of interest, such as the return probability, inverse participation ratio, and level statistics, can be computed numerically in the single-excitation subspace, showing that semilocalization is characterized by an hybrid probability distribution that interpolates between Gaussian orthogonal ensemble and Poisson law and is distributed over multiple sites that can be arbitrarily distant from one another. These properties are then analyzed in more detail in Section 9.2.2 in the absence of nearest-neighbor hopping, in which case d becomes irrelevant. Here, we derive asymptotically exact results in the large-N limit, showing that dark state semilocalization manifests itself as peculiar, diffusive-like transport properties and is associated with a multifractal spectrum characterized by a continuous set of fractal dimensions. Multifractal properties in our minimalistic model turn out to be exactly the same as in Anderson metal-insulator transition on the Bethe lattice.

9.2.1 The Disordered TC Model with Hopping

The effect of strong light–matter coupling on the Anderson transition can be understood in the framework of the disordered TC model with nearest-neighbor hopping. Here, a disordered ensemble of N two-level quantum emitters forming a cubic lattice ($d = 3$) is embedded in a cavity with only one spatially uniform mode (Figure. 9.1a). We refer the reader to Refs. [40, 41, 50] for extensions of the model to a multimode cavity with dissipation. The Hamiltonian is $\hat{H} = \hat{H}_0 + \hat{H}_\mathrm{I}$, with ($\hbar \equiv 1$)

$$\begin{aligned}\hat{H}_0 &= \omega_c \hat{a}^\dagger \hat{a} + \sum_i (w_e + w_i)\hat{\sigma}_i^+ \hat{\sigma}_i^- - J \sum_{\langle i,j \rangle} \hat{\sigma}_i^+ \hat{\sigma}_j^- \\ \hat{H}_\mathrm{I} &= g \sum_i (\hat{a}\hat{\sigma}_i^+ + \hat{a}^\dagger \hat{\sigma}_i^-).\end{aligned} \tag{9.1}$$

$\hat{H}$ commutes with the excitation number operator $\hat{M} = \sum_i \hat{\sigma}_i^+ \hat{\sigma}_i^- + \hat{a}^\dagger \hat{a}$. We restrict our discussion to the single-excitation subspace defined as the eigenspace of $\hat{M}$ with eigenvalue 1, which is of dimension $N+1$. This eigenspace is spanned by the N states $\{|\boldsymbol{i}, 0\rangle\}$ with no photon in the cavity and one excited emitter on lattice site $\boldsymbol{i} \in [1, N]$, and by the state $|G, 1\rangle$ with one photon in the cavity and all emitters in their ground state. Introducing the ground state, $|G, 0\rangle$, the spin lowering and photon annihilation operators in this basis read $\hat{\sigma}_i^- = |G, 0\rangle\langle \boldsymbol{i}, 0|$ and $\hat{a} = |G, 0\rangle\langle G, 1|$. Since $[\hat{H}, \hat{M}] = 0$, the spectrum only depends on the emitter frequencies through the detunings $w_e + w_i - \omega_c$, with ω_c the cavity mode frequency and w_e the average emitter transition. We consider the cavity mode in resonance with the average emitter transition, and set $w_e = \omega_c = 0$ without loss of generality. On-site (diagonal) disorder is introduced by drawing the N independent and identically distributed random variables w_i from a uniform distribution with support $[-W/2, W/2]$. The disorder strength is thus governed by the parameter W.

The last term in $\hat{H}_0$ governs hopping of excitations between nearest neighbor sites $\langle \boldsymbol{i}, \boldsymbol{j} \rangle$ with rate J. Assuming periodic boundary conditions, this term is diagonalized by the Fourier modes $\hat{b}_{\boldsymbol{q}} = \sum_i exp(-i\boldsymbol{q} \cdot \boldsymbol{i})\hat{\sigma}_i^- / \sqrt{N}$, with $\boldsymbol{q}$ the quasi-momentum. The term H_I is the TC emitter-cavity coupling with local strengths g, which are assumed to be spatially uniform. This term can be written

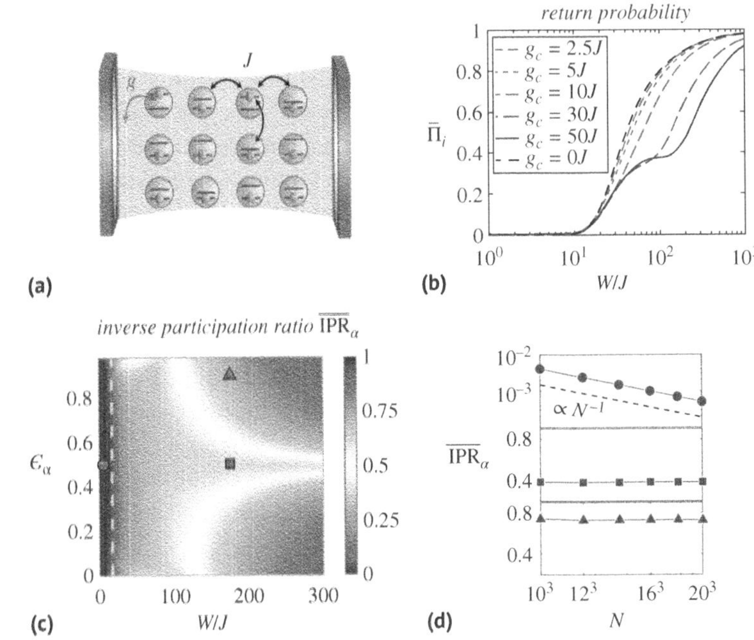

Figure 9.1 (a) An excitation can hop with rate J on a disordered 3D cubic lattice with N sites. Emitters are coupled to a single (homogeneous) cavity mode with local strength g. (b) Disorder-averaged return probability Π_i as a function of W/J for the central site $\boldsymbol{i}$ of a $N = 15^3$ cube. (c) Disorder-averaged inverse participation ratio $\overline{\mathrm{IPR}}_\alpha$ as a function of W/J and the renormalized dark-state energy ϵ_α for $g_c = 30J$. The white dashed line indicates $W = W_c$. (d) Finite-size scaling of $\overline{\mathrm{IPR}}_\alpha$ for the parameters corresponding to the symbols in (c): circles ($W = 5J$, $\epsilon_\alpha = 0.5$); squares ($W = 175J$, $\epsilon_\alpha = 0.5$); triangles ($W = 175J$, $\epsilon_\alpha = 0.9$). We used bins of widths 0.02 and compute the disorder average over ~ 100 realizations. *Source*: Reprinted from Ref. [34].

in the form $H_I = g_c(\hat{a}\hat{b}_{\mathbf{0}}^\dagger + \hat{a}^\dagger\hat{b}_{\mathbf{0}})$, and couples the symmetric bright mode $\hat{b}_{\mathbf{q}=0}$ to cavity photons. Since the scaling of the light–matter coupling with the system size comes from its dependence on the cavity-mode volume V as $\sim 1/\sqrt{V}$, the collective coupling strength $g_c \equiv g\sqrt{N}$ (half of the collective Rabi splitting) is the relevant physical parameter in the thermodynamic limit $N \to \infty$ as it remains independent of N for a fixed density N/V.

In the absence of disorder ($W = 0$), $\hat{H}_I$ has two polariton eigenstates $|\psi_\pm\rangle = (\hat{b}_{\mathbf{q=0}}^\dagger \pm \hat{a}^\dagger)/\sqrt{2}|G,0\rangle$ with energies $E_\pm = \pm g_c$, as well as $N-1$ uncoupled dark states $|\psi_{\alpha\neq\pm}\rangle = \hat{b}_{\mathbf{q\neq0}}^\dagger|G,0\rangle$ with vanishing photon weight, $\langle G,1|\psi_{\alpha\neq\pm}\rangle = 0$. Crucially, finite disorder ($W \neq 0$) leads to a coupling between the bright and the dark states since $\hat{H}_0$ is nondiagonal in quasi-momentum space, as a result of the broken translational invariance. The dark eigenstates, therefore, acquire a finite photon weight and can be thought of as "grey" states.

In the absence of light–matter coupling ($g_c = 0$), $\hat{H} = \hat{H}_0$ corresponds to the Anderson model [47], displaying a W-dependent mobility edge that determines a metal–insulator transition at the critical disorder strength $W_c \simeq 16.5J$ (for energy states lying in the middle of the band) [51, 52]. While for $W \ll W_c$ the eigenstates $|\psi_\alpha\rangle$ resemble extended Bloch states, they are exponentially localized around given sites for $W > W_c$, e.g. $|\langle \boldsymbol{j},0|\psi_\alpha\rangle|^2 \propto e^{-|\boldsymbol{i}-\boldsymbol{j}|/\xi}$ for a state localized on-site $\boldsymbol{i}$, with ξ a localization length that depends on the ratio W/J. In the following, we investigate the spectral and transport properties of the model for $g_c, W, J \neq 0$.

9.2.1.1 Inverse Participation Ratio

Information about the localization of dark eigenstates with energy E_α ($\alpha \neq \pm$) is provided by the inverse participation ratio (IPR),

$$\mathrm{IPR}_\alpha = \sum_{i=1}^{N} |\langle \boldsymbol{i}, 0|\psi_\alpha\rangle|^4 + |\langle G, 1|\psi_\alpha\rangle|^4 \tag{9.2}$$

The IPR is a useful quantity to study Anderson localization–delocalization transitions as it allows to characterize strong fluctuations of the eigenfunctions at all length scales (often referred to as spectrum self-similarity) that are typically observed at the transition point. Introducing the average of an observable O over disorder realizations, $\overline{O} \equiv \frac{1}{W^N} \int \prod_j dw_j\, O(\{w_j\})$, where the N independent variables w_j are all uniformly distributed in the interval $[-W/2, W/2]$, $\overline{\mathrm{IPR}_\alpha}$ corresponds to the disorder average of the second moment of the probability distribution $p_\alpha(\{w_i\})$, which consists of the squared amplitudes of the eigenstates $|\psi_\alpha\rangle$ in the basis of states $|G, 1\rangle$ and $|\boldsymbol{i}, 0\rangle$ for all sites $\boldsymbol{i}$. For a d-dimensional system of linear size L, with a number of sites $N \sim L^d$, $\overline{\mathrm{IPR}_\alpha}$ typically decays as a power-law with L [53],

$$\overline{\mathrm{IPR}_\alpha} \underset{L\to\infty}{\sim} L^{-D_\mathrm{f}},$$

with an exponent D_f that is the fractal dimension of the support of the wavefunction. In particular, for an extended (metallic) eigenstate, all components $|\langle \boldsymbol{i}, 0|\psi_\alpha\rangle|^2$ are of order $O(1/N)$, which leads to $\overline{\mathrm{IPR}_\alpha} = O(1/N)$, and thus, a fractal dimension $D_\mathrm{f} = d$. In contrast, a localized eigenstate has a finite number of components of order $O(1)$, which yields $\overline{\mathrm{IPR}_\alpha} = O(1)$ and thus $D_\mathrm{f} = 0$. Any intermediate value $0 < D_\mathrm{f} < 1$ is a signature of fractality [53].

9.2.1.2 Return Probability

The IPR is also related to the return probability

$$\Pi_i = \lim_{T\to\infty} \frac{1}{T} \int_0^T dt |\langle \boldsymbol{i}, 0|e^{-i\hat{H}t}|\boldsymbol{i}, 0\rangle|^2, \tag{9.3}$$

which is the infinite time-averaged probability for an excitation initially located on site $\boldsymbol{i}$ to be found at the same site. Indeed, by summing the IPR (Eq. 9.2) over all eigenstates one can easily see that $\sum_i \Pi_i = \sum_\alpha \mathrm{IPR}_\alpha$. The IPR can thus be interpreted as the contribution of a given eigenstate to the spatially averaged return probability.

In the single-excitation sector, it is easy to diagonalize the Hamiltonian $\hat{H}$ numerically in a cubic lattice of moderate sizes. We do so for $N = 15^3$ for a given disorder configuration and then compute the disorder-averaged quantities of interest (see Figure 9.1b–d). The disorder-averaged return probability $\overline{\Pi_i}$ for the central site $\boldsymbol{i}$ is shown in Figure 9.1b. For $g_c = 0$ (dashed line), $\overline{\Pi_i}$ increases from 0 (metallic phase) to 1 (localized phase) when increasing the disorder strength W/J. Remarkably, $\overline{\Pi_i}$ exhibits a plateau $\simeq 0.4$ for $g_c > W > W_c$, which persists up to large disorder strengths ($W \sim 100J$ for $g_c = 50J$). The disorder-averaged $\overline{\mathrm{IPR}_\alpha}$ of the dark eigenstates with energies E_α is shown in Figure 9.1c as a function of W/J, together with its scaling with N in Figure 9.1d for $g_c = 30J$. For clarity, we use a renormalized energy scale $\epsilon_\alpha = (E_\alpha - W/2)/W$ with $\epsilon_\alpha \in [0, 1]$. For each disorder realization, we bin different levels into groups with equal energy width and compute the disorder average in each bin. Three different regions can be distinguished: (i) a metallic region (left part) where $\overline{\mathrm{IPR}_\alpha} \sim O(1/N) \approx 0$ ($D_\mathrm{f} = 3$) for $W < W_c$; (ii) a localized region (upper-right and lower-right corners) where $\overline{\mathrm{IPR}_\alpha} \sim O(1) \approx 1$ ($D_\mathrm{f} = 0$) for $W > g_c$; (iii) An extended area (central part) where $\overline{\mathrm{IPR}}$ exhibits an intermediate value ≈ 0.4 ($D_\mathrm{f} = 0$) for $g_c > W > W_c$, and where $\overline{\mathrm{IPR}_\alpha}$ does not scale with N similarly as for localized states. Such a reduced value of the IPR as compared to the localized phase where $\overline{\mathrm{IPR}}_\alpha \approx 1$ suggests that the dark eigenstates in this regime are also localized but on multiple sites. We will see in Section 9.2.2 that this information (e.g., the fractal dimension $D_f = 0$) is, however, not sufficient to fully characterize the dark state spectrum in this regime.

9.2.1.3 Level Statistics

Localization properties of dark eigenstates are also characterized by their level statistics [54], i.e. the probability distribution function $P(s_\alpha)$ of the spacings between adjacent eigenenergies, $s_\alpha = \epsilon_{\alpha+1} - \epsilon_\alpha$. In Figure 9.2a, we compute $P(s_\alpha)$ numerically by binning the eigenstates into groups with equal energy width. As expected, we obtain a Wigner-Dyson distribution, $P(s_\alpha) = \frac{\pi}{2} s_\alpha exp(-\frac{\pi}{4} s_\alpha^2)$, in the metallic region ($W < W_c$, circles), which is the signature of a correlated spectrum featuring level repulsion ($P(0) = 0$) that is well described by Gaussian random matrix models [55]. Conversely, the fully localized region ($W > g_c > J$) is characterized by a Poissonian distribution, $P(s_\alpha) = exp(-s_\alpha)$ [56], where the dark states exhibit independent and identically distributed (uncorrelated) energy levels. We observe that the region with a reduced IPR (as compared to the insulating phase) exhibits a level statistics close to a semi-Poissonian distribution [57], $P(s_\alpha) = 4 s_\alpha exp(-2 s_\alpha)$. We interpret those characteristics as a particular type of localization, which we dubbed semilocalization. It is interesting to point out that such a hybrid distribution with both level repulsion and Poissonian tail was proposed early on as a good candidate to describe the spectrum at the Anderson transition point [58, 59].

A related quantity to characterize the spectrum of the dark states is the disorder-averaged energy shift $\overline{\Delta}_\alpha = N\{E_\alpha - [(w_i + w_{i+1})/2]\}$, which is shown in Figure 9.2b. Here w_i and w_{i+1} refer to the closest bare levels immediately below and above the dark-state eigenenergy E_α. $\overline{\Delta}_\alpha$ is computed in a similar way as the IPR, i.e. by binning energy levels into groups with equal energy window and then computing the disorder average. While in the localized phase $W > J$ (triangle) the

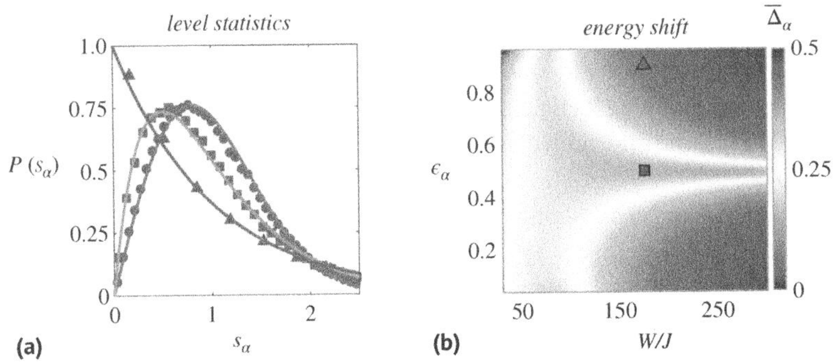

Figure 9.2 (a) Numerically computed probability distributions $P(s_\alpha)$ for the parameters corresponding to the symbols in Figure 9.1(c, d). The solid lines refer to the analytical formulas given in the main text: Wigner-Dyson distribution (circles), Poissonian distribution (triangles), semi-Poissonian distribution (squares). (b) Disorder-averaged level deviations $\overline{\Delta}_\alpha$ as a function of W/J for $g_c = 30J$. Other parameters are the same as in Figure 9.1. *Source*: Reprinted from Ref. [34].

eigenenergies are found to be very close to the bare energy levels w_i and thus follow the same Poissonian statistics, they are pushed in between the bare levels as a consequence of hybridization of the latter (square) when W decreases toward $g_c = 30J$, thus approaching SCs. Note that the dark states become metallic when $W < W_c$ (left side of Figure 9.2b). Exact semi-Poisson statistics arise when assuming that the dark state energy levels are exactly equidistant from the two closest bare levels, i.e., $\Delta_\alpha = 0$, which provides the distribution of level spacings

$$P(s_\alpha) = \int \mathrm{d}x\mathrm{d}y\, \delta\left(s_\alpha - \frac{x+y}{2}\right) \mathrm{e}^{-x}\mathrm{e}^{-y} = 4s_\alpha \mathrm{e}^{-2s_\alpha}$$

In Refs. [34, 35], the distributions of level spacings were found to be close to – but not exactly equal to – that semi-Poisson distribution.

9.2.2 Arrowhead Matrix Model and Dark State Multifractality

Significant insight about semilocalized dark states is obtained by investigating the Hamiltonian (Eq. 9.1) without nearest-neighbor hopping ($J = 0$) [35], in which case the spectrum becomes independent of the spatial dimension d. In fact, the simpler model obtained by setting $J = 0$ exhibits all the phenomenology of *semilocalization*, and it is an ideal starting point for analytical calculations. As in Section 9.2.1, we restrict our analysis to the single-excitation subspace.

9.2.2.1 Perturbation Theory

A first approach is to treat the light–matter coupling $\hat{H}_\mathrm{I}$ in second-order perturbation theory. In this case, one can show that a (trivially) localized eigenstate on site i, $|i, 0\rangle$, for $g_c = 0$ acquires a disorder-averaged amplitude

$$\overline{\frac{g^4}{(w_i - w_j)^2 w_i^2}} \underset{N\to\infty}{\simeq} \frac{4g_c^4[4 - 2log(4)]}{NW^4}$$

on site $j \neq i$ via hybridization with the cavity mode for $g^2 \ll |(w_j - w_i)w_i|$ [34]. This probability amplitude rising with g_c/W is consistent with the IPR analysis suggesting that semilocalized dark states are distributed over multiple sites, which can be arbitrarily distant from one another. Note that a similar behavior was reported for dissipative couplings to a common reservoir [38]. The failure of perturbation theory for energy levels close to the center of the distribution, $w_i \approx w_j \approx 0$ explains the persistence of semilocalized dark states in the middle of the spectrum $\epsilon_\alpha \approx 0.5$ for $W \gg g_c \gg J$ (see Figure 9.1c). Even if $g_c \ll W$ in this regime, the coupling to the cavity is nonperturbative due to the proximity of the energy levels with the cavity mode (resonant situation), and the energy levels can therefore hybridize.

9.2.2.2 Arrowhead Matrix Model

Let us rewrite the state with one photon, $|G, 1\rangle$, as $|N + 1\rangle$, and all the states $|i, 0\rangle$ with an excited emitter on site i as $|i\rangle$. Indexing the emitters from 1 to N, one can then identify the Hamiltonian Eq. (9.1) (with $J = \omega_c = w_e = 0$) with the one of an $N + 1$-level system,

$$\hat{H} = \sum_{i=1}^{N} w_i |i| \langle i\rangle + g \sum_{i=1}^{N} (|N+1| \langle i\rangle + |i| \langle N+1\rangle) = \sum_{1\leq i,j\leq N+1} h_{i,j} |i| \langle j\rangle$$

where the $(N + 1) \times (N + 1)$ Hamiltonian matrix h turns out to be an *arrowhead matrix*,

$$h = \begin{pmatrix} w_1 & & & & g \\ & w_2 & & & g \\ & & \ddots & & \vdots \\ & & & w_N & g \\ g & g & \dots & g & 0 \end{pmatrix} \tag{9.4}$$

This is interesting because it leads to a simple form for the energy spectrum and for the eigenstates, $h\psi_\alpha = E_\alpha \psi_\alpha$,

$$\psi_\alpha = \frac{1}{\mathcal{N}_\alpha} \begin{pmatrix} \frac{g}{E_\alpha - w_1} & \frac{g}{E_\alpha - w_2} & \cdots & \frac{g}{E_\alpha - w_N} & 1 \end{pmatrix}^{\mathrm{T}} \tag{9.5}$$

where $\mathcal{N}_\alpha = \left(1 + \sum_i \frac{g^2}{(E_\alpha - w_i)^2}\right)^{-1/2}$ is a normalization constant, T denotes the transpose, and where the energy E_α is a solution of the secular equation,

$$E_\alpha = \sum_{i=1}^{N} \frac{g^2}{E_\alpha - w_i} \tag{9.6}$$

It can easily be seen (see Ref. [35]) that there are exactly $N + 1$ solutions to this equation, and that they are interlaced with the bare energies w_i: $E_1 < w_1 < E_2 < w_2 < \cdots < E_{N-1} < w_N < E_{N+1}$. Moreover, for SC $g_c \gg W$, the secular equation for the minimal (resp. maximal) energy becomes $E_1 \simeq \frac{Ng^2}{E_1} = \frac{g_c^2}{E_1}$ (resp. $E_{N+1} \simeq \frac{g_c^2}{E_{N+1}}$), so $E_1 \simeq -g_c$ and $E_{N+1} \simeq +g_c$, while all the other energy levels are in the interval $[-W/2, W/2]$. We can thus identify the first (resp. last) eigenstate as the lower polariton (resp. upper polariton), while the $N - 1$ central eigenstates all correspond to dark states.

Many results can be obtained by applying the above formulas for arrowhead matrices in the case of any random distribution of the bare energies w_i with finite support. These results include asymptotically exact formulas (in the limit of large N) for static and dynamical quantities of physical interest, such as the photonic weights of the eigenstates, the spectral function, the energy shift, the time-dependent escape probability from an initial site, and many others. Here, we simply illustrate the general flavor of these results with the energy shift and the escape probability, and refer to Ref. [35] for more details. The escape probability is particularly interesting because it shows that the arrowhead matrix model exhibits a peculiar diffusive-like behavior.

9.2.2.3 Escape Probability

The escape probability from a site i is defined as

$$P_i(t) = 1 - |\langle i|e^{-i\hat{H}t}|i\rangle|^2 = 1 - |[e^{-iht}]_{ii}|^2 \tag{9.7}$$

For finite times $1/g_c \ll t \ll N/g_c$, the escape probability exhibits a diffusive-like behavior, $P_i(t) = \Gamma(w_i)t$, with an escape rate governed by the individual coupling strength g and given by the Fermi golden rule,

$$\Gamma(w_i) = 2\pi g^2 \overline{A}(w_j) \tag{9.8}$$

in the limit of large N. Here $\overline{A}(w_j)$ is the disorder-averaged photon spectral function $A(\epsilon) = -\frac{1}{\pi}\mathrm{Im}[(\epsilon - h + i0^+)^{-1}]_{N+1,N+1}$, which satisfies the normalization property $\int d\epsilon A(\epsilon) = 1$, and can be computed exactly at large N as [35]

$$\overline{A}(\epsilon) = \frac{W}{\pi^2 g_c^2} \frac{1}{1 + \left(\frac{1}{\pi}\log\left|\frac{\epsilon+(W/2)}{\epsilon-(W/2)}\right| - \frac{\epsilon W}{\pi g_c^2}\right)^2} \tag{9.9}$$

Diffusive properties are often characterized by the mean square displacement (growing linearly with time) instead of the escape probability. Nevertheless, one can easily show that these two quantities are in fact proportional. For an excitation initially localized on site i, the disorder-averaged mean square displacement reads

$$\overline{\langle x^2(t)\rangle} = \overline{\sum_{j\neq i} (i-j)^2 \left|\langle i|e^{-i\hat{H}t}|j\rangle\right|^2}$$

and since there is no correlation between random energies and positions, one has

$$\overline{\langle x^2(t)\rangle} = \sum_{j\neq i} \overline{(i-j)^2}\, \overline{\left|\langle i|e^{-i\hat{H}t}|j\rangle\right|^2} \propto \sum_{j\neq i} \overline{\left|\langle i|e^{-i\hat{H}t}|j\rangle\right|^2} = P_i(t)$$

As it is obvious from Figure 9.1b, the large-N limit of the return probability (Eq. 9.3) is not simply connected to the asymptotic formula $P_i(t) = \Gamma(w_i)t$ for the escape probability as $\lim_{N\to\infty} \Pi_i = 1 - \lim_{t\to\infty} P_i(t)$. This stems from the noncommutativity of the limits $N \to \infty$ and $t \to \infty$, which is due to large fluctuations of the wavefunctions that typically are not smooth functions (see Figure 9.3a). However, it can be shown that certain important quantities, such as the Green functions $\langle i|e^{-i\hat{H}t}|j\rangle$ and $\langle N+1|e^{-i\hat{H}t}|N+1\rangle$, are not sensitive to fluctuations of the distribution of bare energies in the large-N limit. They are instead only sensitive to the average density of states N/W, which greatly simplifies analytical calculations in that limit.

For any finite g_c, the energy w_j of the initial site can serve as a knob to control the escape dynamics, which could be exploited in experiments using a narrowband laser, for instance. In particular: (i) for $w_j = 0$, i.e. injection in the middle of the distribution, the escape rate reaches its maximum possible value $2W/(\pi N)$, independently of g_c, (ii) for $w_j = \pm W/2$, i.e. injection on the

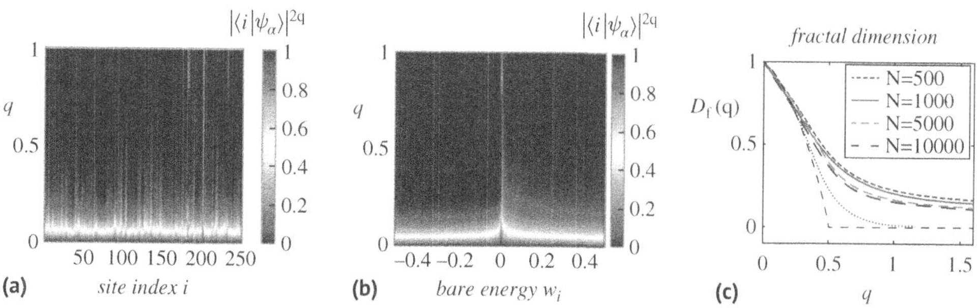

Figure 9.3 Components of $\text{IPR}_\alpha(q)$ for a given eigenstate ψ_α in the middle of the spectrum as a function of the site index i (with no ordering with respect to the bare energy w_i) (a), and as a function of the bare energy w_i (b). The fractal structure visible in (a) is caused by the random ordering of the sites: when the squared amplitudes are ordered according to the bare energy w_i, they are a smooth function of w_i decaying as $\sim 1/(E_\alpha - w_i)^2$. (c) Fractal dimension $D_f(q)$ of the dark states, computed numerically for different system sizes N (lines). The dashed line corresponds to the analytical result (Eq. 9.12). Deviations from the analytical formula are due to finite-size effects. The dotted line shows the numerical extrapolation to $N \to \infty$. Parameters are $g_c/W = 0.5$ and $N = 250$. *Source*: Reprinted from Ref. [35].

edges of the box distribution, the escape rate vanishes for all values of g_c, (iii) in all other cases, the escape rate increases $\propto g_c^4$ for weak couplings $g_c \ll W$, reaches its maximum value $2W/(\pi N)$ for intermediate couplings $g_c \sim W$, and then saturates to a lower, g_c-independent value. Since the maximum escape rate $2W/(\pi N)$ increases with the disorder strength, we conclude that the cavity does not only protect transport against disorder but also turns disorder into an ally that can contribute favorably to transport.

Interestingly, it is the contribution of dark states to this escape dynamics that completely dominates over that of polariton states, consistently with semilocalization where dark states are localized on multiple, noncontiguous sites. The latter are thus connected to each other via the dark states regardless of the distance separating them, which allows long-range transfer of excitations throughout the system of emitters. The coupling to the cavity allows to turn a fully localized phase characterized by the complete absence of diffusion without the cavity into a new phase with diffusive-like properties. We stress, however, that due to the nonlocal character of the coupling to the cavity, the escape dynamics does not correspond to standard diffusion as in Brownian motion, because the initial excitation does not propagate from one site to the next. A more accurate picture is that the excitation gets immediately delocalized over all the emitters, with occupation rising linearly with time, similarly to a diffusive motion. It is worth pointing out that a very similar diffusive-like behavior with continuous energy exchange between polaritonic and dark states has been observed in semiclassical large-scale atomistic molecular dynamics simulations for a realistic setup of Rhodamine molecules coupled to a Fabry–Pérot cavity [60].

9.2.2.4 Multifractality

Multifractality is a generalization of the concept introduced in Section 9.2.1 for a continuous spectrum of fractal dimensions,

$$D_f(q) \equiv \lim_{N\to\infty} \left(\frac{d}{1-q} \frac{\log \overline{\mathrm{IPR}_\alpha(q)}}{\log N} \right) \tag{9.10}$$

i.e., when the second moment of the probability distribution $p_\alpha(\{w_i\})$, $\overline{\mathrm{IPR}_\alpha}$ (Eq. 9.2) is not sufficient to fully characterize the spectrum. Instead, one has to use a generalized IPR providing also higher-order moments of the probability distribution,

$$\mathrm{IPR}_\alpha(q) = \sum_{i=1}^{N} |\langle i,0|\psi_\alpha\rangle|^{2q} + |\langle G,1|\psi_\alpha\rangle|^{2q}$$

with q a continuous parameter. For small q, $\overline{\mathrm{IPR}_\alpha(q)}$ is sensitive to the tails of probability distribution p_α, while for large q it is mostly sensitive to the largest components p_α. A qualitative change of the behavior of $\overline{\mathrm{IPR}_\alpha(q)}$ as a function of q is a signature of multifractality of the wavefunction [49].

Interestingly, one finds that all eigenstates of the arrowhead matrix model exhibit multifractality in that sense [35]. Moreover, multifractality arises in an extremely simple way: for independent, identically distributed bare energies w_j, the probabilities p_α [looking only at the components $\langle i,0|\psi_\alpha\rangle$] look like a random distribution which exhibits a fractal structure, see Figure 9.3a. However, when one sorts the amplitudes in order of increasing w_j, then the probability p_α turns out to be a smooth function of w_j (see Figure 9.3b) which decays as a power-law, $p_\alpha \sim 1/(E_\alpha - w_j)^2$. This power-law behavior implies that the generalized IPR of a dark state behaves as

$$\mathrm{IPR}_\alpha(q) \underset{N\to\infty}{=} \begin{cases} O(N^{1-2q}) & \text{if} \quad q < 1/2 \\ O(1) & \text{if} \quad q > 1/2. \end{cases} \tag{9.11}$$

The fractal dimension of all dark states is thus (see Figure 9.3c)

$$D_f(q)/d = \min\left(\frac{1-2q}{1-q}, 0\right) \tag{9.12}$$

and the nontrivial dependence on q signals multifractality. It is remarkable that multifractal behavior, which is usually associated to much more complex models in the context of localization–delocalization transitions [49], emerges in such an elementary way in the arrowhead matrix model. Moreover, the dark state multifractal dimension (Eq. 9.12) turns out to be exactly the same as the ones for the Anderson model on the Bethe lattice, which is commonly regarded as a mean-field model (i.e. with spatial dimension $d \to \infty$) of the localization–delocalization transition [49, 61]. We see that the random arrowhead matrix model naturally reproduces some aspects of the physics of the mean-field Anderson transition.

9.3 The Influence of Disorder and Semilocalization on Vibrational Dynamics

In this section, we now discuss how the concept of semilocalized eigenstates that has been introduced in the previous section can have an impact on vibrational dynamics of toy-model molecules, summarizing the main findings from [44]. In Section 9.3.1, we will first introduce the arguably most minimalistic model featuring electro-vibrational dynamics for molecules that are electronically coupled to a single cavity-mode, the HTC model. We then briefly discuss how this model can be exactly numerically simulated also for large molecule numbers N, using a matrix-product state method in Section 9.3.2. Importantly, there we also introduce the concept of vibrational entanglement and how it is connected to the possibility of simulating HTC dynamics numerically, in general. In Section 9.3.3, we then summarize the main recent findings about dynamics in disordered HTC models: counter-intuitively, because of inhomogeneous energy re-distribution through semilocalized states, we explain how large disorder can lead to increased vibrational entanglement. This is, on the one hand, important to know, since it leads to a breakdown of commonly used Ehrenfest mean-field methods. On the other hand, disorder-induced vibrational entanglement also leads to more quantum-mechanical effects. This manifests itself by, e.g. non-Gaussian shapes of vibrational wave-packets. Remarkably, only in the presence of disorder, these effects also survive for larger molecule numbers, especially when the incoherent photo-excitation takes place via the cavity mode. Here, we demonstrate robust modifications of nuclear wavepackets for up to $N \sim 150$ toy-model molecules. Disorder may therefore be the key ingredient to explain cavity-modified chemistry in collective coupling setups with small individual coupling strengths.

9.3.1 The Holstein–Tavis–Cummings Model

The fundamental origin of the electro-vibrational Holstein coupling [62] for a single molecule n can be understood in a simple model with two-electronic states and a single nuclear coordinate d, thus leading to two potential energy "surfaces" as function of d, as shown in the sketch in Figure 9.4. We now consider a situation where the nuclear dynamics remain restricted to the minima of the potential energy curves in both the electronic ground and excited states. In this scenario, we can make a harmonic approximation for the nuclear potentials, leading to discretized harmonic oscillator levels with frequency spacing ν_n, which we assume are equal in both electronic states.

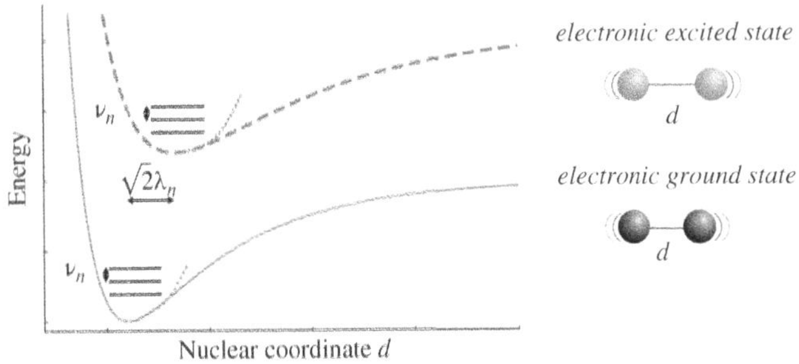

Figure 9.4 Sketch of the origin for the Holstein coupling between electronic and vibrational degrees of freedom in a simple nuclear model.

The electro–vibronic coupling manifests itself by a shift of the energetic minima in the two potential energy curves, here by an amount defined as $\sqrt{2}\lambda_n$ in standard harmonic oscillator units. Physically, this shift represents the fact that, depending on the electronic state of a molecule, the nuclear equilibrium positions differ by a certain amount. This implies that an electronic excitation requires an additional "reorganization energy" of the amount $\lambda_n^2\nu_n$, and the dimensionless quantity λ_n^2 is also known as Huang–Rhys factor. On the Hamiltonian level, the Holstein coupling thus leads to an overall minimal Hamiltonian for describing electro-vibrational dynamics of molecule n of the form [63]:

$$\hat{H}_H^{(n)} = \nu_n \hat{b}_n^\dagger \hat{b}_n + (\omega_n + \lambda_n^2 \nu_n)\hat{\sigma}_n^+ \hat{\sigma}_n^- - \lambda_n \nu_n \left(\hat{b}_n + \hat{b}_n^\dagger\right) \hat{\sigma}_n^+ \hat{\sigma}_n^- . \tag{9.13}$$

Here, $\hat{b}_n$ and $\hat{b}_n^\dagger$ are standard harmonic oscillator ladder operators for the nuclear vibration, and $\hat{\sigma}_n^\pm$ are raising/lowering operators for the excited electronic level at energy ω_n^e.

Next, we can combine this Holstein Hamiltonian with the TC model from Eq. (9.1), to arrive at the arguably most minimalistic model for studying the modification of vibrational dynamics of nucleii in systems with collective electronic SC to a single cavity mode $\hat{a}$, the HTC Hamiltonian [10]. Importantly, here we highlight the importance of electronic disorder in such a model, and thus will from now on focus on the disordered HTC Hamiltonian ($\hbar = 1$):

$$\hat{H}_{\mathrm{HTC}} = \hat{H}_{\mathrm{TC}} + \hat{H}_{\mathrm{vib}} + \hat{H}_{\mathrm{H}} + \hat{H}_{\mathrm{dis}}, \tag{9.14}$$

with

$$\hat{H}_{\mathrm{TC}} = \sum_{n=1}^{N} \Delta \hat{\sigma}_n^+ \hat{\sigma}_n^- + g \sum_{n=1}^{N} \left(\hat{a}\hat{\sigma}_n^+ + \hat{a}^\dagger \hat{\sigma}_n^-\right), \tag{9.15}$$

$$\hat{H}_{\mathrm{vib}} = \nu \sum_{n=1}^{N} \hat{b}_n^\dagger \hat{b}_n , \tag{9.16}$$

$$\hat{H}_{\mathrm{H}} = -\lambda\nu \sum_{n=1}^{N} \left(\hat{b}_n + \hat{b}_n^\dagger\right) \hat{\sigma}_n^+ \hat{\sigma}_n^- ., \tag{9.17}$$

$$\hat{H}_{\mathrm{dis}} = \sum_{n} \epsilon_n \hat{\sigma}_n^+ \hat{\sigma}_n^- . \tag{9.18}$$

Here we have made several assumptions: we considered all N molecules to have a homogeneous Huang–Rhys factor $\lambda_n^2 = \lambda^2$, as well as a homogeneous vibrational frequency $\nu_n = \nu$. We have then defined the problem in a frame rotating at the cavity frequency ω_c with detuning $\Delta = (\omega + \lambda^2\nu) - \omega_c$ from the electronic transition energy at the Franck–Condon point. Here we introduced a mean homogeneous electronic transition frequency, $\omega_n = \omega$. We will consider $\Delta = 0$ from now on. As in Section 9.2, the cavity-coupling strength of a single molecule to the cavity is homogeneous and given by $g \equiv g_c/\sqrt{N}$, where g_c is the collective coupling strength. Lastly, the only disorder is assumed to occur at electronic levels, where $\epsilon_n = \omega_n - \omega$ is a random offset in the electronic transition energy of the n-th molecule (as typically the case for the modeling of organic molecules [32, 64, 65]). Here, we will consider the random energies ϵ_n to be drawn from a normal distribution with zero mean and variance W^2.

9.3.2 Vibrational Entanglement and Numerical Simulations

Our goal is to study the full quantum mechanical dynamics of Hamiltonian (Eq. 9.14) after an incoherent photo-excitation, mimicking the experimental scenarios of early polaritonic chemistry experiments focusing on cavity-modified photochemistry under collective electronic SC [7]. The goal of our analysis is twofold: on the one hand, we want to fundamentally better understand the modifications of the nuclear dynamics due to the cavity coupling, in particular in the limit of large N. The goal here is to shed light on how modifications of nuclear dynamics may also occur for small individual coupling strengths $g \to 0$ and $N \to \infty$ ($1/N$ problem). On the other hand, an important question is to analyze to what extent genuine quantum effects play a role in such modifications. Besides being of fundamental interest, the latter question is also intrinsically connected to numerical simulation methods.

Computing the full collective dynamics for many molecules, taking all photonic, electronic, and vibrational degrees of freedom into account, is an extremely challenging task, even for minimal models as Hamiltonian (Eq. 9.14). In particular, for large systems, strong approximations reducing the "quantum-ness" of the problem are needed. For example, in a cavity Born–Oppenheimer picture, the adiabatic nuclear dynamics on polaritonic and dark potential energy surfaces can be computed [66], neglecting nonadiabatic couplings between the surfaces. Such dynamics remain valid in regimes where the separation between the potential energy surfaces remains large, a situation which is often not realistic in polaritonic chemistry [67–69]. A "mean-field" Ehrenfest approach is a common way to go beyond this limitation [70, 71]. This approach relies on a product state assumption at all times t between the electro-photonic and the nuclear degrees of freedom, which for our model would imply the approximation

$$|\psi_{\mathrm{mf}}(t)\rangle \approx |\phi_{\mathrm{el+ph}}(t)\rangle|\phi_{\mathrm{vib}}(t)\rangle. \tag{9.19}$$

Here, $|\phi_{\mathrm{el+ph}}(t)\rangle$ and $|\phi_{\mathrm{vib}}(t)\rangle$ are two wave functions fully defined on the electro-photonic and the vibrational Hilbert space, respectively. By definition, ansatz (Eq. 9.19) neglects any type of quantum entanglement between the nuclear Hilbert space and the rest of the system, indicated by the fact that the vibrational entanglement entropy vanishes. The latter we define as

$$S_{\mathrm{vib}}(t) = -\mathrm{tr}\,[\hat{\rho}_{\mathrm{vib}} log_2(\hat{\rho}_{\mathrm{vib}})]. \tag{9.20}$$

Here, $\hat{\rho}_{\mathrm{vib}}$ is the reduced density matrix on the vibrational Hilbert space. From a pure full state of the system, $|\psi(t)\rangle$ it is obtained by partially tracing over the electro-photonic Hilbert space,

$\hat{\rho}_{vib} = tr_{el+ph}(|\psi(t)\rangle\langle\psi(t)|)$. Here, as entropy measure we choose the von Neumann entropy (with base 2), but any other entropy (e.g. Renyi entropies) could also be used. Given that the state of the full system is pure, $\hat{\rho} = |\psi(t)\rangle\langle\psi(t)|$, the entropy (Eq. 9.20) quantifies how much information is encoded in the quantum correlations between the nuclear wave function and the rest of the system. Trivially, in the case that the full wave function is factorized, as in ansatz (Eq. 9.19), $|\psi(t)\rangle = |\psi_{mf}(t)\rangle$, it follows that $S_{vib}(t) = 0$ at all times. Therefore, large vibrational entropies indicate a breakdown of the Ehrenfest mean-field assumption, which motivates our study of the dynamics of $\hat{S}_{vib}$ in this section. A key insight of the work in Ref. [44] has been that for large N, in the presence of disorder, $\hat{S}_{vib}$ can become large, therefore leading to the breakdown of Ehrenfest approximations and more genuine quantum effects in the dynamics.

An alternative method to exactly simulate the full quantum dynamics of a many-body model as (Eq. 9.14) is given by matrix-product state (MPS) algorithms [72]. The general concept of this technique emerged from density-matrix renormalization group algorithms, and has been in particular adapted to simulate Hamiltonian time-evolution in the early 2000s [73–75]. Originally conceived for 1D models, a main challenge for matrix-product state (or more generally tensor-network) methods has been an adaptation to higher dimensional geometries. However, in particular, for models with long-range couplings, several solutions for simulating dynamics, e.g. using matrix-product operators are known [72, 76]. Note that our disordered HTC model only has a very small connectivity between the individual molecules, which only interact over the central site (the cavity). This makes simple methods such as introducing swap-gates between individual sites of the MPS a viable option [44]. Generally, in the field of polaritonic chemistry setups, MPS techniques have recently started to be successfully applied [44, 77, 78].

For a detailed introduction into MPS techniques, in particular with long-range connectivity, we refer the reader to the extensive literature in, e.g. [72, 76, 79, 80], for a specific implementation of an MPS algorithm for the disordered HTC model, see Ref. [44]. Here, we will only briefly introduce the fundamental concept: an MPS can be thought of as a generalization of a product state. A product state for N molecules is written as $|\psi\rangle = \otimes_n^N |\phi_n\rangle$, where $|\phi_n\rangle$ are local states for each molecule n (and the cavity, $n = 0$). Here, as local Hilbert space, we consider the electronic and vibrational excitations of a single toy-model molecule and the photonic mode. Note that in principle the local Hilbert space is infinite, however, we are only concerned with small nuclear displacement such that the harmonic oscillator levels can be truncated at relatively small vibrational excitation numbers. Furthermore, we can restrict simulations to small electro–photonic excitation numbers to also truncate the photon mode occupation number. Then, a product state representation only requires on the order of $\mathcal{O}(N)$ complex parameters, but is by definition limited to only nonentangled states. This strong limitation is a natural consequence of the fact that the full Hilbert space has a dimension of order $\propto exp(N)$. Instead, an MPS can be thought of as a numerical decomposition of $|\psi\rangle$ into a product of $\chi \times \chi$ matrices, where the entries of the matrices itself are local states. An MPS then requires $\mathcal{O}(N\chi^2)$ parameters to be described, however, can represent quantum states with a controllable amount of bipartite entanglement entropies, depending on the matrix size χ [81]. For any-bipartition of the state into two separate blocks, von Neumann entanglement entropies are then strictly limited to values smaller than $log_2(\chi)$. In quantum many-body systems, the scaling behavior of entanglement entropies in space and time thus crucially determines the simulability by MPS methods [82, 83]. In particular, doing convergence tests in the matrix (or "bond dimension") χ, an MPS allows to simulate the full dynamics of Hamiltonian (Eq. 9.14) with a finite amount of vibrational entanglement $S_{vib} > 0$ as long as it does not become excessively large.

9.3.3 Dynamics After Photo-excitation

We now discuss the main insights of the effect of disorder on the vibrational dynamics of model (Eq. 9.14). We generally consider a scenario where initially the system is taken by an incoherent photo-absorption. However, as we will see below, it is crucial to distinguish two scenarios, where the photon is either initially absorbed by a molecule, or by the cavity, as sketched in Figure 9.5a, b, respectively. In the two cases, the initial state of the system is then given by $|\psi_0^m\rangle = \hat{\sigma}_1^+|0\rangle_{\text{ph}}|0\rangle_{\text{exc}}|0\rangle_{\text{vib}}$ (molecular absorption), or by $|\psi_0^c\rangle = \hat{a}^\dagger|0\rangle_{\text{ph}}|0\rangle_{\text{exc}}|0\rangle_{\text{vib}}$ (cavity absorption). Here, the states $|0\rangle_{\text{exc,vib,ph}}$ describe the ground states of the free electronic, vibrational and photonic Hamiltonian. Note that the equilibrium states of the unexcited parts of the systems (vibrational Hilbert space of unexcited molecule and cavity-mode) are very well approximated by $|0\rangle_{\text{exc,vib,ph}}$ since thermal occupations of higher levels are negligible at room temperature. Note that, without loss of generality, we identify the initially excited molecules as the first molecule with index 1.

Although we do not claim that our minimal model (Eq. 9.14) exactly describes a real molecular system, to get a general idea about typical number we can motivate our model by an organic setup with Rhodamine 800 under strong electronic coupling [77, 84]. In such a scenario, it is useful to consider nuclear displacements in the HTC model between $0.1 \leq \lambda \leq 0.5$ and $\nu = 0.3g_c$ [44]. For a typical collective vacuum Rabi splitting of $2g_c = 700$ meV [7], this then implies $\nu \sim 0.1$ eV. In the following we are interested in the dynamics over vibrational period, $2\pi/\nu$. This implies simulations up to times of typically $\sim 20g_c^{-1}$, which correspond to timescales on the order of 10–100 fs. We are interested in the coherent dynamics on these ultra-fast timescales that one obtains from evolving the system according to the Schrödinger equation with Hamiltonian (Eq. 9.14), i.e. we compute and analyze the states $|\psi(t)\rangle = \mathrm{e}^{-\mathrm{i}\hat{H}_{\text{HTC}}t}|\psi_0^{(m/c)}\rangle$.

To get an initial idea about the dynamics, in Figure 9.6, we show results on the time-evolution after one out of 100 molecules is photo-excited (i.e. for the initial state $|\psi_0^m\rangle$). We compare a case with strong disorder ($W = g_c/2$, single disorder realization) and without disorder ($W = 0$). Figure 9.6a shows the phase-space evolution of all vibrational oscillators, $i = 1, \ldots, 100$, with $i = 1$ denoting the initially excited molecule. Here, $\langle\hat{x}_i\rangle$ and $\langle\hat{p}_i\rangle$ are the time-dependent expectation values of the harmonic oscillator position and momentum operators of vibrational mode i, $\hat{x}_i = (\hat{b}_i^\dagger + \hat{b}_i)/\sqrt{2}$, $\hat{p}_i = \mathrm{i}(\hat{b}_i^\dagger - \hat{b}_i)/\sqrt{2}$. Due to the initial displacement in the excited state, the initially excited molecule (black solid line marked with a star) exhibits a large oscillation, which is outside of the scale of the plot. In contrast, in the case of $W = 0$, we find that all the other 99 molecules that were initially not excited barely exhibit any oscillation (barely visible gray circle). The almost nonexistence of any vibrational molecular dynamics for the initially unexcited molecules can be seen as a direct manifestation of the $1/N$ problem in polaritonic chemistry: the only mechanism of how an initially unexcited molecule can receive an electronic excitation is via the cavity mode. Since all excitations are equally coupled, perturbatively it is easy to see that the electronic excitation number $\langle\hat{\sigma}_i^+\hat{\sigma}_i^-\rangle(t)$

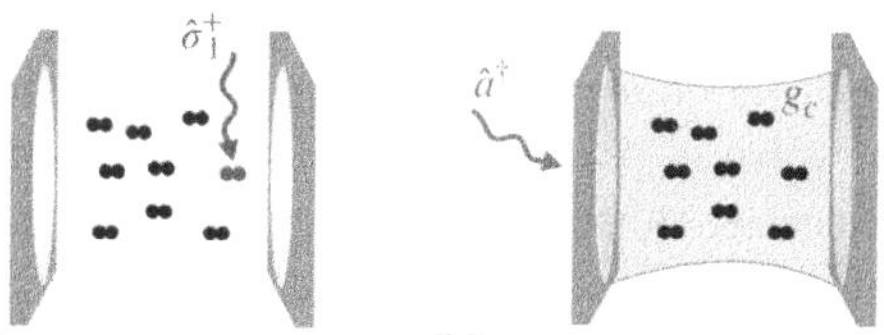

(a) (b)

Figure 9.5 Sketch of the two different initial incoherent photo excitation schemes considered here, either after photo absorption into a molecule (a) or the cavity (b). *Source*: Reprinted from Ref. [44].

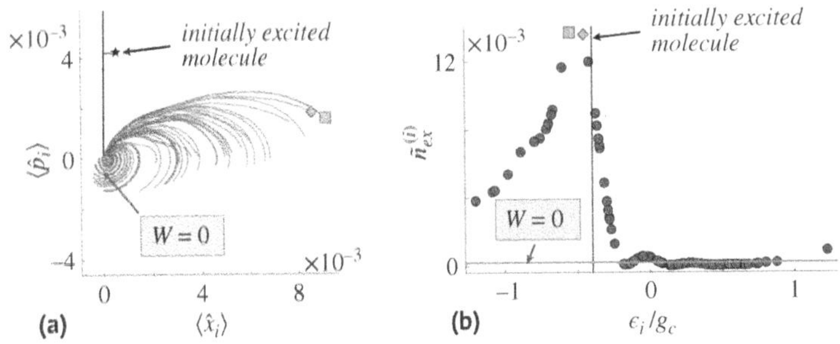

Figure 9.6 (a) Phase space dynamics over a single oscillation period $t \leq 2\pi/\nu$ of $N = 100$ vibrational oscillators (single disorder realization) with $W = g_c/2$ after an incoherent molecular photo excitation. The almost vertical black lines correspond to the initially excited molecule, gray lines to 99 initially unexcited ones. The gray circle (barely visible) corresponds to $W = 0$. (b) Excitation fraction of the initially unexcited 99 molecules as a function of their random energy offset ϵ_i, at time $t = 2\pi/\nu$. The square and diamond highlight two molecules with random energies close to the initially excited molecule in (a) and (b). Parameters: $N = 100$, $\nu = 0.3g_c$, $\lambda = 0.4$. *Source*: Reprinted from Ref. [44].

for molecules $i = 2, \ldots, N$ is proportional to $1/N^2$ [44], here $\sim 10^{-4}$. This is shown in Figure 9.6b, where we plot the electronic excitation number at the end of a vibrational period, $\langle \hat{\sigma}_i^+ \hat{\sigma}_i^- \rangle (t = 2\pi/\nu)$ as function of the initial random electronic energy offset ϵ_i. As expected, for $W = 0$ (grey line) the excitation is diluted among $N - 1$ molecules, leading to a negligible displacement and vibrational evolution for $i > 1$.

This changes drastically in the presence of disorder (colored lines and points). While many molecules still exhibit only small phase-space circles, a few of the molecules now undergo evolution with displacements $\langle \hat{x}_i \rangle$ orders of magnitudes larger than for $W = 0$. This can be understood from the characteristics of semilocalized eigenstates. As described in Section 9.2, those come from the hybridization of bare states that are arbitrarily close in energy. This now leads to a much more efficient excitation transfer from the initially excited molecule (black vertical line in [b]) to a few molecules in the energetic vicinity. Perturbatively, it is clear that this energy transfer leads to electronic excitation numbers $\langle \hat{\sigma}_i^+ \hat{\sigma}_i^- \rangle (t)$ proportional to $1/N$ [44], here $\sim 10^{-2}$, as numerically verified in Figure 9.6b. While the excitation transfer thus still decreases with N, in practice the disorder can here enhance it by an order of magnitude with a direct impact on the subsequent vibrational dynamics. In Figure 9.6a, we furthermore highlight two molecular oscillators (diamond, square) which undergo the largest phase-space evolution. Comparing to Figure 9.6b we observe indeed that those two oscillators correspond to molecules with a random energy offset very close to the initially excited molecule. The small shift toward lower energies $\sim \lambda^2 \nu$ and asymmetry with respect to the energy of the initially excited vibration can be attributed to the additional reorganization energy needed for the energy transfer.

Another important impact of the disorder on the dynamics in Figure 9.6a is the fact that the oscillation of the individual vibrational oscillators becomes highly inhomogeneous. In the case of $W = 0$ all initially nonexcited oscillators ($i > 1$) evolve in phase, which allows to consider them as a single collective phonon-mode of $N - 1$ molecules. However, for $W > 0$ a much larger vibrational Hilbert space is accessed, since $N-1$ modes are now needed to describe the full coherent dynamics of the system. Therefore, we find that disorder leads to a strong increase of the vibrational entanglement, S_{vib} as shown in Figure 9.7.

There, we plot the disorder averaged entropy, S_{vib} as function of time for both scenarios of an initial molecular (Figure 9.7a) and an initial cavity (Figure 9.7b) excitation. In both cases the

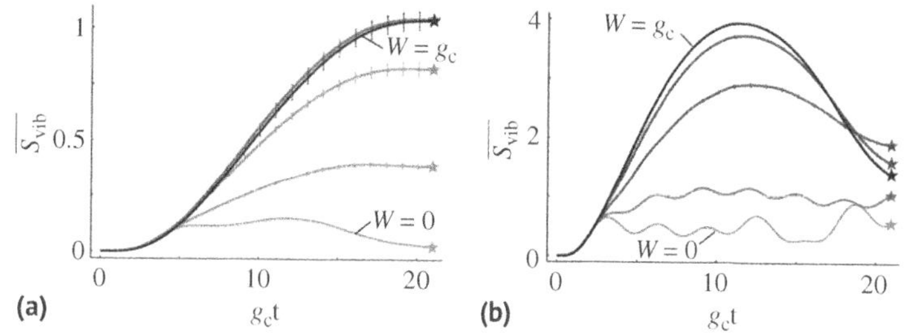

Figure 9.7 Evolution of disorder averaged entanglement entropies $\overline{S_{vib}}$. The disorder strengths W is increased (from light to dark: $W = 0$, $g_c/4$, $g_c/2$, $3g_c/4$, g_c). The initial incoherent photo excitation occurs in a molecule in (a) and the cavity mode in (b). Parameters: $N = 100$, $\nu = 0.3g_c$, $\lambda = 0.4$, 64 disorder realizations. *Source*: Reprinted from Ref. [44].

entanglement increases strongly with W, and in general we find the values of S_{vib} for scenario (b) to be significantly larger (on average) than for case (a).

It is important to point out that disorder-enhanced entanglement entropy is not only an abstract mathematical quantity, but it can also have direct practical physical consequences for the vibrational dynamics as shown in Ref. [44]. Crucially, it can have a strong impact on the shape of nuclear position distribution functions for the molecules i, $P_i(x_i, t)$, for the harmonic oscillators.

Let us first consider just a single molecule n with the initial product state $|\phi(0)\rangle = |\phi(0)\rangle_{el}|0\rangle_{vib}$, where $|\phi(0)\rangle_{el} = \sin(\theta_0/2)|0\rangle_{el} + e^{i\phi_0}\cos(\theta_0/2)|1\rangle_{el}$ in an arbitrary state of the electronic two-level system with Bloch vector angles $0 \leq \theta_0 \leq \pi$ and $0 \leq \phi_0 < 2\pi$. Here, $|0\rangle_{vib}$ denotes the harmonic oscillator ground state. Then, the Holstein coupling Hamiltonian (Eq. 9.17) generates the time-dependent states $|\phi(t)\rangle = e^{-i\hat{H}_H t}|\phi(0)\rangle = e^{-i\lambda\nu\sqrt{2}t\hat{\sigma}_i^+\hat{\sigma}_i^-\hat{x}_i}|\phi(0)\rangle_{el}|0\rangle_{vib}$. Enforcing the product state assumption (Eq. 9.19) at all times would now imply time-dependent states of the form

$$|\phi(t)\rangle \approx \left(\sin(\theta_0/2)|0\rangle_{el} + e^{i\eta(t)}\cos(\theta_0/2)|1\rangle_{el}\right)|i\lambda\nu t\rangle_{vib} \tag{9.21}$$

where we defined the time-dependent phase $\eta(t) = \phi_0 - \sqrt{2}\lambda\nu t$ and the usual harmonic oscillator coherent state $|\alpha\rangle_{vib} = \exp(\alpha\hat{b}_i^\dagger - \alpha^*\hat{b}_i)|0\rangle_{vib}$. We can further parameterize the coherent state as $|i\lambda\nu t\rangle_{vib} = |x(t) + ip(t)\rangle_{vib}$ with the time-dependent real (position) and imaginary (momentum) parts $x(t)$ and $p(t)$, respectively. Since, besides the Holstein coupling term (Eq. 9.17) only the free oscillator Hamiltonian (Eq. 9.16) governs the dynamics of the vibrations, this implies that all oscillators will remain in coherent states with a Gaussian position and momentum distributions at all times, also after a disorder average. In contrast, dropping the product state assumption, the exact quantum state generated by the Holstein coupling for the dynamics of our single molecule takes the form

$$|\phi(t)\rangle = \sin(\theta_0/2)|0\rangle_{el}|0\rangle_{vib} + \cos(\theta_0/2)e^{i\eta(t)}|1\rangle_{el}|i\lambda\nu t\rangle_{vib} \tag{9.22}$$

For the general scenario of $\theta_0 \neq 0$ and $\theta_0 \neq \pi$ this state is an entangled superposition of a coherent state and the electronic state. When tracing over the electronic degree of freedom, the nuclear position distribution $P_i(x_i, t)$ can now take distinctive non-Gaussian shapes.

Indeed, with our numerically exact MPS simulations in Ref. [44], we have shown that disorder leads to distinct non-Gaussian features in the coupled many-body setting. This can be seen in Figure 9.8a where we show the modification of the (disorder-averaged) nuclear position distribution for the initial excited molecule and increasingly large disorder W at a fixed time of $t = 2\pi/\nu$.

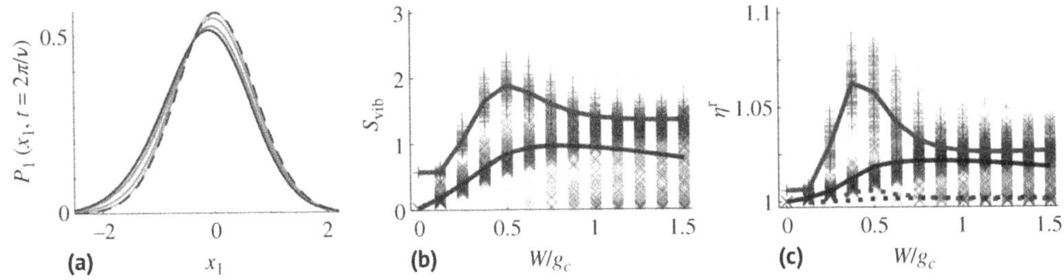

Figure 9.8 (a) Position probability distribution of an initially excited molecular oscillator at time $t = 2\pi/\nu$ (gray dashed line: $g = 0$). From light to dark disorder is increased: $W = 0, g_c/4, g_c/2, 3g_c/4, g_c$. (b) and (c) show vibrational entanglement entropies S_{vib} and right tail-weights η^r as function of W ($t = 2\pi/\nu$). We compare the two different initial states. Molecular excitation: "x" symbols, bottom solid lines; and cavity excitation: "+" symbols, top solid lines. Symbols show 256 individual disorder realizations, lines are their mean value. Dashed lines in (c) show the mean value when using the product state ansatz (Eq. 9.19). Parameters: $N = 100$, $\nu = 0.3g_c$, $\lambda = 0.4$. *Source*: Reprinted from Ref. [44].

Here, at this time, we find a clear asymmetry toward small values of x_1, while at this time the initially nonexcited vibrational oscillators show an asymmetry to larger values of $x_{n>1}$ (not shown). To capture the amount of the asymmetry with a single number, we can define a tail-weight, which for the right tail of the distributions $P_i(x_i, t)$ reads

$$\eta^r(t) = \sum_i \int_{x^r_{thr}}^{\infty} dx_i\, P_i(x_i, t) \tag{9.23}$$

with $x_{thr} \gg 0$ a large threshold value. Here, we choose the threshold in such a way that for the bare (Gaussian) ground state distribution one percent of the probability is captured, which corresponds to $x_{thr} \approx 1.6$. We have also checked that the physics shown in Figure. 9.8c does not strongly depend on the threshold, e.g. when increasing the threshold up to scenarios where only 0.01% probability is in the bare tail. For large values of x_{thr} this tail-weight can be thought of as a probability for extremely large nuclear distances. Although with our simplified minimal model we cannot claim to capture any real chemical processes, we still propose to think about such extreme configurations as potential probabilities, e.g. for breaking bonds [85]. Comparing the dependence of S_{vib} and η^r (at fixed time $t = 2\pi/nu$) on the disorder strength W in Figure 9.8b, c, respectively, it is striking that there is direct correlation between the vibrational entanglement entropy and the tail-weight for both initial states. Furthermore, the enhanced tail probabilities seen in Figure 9.8c clearly disappear when running the many-body simulation with the product state assumption (Eq. 9.19) (dashed lines). Both of these are a very strong indication that disorder can lead to enhance a genuine quantum effect in the HTC model that can have real physical consequences on the evolution of nuclear distribution functions.

The final important question to come back to is whether disorder-induced quantum mechanical modification of nuclear wave-packets can also survive in the $N \to \infty$ limit. For the case of an initial molecular photo-excitation, we have already seen that energy transfer is suppressed $\propto 1/N$ in the disordered scenario (compared to $1/N^2$ for $W = 0$). Therefore, in this scenario we also expect the entanglement entropy as well as the (system-size normalized) tail-weights to be suppressed for large N. Indeed, we find this to be the case in our exact simulations (bottom solid lines) in Figure 9.9, where we show S_{vib} and η^r at fixed time $t = 2\pi/\nu$. Note that comparing Figure 9.9a, b, we find again a direct correlation between the two.

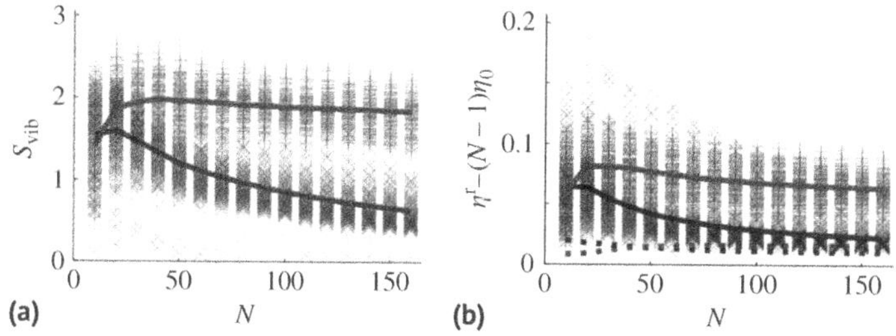

Figure 9.9 Vibrational entanglement entropies S_{vib}, (a), and right tail-weights η^r, (b), as functions of N at fixed $W = g_c/2$ at time $t = 2\pi/\nu$. We compare the two different initial states. Molecular excitation: "x" symbols, bottom solid lines; and cavity excitation: "+" symbols, top solid lines. For a fair comparison, from the extensive quantity η^r we subtract a trivial ground state contribution $(N-1)\eta_0$ (here: $\eta_0 = 0.01$). Parameters: $\nu = 0.3g_c$, $\lambda = 0.4$. *Source*: Reprinted from Ref. [44].

Importantly, however, we find a striking different behavior for an initial cavity excitation (top solid lines) in Figure 9.9. We have performed exact MPS simulations for up to $N = 160$ molecules and found a nearly constant scaling in the large N limit for both S_{vib} and η^r. While it is difficult to confirm this behavior precisely numerically for much larger numbers, we can rationalize the scaling using a perturbative argument: Since disorder leads to the mixing of the cavity mode with all dark states, the energy transfer from the cavity mode to all dark states should be proportional to the overall photonic weight of all dark states from the semilocalized states. We know that $\sum_d |\langle d||1_{ph}\rangle|^2 \approx W^2/g_c^2$, where $|1_{ph}\rangle$ denotes the initial single-photon state of the cavity, and $|d\rangle$ are the $N-1$ semilocalized dark states. The fraction of total transferred energy from the initial state to the individual molecules is therefore independent of N, also in the large N limit. Note again that going beyond the Ehrenfest product state approximations is essential to observe such effects, since the dashed lines in Figure 9.9b fail to reproduce it. In conclusion, to observe a robust modification of nuclear dynamics in the large N limit, it is essential to consider an initial incoherent photo excitation *through the cavity mode*, and *not through individual molecules*. It is worth pointing out that a similar conclusion can be found when considering an open-system dissipative quantum optics model for electron-transfer reactions [86].

9.4 Conclusion and Outlook

This chapter provides an account of the considerable progress achieved in recent years in our understanding of the interplay of disorder and strong collective light–matter interactions from the point of view of theorists. We introduced an approach relying on minimal models of increasing complexity, taking relevant degrees of freedom (electronic, photonic, and nuclear motion) into account in their simplest forms.

A notable consequence of this progress is the demonstration of macroscopic effects of *collective* SC of the dynamics and, in particular, on energy transport in large ensemble of emitters, and especially the importance of disorder for such dynamics. As we have seen, for all finite light–matter coupling strengths excitation transport can be modified due to an *alteration of all the eigenstates of the Hamiltonian* – arising even in situations where the light–matter coupling strength per molecule is very tiny. This effect, which at the single-particle level could naively be regarded as an $1/N$-effect, can lead to a macroscopic enhancement of transport and nuclear dynamics that becomes *visible also at very large scales*. This result, which is exact for rather general models considered here, is a clear manifestation of the subtle, though highly robust, collective quantum effects that can be found to dominate at large scales. This example thus provides a cautionary tale for theories that are limited to the study of disordered systems with just a few particles. This seems particularly important for ab-initio calculations, where the numerical cost of precisely simulating the internal structure of single emitters prohibits the extension of calculations to large ensembles of particles. The results presented here also highlight that care has to be taken in choosing approximations, in order to access large-scale regimes.

In order to further progress in the field of cavity-modified transport and polaritonic chemistry, it seems essential to now devise *theories and numerical predictions that can be directly tested in experiments*. While this should ultimately be done with "real molecular systems," we believe that a promising path forward is constituted by the possibility to realize controlled experimental "quantum simulation" setups that are very close to our minimal cQED models with laser-cooled atoms or molecules.

In this context, cold atom experiments may provide a bridge between models, theories and experiments. For example, two recent works by Brantut [36] and Reichel [87] have used internal electronic excitations of alkali atoms trapped in high-finesse cavities in the regime of cavity QED to implement disordered TC models as discussed in this chapter. Both experiments have directly measured the existence of "grey" states, due to the combined effects of disorder and light–matter coupling. The results in Ref. [36] have further already provided evidence of multi-fractal behavior of low-energy excited states in the spectrum. Future work may address the influence of dephasing and atomic motion in these systems, which would be a breakthrough for experiments. We believe that quantum simulations of light–matter coupled systems could also be crucial to understanding the complex role that possible direct interactions (e.g. dipole–dipole, van der Waals) between the emitters can play in the excitation dynamics due to the competition with light–matter coupling and disorder. This would potentially also open up solid state and chemical cQED setups to fundamental research fields such as quantum thermalization and many-body localization, fields with many open questions. In the first recent work [88] with a simple TC model in the presence of multiple excitations, direct interactions between emitters were shown to favor delocalization in the many-body Hilbert space, while Poisson statistics and multifractality were also simultaneously observed, in contrast to the single-excitation case. Thus, while being integrable, this model was shown to share similarities with other archetypal models for many-body localization, a hot topic of research in the last decade in the field of condensed matter physics.

It can be hoped that more experimental groups in chemical departments will take on the challenge to realize fully controllable molecular systems. Simple diatomic molecules as realized in traditional quantum optics experiments [89] may provide platforms for the first explorations of cQED effects in simple chemistry. One example could be the realization of ground state molecules using *collective dissipative effects in a cavity* as recently proposed in Ref. [90]. Very recent experiments [91] with more complex molecular structures in the gas phase may provide an exciting path forward, bridging the gap between the controlled setups typical of traditional quantum optics toward molecules of chemical interest and in frequency ranges (mid- and far-infrared frequencies) highly relevant for vibrational dynamics and chemical reactions. This is an exciting path forward, and as with all the best stories, it seems that the very best is still to come.

References

1 Raimond, J.M., Brune, M., Haroche, S. Manipulating quantum entanglement with atoms and photons in a cavity. *Rev. Mod. Phys.* **2001**, *73*, 565–582. https://doi.org/10.1103/RevModPhys.73.565.

2 Kimble, H.J. Strong interactions of single atoms and photons in cavity QED. *Phys. Scr.* **1998**, *1998*(T76), 127. https://dx.doi.org/10.1238/Physica.Topical.076a00127.

3 Jaynes, E.T., Cummings, F.W. Comparison of quantum and semiclassical radiation theories with application to the beam maser. *Proc. IEEE* **1963**, *51*(1), 89–109. https://doi.org/10.1109/PROC.1963.1664.

4 Tavis, M., Cummings, F.W. Exact solution for an *n*-molecule—radiation-field hamiltonian. *Phys. Rev.* **1968**, *170*, 379–384. https://doi.org/10.1103/PhysRev.170.379.

5 Raizen, M.G., Thompson, R.J., Brecha, R.J., Kimble, H.J., Carmichael, H.J. Normal-mode splitting and linewidth averaging for two-state atoms in an optical cavity. *Phys. Rev. Lett.* **1989**, *63*, 240–243. https://doi.org/10.1103/PhysRevLett.63.240.

6 Fink, J.M., Göppl, M., Baur, M., Bianchetti, R., Leek, P.J., Blais, A., et al. Climbing the jaynes–cummings ladder and observing its nonlinearity in a cavity qed system. *Nature* **2008**, *454*, 315. https://doi.org/10.1038/nature07112.

7 Hutchison, J.A., Schwartz, T., Genet, C., Devaux, E., Ebbesen, T.W. Modifying chemical landscapes by coupling to vacuum fields. *Angew. Chem. Int. Ed.* **2012**, *51*(7), 1592–1596. https://doi.org/10.1002/anie.201107033.

8 Kowalewski, M., Bennett, K., Mukamel. S. Cavity femtochemistry: manipulating nonadiabatic dynamics at avoided crossings. *J. Phys. Chem. Lett.* **2016**, *7*(11), 2050–2054. https://doi.org/10.1021/acs.jpclett.6b00864.

9 Galego, J., Garcia-Vidal, F.J., Feist, J. Cavity-induced modifications of molecular structure in the strong-coupling regime. *Phys. Rev. X* **2015**, *5*, 041022. https://doi.org/10.1103/PhysRevX.5.041022.

10 Herrera, F., Spano, F.C. Cavity-controlled chemistry in molecular ensembles. *Phys. Rev. Lett.* **2016**, *116*, 238301. https://doi.org/10.1103/PhysRevLett.116.238301.

11 Fregoni, J., Granucci, G., Coccia, E., Persico, M., Corni, S. Manipulating azobenzene photoisomerization through strong light–molecule coupling. *Nat. Commun.* **2018**, *9*, 4688. https://doi.org/10.1038/s41467-018-06971-y.

12 Munkhbat, M., Wersäll, B., Baranov, D.G., Antosiewicz, T.J., Shegai, T. Suppression of photo-oxidation of organic chromophores by strong coupling to plasmonic nanoantennas. *Sci. Adv.* **2018**, *4*(7), eaas9552. https://doi.org/10.1126/sciadv.aas9552.

13 Peters, V.N., Faruk, M.O., Asane, J., Alexander, R., Peters, D.A., Prayakarao S., et al. Effect of strong coupling on photodegradation of the semiconducting polymer p3ht. *Optica* **2019**, *6*(3), 318–325. https://doi.org/10.1364/OPTICA.6.000318.

14 Antoniou, P., Suchanek, F., Varner, J.F., Foley IV, J.J. Role of cavity losses on nonadiabatic couplings and dynamics in polaritonic chemistry. *J. Phys. Chem. Lett.* **2020**, *11*(21), 9063–9069. https://doi.org/10.1021/acs.jpclett.0c02406

15 Mauro, L., Caicedo, K., Jonusauskas, G., Avriller, R. Charge-transfer chemical reactions in nanofluidic Fabry-Pérot cavities. *Phys. Rev. B* **2021**, *103*, 165412. https://doi.org/10.1103/PhysRevB.103.165412.

16 Thomas, A., George, J., Shalabney, A., Dryzhakov, M., Varma, S.J., Moran, J. Ground-state chemical reactivity under vibrational coupling to the vacuum electromagnetic field. *Angew. Chem. Int. Ed.* **2016**, *55*(38), 11462–11466. https://doi.org/10.1002/anie.201605504.

17 Thomas, A., Lethuillier-Karl, L., Nagarajan, K., Vergauwe, R.M.A., George, J., Chervy, T. Tilting a ground-state reactivity landscape by vibrational strong coupling. *Science* **2019**, *363*(6427), 615–619. https://doi.org/10.1126/science.aau7742.

18 Thomas, A., Jayachandran, A., Lethuillier-Karl, L., Vergauwe, R.M.A., Nagarajan, K., Devaux, E. Ground state chemistry under vibrational strong coupling: dependence of thermodynamic parameters on the rabi splitting energy. *Nanophotonics* **2020**, *9*(2), 249–255. https://doi.org/10.1515/nanoph-2019-0340.

19 Feist, J., Galego, J., Garcia-Vidal, F.J. Polaritonic chemistry with organic molecules. *ACS Photonics* **2018**, *5*(1), 205–216. https://doi.org/10.1021/acsphotonics.7b00680.

20 Hertzog, M., Wang, M., Mony, J., Börjesson, K. Strong light–matter interactions: a new direction within chemistry. *Chem. Soc. Rev*. **2019**, *48*, 937–961. http://dx.doi.org/10.1039/C8CS00193F.

21 Garcia-Vidal, F.J., Ciuti, C., Ebbesen, T.W. Manipulating matter by strong coupling to vacuum fields. *Science* **2021**, *373*(6551), eabd0336. https://doi.org/10.1126/science.abd0336.

22 Reitz, M., Sommer, C., Genes, C. Cooperative quantum phenomena in light-matter platforms. *PRX Quantum* **2022**, *3*(1), 010201. https://doi.org/10.1103/PRXQuantum.3.010201.

23 Zhong, X., Chervy, T., Zhang, L., Thomas, A., George, J., Genet, C., Energy transfer between spatially separated entangled molecules. *Angew. Chem. Int. Ed.* **2017**, *56*(31), 9034–9038. https://doi.org/10.1002/anie.201703539.

24 Du, M., Martínez-Martínez, L.A., Ribeiro, R.F., Hu, Z., Menon, V.M., Yuen-Zhou, J. Theory for polariton-assisted remote energy transfer. *Chem. Sci.* **2018**, 9, 6659–6669. http://dx.doi.org/10.1039/C8SC00171E.

25 Lerario, G., Ballarini, D., Fieramosca, A., Cannavale, A., Genco, A., Mangione, F. High-speed flow of interacting organic polaritons. *Light Sci. Appl.* **2017**, *6*, e16212. https://doi.org/10.1038/lsa.2016.212.

26 Reitz, M., Mineo, F., Genes, C. Energy transfer and correlations in cavity-embedded donor-acceptor configurations. *Sci. Rep*. **2018**, *8*(9050), 1–11. https://doi.org/10.1038/s41598-018-27396-z.

27 Schäfer, C., Ruggenthaler, M., Appel, H., Rubio, A. Modification of excitation and charge transfer in cavity quantum-electrodynamical chemistry. *Proc. Natl. Acad. Sci. USA* **2019**, *116*(11), 4883–4892. https://doi.org/10.1073/pnas.1814178116.

28 Feist, J., Garcia-Vidal, F.J. Extraordinary exciton conductance induced by strong coupling. *Phys. Rev. Lett.* **2015**, *114*, 196402. https://doi.org/10.1103/PhysRevLett.114.196402.

29 Schachenmayer, J., Genes, C., Tignone, E., Pupillo, G. Cavity-enhanced transport of excitons. *Phys. Rev. Lett.* **2015**, *114*(19), 196403. https://doi.org/10.1103/PhysRevLett.114.196403.

30 Houdré, R., Stanley, R.P., Ilegems, M. Vacuum-field Rabi splitting in the presence of inhomogeneous broadening: resolution of a homogeneous linewidth in an inhomogeneously broadened system. *Phys. Rev. A* **1996**, *53*(4), 2711–2715. https://doi.org/10.1103/PhysRevA.53.2711.

31 Litinskaia, M., La Rocca, G.C., Agranovich, V.M. Inhomogeneous broadening of polaritons in high-quality microcavities and weak localization. *Phys. Rev. B* **2001**, *64*, 165316. https://doi.org/10.1103/PhysRevB.64.165316.

32 Agranovich, V.M., Litinskaia, M., Lidzey, D.G. Cavity polaritons in microcavities containing disordered organic semiconductors. *Phys. Rev. B* **2003**, *67*(8), 085311. https://doi.org/10.1103/PhysRevB.67.085311.

33 Gonzalez-Ballestero, C., Feist, J., Gonzalo Badía, E., Moreno, E., Garcia-Vidal, F.J. Uncoupled dark states can inherit polaritonic properties. *Phys. Rev. Lett.* **2016**, *117*(15), 156402. https://doi.org/10.1103/PhysRevLett.117.156402.

34 Botzung, T., Hagenmüller, D., Schütz, S., Dubail, J., Pupillo, G., Schachenmayer, J. Dark state semilocalization of quantum emitters in a cavity. *Phys. Rev. B* **2020**, *102*, 144202. https://doi.org/10.1103/PhysRevB.102.144202.

35 Dubail, J., Botzung, T., Schachenmayer, J., Pupillo, G., Hagenmüller, D. Large random arrowhead matrices: multifractality, semilocalization, and protected transport in disordered quantum spins coupled to a cavity. *Phys. Rev. A* **2022**, *105*(2), 023714. https://doi.org/10.1103/PhysRevA.105.023714.

36 Sauerwein, N., Orsi, F., Uhrich, P., Bandyopadhyay, S., Mattiotti, F., Cantat-Moltrecht, T. Engineering random spin models with atoms in a high-finesse cavity. *Nat. Phys.* **2023**, *19*(8), 1128–1134. https://doi.org/10.1038/s41567-023-02033-3.

37 Scholes, G.D. Polaritons and excitons: Hamiltonian design for enhanced coherence. *Proc. R. Soc. A* **2020**, *476*(2242), 20200278. https://doi.org/10.1098/rspa.2020.0278.

38 Chávez, N.C., Mattiotti, F., Méndez-Bermúdez, J.A., Borgonovi, F., Celardo, G.L. Disorder-enhanced and disorder-independent transport with long-range hopping: application to molecular chains in optical cavities. *Phys. Rev. Lett.* **2021**, *126*(15), 153201. https://doi.org/10.1103/PhysRevLett.126.153201.

39 Sommer, C., Reitz, M., Mineo, F., Genes, C. Molecular polaritonics in dense mesoscopic disordered ensembles. *Phys. Rev. Res.* **2021**, *3*(3), 033141. https://doi.org/10.1103/PhysRevResearch.3.033141.

40 Allard, T.F., Weick, G. Disorder-enhanced transport in a chain of lossy dipoles strongly coupled to cavity photons. *Phys. Rev. B* **2022**, *106*(24), 245424. https://doi.org/10.1103/PhysRevB.106.245424.

41 Engelhardt, G., Cao, J. Unusual dynamical properties of disordered polaritons in microcavities. *Phys. Rev. B* **2022**, *105*(6), 064205. https://doi.org/10.1103/PhysRevB.105.064205.

42 Ribeiro, R.F. Multimode polariton effects on molecular energy transport and spectral fluctuations. *Commun. Chem.* **2022**, *5*(48), 1–10. https://doi.org/10.1038/s42004-022-00660-0.

43 Balasubrahmaniyam, M., Simkhovich, A., Golombek, A., Sandik, G., Ankonina, G., Schwartz, T. From enhanced diffusion to ultrafast ballistic motion of hybrid light–matter excitations. *Nat. Mater.* **2023**, *22*, 338–344. https://doi.org/10.1038/s41563-022-01463-3.

44 Wellnitz, D., Pupillo, G., Schachenmayer, J. Disorder enhanced vibrational entanglement and dynamics in polaritonic chemistry. *Commun. Phys.* **2022**, *5*(120), 1–11. https://doi.org/10.1038/s42005-022-00892-5.

45 Ahn, W., Triana, J.F., Recabal, F., Herrera, F., Simpkins, B.S. Modification of ground-state chemical reactivity via light–matter coherence in infrared cavities. *Science* **2023**, *380*(6650), 1165–1168. https://doi.org/10.1126/science.ade7147.

46 Cui, B., Nizan, A. Collective response in light–matter interactions: The interplay between strong coupling and local dynamics. *J. Chem. Phys.* **2022**, *157*(11). https://doi.org/10.1063/5.0101528.

47 Anderson. W. Absence of diffusion in certain random lattices. *Phys. Rev.* **1958**, *109*(5), 1492–1505. https://doi.org/10.1103/PhysRev.109.1492.

48 Abrahams, E., Anderson, P.W., Licciardello, D.C., Ramakrishnan, T.V. Scaling theory of localization: absence of quantum diffusion in two dimensions. *Phys. Rev. Lett.* **1979**, *42*, 673–676. https://doi.org/10.1103/PhysRevLett.42.673.

49 Evers, F., Mirlin, A.D. Anderson transitions. *Rev. Mod. Phys.* **2008**, *80*(4), 1355–1417. https://doi.org/10.1103/RevModPhys.80.1355.

50 Engelhardt, G., Cao, J. Polariton localization and dispersion properties of disordered quantum emitters in multimode microcavities. *Phys. Rev. Lett.* **2023**, *130*, 213602. https://doi.org/10.1103/PhysRevLett.130.213602.

51 MacKinnon, A., Kramer, B. One-parameter scaling of localization length and conductance in disordered systems. *Phys. Rev. Lett.* **1981**, *47*, 1546–1549. https://doi.org/10.1103/PhysRevLett.47.1546.

52 Hofstetter, E., Schreiber, M. Finite-size scaling and critical exponents. A new approach and its application to Anderson localisation. *Eur. Lett. (EPL)* **1993**, *21*(9), 933–939. https://doi.org/10.1209/0295-5075/21/9/010.

53 Evers, F., Mirlin, A.D. Fluctuations of the inverse participation ratio at the Anderson transition. *Phys. Rev. Lett.* **2000**, *84*, 3690–3693. https://doi.org/10.1103/PhysRevLett.84.3690.

54 Zharekeshev, I.K., Kramer, B. Scaling of level statistics at the disorder-induced metal-insulator transition. *Phys. Rev. B* **1995**, *51*, 17239–17242. https://doi.org/10.1103/PhysRevB.51.17239.

55 Mehta, M.L. *Random Matrices*. Elsevier: Amsterdam, **2004**.

56 Haake, F. *Quantum Signatures of Chaos*. Springer-Verlag: Berlin, Heidelberg, **2006**.

57 Bogomolny, E.B., Gerland, U., Schmit, C. Models of intermediate spectral statistics. *Phys. Rev. E* **1999**, *59*, R1315. https://doi.org/10.1103/PhysRevE.59.R1315.

58 Altshuler, B.L., Zharekeshev, I.K., Kotochigova, S., Shklovskii, B. Repulsion between energy levels and the metal-insulator transition. *Sov. Phys. JETP* **1988**, *67*, 625. http://jetp.ras.ru/cgi-bin/e/index/e/67/3/p625?a=list.

59 Shklovskii, B.I., Shapiro, B., Sears, B.R., Lambrianides, P., Shore, H.B. Statistics of spectra of disordered systems near the metal-insulator transition. *Phys. Rev. B* **1993**, *47*, 11487–11490. https://doi.org/10.1103/PhysRevB.47.11487.

60 Sokolovskii, I., Tichauer, R.H., Morozov, D., Feist, J., Groenhof, G. Enhanced excitation energy transfer under strong light-matter coupling: insights from multi-scale molecular dynamics simulations. *arXiv* **2022**. https://doi.org/10.48550/arXiv.2209.07309.

61 Abou-Chacra, R., Thouless, D.J., Anderson, P.W. A selfconsistent theory of localization. *J. Phys. C Solid State Phys.* **1973**, *6*(10), 1734.

62 Holstein, T. Studies of polaron motion: Part I. The molecular-crystal model. *Ann. Phys.* **1959**, *8*(3), 325–342. https://doi.org/10.1016/0003-4916(59)90002-8.

63 Reitz, M., Sommer, C., Genes, C. Langevin approach to quantum optics with molecules. *Phys. Rev. Lett.* **2019**, *122*(20), 203602. https://doi.org/10.1103/PhysRevLett.122.203602.

64 Michetti, P., La Rocca, G.C. Exciton-phonon scattering and photoexcitation dynamics in *J*-aggregate microcavities. *Phys. Rev. B* **2009**, *79*(3), 035325. https://doi.org/10.1103/PhysRevB.79.035325J.

65 Sommer, C., Reitz, M., Mineo, F., Genes, C. Molecular polaritonics in dense mesoscopic disordered ensembles. *Phys. Rev. Res.* **2021**, *3*(3), 033141. https://doi.org/10.1103/PhysRevResearch.3.033141.

66 Galego, J., Garcia-Vidal, F.J., Feist, J. Cavity-induced modifications of molecular structure in the strong-coupling regime. *Phys. Rev. X* **2015**, *5*(4), 041022. https://doi.org/10.1103/PhysRevX.5.041022.

67 Vendrell, O. Collective Jahn-Teller interactions through light-matter coupling in a cavity. *Phys. Rev. Lett.* **2018**, *121*(25), 253001. https://doi.org/10.1103/PhysRevLett.121.253001.

68 Feist, J., Galego, J., Garcia-Vidal, F.J. Polaritonic chemistry with organic molecules. *ACS Photonics* **2018**, *5*(1), 205–216. https://doi.org/10.1021/acsphotonics.7b00680.

69 Fábri, C., Halász, G.J., Cederbaum, L.S., Vibók, Á. Born–Oppenheimer approximation in optical cavities: from success to breakdown. *Chem. Sci.* **2021**, *12*(4), 1251–1258. https://doi.org/10.1039/D0SC05164K.

70 Clàudia, G., Gerrit, C., Feist, J., Morozov, D., Toppari, J.J. Tracking polariton relaxation with multiscale molecular dynamics simulations. *J. Phys. Chem. Lett.* **2019**, *10*(18), 5476–5483. https://doi.org/10.1021/acs.jpclett.9b02192.

71 Zhang, Y., Nelson, T., Tretiak, S. Non-adiabatic molecular dynamics of molecules in the presence of strong light-matter interactions. *J. Chem. Phys.* **2019**, *151*(15), 154109. https://doi.org/10.1063/1.5116550.

72 Schollwöck, U. The density-matrix renormalization group in the age of matrix product states. *Ann. Phys.* **2011**, *326*(1), 96–192. https://doi.org/10.1016/j.aop.2010.09.012.

73 Vidal, G. Efficient simulation of one-dimensional quantum many-body systems. *Phys. Rev. Lett.* **2004**, *93*(4), 040502. https://doi.org/10.1103/PhysRevLett.93.040502.

74 Daley, A.J., Kollath, C., Schollwöck, U., Vidal, G. Time-dependent density-matrix renormalization-group using adaptive effective Hilbert. *J. Stat. Mech.: Theory Exp.* **2004**. https://doi.org/10.1088/1742-5468/2004/04/P040052004(04):P04005.

75 White, S.R, Feiguin, A.E. Real-time evolution using the density matrix renormalization group. *Phys. Rev. Lett.* **2004**, *93*(7), 076401. https://doi.org/10.1103/PhysRevLett.93.076401.

76 Paeckel, S., Köhler, T., Swoboda, A., Manmana, S.R., Schollwöck, U., Hubig, C. Time-evolution methods for matrix-product states. *Ann. Phys.* **2019**, *411*, 167998. https://doi.org/10.1016/j.aop.2019.167998.

77 del Pino, J., Schröder, F.A., Chin, A.W., Feist, J., Garcia-Vidal, F.J. Tensor network simulation of non-markovian dynamics in organic polaritons. *Phys. Rev. Lett.* **2018**. *121*(22), 227401. https://doi.org/10.1103/PhysRevLett.121.227401.

78 Fowler-Wright, P., Lovett, B.W., Keeling, J. Efficient many-body non-markovian dynamics of organic polaritons. *Phys. Rev. Lett.* **2022**, *129*(17), 173001. https://doi.org/10.1103/PhysRevLett.129.173001.

79 Bentsen, G.S., Daley, A.J., Schachenmayer, J. Entanglement dynamics in spin chains with structured long-range interactions. In: *Entanglement in Spin Chains: From Theory to Quantum Technology Applications*. Springer: Cham, **2022**, 285–319. https://doi.org/10.1007/978-3-031-03998-0_11.

80 Schachenmayer, J., Lanyon, B.P., Roos, C.F., Daley, A.J. Entanglement growth in quench dynamics with variable range interactions. *Phys. Rev. X* **2013**, *3*(3), 031015. https://doi.org/10.1103/PhysRevX.3.031015.

81 Schuch, N., Wolf, M.M., Verstraete, F., Cirac, J.I. Entropy scaling and simulability by matrix product states. *Phys. Rev. Lett.* **2008**, *100*(3), 030504. https://doi.org/10.1103/PhysRevLett.100.030504.

82 Amico, L., Fazio, R., Osterloh, A., Vedral, V. Entanglement in many-body systems. *Rev. Mod. Phys.* **2008**, *80*(2), 517. https://doi.org/10.1103/RevModPhys.80.517.

83 Eisert, J., Cramer, M., Plenio, M.B. Colloquium: area laws for the entanglement entropy. *Rev. Mod. Phys.* **2010**, *82*(1), 277–306. https://doi.org/10.1103/RevModPhys.82.277.

84 Valmorra, F., Bröll, M., Schwaiger, S., Welzel, N., Heitmann, D., Mendach, S. Strong coupling between surface plasmon polariton and laser dye rhodamine 800. *Appl. Phys. Lett.* **2011**, *99*(5), 051110. https://doi.org/10.1063/1.3619845.

85 Vendrell, O. Coherent dynamics in cavity femtochemistry: application of the multi-configuration time-dependent Hartree method. *Chem. Phys.* **2018**, *509*, 55–65. https://doi.org/10.1016/j.chemphys.2018.02.008.

86 Wellnitz, D., Pupillo, G., Schachenmayer, J. A quantum optics approach to photoinduced electron transfer in cavities. *J. Chem. Phys.* **2021**, *154*(5), 054104. https://doi.org/10.1063/5.0037412.

87 Baghdad, M., Bourdel, P.A., Schwartz, S., Ferri, F., Reichel, J., Long, R. Spectral engineering of cavity-protected polaritons in an atomic ensemble. *Nat. Phys.* **2023**, *19*(8), 1104–1109. https://doi.org/10.1038/s41567-023-02035-1.

88 Mattiotti, F., Dubail, J., Hagenmüller, D., Schachenmayer, J., Brantut, J.P., Pupillo, G. Multifractality in the interacting disordered Tavis-Cummings model. *arXiv* **2023**. https://doi.org/10.48550/arXiv.2302.14718.

89 Liu, Y., Ni, K.K. Bimolecular chemistry in the ultracold regime. *Annu. Rev. Phys. Chem.* **2022**, 73(1), 73–96. https://doi.org/10.1146/annurev-physchem-090419-043244.

90 Wellnitz, D., Schütz, S., Whitlock, S., Schachenmayer, J., Pupillo, G. Collective dissipative molecule formation in a cavity. *Phys. Rev. Lett.* **2020**, *125*(19), 193201. https://doi.org/10.1103/PhysRevLett.125.193201.

91 Wright, A.D., Nelson, J.C., Weichman, M.L. Rovibrational polaritons in gas-phase methane. *J. Am. Chem. Soc.* **2023**, *145*(10), 5982–5987. https://doi.org/10.1021/jacs.3c00126.

Part III

Applications

10

Engineering Organic Exciton–Polariton Condensates in Microcavities

Sitakanta Satapathy[1] *and Vinod M. Menon*[2,3]

[1] *Quantum Circuits, Inc., New Haven, CT, USA*
[2] *Department of Physics, Center for Discovery and Innovation, The City College of New York, New York, NY, USA*
[3] *The PhD Program in Physics, The Graduate Center of the City University of New York, New York, NY, USA*

10.1 Introduction

In 1924, Luis de Broglie proposed the wave nature of the particles, where the wavelength (λ_{dB}) is directly related to mass (m) of the particles as $\lambda_{\mathrm{dB}} = (2\pi\hbar^2/mk_{\mathrm{B}}T)^{1/2}$ at temperature (T) [1]. Below a critical temperature (T_{c}), the de Broglie wavelength of these bosonic particles becomes comparable to their interatomic distances, where the wavepackets of these bosons starts to overlap. As a result, they experience a phase transition in which all the particles collectively occupy a single quantum state, leading to the formation of a condensate. The condensate state is defined by the emergence of macroscopic phase coherence, with the particles in the condensed state being represented by a single wave function. The critical temperature (T_{c}) corresponding to this quantum degeneracy limit is related to the mass (m) and density of the bosonic particles as $T_{\mathrm{c}} \propto \hbar^2 n^{2/3}/m$. From this, it is evident that achieving higher critical condensation temperatures requires either reducing the mass of the bosonic particles or increasing their density. Bose–Einstein condensates (BEC) are the ideal systems for observing quantum phenomena on a macroscopic level. Ever since the discovery of atomic condensates in 1995 [2], many groups worldwide have carried out fundamental research on them and more recently even found applications in emerging technologies such as quantum computing [3–5]. The primary challenge with atomic condensates lies in the extremely low temperatures required (on the order of nanokelvin) due to the large masses associated with atoms. This would be significantly easier if the condensing bosonic particles had a lower mass. In this context, solid-state systems, such as magnons [6], excitons [7], and exciton–polaritons (EPs) [8], have been proposed as candidates for condensation at elevated temperatures.

As discussed in the previous chapters of this volume, EPs are bosonic quasiparticles which emerge because of strong light–matter coupling between excitons and photons in a cavity [9]. Due to their hybrid construct, these quasiparticles inherit the properties of both excitons and photons. While the photon component makes them nearly massless, with the effective mass of the lower polariton (LP) component being ~10^{-12} times lighter than the mass of Rubidium atom (Rb), the

Polariton Chemistry: Molecules in Cavities, First Edition. Edited by Joel Yuen-Zhou, Noel C. Giebink and Raphael F. Ribeiro. Published 2026 by John Wiley & Sons, Inc.

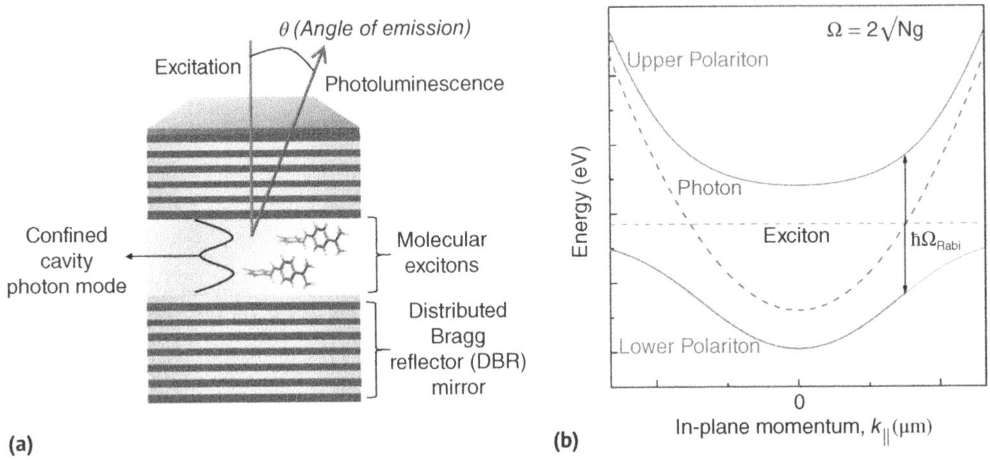

Figure 10.1 (a) Schematic representation of a microcavity device containing a thin film of organic molecules (green) sandwiched between two distributed Bragg reflector mirrors with alternating layers. (b) Strong coupling between microcavity photons (dispersive) and organic molecular excitons (nondispersive) which results in the forming of two polariton branches – the upper (UP) and the lower polariton (LP) branches are shown as top and bottom solid lines, respectively.

exciton component provides interactions and nonlinearity which is otherwise difficult to engineer in purely photon-based devices. The lower effective mass of the LP component close to zero in-plane momentum made EPs attractive for achieving condensation at elevated temperatures, with the one important distinction being their out-of-equilibrium characteristics compared to atomic BECs (Figure 10.1).

Initial investigations into EP condensation (EP condensation, polariton condensation, and polariton lasing are used interchangeably in this chapter) focused on II–VI and III–V inorganic semiconductor quantum-well microcavities [10, 11]. Nevertheless, such systems could only operate at temperatures below approximately 100 K due to the small exciton binding energy of their weakly bound and delocalized Wannier–Mott excitons [12]. Subsequently, the achievement of polariton condensation under ambient conditions was demonstrated in materials such as GaN and ZnO, where the excitons exhibit higher binding energies and oscillator strengths [13–15]. However, these materials typically require demanding fabrication conditions, such as epitaxial growth, to attain high-quality factor microcavities, and they encounter challenges associated with lattice mismatch. Such limitations have sparked interest in exploring materials that possess significantly larger exciton binding energies and offer greater ease of fabrication. For a polariton condensate to form, both the exciton binding energy and the matter–light coupling must surpass the temperature. Organic molecular materials that host large binding energy Frenkel excitons possess the capability to fulfill this requirement even at elevated temperatures [16], thereby facilitating the observation of condensation at room temperature. Furthermore, organic materials offer distinct advantages and unique characteristics that make them appealing for exploring molecular polariton condensation. These include:

- *Room temperature operation*: In molecular solids, the formation of localized excitations, known as Frenkel excitons, arises from the process of electronic excitations hopping between molecules [17]. This stands in contrast to Wannier–Mott excitons found in inorganic materials, where the delocalized electrons and holes are bound [12]. Notably, Frenkel excitons exhibit greater binding energy and oscillator strength compared to Wannier excitons, enabling their observation and study even at room temperature [18].

- *Disorder*: In contrast to inorganic materials, organic molecular solids show notable energetic and positional disorders, particularly in amorphous molecular systems or polymers [19]. Additionally, high-energy vibrational modes exist, which can exhibit strong coupling with the electronic state. The role of disorder in achieving molecular polariton condensates is still subject to debate and lacks comprehensive understanding. However, recent studies have indicated that when the Rabi splitting surpasses the linewidth associated with disorder, lasing can still be achieved [20–30]. This has been demonstrated in various systems, including amorphous organic films, fluorescent proteins (FPs) and organic polymers.
- *Versatile material processing and ease of fabrication*: Organic materials can be easily processed into various forms, including thin films, crystals, and nanostructures. This flexibility enables the fabrication of different device geometries and integration with diverse optical cavity structures, such as microcavities and photonic crystals, to enhance light–matter coupling and facilitate polariton condensation.
- *Tailorable molecular design*: Organic molecules can be chemically synthesized with a high degree of precision, allowing researchers to control their electronic and optical properties. By carefully selecting the molecular structure, it is possible to tune the exciton energy, transition dipole moment, disorder, Stokes shift, and intermolecular interactions, which are crucial for achieving strong coupling and polariton condensation.

There is a large body of work on strong light–matter coupling in molecular systems, including the topic of polariton chemistry, which has been discussed previously, including in earlier chapters of this volume [18, 19, 31–33]. Our objective in this chapter is to provide an overview of the latest progress specifically on EP condensates in organic microcavities from the standpoint of molecular photophysics and microcavity design. The aim is to give a comprehensive understanding of how to explore the role of molecular engineering in tailoring the properties of polariton condensates.

10.2 Mechanism of Polariton Condensation in Organic Microcavities

While the physics of EP condensation in inorganic quantum well systems has been studied in detail, there is much less theoretical work on the underlying mechanism in the context of organic molecular EP condensates. The widely accepted picture representing the mechanism of polariton condensation more broadly is given in Figure 10.2 [11].

In a typical condensation experiment, a pulsed or continuous wave laser is incident onto the strongly coupled microcavity sample, where the sample is excited nonresonantly at some energy above the upper polariton (UP) branch. Typically, the wavelength of excitation for an organic microcavity involving distributed Bragg reflector (DBR) mirror corresponds to the reflective minimum at the blue edge of the mirror. This is done to maximize the light absorption in the microcavity sample. Upon nonresonant excitation, a reservoir of excitons is created. Since the excitons interact with each other as well as with lattice vibrations, they undergo fast relaxation via phonon scattering to create a hot reservoir of dark excitons at some higher momenta across the LP branch. The exciton density of the reservoir increases with an increase in the excitation laser power. Eventually, the quantum degeneracy limit is reached at some critical threshold excitation power, where there will be an onset of the bosonic-stimulated scattering of EPs directly towards the ground state ($k_{\parallel} = 0$) of the LP branch, vide Figure 10.2. Beyond this threshold excitation density, the stimulated relaxation rate is enhanced and the ground state population density of polaritons

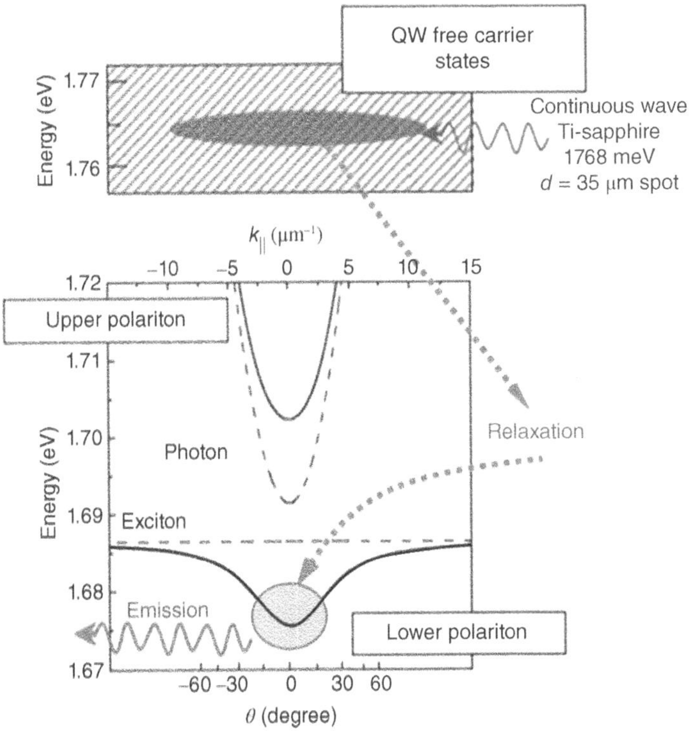

Figure 10.2 Conventional mechanism of polariton condensation via nonresonant pumping of the microcavity system. *Source*: Kasprzak et al. (2006) / Springer Nauture.

shoots up until the population clamps and saturates. The relaxation dynamics of the polaritons are dictated by the interaction between the excitons, polaritons, and the phonons or molecular vibrational modes. The dominant polariton relaxation mechanisms in organic microcavities include: (a) radiative (or intracavity) pumping and (b) vibron (or Raman mode) assisted relaxation.

10.2.1 Radiative Pumping

In the case of inorganic polaritons, fermionic exchange interactions of Wannier–Mott excitons play a major role in relaxation and stimulated scattering of polaritons. In contrast, this mechanism is largely absent in molecular polaritons due to the strong localization of Frenkel excitons, which effectively quenches the fermionic exchange interactions. In the context of polariton condensation in organic microcavities, an important mechanism that promotes the population of polariton modes is known as radiative pumping or intracavity pumping, which is shown in Figure 10.3a. This process, demonstrated in 2010 by Kéna-Cohen and Forrest using an organic single-crystal microcavity, resembles conventional emission, but the final state is a polariton [35]. It involves an electronic transition, such as a vibronic sideband, localized state, or defect, that is not strongly coupled. This transition radiatively pumps the LP, usually preferring a negative detuning of the cavity mode to bring the LP branch resonant with respect to the bare emission of the molecule, as illustrated in Figure 10.3a.

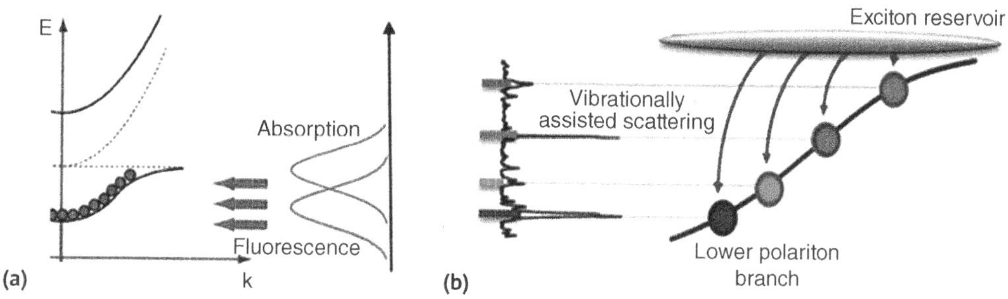

Figure 10.3 Schematic representation showing different mechanisms of polariton scattering in organic microcavities: (a) radiative pumping [35]; (b) vibron-assisted relaxation [36].

10.2.2 Vibron-assisted Relaxation

Organic semiconductors exhibit several vibronic modes with energies typically ranging from 30 to 430 meV, which often coincide with the energy of Rabi splitting. This fortuitous overlap establishes a direct relaxation pathway between excitons and polaritons, almost identical to the LO phonon emission in GaAs quantum well-based microcavity structures, which facilitates the scattering of light-hole excitons into heavy-hole EPs. Lidzey and colleagues demonstrated that in a strongly coupled *J*-aggregated cyanine dye-based microcavity, an exciton reservoir can be created following nonresonant excitation [36]. These exciton states can then scatter into polariton states within the LP branch by emitting energy that caused the excitation of the different vibrational modes of the molecular system. This process is depicted in Figure 10.3b. By designing the cavity in a way that the energy difference between the polariton mode at normal incidence and the uncoupled exciton energy resonates with a vibrational quantum, i.e., $E_{\text{vib}} = E_{\text{ex}} - E_{\text{LP}}\ (k_0)$, polariton states near $k_{\|} = 0$ of the LP branch can be rapidly populated after nonresonant excitation of uncoupled excitons.

10.3 Experimental Signatures of Polariton Condensation in Organic Microcavities

Angle-resolved photoluminescence (PL) is a widely utilized technique for investigating condensation phenomena. This method allows the PL intensity associated with specific in-plane wavevectors, denoted as, $k_{\|}$, to be correlated with the population of polaritons. The relationship is expressed through the equation $N(E) = I(k_{\|})/|C(k_{\|})|^2$, where $N(E)$ represents the relative occupation number of the respective polariton branch. It is determined by normalizing the PL intensity, $I(k_{\|})$, with respect to the photon fraction of the cavity determined at each angle, $|C(k_{\|})|^2$. The angle-dependent energy is determined by the maximum of the PL spectrum. Researchers have employed the polariton occupation number, $N(E)$, to study thermalization processes in organic microcavities. By examining $N(E)$ as a function of $[E(k_{\|}) - E(k_{\|} = 0)]$ across the LP branch, they investigate the nonequilibrium or quasi-equilibrium relaxation dynamics leading to forming of the polariton condensates. Furthermore, by fitting the $N(E)$ according to Maxwell–Boltzmann distribution as $N(E) \propto \exp[-\{E(k_{\|}) - E(k_{\|} = 0)\}/k_B T_{\text{eff}}]$, efforts have been made to determine the effective temperature (T_{eff}) of the organic polariton gas, providing insights into the degree of equilibrium within the investigated system.

Experimentally, polariton condensation is distinguished by several features.

- Dark excitons below the threshold have multiple decay pathways, including scattering into LP states, energy transfers in near-field to other molecules, radiating into weakly coupled modes, and nonradiative decay. However, above the threshold, stimulated scattering towards $k_{\|} = 0$ becomes dominant, surpassing the rates associated with the other decay pathways, and a pronounced nonlinear increase in emission from $k_{\|} = 0$ of the LP branch is observed. Above a certain threshold density, there is a significant enhancement in both spatial and temporal coherence. This is evident through a reduction in linewidth and a shift of the PL towards the lower $k_{\|}$-vectors (or low energy states) of the LP branch.
- The formation of a condensate in organic microcavities is characterized by the spontaneous emergence of spatial coherence at room temperature, which is defined by its long-range order. Experimental evidence for this can be observed through interferometry using a Michelson interferometer, as shown in Figure 10.4a. By comparing the emitted light from the cavity with its inverted image, the first-order spatial coherence, denoted as $g^1(r, r')$, which measures the phase coherence between the points at r and r', can be determined. Below the threshold, the PL intensity of the polariton appears uniformly distributed throughout the entire region. At the threshold, interference fringes start appearing. This signifies the onset of coherent emission coming out from the cavity polaritons. As shown in Figure 10.4b, further increasing the pump intensity leads to the identification of interference fringes across the entire condensate area, demonstrating the existence of long-range order. Importantly, these experiments can be conducted at room temperature using organic microcavities. Above the threshold, the degree of spatial coherence increases with power and can be measured quantitatively. However, it is important to consider the pump intensity distribution when analyzing the spatial decay of correlation functions. Measurements using large spot sizes or using a flat-top pump profile exhibit reduced phase contrast and weaker spatial structure. The appearance of such features in the intensity profile is attributed to the existence of dynamic instabilities [21]. As a consequence, the condensate appears to be fragmented. This is due to an interplay of polariton–exciton repulsive interactions and gain that creates spatial hole burning in the density of the dark exciton reservoir.
- Most experimental observations show a blueshift in the polariton emission energy with increased excitation energy. This blueshift is influenced by the total population near the threshold and arises primarily due to a reduced ground-state occupancy due to pumping. The

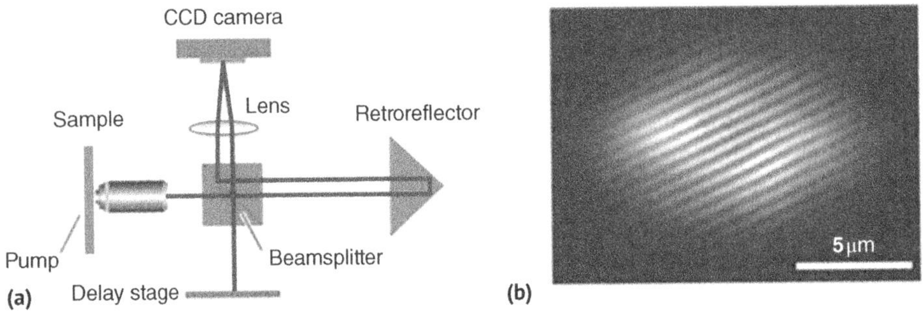

Figure 10.4 (a) Schematic of a Michelson interferometer with a retroreflector configuration to measure the temporal coherence. Both the images are focused onto a charge-coupled device (CCD) using a lens to produce the spatial interference pattern. (b) Spatial interference patterns measured from a molecular polariton condensate. *Source*: Ref. [21], with permission of American Physical Society.

blueshift can provide insights into the coupling strength. Experiments conducted by Lidzey and colleagues investigated the origin of blueshift in such condensates, considering various contributing factors such as intracavity Kerr effect, polariton–polariton interactions, gain-induced frequency pulling, and effects associated with the saturation of optical transitions [25]. The study concludes that blueshifts in organic polariton condensates result from the combined effects of saturation and intermolecular energy transfer. A step-like increase in blueshift occurs at the condensation threshold when the stimulated scattering of optically aligned excitons into the ground polariton state exceeds the rate of intermolecular energy transfer. A notable feature observed in some organic polariton condensates is their spontaneous polarization behavior. Below the threshold, emission from the LP state at $k_{\parallel} = 0$ is unpolarized. However, above the threshold, the polarization of condensate aligns spontaneously with the polarization of the pump. This alignment indicates the stimulated scattering to take place at a rate faster than any depolarization mechanisms (internal conversion and intermolecular energy transfer), resulting in ultrafast decay of fluorescence. Lidzey et al. propose a model based on the interplay between the stimulated scattering of excitons and intermolecular energy transfer. The model successfully predicts a correlation between the experimentally observed step-like blueshift at the condensation threshold with a step-like increase in the degree of linear polarization of the condensate [25]. At a molecular level, below the threshold, the sublinear increase in polarization accounts for intermolecular energy transfer, leading to exciton–exciton annihilation. Above the threshold, the molecules with the transition dipole moments aligned with respect to the laser (pump) polarization undergo stimulated scattering approaching the condensation threshold. Additionally, saturation effects due to photobleaching become dominant with further increase in excitation energy.

10.4 The Molecular Medley for Polariton Condensation

So far, organic polariton condensation has been realized using four classes of molecular materials: (a) single crystalline systems, (b) low molecular weight emitters (LMWEs), (c) polymers, and (d) host–guest systems. In this section, we outline the experimental realizations of organic polariton condensates using different organic materials. Understanding the diverse range of materials that can exhibit organic polariton condensation is crucial for advancing this field of research. Figure 10.5 presents representative examples of materials employed so far to achieve organic polariton condensation. By examining these realizations, researchers have gained insights into the underlying mechanisms and properties of these systems.

10.4.1 Single Crystalline Systems

Organic molecular crystals have emerged as promising candidates for observing polariton condensation due to their well-defined molecular structures and optical properties, showcasing their potential for coherent light emission and low-threshold lasing [35, 37, 38]. Previously, strong coupling between excitons and photons was successfully shown in nanocrystalline films of 3,4,7,8-naphthalenetetracarboxylic dianhydride (NTCDA) and polycrystalline tetracene [39, 40]. Polycrystalline tetracene-exhibited crystalline domains spanning several micrometers, enabling the observation of anisotropy with band structure of excitons. However, the integration of polycrystalline films into high-quality factor microcavities poses challenges due to their significant surface roughness. In 2010, Kéna-Cohen et al. demonstrated polariton condensation in organic semiconductor microcavities using anthracene single crystals at room temperature [35]. The dipole moment

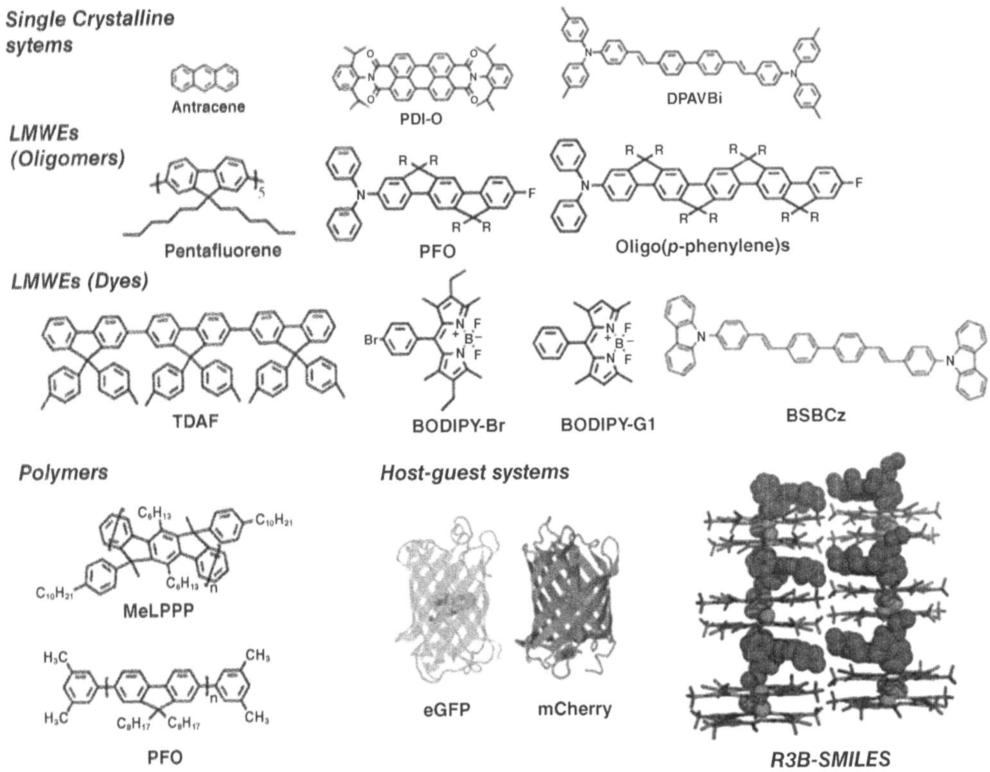

Figure 10.5 Molecular systems investigated so far for studying polariton condensation.

associated with the lowest electronic energy transition in anthracene is along its short axis. The crystal structure reveals that there are two molecules per unit cell. This results in Davydov splitting of this lowest exciton energy transition into bands that are polarized along different directions.

In this work, the microcavity was composed of two DBR mirrors with a melt-grown anthracene single crystal as the active layer. From the cavity mode linewidth (Dω). The anthracene microcavities had a high-quality factor ($Q \sim 6000$), indicating a high level of photon confinement. The *b*-polarized absorption spectra in anthracene reveals three distinct and structured vibronic transitions (0–0, 0–1, and 0–2) which strongly couple to the cavity photon mode, leading to the formation of four polariton branches, as shown in Figure 10.6a. The polariton lifetime, estimated from the linewidth of the PL peak, $DE_{LP} \sim 3.5$–5.5 meV, was found to be $\tau_p = \hbar/DE_{LP} \sim 150$–$200$ fs. Despite being shorter than in inorganic microcavities, it was sufficient for achieving condensation in organic materials. In anthracene microcavities, condensation was observed when the LP minimum at $E_{LP}(k_0) = 2.94$ eV was precisely tuned below the zero-phonon-line ($E_{0\text{-}0}$) energy by one vibrational energy, $E_{vib} = 173$ meV. It is noteworthy to mention here that a previous study by the same group introduced a melt-processing method to synthesize large single crystals of anthracene with precision at the level of nanometers over the thickness. Consequently, this results in strong anisotropic dispersion and birefringence [41]. Increasing the pump power beyond a certain threshold (P_{th}) resulted in the manifestation of condensation signatures as discussed in the previous section, such as a superlinear increase in intensity and a drastic reduction in PL lifetime. This is shown in Figure 10.6a, where the reduction in the condensation threshold can be attributed to

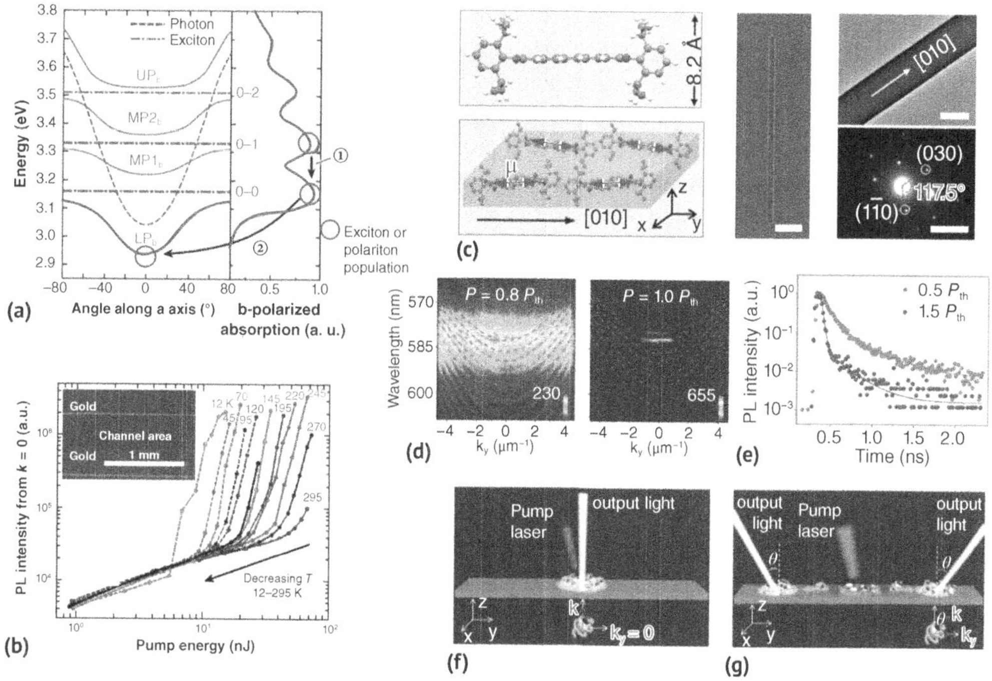

Figure 10.6 (a) Anthracene microcavity dispersion polarized along the b axis showing the respective polariton branches (left). Dashed lines show the bare photon and exciton energies. Absorption spectra of anthracene and representation of the lasing mechanism (right). (b) Power dependent PL at $k = 0$ measured at different temperatures with the inset depicting an optical image of the microcavity structures used. *Source*: Figures adapted with permission from Ref. [35]; with permission from Springer Nature. (c) Chemical structure of the PDI-O molecule and the TEM characterization results for the PDI-O microribbon structure. (d) Power dependent PL dispersion plots below (left) and (right) above the threshold depicting PL in k-space. (e) Time-resolved decay of PL at 0.5 P_{th} and 1.5 P_{th}. Schematic showing the light output from the condensate at (f) low and (g) high pump powers figures adapted with permission from Ref. [37]; with permission from Springer Nature.

enhanced relaxation of the dark excitons from the reservoir to the $k = 0$ of the LP involving vibrational phonons or radiative pumping of the LP by the undressed 0–1 fluorescence peak.

As shown in Figure 10.6b, the lasing threshold of anthracene exhibited temperature dependence, with a significant decrease in threshold when cooled down to 12 K from room temperature. This is possibly due to an increase in the lifetime of the polariton caused by the narrowing of the exciton linewidth. Moreover, nonlinear behavior was observed at energies corresponding to the vibronic replica ($E_{0\text{-}0} - E_{vib}$) away from the $k = 0$ point [19]. Although higher powers were required for these observations compared to $k = 0$ condensation, they underscored the importance of vibronic relaxation processes and the limited scattering efficiency within the LP branch due to the short lifetime of polariton.

In another study, Tang et al. showed that organic single-crystal microribbons (Figure 10.6c) made of *N,N′*-bis(2,6-diisopropyl phenol) 3,4,9,10-perylenetetracarboxylic diimide (PDI-O) exhibit room temperature EP condensation [37]. In contrast to all other configurations which have so far relied on the use of external mirrors (either DBR or metal mirrors), in this study, the regular-shaped self-assembled microribbons served as waveguide Fabry–Pérot microcavity modes that strongly couple to the Frenkel excitons at room temperature. Despite the poor cavity quality, the large transition dipole moments and the high densities of excitons of PDI-O microcrystals resulted in a very high coupling strength ($\hbar\Omega_R = 530$ meV), which aided in polariton condensation and lasing, as shown in Figure 10.6d. Time-resolved PL measurements in Figure 10.6e demonstrate that the polariton PL lifetime drops from 277 to 47 ps when the pump fluence is raised above the threshold, providing evidence in favor of the stimulated scattering process. As seen in Figure 10.6f, the produced ground-state condensates had no wavevector component when the threshold ($P_{th} = 2.3$ μJ cm^{-2}) was achieved, resulting in the lasing output with a zero-emission angle. Polariton–exciton interactions can result in the propagation of polaritons over macroscopic distances at higher pump powers, vide Figure 10.6g. The single-crystal microribbons' exceptional structural homogeneity allowed the polaritons to travel over macroscopic distances during the course of their estimated 50 fs lifetime. This device became a direction-tunable polariton laser that can be controlled via pump power and temperature when it was discovered that such interaction could be controlled to regulate exciton densities upon varying the temperature. Through strong coupling in low-dimensional geometries without the use of external microcavities to create tunable polariton lasers, this work further demonstrated the enormous potential of organic materials as natural microresonators.

Ren et al. presented a cavity made of a single crystalline microbelt of 4,4'-bis[4-(di-p-tolylamino)styryl]biphenyl (DPAVBi) covered by a space-layer (SiO_2) and sandwiched between two silver reflectors [38]. DPAVBi molecules self-assemble in a herringbone pattern along the length of the microbelt. The transition dipole-moment (TDM) of DPAVBi is given by an angle of 36° with respect to the stacking direction along the long axis of the molecule. As a result, the projection of TDM along the length direction (also known as the parallel direction) is substantially larger than the projection along the width direction (also known as the vertical direction) of the microbelt. The molecule shows polarization-dependent absorption and strong coupling, resulting in the formation of five polariton branches. With the help of an effective vibration-assisted relaxation channel and heavy, exciton-like polaritons produced by positive detuning in this cavity, as well as interparticle interactions between polaritons and reservoirs, low-threshold (84 μJ cm^{-2}) polariton lasing was realized in an Ag-Ag cavity.

10.4.2 Low Molecular Weight Emitters

In polariton condensation experiments, disordered molecular thin films employing low molecular weight organic emitters have emerged as significant research topics. The amorphous nature of these films allows for greater control over material properties, such as the induction of localization effects via disorder. At first, it was believed that the presence of significant static disorder in these structures would impede organic polariton BEC. It was anticipated that the disorder would localize polariton wavefunctions, particularly for states close to $k = 0$ [40]. In addition, the presence of molecules with incorrect dipole moment orientations in disordered films would reduce the Rabi energy when coupled to a particular electric field polarization. In addition, these misaligned molecules may serve as a receptacle for the excited population via Forster transfer, which results in exciton–exciton annihilation or bimolecular fluorescence quenching [25]. Although disorder-induced inhomogeneous broadening is significantly greater in amorphous or polycrystalline thin films compared to crystals, recent research has demonstrated polariton lasing using these systems, if the polariton separation exceeds the disorder linewidth [20–30]. In comparison to the production of single crystals, low-molecular-weight organic molecules can be readily processed into disordered or polycrystalline thin films. These systems have demonstrated compelling phenomena such as Anderson localization [42], and disorder-enhanced condensation [21], paving the way for novel methods of manipulating polariton states. For the fabrication of disordered or polycrystalline thin films, the prevalent techniques include spin coating and vacuum sublimation [43]. These fabrication techniques permit the production of pristine (undoped) or doped films in which the organic dye is incorporated into the host matrix. In a number of instances, these films display a degree of orientational alignment, with the molecules primarily lying flat on the substrate [44]. Additionally, these films may contain microscopic crystalline domains. Without directed templating, however, the in-plane alignment of these domains remains arbitrary.

One of the most widely used organic materials from which polariton condensation has been investigated many times is the fluorene oligomer 2,7-Bis[9,9-di(4-methylphenyl)-fluoren-2-yl]-9,9-di(4-methylphenyl)fluorene (TDAF) [20]. TDAF also shows in-plane anisotropy to a certain degree, which consequently results in a large absorption coefficient for its in-plane polarization [43, 44]. This consequently results in a large Rabi energy for the transverse-magnetic (TM) polarization. TDAF films have been employed as active layers in all-metal microcavities to demonstrate ultrastrong coupling behavior with giant Rabi splitting of $\hbar\Omega_R \sim 1$ eV [19]. Figure 10.7a, b depicts the wavevector (k-space) resolved imaging of PL intensity maps below and above the threshold ($P_{th} = 63$ μJ cm^{-2}), respectively. The PL intensity distribution changes when a condensate forms, moving toward $k = 0$ of the LP branch. As demonstrated in Figure 10.7b (see Table 10.1), the condensation threshold in TDAF is 10 times lower than that in anthracene. The threshold might be lowered as a result of less bimolecular annihilation, a relatively longer polariton lifetime, and a higher Rabi energy. One uniqueness of TDAF microcavity in this study was the spontaneous polarization of the condensate which is shown in Figure 10.7c. While below the threshold ($<P_{th}$), the $k = 0$ emission is unpolarized as a result of in-plane disorder, spontaneous polarization of the condensate occurs above the threshold at 1.6 P_{th} along the pump polarization direction. This shows that stimulated scattering becomes faster than internal conversion and any other intermolecular energy transfer process that could lead to ultrafast depolarization of fluorescence. When the LP is pumped resonantly, a threshold as low as $P_{th} = 13$ μJ cm^{-2} could be achieved, which has been shown in follow-up research. Interferometry experiments reveal the presence of well-resolved fringes spread virtually throughout the condensate region above P_{th} (see Figure 10.4b). The coherence time is estimated to be $\tau_c = 0.8$ ps at $1.5P_{th}$, which is basically determined by how fast the dark exciton reservoir depopulates to a certain value below the threshold. Above the threshold, the amount of spatial coherence is found to increase with power. As shown in Figure 10.7d, fragmentation of the condensate can be observed when the gain along with a repulsive polariton–exciton interaction can lead to burning of spatial holes in the dark exciton reservoir [21].

The first demonstration of room-temperature superfluidity using TDAF organic polariton condensates was achieved by Lerario et al. [45]. As illustrated in Figure 10.7e, an artificially introduced defect scatters polaritons with finite momentum upon resonant pumping at low intensities. This scattering caused interference with the incoming polariton flow, creating a spatial fringe pattern in real-space images. Simultaneously, a shadow cone formed behind the defect due to reduced transmission, and a dynamically generated vortex–antivortex pair appeared. In momentum space, the scattering resulted in the characteristic Rayleigh scattering ring. Upon increasing the pump intensity, these features, including the fringes, shadow cone, vortex pair, and scattering ring, nearly vanished (Figure 10.7f). This suppression of scattering under higher pump intensities aligns with the expected behavior of a superfluid. This breakthrough not only enabled the exploration of macroscopic quantum phenomena at room temperature but also shows strong promise for designing polaritonic devices with enhanced resilience to structural defects and disorder.

Reducing the polariton lasing threshold is a key objective in the field of organic EP condensates. Recent endeavors have primarily concentrated on identifying materials with a high photoluminescence quantum yield (PLQY), greater absorption coefficient, and increased polariton relaxation

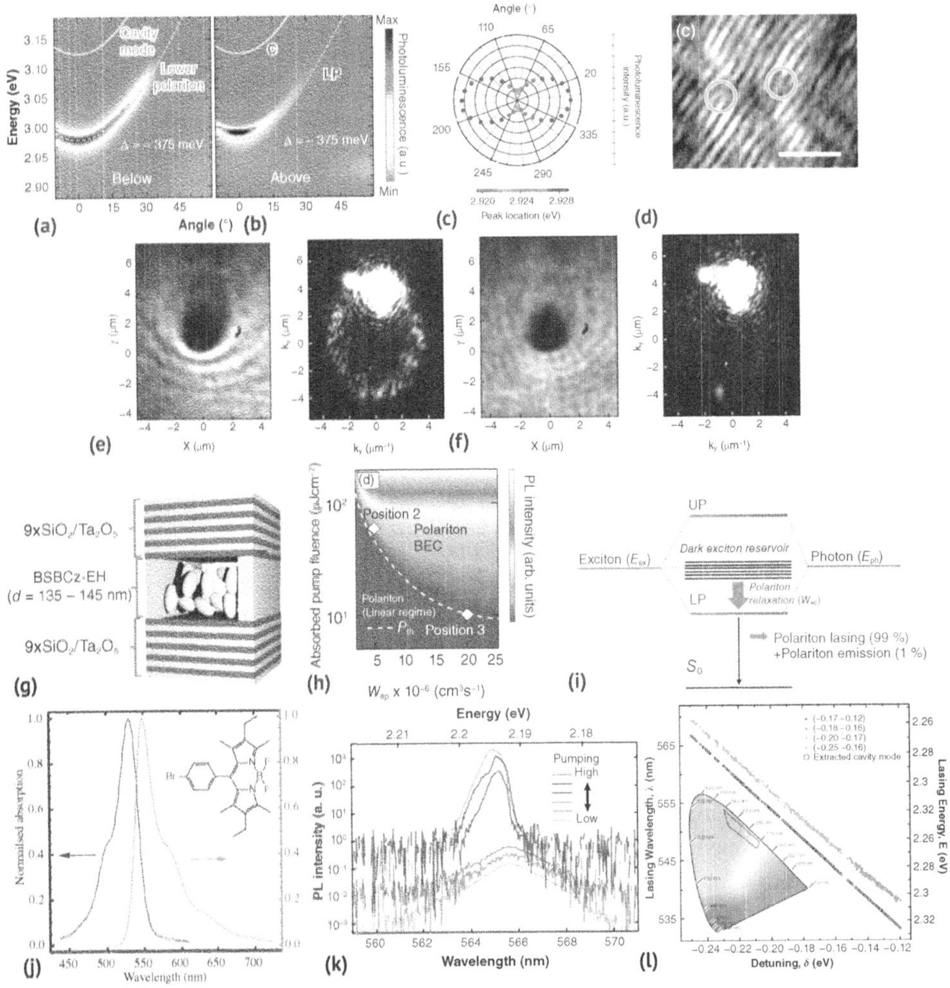

Figure 10.7 Power dependent PL dispersion from the TDAF microcavities (a) below (left) and (b) above (right) the P_{th} demonstrating the PL collapse in k-space above the P_{th}. (c) Polarization plot showing spontaneous polarization of the condensate above the P_{th}. (d) Interferograms from the TDAF microcavity polariton condensate using a small (27 × 25 μm) Gaussian pump showing disorder in condensate as a consequence of the optical hole burning of the exciton reservoir. *Sources*: [19], with permission of Annual Reviews; [21], with permission of American Physical Society. Experimental real space (left) and k-space (right) interferogram images showing the polariton flow occurring across an artificial defect at (e) low and (f) high pump power. *Source*: [45], with permission of Springer Nature. (g) Schematic of the microcavity structure containing melt processed BBSCz-EH capsules. (h) A contour map showing the dependence of the absorbed pump fluence against the simulated PL intensity for different polariton scattering rates from the BBSCz-EH microcavity. The theoretical threshold of a polariton condensate with respect to the scattering rate is shown by the dotted line, The polariton scattering rate and the threshold at two different positions are shown by the white plots. (i) A schematic showing the radiative pumping assisted polariton relaxation mechanism in BBSCz-EH microcavities. *Source*: Figures adapted with permission from Ref. [46]; copyright 2021 Wiley-VCH. (j) BODIPY-Br thin film's absorption and PL spectra. The characteristic excimer (dimer) peak corresponds to 593 nm in the PL spectra. (k) Power dependence of the microcavity emission spectrum obtained from $k = 0$ of the LP branch of BODIPY-Br showing the onset of condensation. *Source*: Figures adapted with permission from Ref. [23]; copyright 2017 Wiley-VCH. (l) Broadband tunable polariton lasing obtained from a wedged cavity with BODIPY-G1 as the active layer with varying exciton–photon detuning. *Source*: Figures adapted with permission from Ref. [49]; copyright 2019 Wiley-VCH.

Table 10.1 Summary of various platforms that have been used to realize exciton–polariton condensation in organic molecular systems along with their Rabi splitting and threshold conditions.

Category	Material	Microcavity bottom/top	λ_{exc} (nm)	Ω_R (meV)[a]	Abs. P_{th} or I_{th}[b]	References
Single crystalline systems	Anthracene	DBR/DBR	420	256	320 μJ cm^{-2}	[35]
	PDI-O	Self-assembled waveguide modes	525	530	2.3 μJ cm^{-2}	[37]
	DPAVBi[c]	Ag-SiO_2/ SiO_2-Ag	448	265	70.0 μJ cm^{-2}	[38]
LMWEs (oligomers)	Pentafluorene	DBR/DBR	437	500	12 μJ cm^{-2}	[24]
	Oligo (p-phynylene)s	DBR/DBR	388	388	63 μJ cm^{-2}	[45]
	TDAF	DBR/DBR	415	600	13–27 μJ cm^{-2}	[20, 21, 45]
LMWEs (dyes)	BODIPY-Br	DBR/DBR	565	91	527 μJ cm^{-2}	[23, 47, 48]
	BODIPY-G1	DBR/DBR	550	114	6 mJ cm^{-2}	[49]
	BSBCz	DBR/DBR	388	246	9.7 ± 0.1 μJ cm^{-2}	[46]
	DPAVB[c]	DBR/DBR	420	227	215 μJ cm^{-2}	[51]
Polymers	MeLPPP	DBR/DBR	475	116	130–500 μJ cm^{-2}	[52–55]
	PFO	DBR/DBR	520	437	27 μJ cm^{-2}	[28, 56]
Host–guest systems	eGFP	DBR/DBR	520	194	300 kW cm^{-2}	[29]
	mCherry	DBR/DBR	640	266	600 kW cm^{-2}	[30, 57, 58]
	R3B-SMILES	DBR/Ag DBR/DBR	566 515	100 110	200 μJ cm^{-2} 130 μJ cm^{-2}	[59, 60]

a. Instead of the pump fluence, threshold irradiance is given in the case of FPs. This is because the exciton density in these cases is related to the product of the peak irradiance with the dark reservoir lifetime. The experiments are performed using long excitation pulses.

b. There is a difference almost by a factor of ~2 between the absorbed and incident fluence in all the reports.

c. DPAVBi and DPAVB are two different derivatives of the same material.

rates. This also ensures that fewer alternative channels for decay interfere with the lasing process. Condensation experiments on a very similar oligofluorene, known as pentafluorene, which exhibits a film PLQY of approximately 90%, have demonstrated a very low lasing threshold of 17 μJ cm^{-2}, with an absorbed fluence of merely 11.7 μJ cm^{-2} [24]. In the present investigation, the researchers employed a buffer layer (LiF) as a protection to mitigate the degradation of the active layer during the sputtering process of the top DBR. Additionally, they effectively decreased the lasing threshold by employing negative detuning, aiming to maximize the relaxation process. Polariton condensation investigations on DBR microcavities composed of melt-processed films of 4,4″-bis((E)-4-(3,6-bis(2-ethylhexyl)-(9H-carbazol-9-yl))styryl)-1,1″-biphenyl (BSBCz-EH) as an active layer (Figure 10.7g) have shown that apart from the resonator's quality factor is critical to realize low threshold polariton condensates. Another important consideration to realize a low threshold polariton condensate is to increase the rate of polariton relaxation [46]. With a decent *Q*-factor (~135) for the loaded cavity under optical excitation, the condensate was characterized with an ultralow threshold absorbed fluence of 9.7 ± 0.1 μJ cm^{-2} at an incident fluence of 15.5 μJ cm^{-2}. As shown in Figure 10.7h, the study also employed a semiclassical model for describing the kinetics, which accounted for the origin of low threshold stemming from the fast relaxation of polariton in BSBCz-EH ($W_{ep} = 2.0 \times 10^{-5}$ cm^3 s^{-1}) due to radiative pumping of the LP branch (Figure 10.7i).

Microcavity polariton condensation results using brominated boron dipyrromethene (BODIPY-Br) dye as the active layer have shown that the most important mechanism responsible to populate the LP branch in this case consists of radiative pumping from a single state near 2.09 eV (593 nm, ref to UV-Vis spectra of BODIPY-Br, vide Figure 10.7j) which corresponds to weakly coupled excimer emission energy as observed in bare films [23]. BODIPY-Br thin films have high PLQY (between 10 and 50%, depending upon the material preparation) that are processed by dispersing the solution in a polystyrene matrix. Figure 10.7k shows the power-dependent PL linecuts near to $k_{||} = 0$ obtained from the BODIPY-Br microcavity below and above threshold. The excited state photochemistry involves the transfer of energy from the dark excitons to the excimer, which then radiatively populates the LP. In contrast with TDAF, which has a different slope below and above the threshold excitation density, the blueshift in BODIPY-Br is found to increase linearly upon increasing the excitation fluence. BODIPY-Br possesses a notable characteristic in terms of its optical properties, as it exhibits the emission of coherent yellow light at a wavelength of $\lambda = 565$ nm. This emission wavelength falls within a range that is typically challenging to attain using inorganic semiconductor lasers. A follow-up study used long excitation pulses of 4 ns to nonresonantly pump a $\lambda/2$ planar microcavity (Q-factor ~ 450) with BODIPY-Br as an active layer [47]. The results demonstrated a quasi-steady state single-mode polariton lasing with an ultralong condensate lifetime of 1.2 ns at a threshold excitation density of 1.5 MW cm^{-2}. The condensate lifetime in this study exceeds the cavity lifetime (~100 fs) by ~4 orders of magnitude. The presence of long-lived condensates, which surpass the typical polariton lifetime by several orders of magnitude, brings the system closer to the dynamic equilibrium. This occurrence implies to a quasi-steady-state operational regime, considering the ultrafast dynamics that are inherent to strongly coupled organic microcavities.

McGhee and colleagues presented a simplified fabrication approach to create a microcavity incorporating the molecular dye BODIPY-Br [48]. In this approach, the bottom mirror of the cavity is formed by depositing a pair of SiO_2 and Nb_2O_5 dielectric mirrors on a silver film. Surprisingly, they observed a similar polariton condensation threshold for this cavity in comparison to a control cavity, which has a bottom mirror made up of 10 quarter-wave dielectric pairs. However, interestingly for the hybrid cavity, the researchers also noticed a partial breakup of the k-space PL above the threshold. They attribute this behavior to spatial disorder propagating throughout the entire structure which mainly occurs at the silver film-single DBR pair interface. Despite these disorder effects, the demonstrated approach offers a simple fabrication method to produce high-performance mirrors suitable for polariton lasers.

Tunable organic polariton lasing over a broad range has also been proposed using two LMWEs [49, 50]. Technically, wavelength tunability can be achieved by making a wedge cavity structure [49], where the cavity detuning could be controlled in order to regulate the polariton energies. However, the limitation of this approach was that condensation could be triggered only at specific detunings that could assist the relaxation process in cavities. BODIPY-G1 consists of a wedge-like gradient in the active layer thickness and is positioned between two DBR mirrors, resulting in a cavity with continuously varying exciton–photon detuning. This structure shows polariton condensation over a broad range of wavelengths, spanning approximately 33 nm across the green and yellow regions of the spectrum, vide Figure 10.7. The second approach involves oligo(p-phenylene)s as the active material, wherein its π-conjugation length can be tuned [50]. By modifying the π-conjugation length, the wavelength of polariton lasing can be altered by approximately 27 nm.

10.4.3 Polymers

The unique properties of polymers, including their tunable optoelectronic characteristics and easy processability, have attracted attention for polariton condensation studies. The photophysics of conjugated polymers incorporates aspects of both organic crystals and small molecules to a certain degree. The dipole–dipole coupling and the superexchange between moieties can cause exciton bands to form along the polymer chain. Inter-chain energy transfer is another event that could occur due to incoherent Förster energy transfer. On the other hand, uneven distribution of chromophores throughout each chain can lead to significant disorder in spin-coated films as a result of bent, twisted, or other chain topologies [19]. This is a typical characteristic of most conjugated polymers, which leads to inhomogeneous broadening of their absorption spectrum in the solid state, where disorder-induced traps can function as effective quenching locations for excitons. One solution to prolong the exciton dephasing times at low temperatures was made possible by using ladder-type conjugated polymers [53]. The rigid one-dimensional (1D) structure of such polymers significantly lessens the inhomogeneous broadening.

The first spin-coated conjugated polymer film which was employed to achieve room temperature polariton condensation is the methyl-substituted ladder-type polymer – poly(paraphenylene) (MeLPPP) [26]. Polariton condensation was observed when the 0–0 and the 0–1 transitions of a 35-nm-thick MeLPPP film were embedded in a DBR-DBR microcavity to observe strong coupling. Interestingly, the wavelength of the condensate emission (474 nm) is just beyond the range where conventional optical gain has been observed (480–573 nm). These microcavities show an interesting temperature dependence. While the room temperature emission maximum occurs near $k = 0$, it shifts to higher k modes upon lowering the temperature. At low temperature, inefficient pumping occurs for the states near $k = 0$, where 10 times increase in threshold occurs upon decreasing the temperature below 50 K. Interestingly, as can be seen from Figure 10.8a, the vibronic PL intensities of MeLPPP thin films are found to be strongly dependent on the temperature. This was also accompanied by a redshift of the peak positions to a significant extent. The spatial interference patterns observed from this material are also different compared to those observed from the other reports. While clear interference fringes occur beyond the threshold, as shown in Figure 10.8b, speckle-like features within the interference patterns imply the presence of significantly high disorder in such spin-coated polymeric films compared to other organic materials. Another reason could be due to the choice of excitation conditions which can also lead to spatial hole burning within the density of the dark exciton reservoir, resulting in the occurrence of such dynamic instabilities for the MeLPPP polariton condensates. Overall, the relaxation dynamics in MeLPPP microcavities remain poorly understood due to these factors.

Zasedatelev et al. reported an organic polariton transistor capable of room temperature operation in 2019 using MeLPPP microcavity by tuning the energy of the pump beam resonant to one vibron above the ground state of LP energy [26]. This is shown in Figure 10.8c, where the detuning was intentionally created to initiate a single-step vibron-mediated relaxation channel which helped efficiently in achieving a condensation threshold of ~82 $\mu J\ cm^{-2}$. This vibron-mediated condensation threshold obtained in this study is much lower than all the previous attempts using MeLPPP. As shown in Figure 10.8d, by incorporating a weak optical control beam that is resonant with the ground state which also serves as a seed for initiating bosonic stimulation, the authors were able to effectively reduce the condensed threshold by one-half. Additionally, this setup enabled a signal amplification process, achieving a net gain of up to 10 dB μm^{-1}, vide Figure 10.8e. This significant advancement created promising prospects for utilizing organic polariton condensates

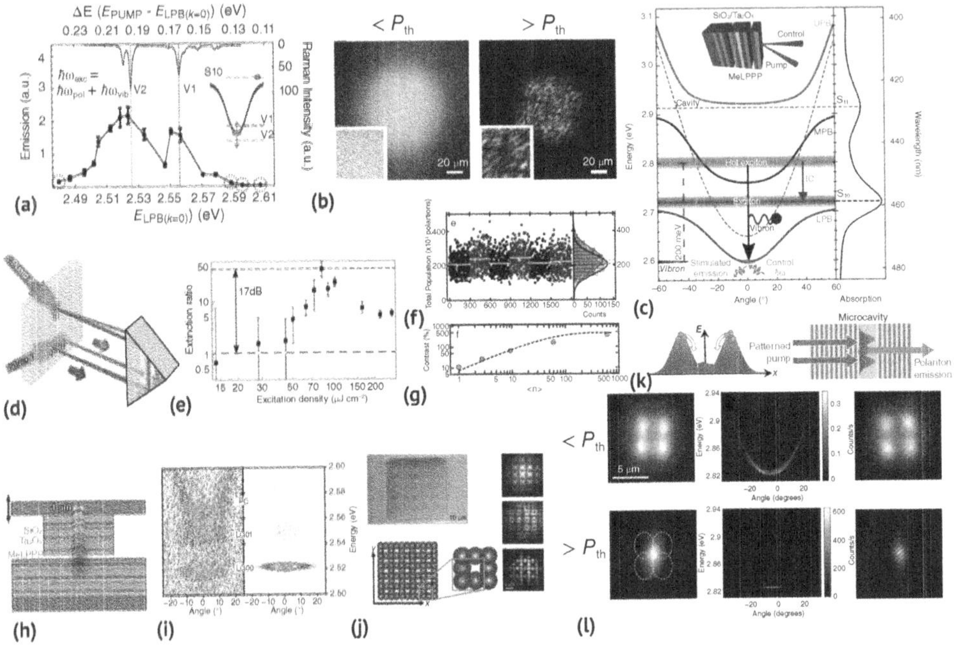

Figure 10.8 (a) Emission intensity of the MeLPPP film with respect to the LP state energy at k_0. The black dots represent the emission intensity at different energies of the LP state, while the top graph shows the Raman spectrum of MeLPPP, (b) spatial coherence of the MeLPPP polariton condensate below (left) and above (right) P_{th}, (c) left panel depicts the microcavity design (inset), and polariton dispersion of the UP, MP, and LP branch. The cavity mode and the two exciton modes are shown as dotted lines. The pump energy is tuned with respect to the hot exciton energy (one vibron high) of MeLPPP. The right panel depicts the absorption spectra of the bare MeLPPP film representing the two exciton modes, (d) schematic representation of the single pump-double probe optical configuration, (e) all optical polariton switch where the extinction ratio (ratio of the PL intensity of "1" state vs. the PL intensity of "0" state) is plotted versus the excitation density of the pump. *Source*: [26], with permission of Springer Nature. (f) Left panel shows pulse-to-pulse analysis of the condensate population showing the realization of the seeded polariton condensate triggered by 1 photon/pulse on an average. Right panel Gaussian distributions show histograms of total polariton population within single-condensate realizations seeded with correspondent photon number state as well histograms of the spontaneously built condensate (black), (g) switching contrast as a function of the average number of photons/seed pulse. *Source*: Reproduced from [53] / with permission of Springer Nature. (h) Schematic of the MeLPPP DBR-DBR microcavity with an etched Gaussian defect on the top DBR mirror, (i) k-space distribution of PL from the patterned microcavity below (left) and above (right) the threshold. *Source*: Reproduced from [54] / with permission of American Chemical Society. (j) A patterned 2D Leib lattice (SEM image on the top and schematic of the unit cell on the bottom) microcavity structure using MeLPPP and the real space mode patterns of the PL. *Source*: Figures adapted with permission from Ref. [55]; copyright 2021 Springer Nature. (k) Schematic of a patterned laser incident on a planar microcavity and the collected PL that comes out of the cavity. (l) Characterization of a single-mode condensate in a 2D trap with the real-space image (left), angle-resolved PL (middle), and interferogram (right) obtained from the cavity which is pumped using four laser spots below (top) and above (bottom) the P_{th}. White dashed ellipses in the real space image $>P_{th}$ indicate the pump spot positions. *Source*: Figures adapted with permission from Ref. [56]; copyright 2022 Springer Nature.

as all-optical transistors and other devices, offering exciting opportunities for future applications. In a follow-up study, Zasedatelev et al. used a similar setup to demonstrate polariton occupancy of the condensate at a very few photon limit [53]. This resulted in an unprecedented gain of ~23000 secondary polaritons per single seed polariton injected resonantly, vide Figure 10.8f. As shown in

Figure 10.8g, Pulse-to-pulse analysis of the single-condensate revealed ultrafast switching of the polariton condensate at a single photon level with a contrast of ~11 and ~20% according to integrated measurements.

Another research utilized an open microcavity approach to demonstrate polariton condensation in MeLPPP [54]. The confined lateral structures were achieved when micrometer-sized concave defects were etched onto one of the mirror substrates. In this work, nanopositioners were used to close the two ends (the planar DBR and film, and the patterned mirror) of the microcavity, as shown in Figure 10.8h. The Gaussian defect-induced confinement led to the forming of zero-dimensional confined Laguerre-Gaussian modes. As shown in Figure 10.8i, the condensation process occurs predominantly in the lowest-lying Laguerre-Gaussian mode. Under this scenario, clear interference fringes with a high degree of spatial coherence were seen within the micrometer-sized confinement region. Building upon these achievements, the group recently realized controllable polariton condensation in a 2D extended Lieb lattice [55]. They fabricated a Lieb lattice structure, as shown in Figure 10.8j, which resulted in a polariton dispersion relation and also the real-space images (Figure 10.8j) of the polariton PL. Through precise engineering of the photonic bandgap, the initially parabolic cavity resonance underwent a division into distinct gapped Bloch energy bands, forming s-, p-, and d-band manifolds. This allowed polaritons to undergo selective condensation into different band-structured modes by adjusting the cavity's length in situ. These experiments utilizing flexible cavities have shown strong potential for exploring a wide range of lattice Hamiltonians and opened up new possibilities for advanced research in this field.

Unlike MeLPPP, which possesses narrow exciton linewidths along with spectrally resolved or structured vibronic replicas, PFO has inhomogeneously broadened excitonic transition, which is a typical characteristic for organic conjugated polymers [28]. Hence, in contrast to MeLPPP, PFO owing to less rigidity dep higher disorders in the thin film state. The measured threshold for polariton lasing is 27.7 μJ cm^{-2}, equivalent to an absorbed pump fluence of 19.1 μJ cm^{-2}. Remarkably, this value is an order of magnitude lower compared to the MeLPPP microcavity. Furthermore, the emitted light exhibits long-range spatial coherence, with a maximum fringe visibility contrast of 72%.

Very recently, Wei et al. used a spatial light modulator (SLM, vide Figure 10.8k) to create both 1D and two-dimensional (2D) confined potential traps within PFO microcavities [56]. Under non-resonant pumping for the organic microcavities with 1D potentials, quantized states with energy spacings inversely proportional to their separation were observed. On the other hand, polariton condensate in case of a square-shaped 2D potential (Figure 10.8l) was found to be spatially distant (~2–3 μm) from the region of pumping. The spatial separation of the condensate region from the exciton formation region leads to a substantial reduction in decoherence, resulting in an order of magnitude lower threshold (87.6 μJ cm^{-2}) and improved temporal coherence in comparison to the control Gaussian pump profile (1.2 mJ cm^{-2}). Furthermore, with an increase in separation distance in the 2D trap, polariton condensates were formed in higher-order modes. This separation of the condensate region from the exciton formation region leads to a significant reduction in decoherence, resulting in threshold lower by an order of magnitude and higher temporal coherence in comparison to the control Gaussian pump profile. Theoretical simulations suggest that at higher pump power, lower-energy condensate modes are generated due to stimulated energy relaxation, resembling the effects observed in phonon scattering. The implications of having a delocalized polariton condensate in organic microcavities allow for the suppression of quenching effects near metal electrodes and thus are of significant importance for advancing electrically driven polaritonic devices.

10.4.4 Host–Guest Systems

The first category of a host–guest organic polariton system to demonstrate condensation ability are FPs [29, 30]. In the majority of FPs, the chromophore is enclosed within a cylindrical β-barrel structure that is made up of several β-sheets composed of elongated polypeptide chains. This is shown in Figure 10.6, where the presence of these β-sheets acts as a protective barrier around the chromophore, effectively isolating it from its surrounding environment and restricting its rotational motion. This confinement serves to decrease any internal conversion and/or intersystem crossing processes, ultimately leading to an enhancement in the fluorescent quantum yield.

In the context of FPs, the first report on polariton lasing (P_{th} = 12 nJ per pulse) was demonstrated by using a green fluorescent protein (GFP) film [29]. GFP, first extracted from the jellyfish Aequorea Victoria has been significantly used in fluorescence microscopy due to properties such as low toxicity, high luminescent yield, and photostability. The microcavity consisted of a spin-coated 500 nm-thick film of enhanced GFP (eGFP) on DBR. A second DBR was laminated via mechanical transfer on the top of the film to complete the cavity structure. Interestingly, this induced cavity thickness variation, ranging from 1500 to 4500 nm, which ultimately resulted in a multimode structure. A noteworthy achievement of this study is that the authors used longer nanosecond pulses (7 ns) to optically pump the cavity sample. This was possible only because of the significantly reduced exciton-exciton annihilation between the β-sheet-protected chromophores. A second threshold (P_2) is observed for this sample under high excitation fluence, which is attributed to conventional photon lasing. Among all the other organic microcavities explored so far, this is the only report to showcase a two-threshold behavior.

Another study by the same group employed a highly monomeric red fluorescent protein (RFP), known as mCherry to observe polariton condensate in a three-dimensional (3D) confined geometry [30]. This is shown in Figure 10.9a, where the patterned cavity was obtained via etching out small hemispherical indentations on a glass substrate that is used to support one of the mirrors, resulting in very high-quality factors (unloaded $Q = 2.5 \times 10^4$ and loaded $Q = 7.5 \times 10^3$). This leads to a remarkably long polariton lifetime of 2 ps with a significantly low threshold of 0.78 nJ per pulse. Moreover, employing long pulses enabled an interferometric assessment of the temporal coherence of polariton, without being constrained by the condensate emission duration. The authors reported a coherence time value of $\tau_c = 156$ ps at an excitation power of 1.7 P_{th} which increased the lifetime of polariton by two orders of magnitude. After that work, they also carried out another work where mCherry was strongly coupled to a 1D periodic array of lattices (Figure 10.9b, c) to demonstrate low threshold ($P_{th} = 1.5$ nJ per pulse) polariton condensation under ambient conditions [57]. The complete band structure of polaritonic Bloch bands was observed, which emerged in the tight binding regime. Additionally, they were able to control the loading of condensates into Bloch bands which possess specific orbital symmetry, vide Figure 10.9c. Of particular interest, they observed intriguing phenomena such as spatially increased expansion and self-localized condensates in such lattice structures.

Based on these results, follow-up works by the same group demonstrated the realization of topological organic polariton condensates [58]. As shown in Figure 10.9d, a patterned mirror cavity using mCherry was used to emulate a model of a linear Su-Schrieffer-Heeger chain. The chain can host topological defects at domain boundaries and at the edge. A distinct polariton mode (Figure 10.9e) within the topological optical bandgap was witnessed in the dispersion of the SSH domain boundary defect along with polariton lasing at an ultralow threshold of 0.10 nJ per pulse. The real-space images are shown in Figure 10.9f, where the defect polariton mode is found to be exponentially localized around the defect site. The edge defect, identical to the domain boundary

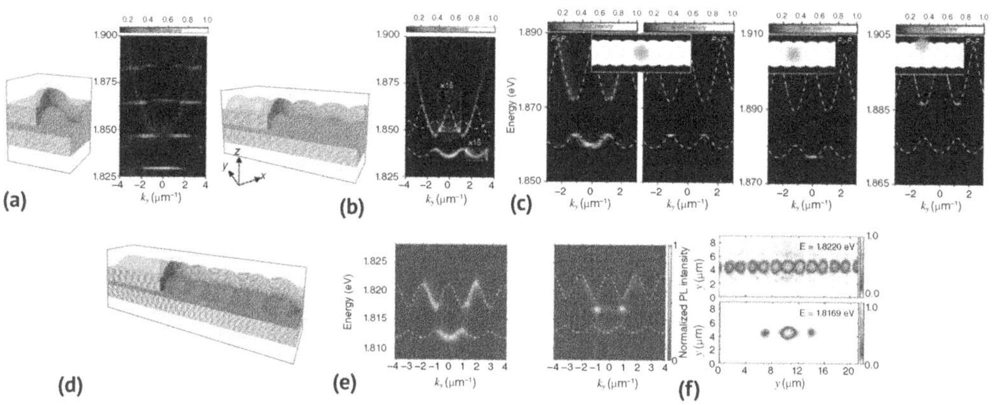

Figure 10.9 (a) Schematic of the patterned mCherry microcavity structure with hemispherical indentation (left) and the *k*-space PL (right). *Source*: [30] / with permission of American Chemical Society. (b) Patterned 1D microcavity lattice array structure (left) and the resulting band structure from the PL emission below P_{th} (right). White lines denote the calculated band structure, (c) angle-dependent PL spectra above P_{th} for on-site pumping (left) and off-site pumping (middle and right) in the 1D lattice. The position of the pump spot which is aligned with respect to the polariton potential is shown. *Source*: [57] / with permission of American Chemical Society. (d) Schematic illustration of a 1D SSH linear chain cavity. (e) Angle-dependent PL spectra with the laser excitation spot being centered on the bulk of the chain (left) and on the domain boundary defect (right). (f) Real-space mode tomographies of the antibinding band (top) and the mode within the gap (bottom). *Source*: Reproduced from [58] / with permission of American Chemical Society.

defect, is also found to be exponentially localized. These distinctive features make this material system a highly promising candidate for room-temperature topological insulators, offering topological robustness even in the presence of structural defects. It is worth noting that while the band structure appeared somewhat obscured in the MeLPPP cavity, a distinctly visible full band structure was achieved using mCherry configurations. This drawback in case of MeLPPP could be attributed to the limitations in the quality factor ($Q = 600$) of the cavity, reduced bandgap, and intrinsic disorder within the thin film material. In contrast, mCherry microcavities possess large quality factors ($Q \approx 10^4$) and well-defined periodic potentials with a significant forbidden bandgap. In the context of the FP microcavities reported so far, only multimode configurations have been explored. This is because most of the FP films are processed from aqueous solutions, which makes it pretty challenging to control the film thickness. While multimode cavities could significantly increase the cavity Q-factor, the effective light–matter interaction strength gets decreased compared to that for single-mode cavity structures.

From the above discussions, it is very clear that exciton–exciton annihilation and photobleaching in organic solids is one of major bottleneck for achieving a stable polariton condensate. This is also one of the most common problem existing for more than 150 years for >10000 organic dyes [61, 62]. A universal solution toward this challenging problem was proposed in 2020, shown in Figure 10.10a, where the anionic counterpart of the organic dye undergoes co-crystallization in the presence of a colorless host macrocycle which then helps the electrostatically bound emissive cationic dye part to reorganize in a checkerboard like lattice pattern [63]. The host–guest complex known as SMILES (small molecule ionic isolation lattices) helped in maintaining the solution-state PLQY of the dyes in the thin film state. The co-crystallization approach was attempted for some of the major classes of commercially available organic dyes in solid thin film state, which represent more than 1000 commercial homologs and synthetic dyes, making them the world's brightest

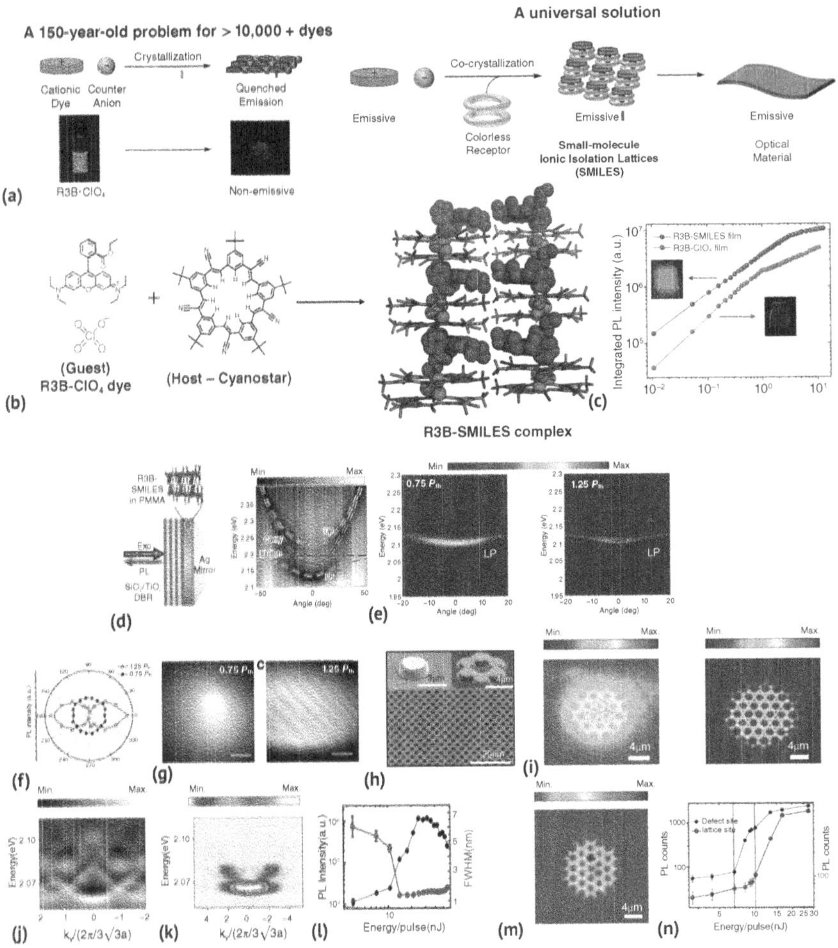

Figure 10.10 (a) Aggregation-induced bimolecular quenching for organic solids with R3B-ClO4 dye as an example for comparison in solution and solid state, (b) schematic depicting the formation of highly emissive SMILES dyes via co-crystallization approach, (c) chemical structure of the reactant precursors and R3B-SMILES material, (d) power-dependent photoluminescence (PL) intensity comparison for thin PMMA films incorporating R3B alone (10 mM in R3B) and R3B-SMILES (10 mM in R3B). Inset shows the PL from the respective films, showing a much brighter SMILES film, (e) momentum-resolved emission showing the lower polariton dispersion below (0.75 P_{th}) and above threshold (1.25 P_{th}), (f) polar plot depicting the emission intensity with respect to the angle of a polarizer in front of the detector at 0.75 P_{th} and 1.25 P_{th} excitation power, (g) real space PL maps showing homogeneous distribution of PL density for pump power < P_{th} and interference patterns for pump power > P_{th} recorded using a Michelson interferometer configuration. *Source*: Reproduced from [59] / with permission of American Chemical Society. (h) SEM image of a single micropillar, honeycomb unit cell, and lattice patterned on the planar cavity using FIB, (i) comparison of real space PL from the 2D honeycomb lattice of SMILES below (left) and above (right) threshold at room temperature, energy resolved momentum space PL (j) below condensation threshold along with tight binding model fits to the observed dispersion and (k) above condensation threshold from the 2D lattice, (l) integrated PL intensity and full width at half-maximum (FWHM) of condensate lattice PL as a function of pulse energy, (m) real-space PL image of defect lattice with a missing hole at input power at $p = 1.25\ P_{th}$ collected with integration time 1 s. Preferential condensation at the lattice defect site is observed, and (n) condensate lattice PL counts at defect sites and lattice citation as a function of pulse energy. The vertical dotted lines show threshold energies of P_{th} ~7 and ~10 nJ for the defect site and the defect-free lattice, respectively. *Source*: Reproduced from [60] / with permission of American Chemical Society.

emitters with a volume-normalized brightness >7000 M^{-1} cm^{-1} nm^{-3}. A comparative study of these dyes without the host confirmed that along with the highly monomeric nature and ease of film processing, the host–guest approach led to a 10× enhancement in PLQY and strong resistance to exciton–exciton annihilation and bimolecular quenching for the thin films.

Recently, a platform for realizing room temperature polariton condensates was proposed using commercially available organic dyes based on SMILES materials [59]. Toward this, rhodamine-3B perchlorate (R3B) was used as an archetypical organic laser dye, vide Figure 10.10b. Thin PMMA films of R3B-SMILES were found to be highly photostable, showing strong greater tolerance (5× higher than the bare R3B films, Figure 10.10c) to pump fluence (10 kHz pulsed laser at 530 nm wavelength) along with in-plane dipole orientation, facilitating better coupling to the electromagnetic field in the cavity (Figure 10.10d). Polariton condensation, as shown in Figure 10.10e, was characterized at room temperature using a DBR-Ag cavity (Q-factor ~ 150) at a threshold of ~200 μJ cm^{-2}. Interestingly, as shown in Figure 10.10f, the condensate was found to be linearly polarized with a degree of ~73% along the direction of laser polarization. In subsequent research, highly photostable host–guest Frenkel excitons in SMILES were utilized to achieve room temperature polariton condensate lattices using a top-down fabrication approach with focused ion beam (FIB) etching on the R3B-SMILES microcavity [60] (Figure 10.10h). Figure 10.10h shows some of the lattices made via this approach. Experimental results showed that polaritons were confined, resulting in discrete states and condensation to the ground state energy, as observed in the energy- and momentum-resolved spectroscopy of single micropillars (Figure 10.10i–l). By patterning a honeycomb lattice on the SMILES microcavity, a 2D polariton condensate lattice composed of coupled micropillars was successfully realized at room temperature for the first time, vide Figure 10.10i. Additionally, the study of deterministically placed defects in the condensate lattice revealed that condensates prefer high refractive index sites for localization, as shown in Figure 10.10m, which also reduces the condensation threshold pump power, Figure 10.10n. This top-down approach offers a versatile platform for exploring arbitrary polariton condensate lattices and polaritonic circuits at room temperature across large areas. More recently, a DBR-DBR SMILES cavity has been realized using a pick and place technique where the Q factors exceeded 1000 and resulted in highly coherent polariton condensates with temporal coherence exceeding 30 ps and linewidth less than 175 μeV.

As of now, all realizations of organic polariton condensation have remained significantly out of equilibrium, necessitating further efforts to advance them towards a thermal equilibrium limit. The short lifetime of organic polaritons is a bottleneck for realizing thermalization unless there is an efficient scattering mechanism or long-lived polaritons that overcomes this hurdle. Efficient thermalization of organic polaritons approaching thermal equilibrium under ambient conditions was shown using a new class of FP – mScarlet in a DBR-Ag microcavity (Q ~ 330). mScarlet is highly monomeric. It is also a β-barrel structure-based RFP with a record 3.5 times higher oscillator strength than mCherry [64]. Its fluorescence lifetime is ~3.9 ns with a PLQY of ~70% in solid-state. The study presents the first experimental demonstration of strongly thermalized polaritons with a drop-in effective temperature, $T_{eff} \sim 350 \pm 35$ K, which is close to the lattice temperature.

The polariton population distribution analysis showed direct evidence for dramatic switching from an equilibrium state into a metastable state. A significant deviation from the Maxwell–Boltzmann statistics at $k_{\|} = 0$ was also observed above the threshold excitation fluence of 15.6 mJ cm^{-2}. The strong thermalization response was also accompanied by a threshold behavior depicting all characteristic hallmarks of optical nonlinearity. Interestingly, however, the k-space results were inconclusive to characterize a proper collapse toward $k_{\|} = 0$ as it occurs in the case of a typical polariton condensate.

10.5 Summary

This book chapter explores the topics of polariton condensation and lasing using organic molecular materials. Table 10.1 lists all the different organic molecular material platforms that have been used to realize exciton–polariton condensation. The utilization of organic polariton condensates offers a remarkable material platform to investigate a diverse range of quantum phenomena, which were previously restricted to cryogenic temperatures. Despite the realization of room-temperature polariton condensates using dozens of molecular solids, which include both crystalline and amorphous materials, the basic design principle to realize polariton condensates in organic microcavities remain unclear. Both theoretical predictions and experimental findings converge on the notion that achieving optimal lasing conditions involves maximizing PLQY, ensuring sufficient exciton–vibron coupling for thermalization, and minimizing cavity losses. A theoretical description of polariton-mediated excited-state dynamics in planar microcavities present significant challenges due to several factors. These challenges arise from the intrinsic disorder of thin films and the complex interplay of radiative and nonradiative relaxation channels inherent in organic systems. Furthermore, selecting suitable molecular emitters for achieving condensation presents significant challenges. This difficulty arises from the coexistence of both positive and negative disorder effects. Positive effects, such as disorder potentials, can facilitate condensation. However, negative effects, including incorrect orientation of transition dipole moments, Förster energy transfer caused by misaligned molecules, and energy or angular broadening of polariton states, hinder the process. A substantial amount of research is currently focused on reducing the polariton condensation threshold. This is indeed essential in the field of electrically pumped organic polariton lasers. However, a targeted chemistry behind the optimal design of an emitter is yet to be unraveled. To mitigate the multiple annihilation and absorption losses stemming from polarons, singlet and triplet excitons, and other species, it is crucial to carefully screen and select suitable material systems. Many organic molecules are known to suffer from issues such as bleaching, low thermal conductivity and charge-carrier mobility, and instability, which are indeed the same elements affecting the prospects of electrical pumping. Addressing and improving these material properties will play a vital role in enabling operation at large polariton densities and enhancing the performance of organic polariton devices.

Additionally, there are several other key questions remain to be addressed in the study of organic polariton condensates. It is unclear if single-crystal materials could help in significantly reducing the condensation threshold or not. Additionally, there is a need to explore methods for realizing stronger nonlinear interactions. In organic materials, nonlinear polariton interactions are restricted due to their localized nature and also due to small Bohr radius of Frenkel excitons. Enhancing these interactions is crucial for future quantum technological applications. This can be achieved through strategies like assembling organic molecules in an orderly fashion at high densities to increase polariton and exciton densities. Additionally, hybridizing with Wannier–Mott systems can be employed to engineer strong nonlinearity [18, 19]. One such exciting quantum phenomenon is the polariton blockade effect, which, although highly intriguing, currently remains a challenging goal to achieve using organic materials. Additionally, achieving thermal-equilibrium organic condensates poses a considerable challenge and more efforts need to be carried out in order to explore materials for this purpose.

Overall, the study of organic polariton condensates holds great promise for unlocking a host of intriguing quantum phenomena and exploring the potential for practical applications in various fields. The flexibility and ease of processing in organic semiconductors offer the potential for various materials to demonstrate exciton-photon coupling, leading to the emergence of possible applications. Exploring novel directions, such as modifying photophysics through strong light coupling to organic materials, can lead to crucial insights. By understanding and manipulating aspects like vibrational modes, molecular chromophores' roles, and vibrational relaxation processes, it may be possible to overcome challenges inherent to conventional organic lasers. Furthermore, the coherent population of polaritons in a condensate may further modify these material properties, offering exciting opportunities for advancing organic optoelectronic devices, and even modify molecular energy transfer processes. Continued research and innovative approaches will undoubtedly shed light on new opportunities and open new directions for the practical use of molecular polariton condensates.

References

1 de Broglie, L. A tentative theory of light quanta. *The London, Edinburgh, and Dublin Philosophical Magazine and J. Sci.* **1924**, ***47***, 446–458.

2 Anderson, M. H., Ensher, J. R., Matthews, M. R., Wieman, C. E., Cornell, E. A. Observation of Bose-Einstein condensation in a dilute atomic vapor. *Science* **1995**, ***269***, 198–201.

3 Ladd, T. D., Jelezko, F., Laflamme, R., Nakamura, Y., Monroe, C., O'Brien, J. L. Quantum computers. *Nature* **2010**, ***464***, 45–53.

4 Byrnes, T., Wen, K., Yamamoto, Y. Macroscopic quantum computation using Bose-Einstein condensates. *Phys. Rev. A* **2012**, ***85***, 040306.

5 Byrnes, T., Rosseau, D., Khosla, M., Pyrkov, A., Thomasen, A., Mukai, T. et al. Macroscopic quantum information processing using spin coherent states. *Opt. Commun.* **2015**, ***337***, 102–109.

6 Demokritov, S. O., Demidov, V. E., Dzyapko, O., Melkov, G. A., Serga, A. A., Hillebrands, B. et al. Bose-Einstein condensation of quasi-equilibrium magnons at room temperature under pumping. *Nature* **2006**, ***443***, 430–433.

7 Wang, Z., Rhodes, D. A., Watanabe, K., Taniguchi, T., Hone, J. C., Shan, J. et al. Evidence of high-temperature exciton condensation in two-dimensional atomic double layers. *Nature* **2019**, ***574***, 76–80.

8 Byrnes, T., Kim, N. Y., Yamamoto, Y. Exciton–polariton condensates. *Nat. Phys.* **2014**, ***10***, 803–813.

9 Weisbuch, C., Nishioka, M., Ishikawa, A., Arakawa, Y. Observation of the coupled exciton-photon mode splitting in a semiconductor quantum microcavity. *Phys. Rev. Lett.* **1992**, ***69***, 3314–3317.

10 Balili, R., Hartwell, V., Snoke, D., Pfeiffer, L., West, K. Bose-Einstein condensation of microcavity polaritons in a trap. *Science* **2007**, ***316***, 1007–1010.

11 Kasprzak, J., Richard, M., Kundermann, S., Baas, A., Jeambrun, P., Keeling, J. M. J. et al. Bose–Einstein condensation of exciton polaritons. *Nature* **2006**, ***443***, 409–414.

12 Deng, H., Haug, H., Yamamoto, Y. Exciton-polariton Bose-Einstein condensation. *Rev. Mod. Phys.* **2010**, ***82***, 1489–1537.

13 Guillet, T., Mexis, M., Levrat, J., Rossbach, G., Brimont, C., Bretagnon, T. et al. Polariton lasing in a hybrid bulk ZnO microcavity. *App. Phys. Lett.* **2011**, ***99***, 161104.

14 Li, F., Orosz, L., Kamoun, O., Bouchoule, S., Brimont, C., Disseix, P. et al. From excitonic to photonic polariton condensate in a ZnO-based microcavity. *Phys. Rev. Lett.* **2013**, ***110***, 196406.

15 Zuniga-Perez, J., Mallet, E., Hahe, R., Rashid, M. J., Bouchoule, S., Brimont, C. et al. Patterned silicon substrates: a common platform for room temperature GaN and ZnO polariton lasers. *Appl. Phys. Lett.* **2014**, ***104***, 241113.

16 Agranovich, V. *Excitations in Organic Solids*. Oxford University Press: Oxford, 2008.

17 Nematiaram, T., Padula, D., Troisi, A. Bright Frenkel excitons in molecular crystals: a survey. *Chem. Mater.* **2021**, ***33***, 3368–3378.

18 Jiang, Z., Ren, A., Yan, Y., Yao, J., Zhao, Y. S. Exciton-polaritons and their Bose–Einstein condensates in organic semiconductor microcavities. *Adv. Mater.* **2022**, ***34***, 2106095.

19 Keeling, J., Kéna-Cohen, S. Bose–Einstein condensation of exciton-polaritons in organic microcavities. *Ann. Rev. Phys. Chem.* **2020**, ***71***, 435–459.

20 Daskalakis, K. S., Maier, S. A., Murray, R., Kéna-Cohen, S. Nonlinear interactions in an organic polariton condensate. *Nat. Mater.* **2014**, ***13***, 271–278.

21 Daskalakis, K. S., Maier, S. A., Kéna-Cohen, S. Spatial coherence and stability in a disordered organic polariton condensate. *Phys. Rev. Let.* **2015**, ***115***, 035301.

22 Grant, R. T., Michetti, P., Musser, A. J., Gregoire, P., Virgili, T., Vella, E. et al. Efficient radiative pumping of polaritons in a strongly coupled microcavity by a fluorescent molecular dye. *Adv. Opt. Mater.* **2016**, ***4***, 1615–1623.

23 Cookson, T., Georgiou, K., Zasedatelev, A., Grant, R. T., Virgili, T., Cavazzini, M. et al. A yellow polariton condensate in a dye filled microcavity. *Adv. Opt. Mater.* **2017**, ***5***, 1700203.

24 Rajendran, S. K., Wei, M., Ohadi, H., Ruseckas, A., Turnbull, G. A., Samuel, I. D. W. Low threshold polariton lasing from a solution-processed organic semiconductor in a planar microcavity. *Adv. Opt. Mater.* **2019**, ***7***, 1801791.

25 Yagafarov, T., Sannikov, D., Zasedatelev, A., Georgiou, K., Baranikov, A., Kyriienko, O. et al. Mechanisms of blueshifts in organic polariton condensates. *Commun. Phys.* **2020**, ***3***, 18.

26 Plumhof, J. D., Stöferle, T., Mai, L., Scherf, U., Mahrt, R. F. Room-temperature Bose–Einstein condensation of cavity exciton–polaritons in a polymer. *Nat. Mater.* **2014**, ***13***, 247–252.

27 Scafirimuto, F., Urbonas, D., Scherf, U., Mahrt, R. F., Stöferle, T. Room-temperature exciton-polariton condensation in a tunable zero-dimensional microcavity. *ACS Photonics* **2018**, ***5***, 85–89.

28 Wei, M., Rajendran, S. K., Ohadi, H., Tropf, L., Gather, M. C., Turnbull, G. A. et al. Low-threshold polariton lasing in a highly disordered conjugated polymer. *Optica* **2019**, ***6***, 1124.

29 Dietrich, C. P., Steude, A., Tropf, L., Schubert, M., Kronenberg, N. M., Ostermann, K. et al. An exciton-polariton laser based on biologically produced fluorescent protein. *Sci. Adv.* **2016**, ***2***, e1600666.

30 Betzold, S., Dusel, M., Kyriienko, O., Dietrich, C. P., Klembt, S., Ohmer, J. et al. Coherence and interaction in confined room-temperature polariton condensates with Frenkel excitons. *ACS Photonics* **2020**, ***7***, 384–392.

31 Ribeiro, R. F., Martínez-Martínez, L. A., Du, M., Campos-Gonzalez-Angulo, J., Yuen-Zhou, J. Polariton chemistry: controlling molecular dynamics with optical cavities. *Chem. Sci.* **2018**, ***9***, 6325–6339.

32 Bhuyan, R., Mony, J., Kotov, O., Castellanos, G. W., Gómez Rivas, J., Shegai, T. O. et al. The rise and current status of polaritonic photochemistry and photophysics. *Chem. Rev.* **2023**, ***123***, 10877–10919.

33 Feist, J., Galego, J., Garcia-Vidal, F. J. Polaritonic chemistry with organic molecules. *ACS Photonics* **2018**, ***5***, 205–216.

34 Porras, D., Ciuti, C., Baumberg, J. J., Tejedor, C. Polariton dynamics and Bose-Einstein condensation in semiconductor microcavities. *Phys. Rev. B* **2002**, ***66***, 085304.

35 Kéna-Cohen, S., Forrest, S. R. Room-temperature polariton lasing in an organic single-crystal microcavity. *Nat. Photonics* **2010**, ***4***, 371–375.

36 Coles, D. M., Michetti, P., Clark, C., Tsoi, W. C., Adawi, A. M., Kim, J.-S. et al. Vibrationally assisted polariton-relaxation processes in strongly coupled organic-semiconductor microcavities. *Adv. Funct. Mater.* **2011**, ***21***, 3691–3696.

37 Tang, J., Zhang, J., Lv, Y., Wang, H., Xu, F. F., Zhang, C. et al. Room temperature exciton–polariton Bose–Einstein condensation in organic single-crystal microribbon cavities. *Nat. Commun.* **2021**, ***12***, 3265.

38 Ren, J., Liao, Q., Huang, H., Li, Y., Gao, T., Ma, X. et al. Efficient bosonic condensation of exciton polaritons in an H-aggregate organic single-crystal microcavity. *Nano Lett.* **2020**, ***20***, 7550–7557.

39 Holmes, R. J., Forrest, S. R. Strong exciton-photon coupling and exciton hybridization in a thermally evaporated polycrystalline film of an organic small molecule. *Phys. Rev. Lett.* **2004**, ***93***(18), 186404.

40 Agranovich, V. M., Litinskaia, M., Lidzey, D. G. Cavity polaritons in microcavities containing disordered organic semiconductors. *Phys. Rev. B* **2003**, ***67***, 085311.

41 Kéna-Cohen, S., Davanço, M., Forrest, S. R. Strong exciton-photon coupling in an organic single crystal microcavity. *Phys. Rev. Lett.* **2008**, ***101***, 116401.

42 Sturges, T. J., Anderson, M. D., Buraczewski, A. A., Navadeh-Toupchi, M., Adiyatullin, A. F., Jabeen, F. et al. Anderson localisation in steady states of microcavity polaritons. *Sci. Rep.* **2019**, ***9***, 19396.

43 Lin, H.-W., Lin, C.-L., Chang, H.-H., Lin, Y.-T., Wu, C.-C., Chen, Y.-M. et al. Anisotropic optical properties and molecular orientation in vacuum-deposited ter(9,9-diarylfluorene)s thin films using spectroscopic ellipsometry. *J. Appl. Phys.* **2004**, ***95***, 881–886.

44 Wu, C.-C., Lin, Y.-T., Wong, K.-T., Chen, R.-T., Chien, Y.-Y. Efficient organic blue-light-emitting devices with double confinement on terfluorenes with ambipolar carrier transport properties. *Adv. Mater.* **2004**, ***16***, 61–65.

45 Lerario, G., Fieramosca, A., Barachati, F., Ballarini, D., Daskalakis, K. S., Dominici, L. et al. Room-temperature superfluidity in a polariton condensate. *Nat. Phys.* **2017**, ***13***, 837–841.

46 Ishii, T., Miyata, K., Mamada, M., Bencheikh, F., Mathevet, F., Onda, K. et al. Low-threshold exciton-polariton condensation via fast polariton relaxation in organic microcavities. *Adv. Funct. Mater.* **2022**, ***10***, 2102034.

47 Putintsev, A., Zasedatelev, A., McGhee, K. E., Cookson, T., Georgiou, K., Sannikov, D. et al. Nano-second exciton-polariton lasing in organic microcavities. *Appl. Phys. Lett.* **2020**, ***117***, 122302.

48 McGhee, K. E., Putintsev, A., Jayaprakash, R., Georgiou, K., O'Kane, M. E., Kilbride, R. C. et al. Polariton condensation in an organic microcavity utilising a hybrid metal-DBR mirror. *Sci. Rep.* **2021**, ***11***, 20879.

49 Sannikov, D., Yagafarov, T., Georgiou, K., Zasedatelev, A., Baranikov, A., Gai, L. et al. Room temperature broadband polariton lasing from a dye-filled microcavity. *Adv. Opt. Mater.* **2019**, ***7***, 1900163.

50 Wei, M., Fang, M., Rajendran, S. K., Lai, W.-Y., Turnbull, G. A., Samuel, I. D. W. Room temperature polariton lasing in ladder-type Oligo(p-Phenylene)s with different π-conjugation lengths. *Adv. Photonics Res.* **2021**, *2*, 2000044.

51 McGhee, K. E. R., Jayaprakash, K., Georgiou, Burg, S. L., Lidzey, D. G. Polariton condensation in a microcavity using a highly-stable molecular dye. *J. Mater. Chem. C* **2022**, ***10***(11), 4187–4195.

52 Zasedatelev, A. V., Baranikov, A. V., Urbonas, D., Scafirimuto, F., Scherf, U., Stöferle, T. et al. A room-temperature organic polariton transistor. *Nat. Photonics* **2019**, ***13***, 378–383.

53 Zasedatelev, A. V., Baranikov, A. V., Sannikov, D., Urbonas, D., Scafirimuto, F., Shishkov, V. Y. et al. Single-photon nonlinearity at room temperature. *Nature* **2021**, ***597***, 493–497.

54 Urbonas, D., Stöferle, T., Scafirimuto, F., Scherf, U., Mahrt, R. F. Zero-dimensional organic exciton–polaritons in tunable coupled Gaussian defect microcavities at room temperature. *ACS Photonics* **2016**, ***3***, 1542–1545.

55 Scafirimuto, F., Urbonas, D., Becker, M. A., Scherf, U., Mahrt, R. F., Stöferle, T. Tunable exciton–polariton condensation in a two-dimensional Lieb lattice at room temperature. *Commun. Phys.* **2021**, ***4***, 39.

56 Wei, M., Verstraelen, W., Orfanakis, K., Ruseckas, A., Liew, T. C. H., Samuel, I. D. W. et al. Optically trapped room temperature polariton condensate in an organic semiconductor. *Nat. Commun.* 2022, ***13***, 7191.

57 Dusel, M., Betzold, S., Egorov, O. A., Klembt, S., Ohmer, J., Fischer, U. et al. Room temperature organic exciton–polariton condensate in a lattice. *Nat. Commun.* **2020**, ***11***, 2863.

58 Dusel, M., Betzold, S., Harder, T. H., Emmerling, M., Beierlein, J., Ohmer, J. et al. Room-temperature topological polariton laser in an organic lattice. *Nano Lett.* **2021**, ***21***, 6398–6405.

59 Deshmukh, P., Satapathy, S., Michail, E., Olsson, A. H., Bushati, R., Yadav, R. K. et al. Plug-and-play molecular approach for room temperature polariton condensation. *ACS Photonics* **2024**, ***11***, 348–355.

60 Yadav, R. K., Satapathy, S., Deshmukh, P., Datta, B., Sharma, A., Olsson, A. H. et al. Direct writing of room temperature polariton condensate lattice. *Nano Lett.* **2024**, ***24***, 4945.

61 Barford, W. *Electronic and Optical Properties of Conjugated Polymers*. Oxford University Press: Oxford, 2013.

62 Müller, J. G., Lemmer, U., Raschke, G., Anni, M., Scherf, U., Lupton, J. M. et al. Linewidth-limited energy transfer in single conjugated polymer molecules. *Phys. Rev. Lett.* **2003**, ***91***(26), 267403.

63 Benson, C. R., Kacenauskaite, L., VanDenburgh, K. L., Zhao, W., Qiao, B., Sadhukhan, T. et al. Plug-and-play optical materials from fluorescent dyes and macrocycles. *Chem* **2020**, ***6***, 1978–1997.

64 Satapathy, S., Liu, B., Deshmukh, P., Molinaro, P. M., Dirnberger, F., Khatoniar, M. et al. Thermalization of fluorescent protein exciton–polaritons at room temperature. *Adv. Mater.* **2022**, ***34***, 2109107.

11

Kinetic Models for Polariton Relaxation in Organic Microcavities and Comparison to Experiments

Tomohiro Ishii[1], Stéphane Kéna-Cohen[1], Felipe Herrera[2,3] and Chihaya Adachi[4,5]

[1] *Department of Engineering Physics, École Polytechnique de Montréal, Montréal, QC, Canada*
[2] *Department of Physics, Universidad de Santiago de Chile, Santiago, Chile*
[3] *Millennium Institute for Research in Optics MIRO, Concepción, Chile*
[4] *Center for Organic Photonics and Electronics Research (OPERA), Kyushu University, Nishi, Fukuoka, Japan*
[5] *International Institute for Carbon Neutral Energy Research (WPI-I2CNER), Kyushu University, Nishi, Fukuoka, Japan*

11.1 Introduction

Due to their bosonic nature, exciton–polaritons can form a macroscopic quantum state termed a Bose–Einstein condensate (BEC) at high densities. BEC of exciton–polaritons was first demonstrated in microcavities containing GaAs and CdTe quantum wells [1–4]. However, the small exciton binding energies typical of these semiconductors limit the observation of BEC to cryogenic temperatures. As a potential way to overcome this limitation, organic semiconductors were investigated as an alternative, which culminated in the realization of polariton BEC in an anthracene single-crystal microcavity [5]. However, this achievement required short optical pulses and a high fluence ($\sim$ 300 μJ cm^{-2}), which limits practical applications. Since then, there has been tremendous interest in further lowering the condensation threshold and moving toward electrical pumping [6–18] given the potential of organic polariton BECs for low-threshold lasing and quantum simulations.

These efforts have mainly focused on identifying and developing organic molecules that show polariton condensation at lower pump intensities. Several types of light-emitting materials, including crystalline anthracene, oligofluorenes, π-conjugated polymers, and fluorescent proteins, have been reported to show polariton BEC [5, 19–33]. Reported thresholds for condensation under optical excitation in these systems range from $P_{\mathrm{th}} = 0.36$–500 μJ cm^{-2}, which is still too high for most optoelectronic applications. For example, it has been reported that a current density of 600 A cm^{-2} is required to electrically drive an organic semiconductor laser with an optical pumping threshold of 0.06 μJ cm^{-2} [34, 35]. Table 11.1 summarizes the parameters of various organic polariton condensates reported to date. Achieving a significant decrease in the BEC threshold for polariton BEC continues to be a major challenge.

When organic microcavities are excited nonresonantly, the so-called "dark" exciton reservoir (DS) is initially populated [36]. These essentially correspond to vibrationally relaxed electronic excited states with very weak photonic character (in the Frank–Condon region or on a timescale

Polariton Chemistry: Molecules in Cavities, First Edition. Edited by Joel Yuen-Zhou, Noel C. Giebink and Raphael F. Ribeiro. Published 2026 by John Wiley & Sons, Inc.

Table 11.1 Parameters from reported organic polariton BEC experiments.

Material	λ (nm)	Δ (meV)	$\hbar\Omega$ [meV]	DBR pairs[a]	τ_{cav}[b] (fs)	Film morphology	I_{pump} [$\mu J\ cm^{-2}$][c]	I_{abs} [$\mu J\ cm^{-2}$][d]	Reference
Anthracene	420	−120	256	12/12	85	Single crystal	–	320	[5]
TDAF	415	−440	600	9/9	150	Amorphous	–	13	[20]
MeLPPP	475	−77	116	6.5/9.5	160	Amorphous	500	–	[19]
BODIPY-Br	565	−165	91	8/10	130	Amorphous	527	–	[22]
BODIPY-G1	550	−248	114	8/10	100	Amorphous	6000	–	[29]
Pentafluorene	437	−450	500	6.5/10.5	60	Amorphous	17	12	[21]
BSBCz-EH	465	−470	350	9/9	33	Amorphous	15.5	9.7	[31]
BSFCz	450	−394	430	7.5/10.5	123	Amorphous	13.5	3.3	[32]
BSTFCz	442	−421	384	7.5/10.5	115	Amorphous	9.7	2.2	[32]
PFO (glassy phase)	466	−465	500	7.5/10.5	132	Amorphous	8	3.8	[33]
PFO (β-phase)	464	−470	700	10.5/10.5	132	Liquid crystal	0.71	0.36	[33]
eGFP	520	−60	194	14/14	100	Amorphous	0.3 MW cm^{-2}	–	[23]
mCherry	640	−66	266	10.5/10.5	2000	Amorphous	0.6 MW cm^{-2}	–	[26]
bPDI-3	585	−207	94	9/11.5	200	Amorphous	560	265	[28]

a. Number of top/bottom Distributed Bragg Reflector (DBR) mirror pairs.
b. Lifetime of the cavity photon.
c. Incident pump fluence.
d. Absorbed pump fluence.

comparable to the nuclear dynamics, stronger admixtures can occur due to the presence of disorder [37, 38]). These DS excitons, which are in practice localized on individual molecules [39, 40], must then populate lower-lying lower polariton (LP) states, which consist of a superposition between a photon mode and a delocalized electronic excitation with the same wavevector as the corresponding photon. In a rate equation model for the kinetics, this population transfer can be described by the polariton relaxation rate constant, W_{ep}. In a simple two-state model, consisting of the DS excitons and the bottom the LP branch, polariton condensation occurs when the polariton relaxation rate is faster than the inverse polariton lifetime [30, 31, 36, 41–48]. There are two obvious strategies for reducing BEC thresholds. One is to increase the polariton lifetime by using higher-quality resonators [26]. The second is to use materials with enhanced polariton relaxation rates [30, 31]. However, at present, a comprehensive understanding of the influence of molecular structure and thin film morphology on polariton relaxation is lacking. In this chapter, we will discuss experimental and theoretical work aimed at better understanding and ultimately accelerating polariton relaxation.

11.2 Modeling Polariton Kinetics in the Linear and Nonlinear Regime

Microscopic kinetic models that ignore coherences and describe scattering between states have been extremely successful in describing polariton condensation in inorganic semiconductor microcavities [49, 50]. While strong exciton–phonon coupling can invalidate the applicability of such models in some organic semiconductors, they nevertheless seem to reproduce many of the salient features of polariton luminescence and condensation.

11.2.1 Polariton Kinetics in the Linear Regime

Since the lifetimes of organic polaritons reported to date are very short (<1 ps) and scattering between polaritons is expected to be slow due to the single-molecule local interactions giving rise to population transfer, it appears reasonable to neglect scattering between delocalized polariton states in treating the kinetics. Furthermore, if polaritons are energetically separated from the reservoir density of states, one can also safely ignore the transfer of polaritons which decay to the DS. Ignoring small changes in lifetime between polariton states at different wavevectors, we can then model the kinetics using a minimal two-state master equation for the population of dark excitons in the reservoir and the ensemble of LPs [30, 43].

$$\frac{dn_R}{dt} = -\left(k_{rad}^{DS} + k_{nr} + W_{ep}\right) n_R \tag{11.1}$$

$$\frac{dn_{LP}}{dt} = W_{ep} n_R - \frac{n_{LP}}{\tau_{LP}} \tag{11.2}$$

where n_{LP} and n_R are the surface densities of LPs and the DS excitons, respectively. k_{rad}^{DS} (k_{nr}) is the radiative (nonradiative) decay rate of the dark excitons, W_{ep} is the rate of excitation transfer from the exciton reservoir to the ensemble of LPs, which consists of the sum of ${W_{ep}}^{(q)}$ for each in-plane wavevector ($W_{ep} = \sum_q W_{ep}^{(q)}$) and τ_{LP} is the lifetime of LPs (Figure 11.1a). The solution of Eq. (11.1) is $n_R = n_{e0}e^{-k_{tot}t}$, where n_{e0} is an exciton density at time $t = 0$ and k_{tot} is a total decay rate, $k_{tot} = k_{rad}^{DS} + k_{nr} + W_{ep}$. Since the LPs are rapidly de-excited to the ground state (τ_{LP} is typically <200 fs), the decay of the LP population after exiting the DSs is fixed by the slower

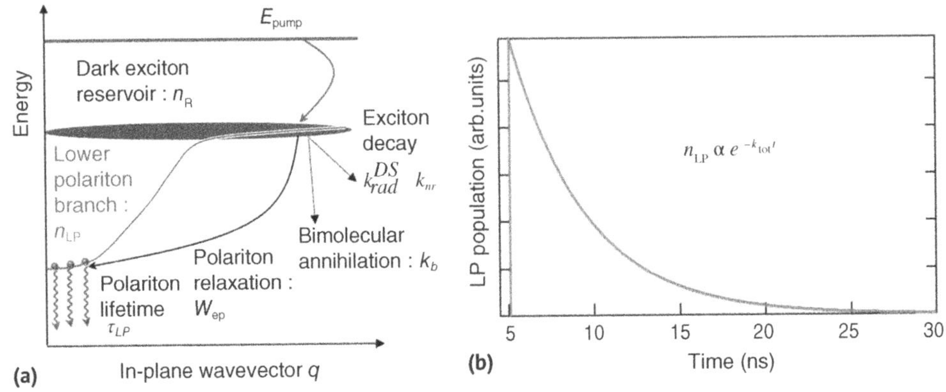

Figure 11.1 (a) Schematic image of the energy diagram of lower polariton (LP) branch. (b) Time dependence of the LP populations calculated by Eqs. (11.1) and (11.2).

reservoir lifetime k_{tot} (Figure 11.1b). The photoluminescence quantum yield (PLQY) of LPs, Φ_{LP}, after DS excitation, can be written as $\Phi_{LP} = W_{ep}/(k_{rad}^{DS} + k_{nr} + W_{ep})$. These two elements allow for the polariton relaxation rate, W_{ep}, to be calculated from the PLQY of LPs and the total decay rate constant, using $W_{ep} = \Phi_{LP}k_{tot}$.

Based on this simple model, it is possible to experimentally estimate W_{ep} using a combination of PLQY and transient PL measurements. This was first recently demonstrated in microcavities containing liquid crystalline perylene bisimide emitters [30]. The structure of the liquid crystalline perylene bisimide emitter (PBLC) is given in Figure 11.2a. It consists of a 1,6,7,12-tetra-4-nonyl phenoxy perylene bisimide (PBI) with two dodecyl chains at the imide positions. This discotic emitter was specifically designed to possess a high PLQY (~25%) and a columnar organization in the solid state, maintained at room temperature. The liquid crystal phase of PBLC can be controlled to the homeotropic and planar alignment by a thermal treatment and surface treatment to quartz substrate using octadecyl trichlorosilane (OTS). The absorption spectrum in the neat film showed a broad $S_0 - S_1$ band with an absorption maximum wavelength at 577 nm and two other bands at around 539 and 449 nm, respectively (Figure 11.2b). The emission spectrum is approximately the mirror image of the $S_0 - S_1$ absorption band with a maximum wavelength at 616 nm and a Stokes shift of about 39 nm.

To explore the polariton properties of PBLCs, microcavities containing oriented thin films with homeotropic and planar alignment were fabricated from the melt between two Distributed Bragg Reflector (DBR) mirrors as displayed in Figure 11.2c, d. The microcavities containing PBLC with homeotropic and planar alignment will be called hereafter MC-HO and MC-PL, respectively. The Rabi splittings for MC-HO and MC-PL are $\hbar\Omega_{HO}^{TM} = 160 \pm 5$ meV and $\hbar\Omega_{PL}^{TM} = 120 \pm 5$ meV. These values can be explained by the anisotropic orientation of the transition dipole moments. Dipole moments are in-plane for homeotropic alignment and are isotropic for planar alignment [30].

The PLQY and lifetime of LPs were measured using an integrating sphere and streak camera, respectively. Figure 11.3b shows the time-resolved LP emission for MC-HO and MC-PL under 400 nm excitation. The emission follows single-exponential decay, with $\tau^{MC\text{-}HO} = 3.8\ ns$, $\tau^{MC\text{-}PL} = 4.2\ ns$, respectively. Care must be taken when estimating the LP PLQY because emission from the DS excitons can also occur through the leaky modes of the DBR (beyond the stop band). The PLQY of the LPs was estimated as shown in Figure 11.3a. The reference line shows the integrated excitation spectrum for an empty sphere. The PLQY is given by

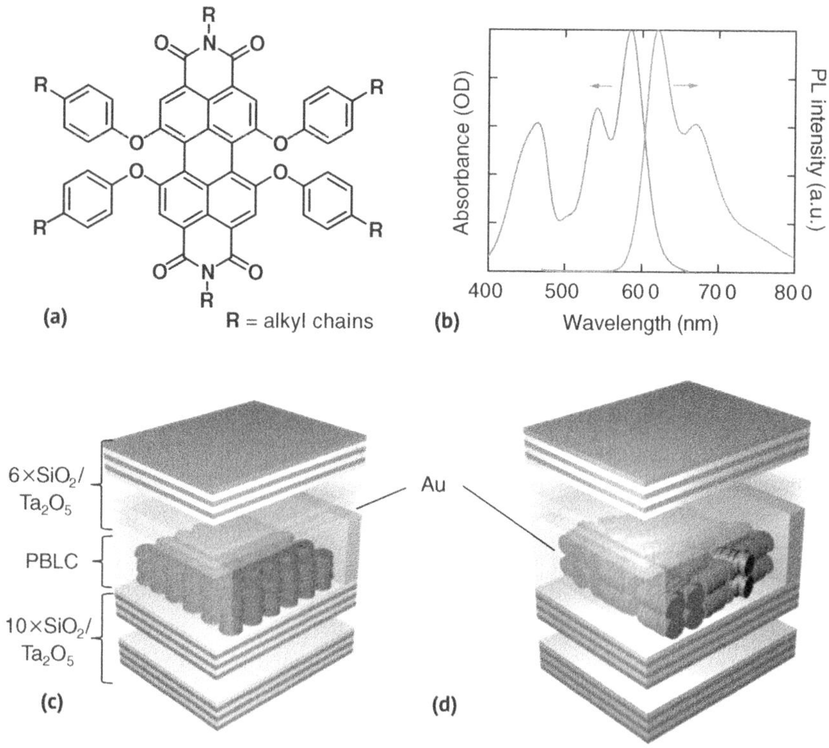

Figure 11.2 (a) Chemical structure of PBLC (1,6,7,12-tetra-4-nonylphenoxy perylene bisimide [PBI] with two dodecyl chains at the imide positions). (b) Emission and absorption spectra for the spin-coated neat film. (c, d) Schematic representation for the microcavities, which consists of two dielectric mirrors of alternating Ta_2O_5 and silicon dioxide (SiO_2) pairs enclosing a single layer of (c) homeotropic and (d) planar-aligned PBLC. *Source*: Ref. [30]/John Wiley & Sons.

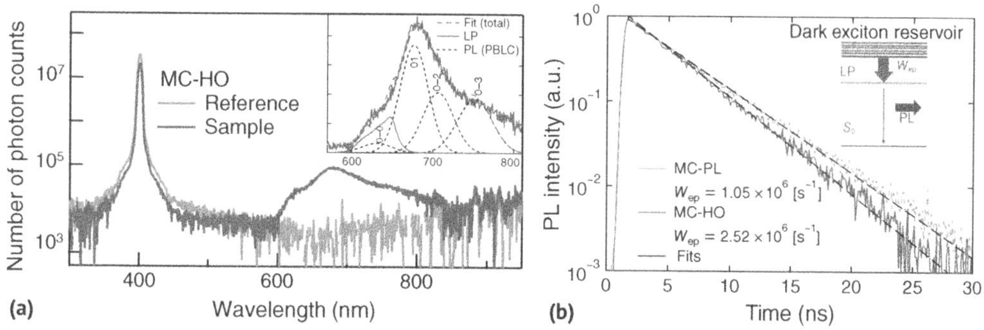

Figure 11.3 (a) Excitation spectrum and photoluminescence of MC-HO. Insets show the emission spectrum of samples and fitting results of each oscillator (black dot lines), the lower polaritons (LP) (black solid line), and the superposition of all functions (black dot lines) [30]. (b) The time-resolved LP emission for MC-HO and MC-PL under 400 nm excitation. At this excited energy regime, the emission follows single-exponential decay in the MC-HO and MC-PL, with $\tau_{MC\text{-}HO}$ = 3.8 ns, $\tau_{MC\text{-}PL}$ = 4.2 ns, respectively. *Source*: Ref. [30]/John Wiley & Sons.

$$\Phi_{LP} = \frac{\int I_{LP}(\lambda)\,d\lambda}{\int I_{int}(\lambda) - I'_{int}(\lambda)\,d\lambda} \tag{11.3}$$

where $I_{LP}(\lambda)$ is the emission spectrum of the LP and $I_{int}(\lambda)$ ($I'_{int}(\lambda)$) is the excitation light spectrum in the absence of the sample (in the presence of the sample). Here, $\int(I_{int}(\lambda) - I'_{int}(\lambda))d\lambda$ corresponds to the number of absorbed photons. $I_{LP}(\lambda)$ was extracted by separating the PL spectrum of MC-HO and MC-PL from the DS exciton PL spectrum, which is nearly that of the bare PBLC.

$$I_{LP}(\lambda) = I_{total}(\lambda) - (I_{0-0}(\lambda) + I_{0-1}(\lambda) + I_{0-2}(\lambda) + I_{0-3}(\lambda)) \tag{11.4}$$

Here, the 0–ν (ν = 0, 1, 2, and 3) emission bands are taken to be Gaussians, with parameters obtained from the shape of the neat film PL and are used to account for DS PL emitted through the leaky modes. Meanwhile, the LP spectrum was taken to be an asymmetric Gaussian distribution (Figure 11.3a inset) with a cut-off at the $k = 0$ LP energy (the cut-off wavelength is $\lambda = 650$ nm). The LP PLQY is estimated to be $\Phi_{LP}^{HO} = 0.96 \pm 0.20\%$ and $\Phi_{LP}^{PL} = 0.43 \pm 0.15\%$ for MC-HO and MC-PL, respectively. From the lifetime, the total decay rates $k_{tot}^{MC-HO} = \frac{1}{\tau_{MC\text{-}HO}} = 2.63 \times 10^8\,[s^{-1}]$ and $k_{tot}^{MC-PL} = \frac{1}{\tau_{MC\text{-}PL}} = 2.38 \times 10^8\,[s^{-1}]$ for MC-HO and MC-PL, respectively. Polariton relaxation rates $W_{ep}^{MC-HO} = 2.5 \pm 0.5 \times 10^6\,[s^{-1}]$ and $W_{ep}^{MC-PL} = 1.0 \pm 0.3 \times 10^6\,[s^{-1}]$ were found for MC-HO and MC-PL. The polariton relaxation rate of the homeotropically aligned MC-HO is twice as fast as that of MC-PL. This follows the trend in Rabi splitting and contrasts with the fact that both bare materials have nearly identical lifetimes and PLQYs.

The importance of dipole orientation for polariton condensation was also recently pointed out [30, 33, 51, 52]. That study used microcavities made of glassy-phase and β-phase PFO sandwiched between DBRs. Since PFO in β-phase has an in-plane orientation of the dipole moment, the absorption coefficient is twice as large as that of glassy-phase. The microcavities for glassy-phase and β-phase PFO show anisotropic Rabi splitting energies of $\hbar\Omega = 0.5$ and 0.7 eV, respectively. The thresholds for condensation in PFO were found to be 3.8 and 0.36 μJ cm^{-2} for glassy and β-phase, corresponding to a 10-fold reduction in the threshold for β-phase [33]. This highlights that the increase in Rabi splitting due to the control of molecular orientation can play an important role in lowering the BEC threshold.

The polariton relaxation rate was also measured recently in BSB-Cz derivatives [31, 53]. These molecules are characterized by a large radiative decay rate of $k_r = 1\times10^8$ [s^{-1}], a high PLQY of 80%, and a low amplified spontaneous emission (ASE) threshold of $P_{th} = 0.6$ μJ cm^{-2} [54]. Importantly, since the emission spectrum of BSBCz has a small overlap with the triplet absorption spectrum, triplet–triplet annihilation is suppressed in the high-density exciton regimes [55, 56]. This class of materials is thus a good candidate for low-threshold polariton condensates.

Among BSBCz derivatives, BSBCz-EH is composed of bi-stilbene (BSB) and carbazole (Cz) in the main p-conjugation scaffold with the solubilizing 2-ethylhexyl (EH) groups (Figure 11.4a top). Since the overlap between the HOMO and LUMO is large because of the small donor–acceptor interaction between the Cz and BSB units, BSBCz-EH has a large transition dipole moment of $\mu = 12.7$ Debye (Figure 11.4a bottom). The photophysical properties in the neat film of BSBCz-EH are shown in Figure 11.4b. BSBCz-EH shows an absorption peak at 388 nm (3.19 eV), a high PLQY value of 70%, and lifetimes of 1.6 ns.

To explore the polariton properties, microcavities were fabricated by spin-coating BSBCz-EH between dielectric mirrors (Figure 11.4c). Three film thicknesses were used, resulting in different detunings, d = 122 nm for $\Delta = -256$ meV, d = 140 nm for $\Delta = -356$ meV, and d = 155 nm for

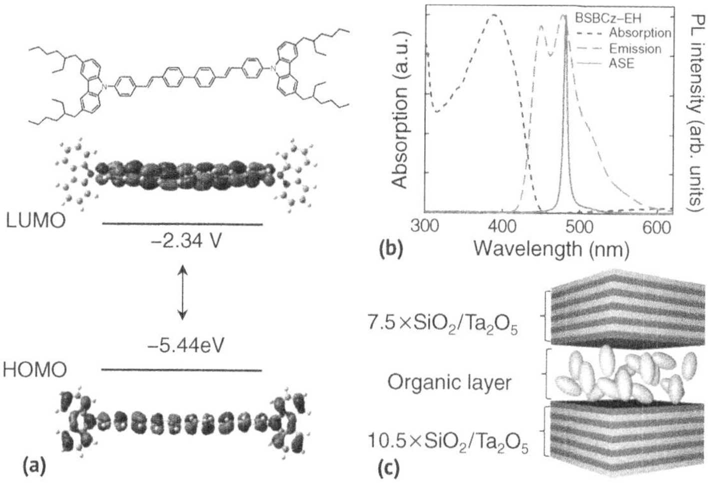

Figure 11.4 (a) Chemical structures of BSBCz-EH, and calculated molecular orbitals for BSBCz [54]. (b) The absorption, PL, and ASE spectrum for a neat film of BSBCz-EH. (c) Schematic image of the planar microcavity containing BSBCz-EH. *Source*: Adapted from Ref. [53].

Table 11.2 Parameters used to estimate polariton relaxation rate, W_{ep}.

Detuning parameter (meV)[a]	$\Gamma_0 = 1/\tau$ (s^{-1})[b]	Φ_{LP} (%)[c]	W_{ep} (s^{-1})[d]
−150	8.84×10^8	0.04 ± 0.03	$3.95 \pm 3.07 \times 10^5$
−256	1.14×10^9	0.27 ± 0.06	$3.03 \pm 1.21 \times 10^6$
−356	1.58×10^9	1.85 ± 0.73	$2.93 \pm 1.15 \times 10^7$
−416	1.96×10^9	2.30 ± 0.86	$4.50 \pm 1.68 \times 10^7$

a. Detuning parameter, $\Delta(0) = E_{ph}(0) - E_{ex}$.
b. $\Gamma_0 = 1/\tau$ is the total decay rate.
c. Φ_{LP} is PLQY of lower polaritons.
d. W_{ep} is polariton relaxation rates.

$\Delta = -416$ meV. The detuning is defined as the difference between the 0-0 absorption band of BSBCz-EH and the $q = 0$ bare cavity mode, where q is the in-plane wavevector. The Rabi energies were estimated to be $\hbar\Omega = 369$, 345, and 322 meV for $\Delta = -256$, -356, and -416 meV, by fitting the LP and UP bands measured by angle-resolved reflectance with a harmonic oscillator model.

The same technique described above for measuring PLQY and lifetimes was used to obtain the polariton relaxation rate [53]. The data are summarized in Table 11.2. The polariton relaxation rates were found to be $W_{ep} = 3.95 \pm 3.07 \times 10^5$, $3.03 \pm 1.21 \times 10^6$, $2.93 \pm 1.15 \times 10^7$, and $4.50 \pm 1.68 \times 10^7\ s^{-1}$ for $\Delta = -150$, -256, -356, and -416 meV, respectively. The rate consistently increases for more negative detuning, which corresponds to increasingly photonic LPs. The origin of this increase will be discussed in Section 11.4.

11.2.2 Polariton Kinetics in the Nonlinear Regime

In the kinetic model, condensation occurs due to the stimulated scattering terms that must be included for a correct description of the thermodynamics of Bose gases. When these terms are included, the occupation of a [LP] with in-plane wavevector q is described by

$$\frac{d\,[\mathrm{LP}]_q}{dt} = \left(1 + [\mathrm{LP}]_q\right) \sum_{q'} W_{q' \to q} [\mathrm{LP}]_{q'} + \left(1 + [\mathrm{LP}]_q\right) W_{\mathrm{DS} \to q} [\mathrm{DS}] \tag{11.5}$$

where $[\mathrm{LP}]_q$ is the density of LPs with wavevector q and $W_{q' \to q}$ is the rate of polariton relaxation rate from LPs with wavevector q' to q. Similarly, $W_{\mathrm{DS} \to q}$ is the scattering rate from dark states to LPs. As in Section 11.2.1, the contribution from the sum on the right-hand side can be ignored when $W_{q' \to q}$ is small and polaritons are short-lived. This equation shows that when the occupancy at $[\mathrm{LP}]_q$ exceeds 1, the LP density is amplified exponentially with the pumping of the DSs until saturation. This corresponds to the onset of condensation.

Similarly to Section 11.2.1, a minimal two-state model can be written to describe the time evolution of the exciton reservoir and the population of LP states that form the condensate [20, 31]. Here, we remind the reader that due to the finite pump spot, the condensate must occupy several modes in reciprocal space. Accounting for pumping, bimolecular annihilation, and spontaneous relaxation into the LPs, a master equation for the volume density of reservoir excitons, $n_{\mathrm{R}}(t)$, and the surface density of polaritons, $n_{LP}(t)$, can be written as:

$$\frac{dn_{\mathrm{R}}}{dt} = \left(1 - \frac{n_{\mathrm{R}}}{N_0}\right) P(t) - \frac{n_{\mathrm{R}}}{\tau_{\mathrm{R}}} - k_{\mathrm{b}} n_{\mathrm{R}}^2 - \frac{W_{\mathrm{ep}}^{(q\sim 0)}}{d} n_{\mathrm{R}} n_{\mathrm{LP}} \tag{11.6}$$

$$\frac{dn_{\mathrm{LP}}}{dt} = W_{\mathrm{ep}}^{(q\sim 0)} n_{\mathrm{R}} n_{\mathrm{LP}} + df \frac{n_{\mathrm{R}}}{\tau_{\mathrm{R}}} - \frac{n_{\mathrm{LP}}}{\tau_{\mathrm{LP}}} \tag{11.7}$$

where $\tau_{\mathrm{R}}(\tau_{\mathrm{LP}})$ is the bare exciton (LP) lifetime, $W_{\mathrm{ep}}{}^{(q\sim 0)}$ is the dark exciton reservoir to LP scattering rate summed over the ensemble of states being considered, $P(t)$ is a Gaussian pump term FWHM, $P(t) = P_0' \exp(-t^2/2\sigma^2)$ with $P_0' = P_0/(dS\hbar\omega_{\mathrm{pump}})$ where d is the cavity thickness, S is the excited pump area, $\hbar\omega_{\mathrm{pump}}$ is the energy of the pump photons and P_0 is the pump energy density per pulse. Given the high excitation densities, this model includes a saturation factor $(1\text{–}n_{\mathrm{R}}/N_0)$ due to ground-state depletion, where N_0 is the molecular density of organic films, and k_{b} is the bimolecular annihilation rate. The constant fraction f, analogous to the spontaneous emission factor, represents the fraction of relaxation from the DSs that spontaneously populates the LP states being considered.

To reveal the role of the parameter of $W_{ep}^{(q\sim 0)}$ and f on the PL power dependence, we show the time-integrated surface density of polaritons $\int_0^\infty n_{\mathrm{LP}}(t)dt$ from solutions of Eqs. (11.6) and (11.7) for various values of $W_{ep}^{(q\sim 0)}$ and f. Figure 11.5a shows the calculation results for polariton relaxation rates of $W_{ep}^{(q\sim 0)} = 1 \times 10^{-5}$, 1.5×10^{-5}, 2×10^{-5}, $3 \times 10^{-5}\ cm^{-3}\ \mathrm{s}^{-1}$ (f was fixed at $f = 0.48$). As expected, the threshold (P_{th}) decreases with the increasing polariton relaxation rate. Figure 11.5b shows the calculation results for the varying "f": $f = 1 \times 10^{-1}$, 1×10^{-2}, 1×10^{-3} *and* 1×10^{-4} ($W_{ep}^{(q\sim 0)}$ was fixed at $W_{ep}^{(q\sim 0)} = 2.0 \times 10^{-5}\ \mathrm{cm}^3\ \mathrm{s}^{-1}$). As can be seen from the figure, the surface density of LPs below the threshold is not changed for all f values, but the surface density of LPs above the threshold is increased with increasing f values. Its inverse, $1/f$, effectively corresponds to the size of the jump in LP density across the threshold.

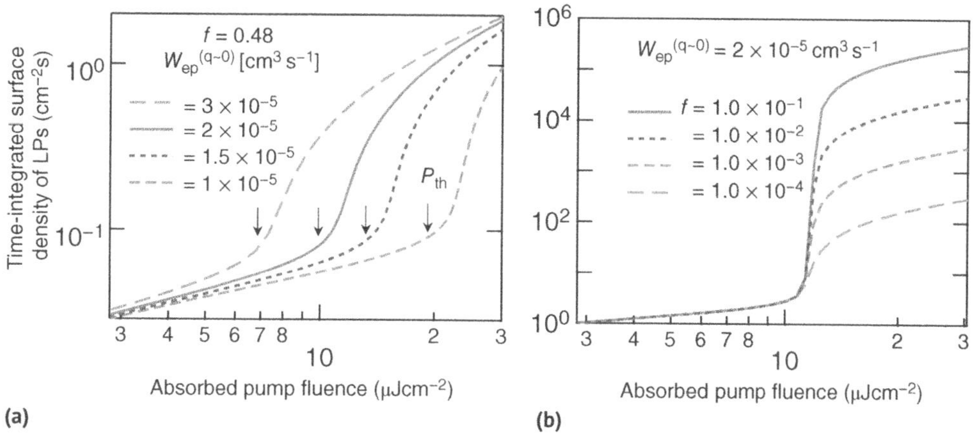

Figure 11.5 Time-integrated surface density of polaritons $\int_0^\infty n_{LP}(t)dt$ calculated from solutions of Eqs. (11.6) and (11.7) for various values of $W_{ep}^{(q\sim0)}$ and f. The used parameters in the simulations are τ_R = 1.6 ns, τ_{LP} = 33 fs, d = 145 nm, $N_0 = 6.76 \times 10^{20}$ cm^{-3} and $k_b = 8.0 \times 10^{-9}$ cm^3 s^{-1}. *Source*: Ref. [31]/John Wiley & Sons.

This minimal model makes it possible to experimentally estimate $W_{ep}^{(q\sim0)}$ by measuring the absorbed pump fluence dependence of the integrated PL. This was first used in microcavities containing anthracene and the fluorene oligomer TDAF emitters [20, 43]. A variation of this approach was also used to understand the role of intermolecular energy transfer in the blueshifts observed upon condensation [27]. In TDAF microcavities with mirrors consisting of six and nine DBR pairs, the parameters $W_{ep}^{(q\sim0)} = 2\times10^{-7}$ *and* 1.3×10^{-7} *cm*^{3} s^{-1} were extracted using this approach. This should be compared with condensation thresholds of P_{th} = 60 and 30 μJ cm^{-2}, respectively, for the same microcavities. This clearly highlights that the lowered threshold for the cavity with nine-pair mirrors is a result of the increased polariton lifetime (55 vs. 150 fs, respectively), while the polariton relaxation rate is relatively independent of the mirror composition, as expected.

The nonlinear power dependence at the threshold was also recently used to obtain the polariton relaxation rate in BSBCz-EH, which allows for a comparison with the PLQY-based approach [31]. These microcavities were fabricated using the melt-growth process as shown in Figure 11.6a. The advantage of the melt-processing method is that it allows the fabrication of organic layers without plasma damage during sputtering to the organic layers [57, 58]. Figure 11.6b shows the reflectance spectra of the melt-grown microcavity for $\Delta = -390$, -430, and -470 meV, respectively. The Rabi splitting energy was estimated to be $\hbar\Omega$ = 301 meV at $\Delta = -390$ meV (Figure 11.6c). The pump-fluence dependence of the PL intensity at $\theta = 0$ was measured for microcavities with $\Delta = -430$ and -470 meV as shown in Figure 11.7. Thresholds corresponding to absorbed fluences of P_{th} = 9.7 ± 0.1 μJ cm^{-2} and 62 μJ cm were found for the two detuning parameters, respectively. The integrated PL results were simulated using the minimal model in Eqs. (11.6) and (11.7). The black lines show the integration of $n_{LP}(t)$ for each absorbed energy fluence ($I_{PL} \propto \int_0^\infty n_{LP}(t)dt$). Here, $W_{ep}^{(q\sim0)}$, k_b, and f are chosen as free parameters. The numerical simulation successfully accounts for the PL data in the region below and above the threshold. From the fits, the polariton relaxation rates were estimated to be $W_{ep}^{(q\sim0)} = 5.2 \times 10^{-6}$ and 2.0×10^{-5} cm^3 s^{-1} for $\Delta = -430$ and -470 meV, respectively. The f and k_b values used for $\Delta = -430$ and -470 meV are f = 0.88, $k_b = 5.5 \times 10^{-10}$ cm^3 s^{-1} and f = 0.55, $k_b = 7.5 \times 10^{-9}$ cm^3 s^{-1}, respectively. Other parameters used in the simulations are shown in Table 11.3. The reduction of

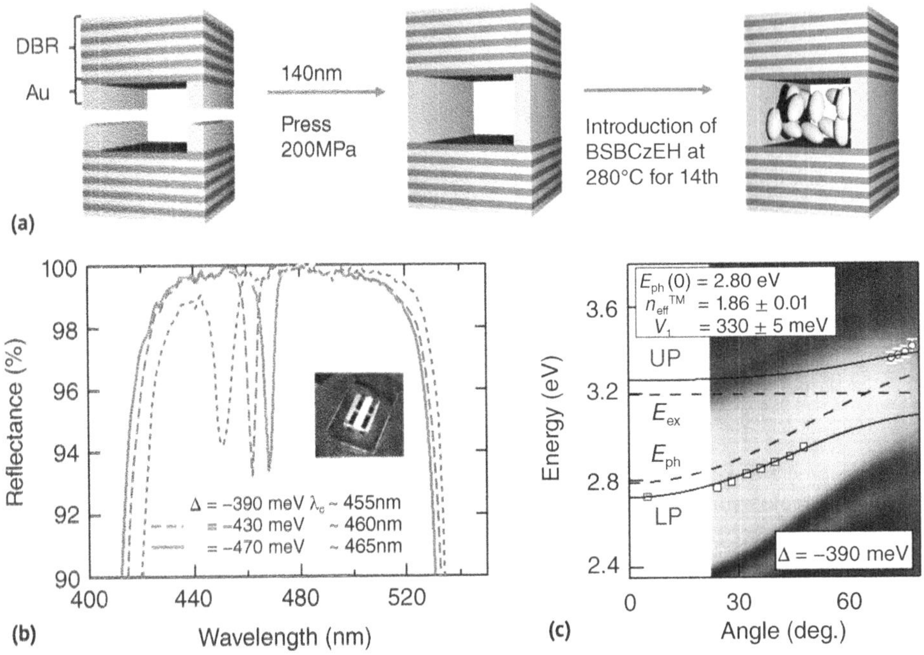

Figure 11.6 (a) Fabrication by melt processing of microcavity containing BSBCz-EH between two clean DBR mirrors. (b) The reflectance spectrum of λ/2 microcavity (DBR/BSBBCz-EH/DBR) for Δ = −390, −430, and −470 meV (Inset: the sample picture of melt-processed microcavity). (c) Contour maps of the angle-resolved reflectivity for transverse magnetic polarization for Δ = −390 meV. *Source*: Ref. [31]/John Wiley & Sons.

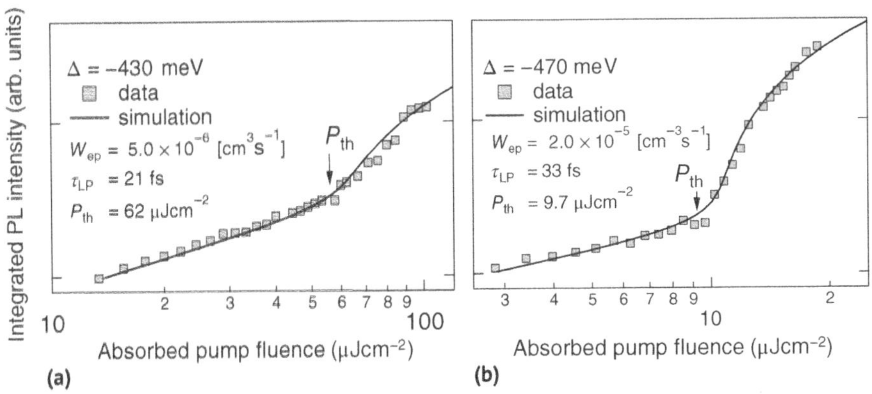

Figure 11.7 Absorbed pump fluence dependence of the integrated photoluminescence data for Δ = −430 meV and Δ = −470 meV and time-integrated surface density of polaritons $\int n_p(t)dt$ calculated from the solution of Eqs. (11.6) and (11.7). *Source*: Ref. [31]/John Wiley & Sons.

the threshold at $\Delta = -470$ meV (P_{th} ~ 9.7 μJ cm^{-2}) compared with that of $\Delta = -430$ meV (P_{th} ~ 62 μJ cm^{-2}) is caused mostly by an increase in the polariton relaxation rate, W_{ep} [31]. The trend with detuning follows that observed in the linear regime, while the rates themselves are smaller as expected since they were now just considering scattering to a small fraction of the LPs near $q = 0$.

In Table 11.4, we summarize the $W_{ep}^{(q\sim0)}$, τ_{LP}, and thresholds for anthracene single crystals, TDAF, and BSBCz-EH. Even though the polariton lifetime for the BSBCz-EH microcavity is shorter

Table 11.3 Parameters used to calculate the integrated photoluminescence by the semiclassical model.

Detuning parameter (meV)	d (nm)	τ_R (ns)	τ_{LP} (fs)	f	k_b (cm^3 s^{-1})	$W_{ep}^{(q\sim0)}$ (cm^3 s^{-1})
$\Delta = -430$	140	1.61	22	0.88 ± 0.05	$5.5 \pm 2.0 \times 10^{-10}$	$5.2 \pm 0.4 \times 10^{-6}$
$\Delta = -470$	145	1.61	33	0.55 ± 0.05	$7.5 \pm 1.0 \times 10^{-9}$	$2.0 \pm 0.1 \times 10^{-5}$

Table 11.4 Parameters for the $W_{ep}^{(q\sim0)}$, τ_{LP}, and P_{th} for anthracene, TDAF, and BSBCz-EH.

Material	Δ (meV)	τ_{LP} (fs)	$W_{ep}^{(q\sim0)}$ (cm^3 s^{-1})	P_{th} (μJ cm^{-2})	Reference
Anthracene	−120	85	2.9×10^{-9}	300	[5]
TDAF	−440	150	1.3×10^{-7}	30	[20]
BSBCz-EH	−470	33	2.0×10^{-5}	9.7	[31]

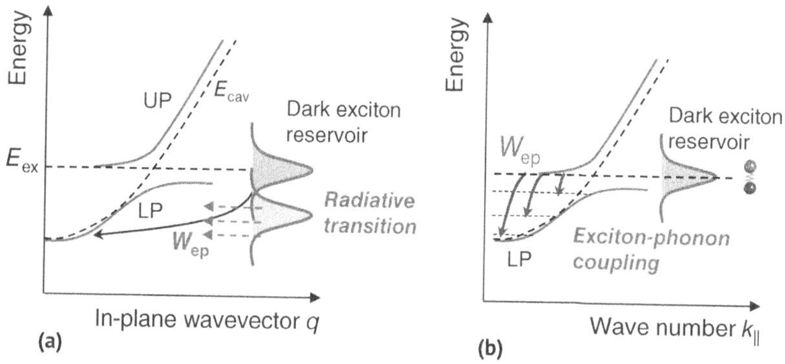

Figure 11.8 Schematic illustrations of the polariton relaxation process. (a) radiative transition: photoluminescence from the dark exciton reservoir pumps the lower polaritons (LPs), (b) nonradiative transition: exciton–phonon coupling pumps the LP states.

than that in anthracene and TDAF microcavities, it is the material showing the lowest threshold. This highlights that accelerating W_{ep} is important for lowering the threshold.

11.3 Polariton Relaxation Mechanisms

In inorganic semiconductors such as GaAs and CdTe, the relaxation from the reservoir to LPs has been proposed to occur mainly via the emission of acoustic phonons [49, 50, 59]. In organic microcavities, radiative pumping from dark states and nonradiative relaxation via molecular vibrations have been proposed as possible mechanisms for relaxation (Figure 11.8) [36, 42, 43].

11.3.1 Radiative Pumping Process (1): Initial Experiments

Mazza et al. proposed that scattering to polaritonic states can be mediated by photon emission accompanied by the creation of a vibrational quantum in the ground state [42, 43]. The energy loss when the energy of the exciton reservoir is scattered into the LP state corresponds exactly to the energy of the vibrational quantum as shown in Figure 11.9. Based on the model, the polariton populations are expected to follow the vibronic progression observed in PL.

Signatures of this mechanism were first observed in the PL from anthracene microcavities [5, 60, 61]. Figure 11.10a shows the PL and absorption spectra for the anthracene single crystal [61]. Peaks in the spectra correspond to transitions from the ground state to the vibronic resonances of the first excited singlet state. Each peak corresponds to the 0–0, 0–1, and 0–2 transitions, respectively. In the microcavity containing anthracene, the polariton states were identified by the angle-resolved reflectance measurements, which show the LP and middle polaritons (MP1 and MP2), respectively [61]. To investigate the polariton population distribution, angle-resolved photoluminescence (PL) was measured for the microcavity as shown in Figure 11.10b [60]. The PL shows a clear maximum at 2.94 eV, whose energy corresponds to the transition from the first electronically excited state to the first vibrational level of the electronic ground state (0–1 transition). However, one should note that in materials with small Stokes shifts, the energies of vibronic replicas also correspond to those that would be attributed to the nonradiative emission of a molecular phonon. One important signature of radiative pumping is its rate dependence on the photon fraction of the final LP state. This highlights that the polariton relaxation mechanism can be distinguished most readily by the dependence on photon and exciton fractions, not by the overlap with the emission spectrum or vibrational modes.

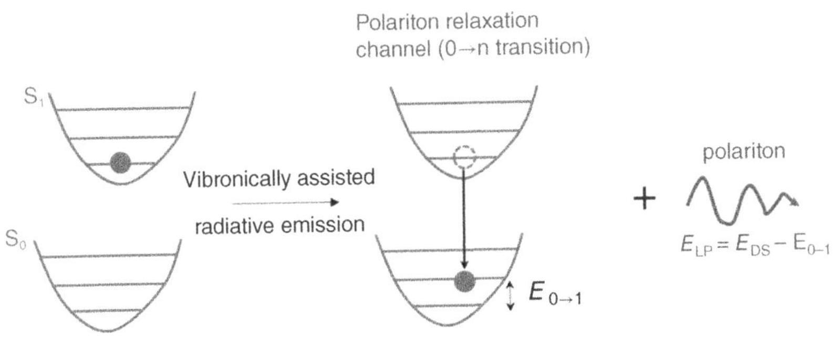

Figure 11.9 Schematic of the polariton relaxation mechanism via molecular emission with the creation of a vibrational quantum in the ground state. E_{0-1}, E_{DS}, and E_{LP} are energies for 0–1 transition in the ground state, dark states, and lower polaritons, respectively.

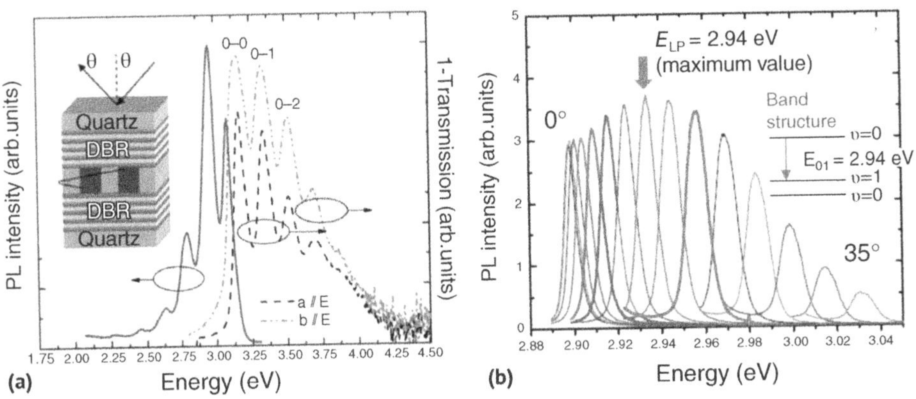

Figure 11.10 (a) Photoluminescence (solid line) and linear transmittance spectra (dashed line) for the microcavity containing anthracene single crystal. Inset: microcavity structure. *Source:* Adapted from Ref. [61]. Copyright (2008) by the American Physical Society. (b) Angle-resolved PL from the a-polarized LP branch. The maximum peak intensity is identified at ~2.94 eV. *Source*: Adapted from Ref. [60].

Radiative pumping has also been studied in microcavities containing semiconducting single-walled carbon nanotubes (SWCNTs) [47]. SWCNTs have attracted considerable interest for their tunability, high-charge carrier mobilities (10000 $cm^2\ V^{-1}\ s^{-1}$), and near-infrared exciton energies (Figure 11.11a). SWCNTs with (6,5) chirality show an intense absorption peak from the E_{11} exciton at 1.24 eV (998 nm) with an FWHM of only 20.5 meV. The G_1 feature in PL results from the decay of E_{11} excitons to the ground state with the emission of a G_0 phonon (Figure 11.11a). Figure 11.11b shows that the LP population approximately tracks the SWCNT emission spectrum. This resemblance strongly suggests that radiative emission populates the LPs. This was further tested by creating luminescent sp^3 defects with a red-shifted emission and observing the redistribution of the polariton population (Figure 11.11c) [47].

Strong signatures that radiative emission dominates the kinetics in BODIPY-Br microcavities have also been observed (Figure 11.11d) [48]. The detuning dependence of the PLQY clearly shows that microcavities possessing more negative detunings emit luminescence with greater efficiency than cavities having a smaller detuning. In addition, the angular dependence of the LP PL shows the most efficient pumping of LPs when the minimum energy of the LPB overlaps with the 0–1 vibronic transition (Figure 11.11e) [48].

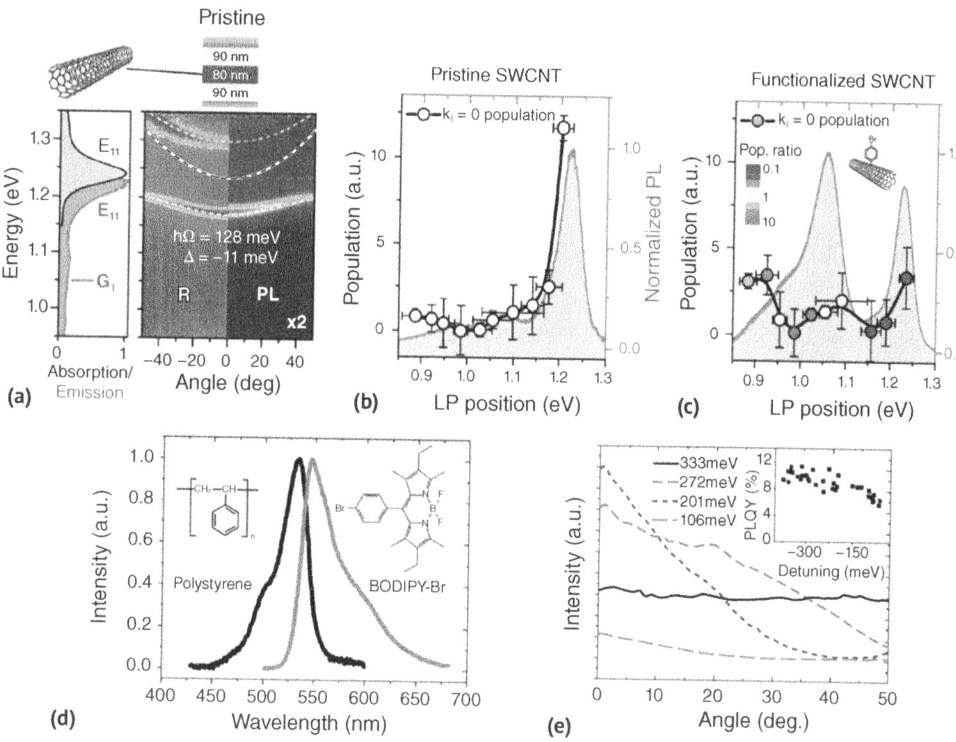

Figure 11.11 (a) Emission and absorption spectra for a pristine (6,5) SWCNT reference film (left). And, angle-resolved reflectivity and photoluminescence for a metal-clad microcavity containing pristine (6,5) SWCNT. (b, c) Calculated polariton population for SWCNT-filled microcavities at $q = 0$ as a function of the LP position for pristine SWCNTs, and functionalized SWCNTs. *Source*: Ref. [47]/American Chemical Society. (d) Absorption and PL emission of BODIPY-Br (inset right) doped into a polystyrene (inset left) film at a ratio of 1 : 100 by mass. (e) Angle dependence of the measured PL intensities at 298 K for four DBR-DBR microcavities having detunings of $\Delta = -333, -272, -201$, and -106 meV. Inset: Detuning dependence of PLQY for a series of DBR-metal microcavities. *Source*: Adapted from Ref. [48]. Copyright (2016) by John Wiley & Sons.

11.3.2 Radiative Relaxation

Mazza et al. were the first to consider radiative pumping of the LP through the luminescence of vibronic replicas. In that work, transition dipole moments between different vibronic states were explicitly modified to account for the Poissonian distribution of the displaced oscillators. The authors argued that light–matter interaction of states that involve a vibrationally excited electronic ground state could be treated perturbatively. Using Fermi's golden rule, one can then obtain scattering rates between states in a way that is completely analogous to conventional spontaneous emission, with amplitudes that follow the vibronic progress in PL, with the notable absence of the 0–0 transition. A key feature of the model is that the scattering rates are proportional to the photon fraction of the final state. The argument for treating certain vibronic transitions perturbatively is hard to justify a priori, however.

Treating collective electronic strong coupling, while formally considering nuclear motion, can be done within the framework of the Holstein-Tavis-Cummings Hamiltonian for a single harmonic potential or generalized approaches for more complex potential energy surfaces. Polaritonic eigen states thus have mixed electronic/photon/vibrational character. Insight from theory has shown that for a large number of molecules, going from the vibrationally relaxed electronic excited state to the ground state with the creation of a photon can indeed be treated perturbatively, with the light–matter coupling g acting as the perturbation.

Here, we treat radiative relaxation as being due to the emission of a photon into cavity mode q by a DS exciton, which drives the incoherent transition between the ground manifold state $|\varepsilon'\rangle$ and an excited (LP) polariton state $|\varepsilon\rangle$. The rate of this process is proportional to the normalized emission spectrum of the molecule, evaluated at the transition eigenfrequency $\omega_{\varepsilon\varepsilon'} \sim \omega_{LP}$, and is proportional to the photonic fraction of the excited polariton state. The scattering rate per molecule from the HTC eigenstate $|\varepsilon'\rangle$ to $|\varepsilon\rangle$ is given by:

$$W_{\varepsilon\varepsilon'} = \sum_{k} |g'_n(q)|^2 |\langle\varepsilon'|\hat{a}_q|\varepsilon\rangle|^2 S_n(\omega_{\varepsilon\varepsilon'}) \tag{11.8}$$

$$S_n(\omega) = \int_0^{\infty} d\tau e^{i\omega\tau} \langle e_n|\hat{\sigma}_n^+(\tau)\hat{\sigma}_n^-(0)|e_n\rangle \tag{11.9}$$

where $g'_n(q)$ is the single-molecule Rabi frequency of the nth molecule with cavity mode at wavevector q, $\hat{a}_q$ is the annihilation operator for a cavity mode with wavevector q. $\hat{\sigma}_n^- = |g_n\rangle\langle e_n|$ are the electronic transition operator and $S_n(\omega)$ is the normalized molecular emission spectrum of the nth molecule.

We can account for the spatial and wavevector dependence of the single molecule Rabi frequency using the microscopic theory of organic exciton–polaritons [62]. The square of the Rabi frequency between one molecule and the cavity mode for TE and TM modes is given by

$$|g'_{\mathrm{TE}}(q,z)|^2 = \frac{1}{\hbar^2}\left(\frac{32\hbar\omega_{\mathrm{cav}}(q)}{A\epsilon\pi\beta_{\mathrm{TE}}(q)}\right)\mu_y^2|E_{\mathrm{nom}}^y(q,z)|^2 \tag{11.10}$$

$$|g'_{\mathrm{TM}}(q,z)|^2 = \frac{1}{\hbar^2}\left(\frac{32\hbar\omega_{\mathrm{cav}}^2(0)}{A\epsilon\pi\omega_{\mathrm{cav}}(q)\beta_{\mathrm{TM}}(q)}\right)\mu_x^2|E_{\mathrm{nom}}^x(q,z)|^2 \tag{11.11}$$

where $\omega_{\mathrm{cav}}(q)$ is the cavity-mode frequencies, $\omega_{cav}(q(\theta)) = \omega_{cav}(0)\left(1-(\sin\theta/n_{eff})^2\right)^{-1/2}$, n_{eff} is the effective refractive index, A is the in-plane quantization area, ϵ is the background dielectric constant of the medium between the mirrors, and $\mu = (\mu_x, \mu_y, \mu_z)$ is the transition dipole moment. The factor $\beta_\lambda(q)$ ($\lambda = TE$ *and* TM *mode*) is given by

$$\beta_{\mathrm{TE}}(q) = \int_{z_{\mathrm{inf}}}^{z_{\mathrm{fin}}} |E^{y}_{\mathrm{nom}}(q, z)|^2 \mathrm{d}z \tag{11.12}$$

$$\beta_{\mathrm{TM}}(q) = \int_{z_{\mathrm{inf}}}^{z_{\mathrm{fin}}} \left[\frac{k_z^2}{k^2}|E^{x}_{\mathrm{nom}}(q, z)|^2 + \frac{k_x^2}{k^2}|E^{z}_{\mathrm{nom}}(q, z)|^2\right] \mathrm{d}z \tag{11.13}$$

where $E_{\mathrm{nom}}(q, z)$ is the normalized electric-field distribution for the x- and y-axis, and k is the total wavevector $k^2 = k_z^2 + q^2$, where k_z is the wavevector for the z-axis. The definition of the x-, y-, and z-axes is shown in Figure 11.12a.

Then, the polariton relaxation rate W_{ep} per molecule is determined by integrating over all wavevectors and spatially averaging

$$W_{\mathrm{ep}} = \frac{1}{L_{\mathrm{c}}} \int_{-\frac{L_{\mathrm{c}}}{2}}^{\frac{L_{\mathrm{c}}}{2}} \mathrm{d}z \frac{A}{(2\pi)^2} \int \mathrm{d}q W_{\mathrm{ep}}(q, z) D(q) \tag{11.14}$$

where the photon density of states in two dimensions is $D(q) = 2\pi q$, and L_{c} is the cavity length. The integral of W_{ep} in the range $[q_{\mathrm{min}}, q_{\mathrm{max}}]$ is:

$$W_{\mathrm{ep}}^{\mathrm{TE}} = \frac{1}{\hbar^2} \frac{32\hbar\mu_y^2}{A\epsilon\pi} \frac{A}{(2\pi)^2} \int_{q_{\mathrm{min}}}^{q_{\mathrm{max}}} dq\, 2\pi q \cdot \frac{\alpha_{\mathrm{TE}}(q)}{\beta_{\mathrm{TE}}(q)} \omega_{\mathrm{cav}}(q) S(\omega_{\varepsilon\varepsilon'}) |C(q)|^2 \tag{11.15}$$

$$W_{\mathrm{ep}}^{\mathrm{TM}} = \frac{1}{\hbar^2} \frac{32\hbar\mu_x^2}{A\epsilon\pi} \frac{A}{(2\pi)^2} \int_{q_{\mathrm{min}}}^{q_{\mathrm{max}}} dq\, 2\pi q \cdot \frac{\alpha_{\mathrm{TM}}(q)}{\beta_{\mathrm{TM}}(q)} \frac{\omega_{\mathrm{cav}}^2(0)}{\omega_{\mathrm{cav}}(q)} S(\omega_{\varepsilon\varepsilon'}) |C(q)|^2 \tag{11.16}$$

with the wavevectors of q_{min} and q_{max} corresponding to $\theta = 0$ and 90 deg, respectively, and $|C(q)|^2 = |\langle\varepsilon'|\hat{a}_q|\varepsilon\rangle|^2$ is the Hopfield coefficient of the photonic part. The definitions of $a_\lambda(q)$ ($\lambda = \mathrm{TE}$ and TM

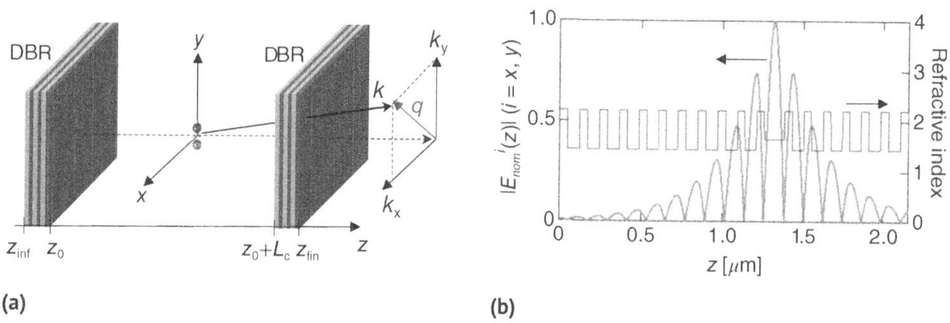

Figure 11.12 (a) Schematic image of the definition of the x-, y-, and z-axis. (b) Normalized electric field intensity distribution $|E_{\mathrm{nom}}{}^{i}(z)|$ ($i = x, y$) for a microcavity consisting of DBR (10.5 pairs)/emitter (122 nm)/DBR (6.5 pairs). Note that electric field at $\theta = 0$ satisfy the below relationship: $|E_{\mathrm{nom}}{}^{x}(z)| = |E_{\mathrm{nom}}{}^{y}(z)|$, and $|E_{\mathrm{nom}}{}^{z}(z)| = 0$.

mode) are given by $\alpha_{TE}(q) = \frac{1}{Lc}\int_{-Lc/2}^{Lc/2} |E_y(q,z)|^2 dz$ and $\alpha_{TM}(q) = \frac{1}{L_c}\int_{-L_c/2}^{Lc/2} |E_x(q,z)|^2 dz$, respectively. The $a_\lambda(q)$ represents the spatial average of the normalized electric field for the y- and x-axis inside a microcavity (Figure 11.12b). For DBR microcavities, the magnitude of the electric field is reduced by the penetration depth: $E(z,q) \propto \sqrt{L_c/L_c + L_{DBR}(q)}$.

Here, we make the quadratic approximation ($E = E_{LP}(0) + \frac{\hbar^2 q^2}{2m_{LP}}$) and integrate with energy over the spectral region of LPs $E \in [E_{inf}(q_{min}), E_{sup}(q_{max})]$. The values of W_{ep} are then given by:

$$W_{ep}^{TE} = \frac{16\mu^2 m_{LP}}{3\pi^2 n_{eff}^2 \hbar^2} \int_{E_{inf}}^{E_{sup}} dE \frac{\alpha_{TE}(E')}{\beta_{TE}(E')} \omega_{cav}(E') |C(E)|^2 S'(E) \tag{11.17}$$

$$W_{ep}^{TM} = \frac{16\mu^2 m_{LP}}{3\pi^2 n_{eff}^2 \hbar^2} \int_{E_{inf}}^{E_{sup}} dE \frac{\alpha_{TM}(E')}{\beta_{TM}(E')} \frac{\omega_{cav}^2(0)}{\omega_{cav}(E')} |C(E)|^2 S'(E) \tag{11.18}$$

where $S'(E)$ is $S' = S(\omega)/\hbar$, and $\omega_{cav}(E')$ is $\omega_{cav}(E') = \omega_{cav}(0)\left(1 - E'^2\right)^{-1/2}$, where E' is $E' = \sin\left\{\tan^{-1}\left(\frac{m_{LP}^{1/2}\lambda_c}{\sqrt{2}n_{eff}\pi\hbar}\sqrt{E - E_{LP}(0)}\right)\right\}$, λ_c is the resonance wavelength, μ is the total dipole moment $\left(\frac{\mu^2}{3} = \mu_x^2 = \mu_y^2 = \mu_z^2\right)$, n_{eff} is the effective refractive index $\left(n_{eff} = \sqrt{\epsilon}\right)$.

In this model, the polariton relaxation rate W_{ep} is scaled by the spectral overlap between the LP and the molecular PL, including all vibronic modes.

We will discuss the theoretical analysis between the Mazza model [42, 43] and the dipole fluctuation model and the comparison with the experimental data in Section 11.4.

11.3.3 Nonradiative Relaxation

The possibility of polariton relaxation occurring through the emission of molecular phonons has also been proposed by Litinskaya and Agranovich [36]. For linear exciton–phonon coupling, the Hamiltonian is given by

$$H_{ex-phon} = gE_{vib} \sum_n b_n^\dagger b_n \left(\upsilon_n^\dagger + \upsilon_n\right) \tag{11.19}$$

where g is the coupling strength ($g = \sqrt{S}$), E_{vib} is the energy of an intermolecular vibrational quantum, $b_n^\dagger$ (b_n) is the creation (annihilation) operator of electronic excitation at site n and $\upsilon_n^\dagger$ (υ_n) is the creation (annihilation) operator of an intramolecular phonon. If the coupling can be treated perturbatively, the scattering rate from one molecular exciton (labeled by j) to a polariton state (labeled by k) is given by

$$W^{j\to k} = \frac{2\pi}{\hbar}\frac{g^2 E_{vib}}{N} |X_j|^2 |X_k|^2 \delta\left[E_0 - E_{LP} - E_{vib}\right] \tag{11.20}$$

where X is the Hopfield coefficient of exciton components. The dependence on the final state exciton fraction distinguishes this mechanism from the radiative one.

This mechanism has also been studied using multiscale molecular dynamics simulations of tetracene molecules strongly coupled to the confined light modes of an optical cavity [63]. It was suggested that polariton relaxation occurs most efficiently for displacements of vibrational modes that overlap with the nonadiabatic coupling vector $\boldsymbol{d}_{DS_0,LP}$:

$$\boldsymbol{d}_{DS_0,LP} = \frac{\langle \psi^{LP} | \nabla_{\mathbf{R}} \hat{H}^{TC} | \psi^{DS_0} \rangle}{E_{DS_0} - E_{LP}} \tag{11.21}$$

where $|\psi^{LP}\rangle$ and $|\psi^{DS_0}\rangle$ are the eigenstates of the multimode Tavis-Cummings Hamiltonian ($\hat{H}^{TC}$) for the LP and the lowest energy dark states (DS_0), $\nabla_{\mathbf{R}}$ is the gradient vector for the displacements of the molecule in the S_1 minimum geometry, and $E_{DS_0} - E_{LP}$ is the energy gap between DS_0 and the LP. The nonadiabatic coupling vector is along the direction of transfer between adiabatic states, and the magnitude determines how much population is transferred. The nonadiabatic coupling ($V_{DS_0,\,LP}$) is given by the overlap between the nonadiabatic coupling vector $\boldsymbol{d}_{DS0,LP}$, and the velocities of the atoms $\dot{\boldsymbol{R}}$:

$$V_{DS_0,\,LP} = \boldsymbol{d}_{DS_0,\,LP} \cdot \dot{\boldsymbol{R}} \tag{11.22}$$

First, we note that $V_{DS_0,\,LP}$ is scaled by $1/\sqrt{N}$, where N is the number of molecules per photon mode [38, 64, 65]. This means that the nonadiabatic transition from DS_0 to LP is reduced in the collective strong coupling regime (typically $N \sim 10^6$). Second, note that nonadiabatic transitions occur when the energy spacing between the eigenstates of the vibrational modes coincides with the energy gap between the LP and the DS_0, $\hbar\omega_{vib} = E_{DS_0} - E_{LP}$. This means that the work performed by the Pechukas force dissipates the corresponding energy into the displacements of the molecules [66]. Third, note that the nonadiabatic transitions are promoted by the overlap between the nonadiabatic coupling vector and the vibrational modes. Figure 11.13a shows the overlap between the $\boldsymbol{d}_{DS0,LP}$, and the vibrational modes of tetracene molecules as a function of energies and total photonic weight at 1 fs in simulations of 512 tetracene molecules coupled to a single cavity mode. As can be seen from these simulations, photonic fractions develop efficiently in proportion to the magnitude of the overlap between the nonadiabatic vector and molecular vibrational modes. In addition, the energy at which the polariton population is enhanced coincides with the peak energy of the Raman vibration (Figure 11.13a). This suggests that the nonadiabatic transitions from the dark state manifold to the LP are mediated by Raman active vibrations.

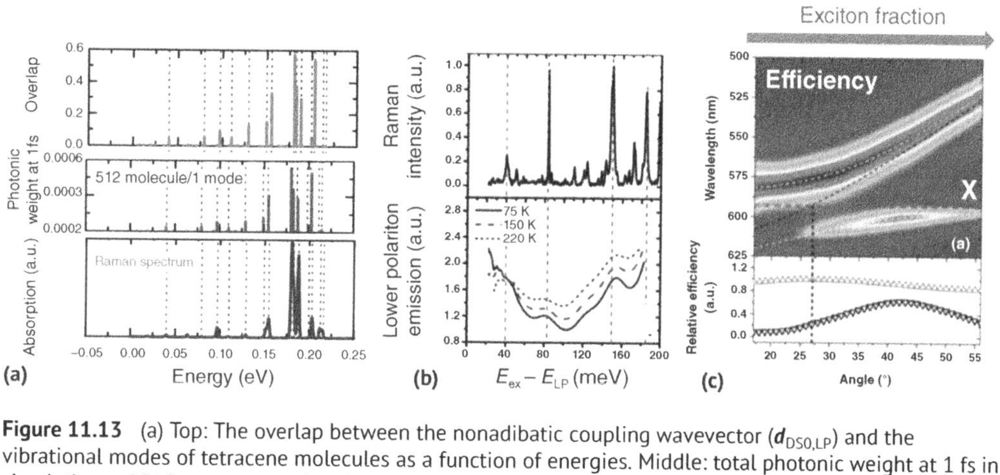

Figure 11.13 (a) Top: The overlap between the nonadibatic coupling wavevector ($\boldsymbol{d}_{DS0,LP}$) and the vibrational modes of tetracene molecules as a function of energies. Middle: total photonic weight at 1 fs in simulations of 512 tetracene molecules coupled to a single cavity mode. Bottom: Theoretically calculated Raman spectra for tetracene molecules. *Source*: Adapted from Ref. [63]. Copyright (2022) by the American Chemical Society. (b) Top: Raman spectrum of TDBC (5,6-dichloro-2-[[5,6-dichloro-1-ethyl-3-(4-sulphobutyl)-benzimidazol-2-ylidene]-propenyl]-1-ethyl-3-(4-sulphobutyl)- benzimidazolium hydroxide). Bottom: LPB emission as a function of energy separation between the exciton reservoir and the lower polaritons (LP). *Source*: Adapted from Ref. [45]. Copyright (2011) by John Wiley & Sons. (c) Polariton scattering efficiency into the LP states for the microcavity including TDBC. *Source*: Adapted from Ref. [44]. Copyright (2013) by the American Physical Society.

Signatures of this mechanism have experimentally been reported in a DBR microcavity containing TDBC *J*-aggregates in a PVA matrix [45]. Figure 11.13b (bottom) shows the peak PL emission intensity as a function of the energy separation between the exciton reservoir and the LP branch (E_{ex}–E_{LP}) for the temperature of 220, 150, and 75 K. The top panel shows the Raman spectrum of TDBC at room temperature, in which the Raman intensity spectrum of TDBC has several sharp Raman modes at energies of 40, 84, and 150 meV. The PL intensity on the LPB is enhanced at the energy point where E_{ex}–E_{LP} matches the energy of Raman vibration modes. This indicates that these Raman active molecular vibrations are populating LP states and play a role in the polariton relaxation mechanism.

The relaxation efficiency into the LP was mapped by dividing the photoluminescence signal by the absorption [44]. The efficiency map is shown in Figure 11.13c. As can be seen from the figure, the LP scattering efficiency decreases as the photon fraction of the polariton state increases. This enhancement of photon fraction suppresses their ability to scatter into LP states, indicating that the exciton component of the final state drives the scattering.

11.4 Comparison Between the Experimentally and Theoretically Estimated W_{ep} in BSBCz-EH System

To understand these polariton relaxation rates in more detail, we theoretically analyzed the polariton relaxation rate (W_{ep}) using the dipole fluctuation model and Mazza model (Section 11.3.2) [42, 43, 53]. Figure 11.14a shows the experimentally and theoretically estimated W_{ep} in the linear and nonlinear regime for several detuning parameters. For a large negative detuning regime above $\Delta = -356$ meV, the experimentally estimated W_{ep} shows good agreement with theoretical values estimated by the dipole fluctuation model, which indicates that the polaritons were indeed populated by the simple one-photon radiative pumping mechanism. In particular, the spectral density

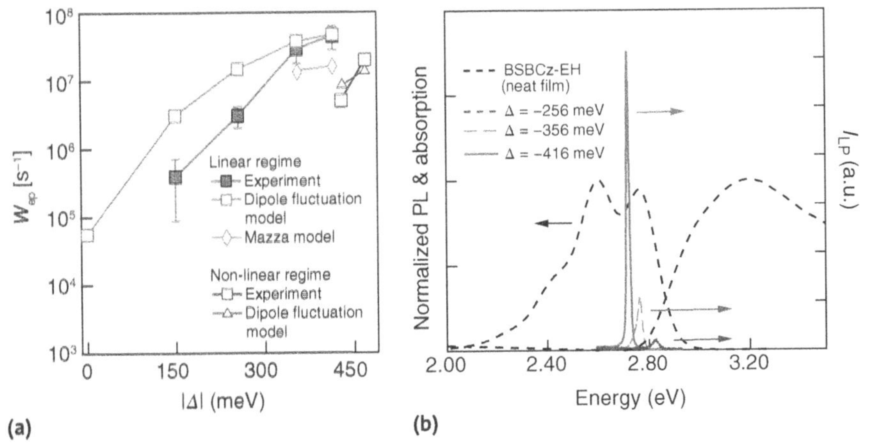

Figure 11.14 (a) The magnitude of detuning parameter dependence of the polariton relaxation rate (W_{ep}). show the experimentally estimated W_{ep} for the linear and nonlinear regimes. Show the theoretically estimated W_{ep} by the dipole fluctuation model and Mazza model [43, 53]. (b) Photoluminescence and absorption spectra of BSBCz-EH in the neat films (black dot lines) and PL spectra of LP at θ = 0 deg for Δ = −256, −356, and −416 meV. *Source*: Adapted from Ref. [53].

takes into account the 0–0, 0–1, and 0–2 vibrational modes, which suggests that all vibronic transitions contribute to this mechanism. Note that the theoretical orders of W_{ep} values estimated by the Mazza model are consistent with the experimental values, but the values are slightly smaller, suggesting that the 0–1 transition alone is difficult to explain the radiative pumping process. On the other hand, in the regime where the detuning is more positive ($\Delta = -256$ meV), the experimental and theoretical values of W_{ep} deviate significantly. As the experimental values are smaller than the theoretical values, this suggests the existence of a mechanism that reduces the efficacy of radiative pumping in this regime. In addition, we estimated the polariton relaxation rate for the microcavity with $\Delta = -150$ meV to compare the theoretical and experimental values in the more positive detuning regime. The results show that the difference between the theoretical and experimental W_{ep} of $\Delta = -150$ meV increases compared to that of $\Delta = -256$ meV, which indicates that an effect suppressing polariton relaxation becomes more pronounced with increasing detuning.

To understand the difference between the experimentally and theoretically estimated W_{ep}, we show the absorption and emission spectra for BSBCz-EH neat film, and the LP spectra for $\Delta = -256$, -356, and -416 meV (Figure 11.14b). For the large negative detuning regime, where the experimental and theoretical W_{ep} values are in good agreement, the LPs do not overlap with the absorption spectrum. On the other hand, for small negative detuning, where the difference between the theoretical and experimental W_{ep} is more pronounced, the LP spectrum largely overlaps with the absorption spectrum, indicating that relaxation of polaritons back to the DS and disorder might both contribute to reducing the efficiency of polariton relaxation [39, 53].

11.5 Conclusion

In this chapter, we described kinetic models for polariton relaxation in organic microcavities. Improving the polariton relaxation is essential to achieve room-temperature polariton BEC devices. For organic microcavities, radiative relaxation through luminescence, and nonradiative relaxation via the molecular vibrations play an active role in the polariton relaxation mechanism. For radiative pumping, the polariton relaxation rate can be obtained from the Rabi energy, the photon Hopfield coefficient, and the spectral overlap of the LPs with the emission spectrum. Thus, decreasing the intramolecular charge transfer property, increasing in-plane molecular orientation, fine-tuning of Stokes shift, and a high PLQY are necessary for fast polariton relaxation. On the other hand, for the nonradiative process, the polariton relaxation rate was calculated to be $W_{ep} = 10^{10}–10^{11}\ [s^{-1}]$ by Litinskaya in the low N regime [36]. This means that the polariton relaxation rate can be 1000 times faster than the currently reported relaxation rates ($W_{ep} = 10^5 - 10^7\ [s^{-1}]$) [20, 30, 31, 43, 53]. If structures showing high Q in the low N regime can be developed, further reduction of the condensation threshold is expected. We hope that the research areas of chemistry and physics will develop in a complementary manner in the future to solve the challenges of creating and controlling polariton BEC states.

References

1 Deng, H., Weihs, G., Santori, C., Bloch, J., Yamamoto, Y. Condensation of semiconductor microcavity exciton polaritons. *Science* **2002**, *298*(5591), 199–202.

2 Deng, H., Weihs, G., Snoke, D., Bloch, J., Yamamoto, Y. Polariton lasing vs. photon lasing in a semiconductor microcavity. *Proc. Natl. Acad. Sci. USA* **2003**, *100*(26), 15318–15323.

3 Kasprzak, J., Richard, M., Kundermann, S., Baas, A., Jeambrun, P., Keeling, J. M. J. et al. Bose–Einstein condensation of exciton polaritons. *Nature* **2006**, *443*(7110), 409–414.

4 Richard, M., Kasprzak, J., Romestain, R., André, R., Dang, L. S. Spontaneous coherent phase transition of polaritons in CdTe microcavities. *Phys. Rev. Lett.* **2005**, *94*(18), 187401.

5 Kéna-Cohen, S., Forrest, S. R. Room-temperature polariton lasing in an organic single-crystal microcavity. *Nat. Photonics* **2010**, *4*(6), 371–375.

6 Berloff, N. G., Silva, M., Kalinin, K., Askitopoulos, A., Töpfer, J. D., Cilibrizzi, P. et al. Realizing the classical XY Hamiltonian in polariton simulators. *Nat. Mater.* **2017**, *16*(11), 1120–1126.

7 Zasedatelev, A.V., Baranikov, A.V., Urbonas, D., Scafirimuto, F., Scherf, U., Stöferle, T. et al. A room-temperature organic polariton transistor. *Nat. Photonics* **2019**, *13*(6), 378–383.

8 Zasedatelev, A. V., Baranikov, A. V., Sannikov, D., Urbonas, D., Scafirimuto, F., Shishkov, V. Y. et al. Single-photon nonlinearity at room temperature. *Nature* **2021**, *597*(7877), 493–497.

9 Dusel, M., Betzold, S., Egorov, O. A., Klembt, S., Ohmer, J., Fischer, U. et al. Room temperature organic exciton–polariton condensate in a lattice. *Nat. Commun.* **2020**, *11*(1), 2863.

10 Wei, M., Verstraelen, W., Orfanakis, K., Ruseckas, A., Liew, T. C., Samuel, I. D. et al. Optically trapped room temperature polariton condensate in an organic semiconductor. *Nat. Commun.* **2022**, *13*(1), 7191.

11 Pannir-Sivajothi, S., Campos-Gonzalez-Angulo, J. A., Martínez-Martínez, L. A., Sinha, S., Yuen-Zhou, J. Driving chemical reactions with polariton condensates. *Nat. Commun.* **2022**, *13*(1), 1645.

12 Klembt, S., Harder, T. H., Egorov, O. A., Winkler, K., Ge, R., Bandres, M. A. Exciton-polariton topological insulator. *Nature* **2018**, *562*(7728), 552–556.

13 Harder, T. H., Sun, M., Egorov, O. A., Vakulchyk, I., Beierlein, J., Gagel, P. Coherent topological polariton laser. *ACS Photonics* **2021**, *8*(5), 1377–1384.

14 Ballarini, D., De Giorgi, M., Cancellieri, E., Houdré, R., Giacobino, E., Cingolani, R. et al. All-optical polariton transistor. *Nat. Commun.* **2013**, *4*(1), 1778.

15 Marsault, F., Nguyen, H. S., Tanese, D., Lemaître, A., Galopin, E., Sagnes, I. et al. Realization of an all optical exciton-polariton router. *Appl. Phys. Lett.* **2015**, *107*(20), 201115.

16 Graf, A., Held, M., Zakharko, Y., Tropf, L., Gather, M. C., Zaumseil, J. et al. Electrical pumping and tuning of exciton-polaritons in carbon nanotube microcavities. *Nat. Mater.* **2017**, *16*(9), 911–917.

17 Schneider, C., Rahimi-Iman, A., Kim, N. Y., Fischer, J., Savenko, I. G., Amthor, M. et al. An electrically pumped polariton laser. *Nature* **2013**, *497*(7449), 348–352.

18 Tischler, J. R., Bradley, M. S., Bulović, V., Song, J. H., Nurmikko, A. Strong coupling in a microcavity LED. *Phys. Rev. Lett.* **2005**, *95*(3), 036401.

19 Plumhof, J. D., Stöferle, T., Mai, L., Scherf, U., Mahrt, R. F. Room-temperature Bose–Einstein condensation of cavity exciton–polaritons in a polymer. *Nat. Mater.* **2014**, *13*(3), 247–252.

20 Daskalakis, K. S., Maier, S. A., Murray, R., Kéna-Cohen, S. Nonlinear interactions in an organic polariton condensate. *Nat. Mater.* **2014**, *13*(3), 271–278.

21 Rajendran, S. K., Wei, M., Ohadi, H., Ruseckas, A., Turnbull, G. A., Samuel, I. D. Low threshold polariton lasing from a solution-processed organic semiconductor in a planar microcavity. *Adv. Opt. Mater.* **2019**, *7*(12), 1801791.

22 Cookson, T., Georgiou, K., Zasedatelev, A., Grant, R. T., Virgili, T., Cavazzini, M. A yellow polariton condensate in a dye filled microcavity. *Adv. Opt. Mater.* **2017**, *5*(18), 1700203.

23 Dietrich, C. P., Steude, A., Tropf, L., Schubert, M., Kronenberg, N. M., Ostermann, K. et al. An exciton-polariton laser based on biologically produced fluorescent protein. *Sci. Adv.* **2016**, *2*(8), e1600666.

24 Wei, M., Rajendran, S. K., Ohadi, H., Tropf, L., Gather, M. C., Turnbull, G. A. Low-threshold polariton lasing in a highly disordered conjugated polymer. *Optica* **2019**, *6*(9), 1124–1129.

25 Putintsev, A., Zasedatelev, A., McGhee, K. E., Cookson, T., Georgiou, K., Sannikov, D. Nano-second exciton-polariton lasing in organic microcavities. *Appl. Phys. Lett.* **2020**, *117*(12), 123302.

26 Betzold, S., Dusel, M., Kyriienko, O., Dietrich, C. P., Klembt, S., Ohmer, J. Coherence and interaction in confined room-temperature polariton condensates with frenkel excitons. *ACS Photonics* **2020**, *7*(2), 384–392.

27 Yagafarov, T., Sannikov, D., Zasedatelev, A., Georgiou, K., Baranikov, A., Kyriienko, O. et al. Mechanisms of blueshifts in organic polariton condensates. *Commun. Phys.* **2020**, *3*, 18.

28 Sabatini, R. P., Maasoumi, F., Prasad, S. K., Zhang, B., Clark, C., Schmidt, T. W. et al. Organic polariton lasing with molecularly isolated perylene diimides. *Appl. Phys. Lett.* **2020**, *117*(4), 041103.

29 Sannikov, D., Yagafarov, T., Georgiou, K., Zasedatelev, A., Baranikov, A., Gai, L. et al. Room temperature broadband polariton lasing from a dye-filled microcavity. *Adv. Opt. Mater.* **2019**, *7*(17), 1900163.

30 Ishii, T., Bencheikh, F., Forget, S., Chénais, S., Heinrich, B., Kreher, D. et al. Enhanced light–matter interaction and polariton relaxation by the control of molecular orientation. *Adv. Opt. Mater.* **2021**, *9*(22), 2101048.

31 Ishii, T., Miyata, K., Mamada, M., Bencheikh, F., Mathevet, F., Onda, K. et al. Low-threshold exciton-polariton condensation via fast polariton relaxation in organic microcavities. *Adv. Opt. Mater.* **2022**, *10*(3), 2102034.

32 Wei, M., Ruseckas, A., Mai, V. T., Shukla, A., Allison, I., Lo, S. C. et al. Low threshold room temperature polariton lasing from fluorene-based oligomers. *Laser. Photonics Rev.* **2021**, *15*(8), 2100028.

33 Le Roux, F., Mischok, A., Bradley, D. D., Gather, M. C. Efficient anisotropic polariton lasing using molecular conformation and orientation in organic microcavities. *Adv. Funct. Mater.* **2022**, *32*(46), 2209241.

34 Sandanayaka, A. S., Matsushima, T., Bencheikh, F., Terakawa, S., Potscavage, W. J., Qin, C. et al. Indication of current-injection lasing from an organic semiconductor. *Appl. Phys. Exp.* **2019**, *12*(6), 061010.

35 Sandanayaka, A. S., Matsushima, T., Bencheikh, F., Yoshida, K., Inoue, M., Fujihara, T. et al. Toward continuous-wave operation of organic semiconductor lasers. *Sci. Adv.* **2017**, *3*(4), e1602570.

36 Litinskaya, M., Reineker, P., Agranovich, V. M. Fast polariton relaxation in strongly coupled organic microcavities. *J. Lumin.* **2004**, *110*(4), 364–372.

37 Wellnitz, D., Pupillo, G., Schachenmayer, J. Disorder enhanced vibrational entanglement and dynamics in polaritonic chemistry. *Commun. Phys.* **2022**, *5*(1), 120.

38 Pérez-Sánchez, J. B., Koner, A., Stern, N. P., Yuen-Zhou, J. Simulating molecular polaritons in the collective regime using few-molecule models. *Proc. Natl. Acad. Sci. USA* **2023**, *120*(15), e2219223120.

39 Agranovich, V. M., Litinskaia, M., Lidzey, D. G. Cavity polaritons in microcavities containing disordered organic semiconductors. *Phys. Rev. B* **2003**, *67*(8), 085311.

40 Fontanesi, L., Mazza, L., La Rocca, G. C. Organic-based microcavities with vibronic progressions: linear spectroscopy. *Phys. Rev. B* **2009**, *80*(23), 235313.

41 Lidzey, D. G., Fox, A. M., Rahn, M. D., Skolnick, M. S., Agranovich, V. M., Walker, S. Experimental study of light emission from strongly coupled organic semiconductor microcavities following nonresonant laser excitation. *Phys. Rev. B* **2002**, *65*(19), 195312.

42 Mazza, L., Fontanesi, L., La Rocca, G. C. Organic-based microcavities with vibronic progressions: photoluminescence. *Phys. Rev. B.* **2009**, *80*(23), 235314.

43 Mazza, L., Kéna-Cohen, S., Michetti, P., La Rocca, G. C. Microscopic theory of polariton lasing via vibronically assisted scattering. *Phys. Rev. B* **2013**, *88*(7), 075321.

44 Coles, D. M., Grant, R. T., Lidzey, D. G., Clark, C., Lagoudakis, P. G. Imaging the polariton relaxation bottleneck in strongly coupled organic semiconductor microcavities. *Phys. Rev. B* **2013**, *88*(12), 121303.

45 Coles, D. M., Michetti, P., Clark, C., Tsoi, W. C., Adawi, A. M., Kim, J. S. et al. Vibrationally assisted polariton-relaxation processes in strongly coupled organic-semiconductor microcavities. *Adv. Funct. Mater.* **2011**, *21*(19), 3691–3696.

46 Zakharko, Y., Rother, M., Graf, A., Hähnlein, B., Brohmann, M., Pezoldt, J. et al. Radiative pumping and propagation of plexcitons in diffractive plasmonic crystals. *Nano. Lett.* **2018**, *18*(8), 4927–4933.

47 Lüttgens, J. M., Berger, F. J., Zaumseil, J. Population of exciton–polaritons via luminescent sp3 defects in single-walled carbon nanotubes. *ACS Photonics* **2021**, *8*(1), 182–193.

48 Grant, R.T., Michetti, P., Musser, A. J., Gregoire, P., Virgili, T., Vella, E. et al. Efficient radiative pumping of polaritons in a strongly coupled microcavity by a fluorescent molecular dye. *Adv. Opt. Mater.* **2016**, *4*(10), 1615–1623.

49 Deng, H., Haug, H., Yamamoto, Y. Exciton-polariton Bose-Einstein condensation. *Rev. Mod. Phys.* **2010**, *82*(2), 1489–1537.

50 Tassone, F., Piermarocchi, C., Savona, V., Quattropani, A., Schwendimann, P. Bottleneck effects in the relaxation and photoluminescence of microcavity polaritons. *Phys. Rev. B* **1997**, *56*(12), 7554–7563.

51 Le Roux, F., Bradley, D. D. C. Conformational control of exciton-polariton physics in metal-poly(9,9-dioctylfluorene)-metal cavities. *Phys. Rev. B* **2018**, *98*(19), 195306.

52 Le Roux, F., Taylor, R. A., Bradley, D. D. C. Enhanced and polarization-dependent coupling for photoaligned liquid crystalline conjugated polymer microcavities. *ACS Photonics* **2020**, *7*(3), 746–758.

53 Ishii, T., Engineering of Exciton-Polariton and Bose-Einstein Condensate in Organic Microcavities. Ph.D. thesis, Kyushu University, 2021.

54 Mamada, M., Fukunaga, T., Bencheikh, F., Sandanayaka, A. S., Adachi, C. Low amplified spontaneous emission threshold from organic dyes based on bis-stilbene. *Adv. Funct. Mater.* **2018**, *28*(32), 1802130.

55 Nakanotani, H., Adachi, C., Watanabe, S., Katoh, R. Spectrally narrow emission from organic films under continuous-wave excitation. *Appl. Phys. Lett.* **2007**, *90*(23), 231109.

56 Oyama, Y., Mamada, M., Shukla, A., Moore, E. G., Lo, S. C., Namdas, E. B. et al. Design strategy for robust organic semiconductor laser dyes. *ACS Mater. Lett.* **2020**, *2*(2), 161–167.

57 Lee, S. D., Sarmadi, M., Denes, F., Shohet, J. L. Surface modification of polypropylene under argon and oxygen-RF-plasma conditions. *Plasma. Pol.* **1997**, *2*(3), 177–198.

58 Fujimoto, H., Miyayama, T., Sanada, N., Adachi, C. Plasma-tolerant structure for organic light-emitting diodes with aluminum cathodes fabricated by DC magnetron sputtering: using a Li-doped electron transport layer. *Org. Electron.* **2013**, *14*(11), 2994–2999.

59 Sanvitto, D., Kéna-Cohen, S. The road towards polaritonic devices. *Nat. Mater.* **2016**, *15*(10), 1061–1073.

60 Kéna-Cohen, S. Strong exciton-photon coupling in organic semiconductor microcavities. Ph.D. thesis, Princeton University, 2010.

61 Kéna-Cohen, S., Davanço, M., Forrest, S. R. Strong exciton-photon coupling in an organic single crystal microcavity. *Phys. Rev. Lett.* **2008**, *101*(11), 116401.

62 Zoubi, H., La Rocca, G. C. Microscopic theory of anisotropic organic cavity exciton polaritons. *Phys. Rev. B.* **2005**, *71*(23), 235316.

63 Tichauer, R. H., Morozov, D., Sokolovskii, I., Toppari, J. J., Groenhof, G. Identifying vibrations that control non-adiabatic relaxation of polaritons in strongly coupled molecule–cavity systems. *J. Phys. Chem. Lett.* **2022**, *13*(27), 6259–6267.

64 Eizner, E., Martínez-Martínez, L. A., Yuen-Zhou, J., Kéna-Cohen, S. Inverting singlet and triplet excited states using strong light-matter coupling. *Sci. Adv.* **2019**, *5*(12), eaax4482.

65 Martínez-Martínez, L. A., Eizner, E., Kéna-Cohen, S., Yuen-Zhou, J. Triplet harvesting in the polaritonic regime: a variational polaron approach. *J. Chem. Phys.* **2019**, *151*(5), 054106.

66 Schwartz, T., Hutchison, J. A., Léonard, J., Genet, C., Haacke, S., Ebbesen, T. W. Polariton dynamics under strong light–molecule coupling. *ChemPhysChem* **2013**, *14*(1), 125–131.

12

Reactions and Assembly Under Vibrational Strong Coupling

Kenji Hirai[1] *and Hirohi Uji-i*[1,2]

[1] *Research Institute for Electronic Science, Hokkaido University, Sapporo, Hokkaido, Japan*
[2] *Department of Chemistry, KU Leuven, Heverlee, Belgium*

12.1 Introduction

A molecule absorbs a photon to be transited to the excited state. Subsequently, the molecule emits a photon and returns to its ground state, which is called spontaneous emission. Spontaneous emission seems to be an intrinsic property of molecules; however, this is explained by the interaction of the emitter (molecule at the excited state) and electromagnetic mode in the vacuum state, which is the lowest energy state. Free space is full of electromagnetic modes that can interact with molecules. The interaction between the molecules and electromagnetic mode is significantly changed by the use of an optical cavity. Ideally, the optical cavity possesses only electromagnetic modes that meet the resonant conditions, that is, the so-called cavity mode. Because of the defined electromagnetic modes in an optical cavity, the interaction between the electromagnetic mode and molecule is largely different from the interaction in free space. In contrast to free space, molecules in an optical cavity can interact with the modes existing in the optical cavity.

The interaction of electromagnetic mode and the molecule and is often described by three parameters (Figure 12.1, κ: rate of resonant decay, γ: rate of nonresonant decay, g: strength of light–matter coupling). Generally, γ is the inverse of lifetime of the emission rate or excited state in free space and κ is the inverse of the lifetime of the photon in an empty cavity. When $g \ll k$ or γ, the interaction of molecules and electromagnetic mode is in the regime of weak coupling. In the regime of weak coupling, spontaneous emission is enhanced, which is called the Purcell effect. The control of spontaneous emission by the Purcell effect was proposed in 1946 [1]. On the other hand, when $g \ll k$ and γ, the interaction of molecules and electromagnetic mode is in the regime of strong coupling [2–4]. This condition implies that the rate of energy exchange between the molecules and cavity photons is larger than that of molecular dephasing and photon leakage. In the regime of strong coupling, electromagnetic mode and matter can interact coherently to form polaritonic states, that is, the hybridized state of photons and matter. The polaritonic states are split into upper and lower branches, which is called the Rabi splitting. Strong coupling not only changes the radiative decay

Polariton Chemistry: Molecules in Cavities, First Edition. Edited by Joel Yuen-Zhou, Noel C. Giebink and Raphael F. Ribeiro. Published 2026 by John Wiley & Sons, Inc.

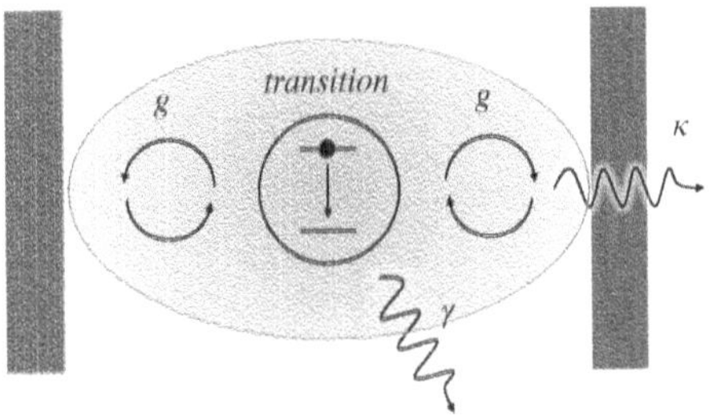

Figure 12.1 Schematic illustration of the interaction between electromagnetic mode and matter. κ: rate of resonant decay, γ: rate of nonresonant decay, g: strength of light–matter coupling.

rates but also forms new energy levels. The superstrong coupling region refers to when the coupling exceeds approximately 10% of the resonance frequency, while the deep strong coupling region occurs when the coupling surpasses the resonance frequency itself.

Initial studies on strong coupling of optical cavities primarily explored exciton and photon interactions of inorganic structures. A notable example is the observation of polaritonic states when GaAs was introduced into an optical cavity [5]. Later, strong coupling was investigated for coordination compound with organic ligands (tetra-(2,6-*t*-butyl)phenol-porphyrin zinc) placed in an optical microcavity [6]. More recently, strong coupling has been demonstrated not only for electron transitions but also for molecular vibration. The phenomenon, involving the coupling between a molecular vibrational mode and a cavity mode, is called vibrational strong coupling (VSC) [7]. In 2015, the VSC of a C=O stretching vibration was investigated for polyvinyl acetate placed in a cavity [8]. This coupling of the C=O stretching vibration to the cavity mode resulted in the generation of vibropolaritonic states. Subsequently, it was experimentally shown that the reactivity of an organic silicon molecule changed under the VSC of the silyl group [9, 10]. A wide variety of chemical reactions, including organic reactions [9–19] and enzymatic reactions [20–23], have been modulated using the VSC of functional groups related to the reaction of interest. Site-selective reactions can be achieved by altering the reactivity of specific functional groups. Most recently, molecular self-assembly [24–28] has also been manipulated under VSC. Chemistry under VSC has attracted attention because this phenomenon is of fundamental scientific interest and has a variety of potential technological applications [29–34]. In this chapter, we introduce a series of experimental reports on chemical reactions under VSC.

12.2 Vibrational Strong Coupling

A Fabry–Pérot (FP) cavity, in which two reflective mirrors face each other, is widely used for VSC experiments. The reflective mirrors are prepared by sputtering a metal (e.g., Au) on a mid-infrared-transparent substrate (e.g., CaF_2, BaF_2, and ZnSe). To avoid direct contact between molecules in the FP cavity and the metal surface, the metal layer is often coated with an inert layer (e.g., SiO_2 and resins). The two mirrors are arranged in parallel with a spacer consisting of a commercially available film (e.g., Mylar and Kapton) between them. Typically, a solution is placed between the reflective mirrors and the mirror distance is tuned to couple the cavity mode to the molecular vibrational mode of interest (Figure 12.2). The formation of vibropolaritonic states can be detected using infrared (IR) spectroscopy (Figure 12.3). The vibropolaritonic states are split into upper and lower branches, called Rabi splitting, and the Rabi splitting energy ($\hbar\Omega_R$) is proportional to the strength of the interaction. As a number of molecules can collectively interact with a single cavity mode,

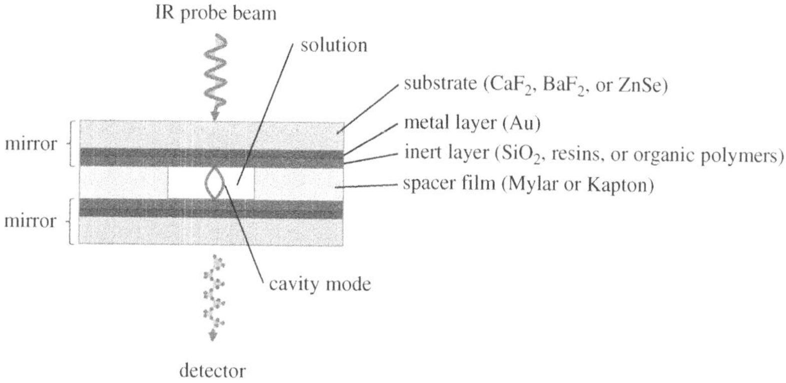

Figure 12.2 Schematic illustration of a Fabry–Pérot cavity. The reflective mirrors are prepared by metal sputtering on the substrates, followed by coating of the inert layers. The formation of vibropolaritonic states is observed by infrared spectroscopy.

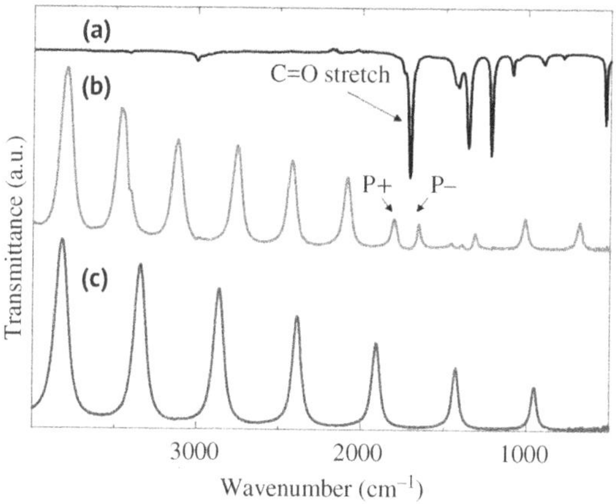

Figure 12.3 (a) Infrared (IR) absorbance of acetone. The vibrational mode of C=O stretch is observed at 1715 cm^{-1}. (b) IR transmittance of acetone in a FP cavity. The periodic peaks indicate the cavity modes. The C=O stretch and fifth cavity mode are coupled at 1715 cm^{-1}, forming polaritonic states. The upper polariton (P+) and lower polariton (P−) branches are observed. The energy gap between P+ and P− is vacuum Rabi splitting. (c) IR transmittance of an empty FP cavity. The periodic peaks indicate the cavity modes.

the Rabi splitting energy is also proportional to the square root of the number of molecules in the FP cavity ($\hbar\Omega_R) \propto \sqrt{N/V}$, where V is the mode volume and N is the number of molecules). This collective coupling results in $N-1$ dark states (DS) in addition to the two vibropolaritonic states, P+ and P− (Figure 12.4). VSC has been investigated in various types of molecules, such as polymers [35], discrete organic molecules [36], organometallic complexes [37, 38], proteins [39], liquid crystals [40], and solvents [41].

The absorption band of polaritonic states is typically sharp, even when the absorption band of bare molecules is broad [42–44]. One example of broad absorbance is the hydroxyl stretch of water molecules. When the absorption of bare molecules is broad, the energy distribution of $N-1$ DS can be broad under a strong coupling. In other words, the broad absorbance does not form the broad absorbance of polaritonic states. It is important to also mention that not entire molecules in the FP cavity are in a strong coupling regime, especially for liquid samples. When the transition dipole of

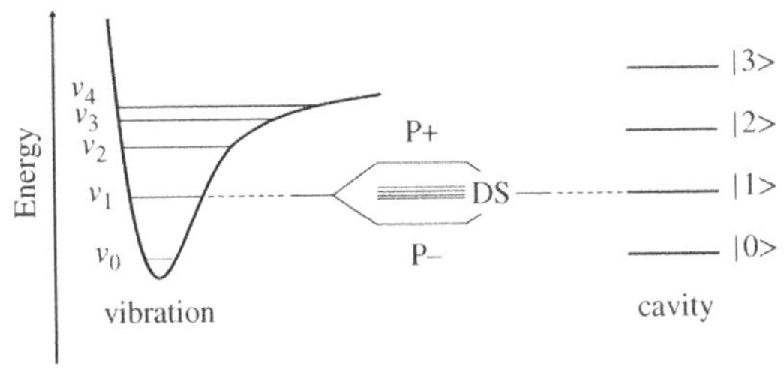

Figure 12.4 Energy diagram of vibropolaritonic states generated through the strong coupling between a cavity mode and a vibrational transition. The gap between the upper polariton (P+) and lower polariton (P–) branches is called the Rabi splitting (DS: dark state).

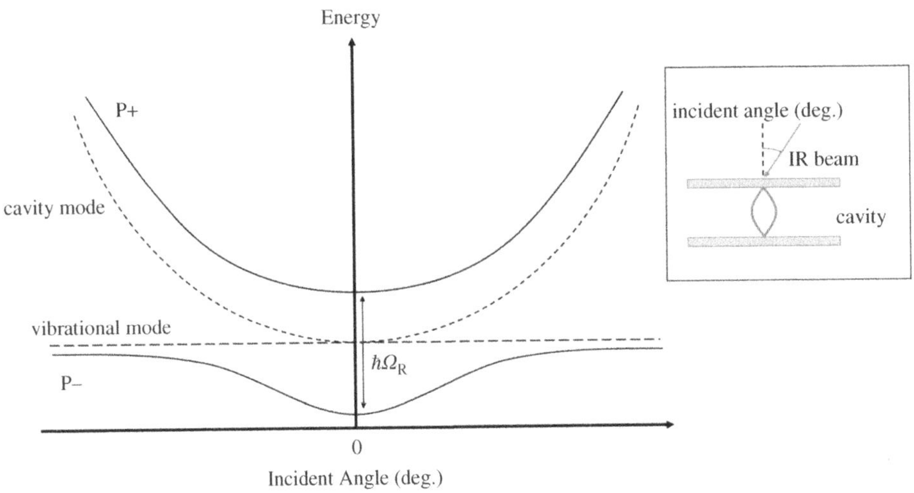

Figure 12.5 Angular dispersion of polaritonic states. The dashed horizontal and parabolic lines represent the molecular vibrational energy and the dispersive photonic mode, respectively. The black solid lines denote the upper (P+) and lower (P–) polaritons generated by vibrational strong coupling (VSC). The gap between P+ and P– at an incident angle at 0° corresponds to the Rabi splitting ($h\Omega_R$). The inset image is an illustration of an FP cavity for VSC. The normal incidence is defined as 0°.

molecules is perpendicular to the amplitude direction of the electronic field, the molecules cannot interact with the photons. The direction of the molecules is not fixed in the liquid samples. A small number of molecules are not coupled to the cavity mode because of the direction of the transition dipole. Furthermore, there are nodes of electric waves where the intensity of the electric field is zero. The molecules around the node of the electric wave do not sense an electric field, indicating that there is no interaction with light [45, 46].

One method for analyzing VSC states is to estimate the angular dispersion. In a typical experiment under VSC, the IR probe beam is perpendicular to the mirror surfaces. As the incident angle is increased from 0° (perpendicular to the mirror surfaces), the energy of the cavity mode increases. In Figure 12.5, the parabolic dashed line indicates the energy of the cavity mode. The energy of the cavity mode is tuned to couple with the vibrational mode (horizontal dashed line, Figure 12.5) at normal incidence (0°). The Rabi splitting is defined as the minimum energy gap between the newly formed polaritonic states, specifically the energy gap ($h\Omega_R$) between P+ (upper solid line) and P– (lower solid line) at normal incidence, as shown in Figure 12.5. At higher incident angles – when the angle is tuned away from resonance – P+ and P– become increasingly photon-like and molecule-like, respectively. These distinctive angular dispersion behaviors serve as clear evidence of VSC.

12.3 Chemical Reactions Under VSC

The VSC of functional groups involved in a chemical reaction can accelerate or decelerate the reaction rate. Additionally, the VSC of solvent molecules can also influence the chemical reaction. In this section, we introduce various examples of chemical reactions influenced by VSC. VSC effects have been observed in various types of reactions, including organic reactions, enzymatic reactions, and self-assembly, as summarized in Tables 12.1 and 12.2.

Table 12.1 Chemical reactions under vibrational strong coupling (VSC).

No.	Reaction	Vibration	Effect
	Organic reactions		
1	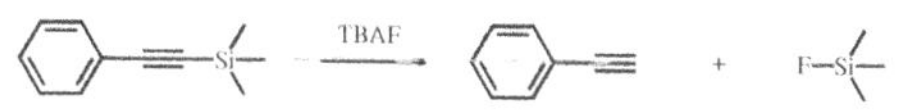	Si—C stretch	Decelerated
2	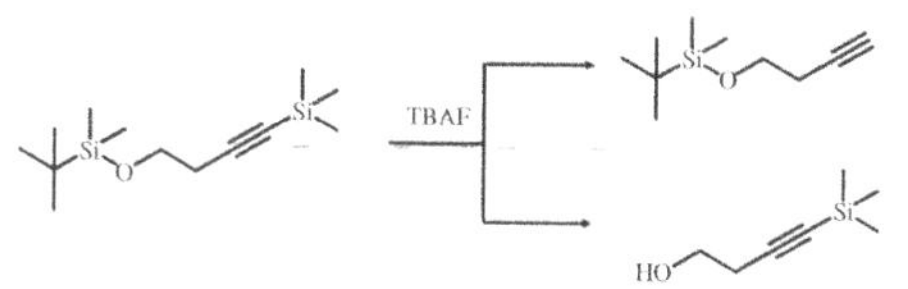	Si—CH_3 stretch Si—C stretch Si—O stretch	Site-selective
3	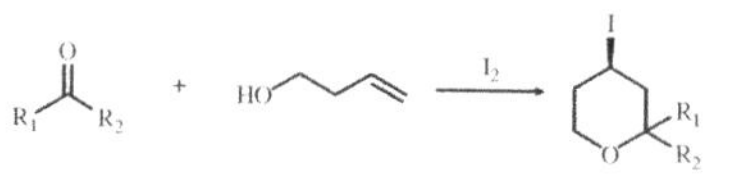	C=O stretch	Decelerated
4	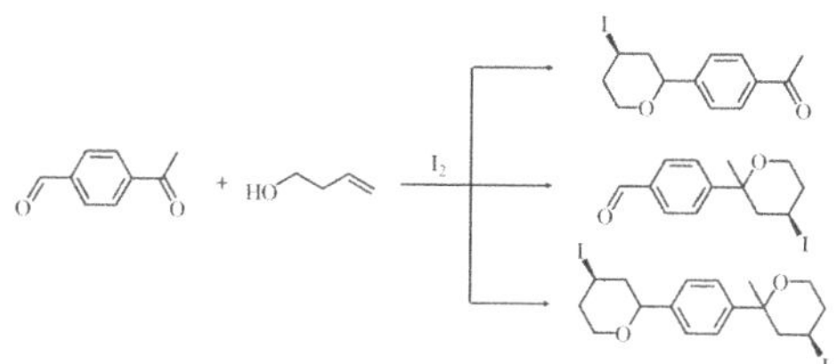	C=O stretch	Site-selective
5	$2H_2O + OCN^- \rightarrow CO_3^{2-} + NH_4^+$ $2D_2O + OCN^- \rightarrow CO_3^{2-} + ND_4^+$ $2H_2O + NH_3BH_3 \rightarrow NH_4^+ + BO_2^- + 3H_2$	O—H stretch O—D stretch	Accelerated
6	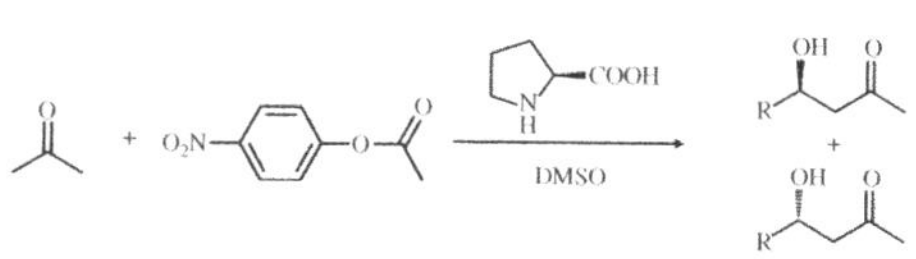	C=O stretch	Accelerated No change of *ee*

(Continued)

Table 12.1 (Continued)

No.	Reaction	Vibration	Effect
	Organic reactions under cooperative VSC		
7	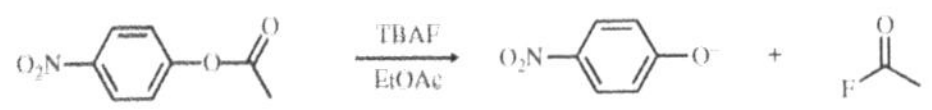	C=O stretch	Accelerated
8	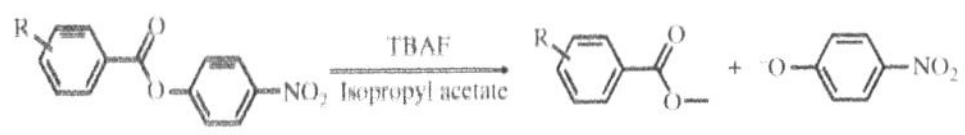	C=O stretch	Accelerated
9	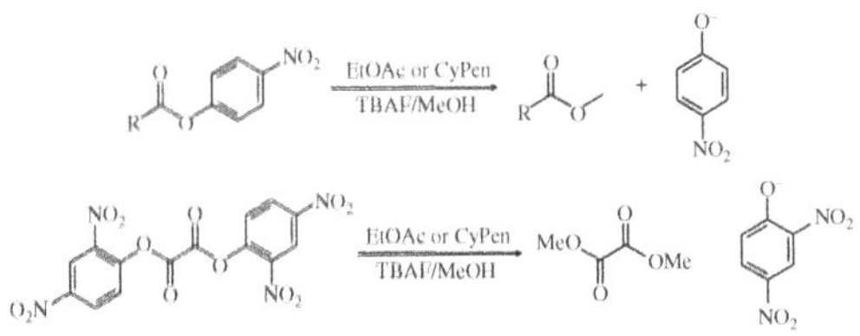	C=O stretch	Accelerated
10		NCO stretch C=O stretch C—H vibration	Decelerated
	Enzymatic reactions		
11	$R_1NHCOR_2 \rightarrow R_1NH_3^+ + R_2COOH$	O—H stretch	Decelerated
12	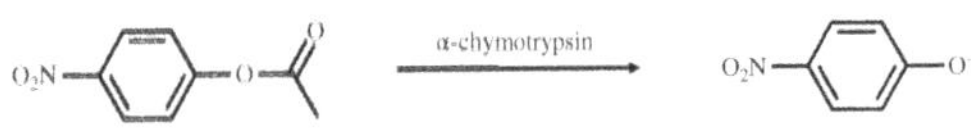	O—H stretch N—H stretch	Accelerated
13	ATP $\xrightarrow[H_2O]{ATPase}$ ADP	O—H stretch	Accelerated
14	Sucrose $\xrightarrow[H_2O]{Sucrase}$ α-glucose + β-fructose	O—H stretch	Accelerated
	Symmetry of molecular vibrations		
15		C=C stretch C—H stretch	Modulation of K_{eq}
16	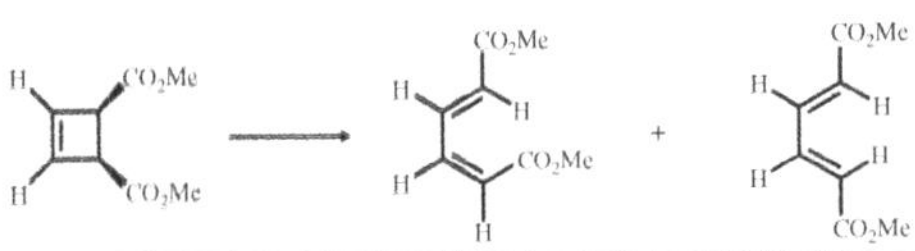	C=O stretch CH_3 bend	Stereoselective

12.3.1 Organic Reactions

12.3.1.1 Organic Reactions Under VSC

For organic reactions, the VSC effect was first demonstrated for the deprotection of an alkynyl silane (No. 1, Table 12.1) [9]. During the nucleophilic substitution reaction in 1-phenyl-2-trimethylsilylacetylene (PTA) and tetra-*n*-butylammonium fluoride (TBAF), F^- from TBAF attacks Si of PTA and the Si—C bond simultaneously dissociates. The introduction of the solution into an FP cavity enabled the VSC state of the Si—C bond in PTA (Figure 12.6a). Under the VSC of the Si—C vibration, the reaction rate was reduced to one-fifth of its original value. It is noted that the cavity mode was adjusted to match the absorption peak of the Si—C vibrational mode, the

Table 12.2 Self-assembly and crystallization modulated by vibrational strong coupling.

No.	Reaction	Vibration
1	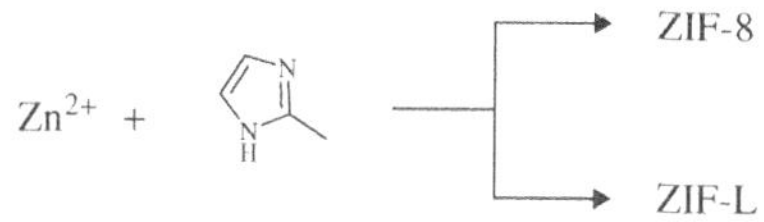	O—H stretch
2	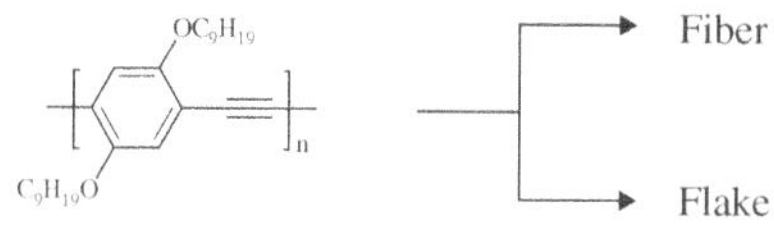	C=C stretch
3	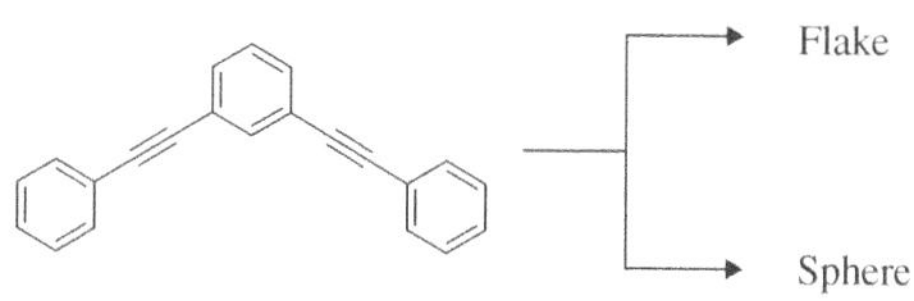	C=C stretch
4	DNA origami	Pyridine
5	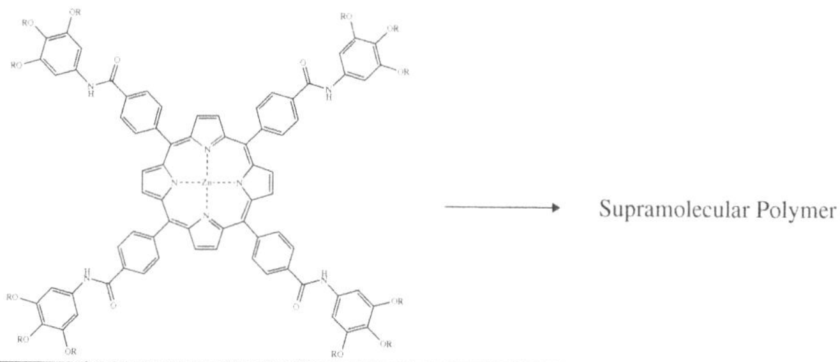	C—H stretch

reaction rate showed the lowest values (Figure 12.6b). Eyring plot analysis revealed that the activation enthalpy ($\Delta H^{\ddagger}$) increased under VSC and that the activation entropy ($\Delta S^{\ddagger}$) was positive (Eq. 12.1, Figure 12.7). The deprotection of alkynyl silanes by TBAF typically proceeds via an S_N2 mechanism, in which the activated state is a five-coordinate intermediate formed by fluoride attack on the silicon atom. Thus, for this S_N2 reaction, $\Delta S^{\ddagger}$ is usually negative. For the reaction under VSC, the sign inversion of $\Delta S^{\ddagger}$ from positive to negative suggested that the modified Si—C reactivity caused a change from an S_N2 mechanism to an S_N1 mechanism. This inhibition of the dissociation reaction for alkynyl silanes was further extended to the site-selective reactions [11].

A site-selective reaction was demonstrated with *tert*-butyldimethyl{[4-(trimethylsilyl)but-3-yn-1-yl]oxy}silane, which contains two silyl bonds (Si—C and Si—O). Both the Si—C and Si—O bonds react with TBAF (No. 2, Table 12.1), with dissociation of the Si—C bond forming (but-3-yn-1-yloxy)(*tert*-butyl)dimethylsilane and dissociation of the Si—O bond forming 4-(trimethylsilyl)but-3-yn-1-ol. Under VSC, the dissociation of the Si—C bond was inhibited, resulting in an increased formation ratio of 4-(trimethylsilyl)but-3-yn-1-ol.

$$\ln \frac{k}{T} = -\frac{\Delta H^{\ddagger}}{R}\frac{1}{T} + \ln \frac{k_B}{h} + \frac{\Delta S^{\ddagger}}{R} \tag{12.1}$$

k: reaction constant, $\Delta H^{\ddagger}$: activation enthalpy, $\Delta S^{\ddagger}$: activation enthalpy, k_B: Boltzmann constant, T: temperature, R: gas constant, h: Planck constant. The plot of k against $1/T$ gives $\Delta H^{\ddagger}$ in the gradient and $\Delta S^{\ddagger}$ in the intersection.

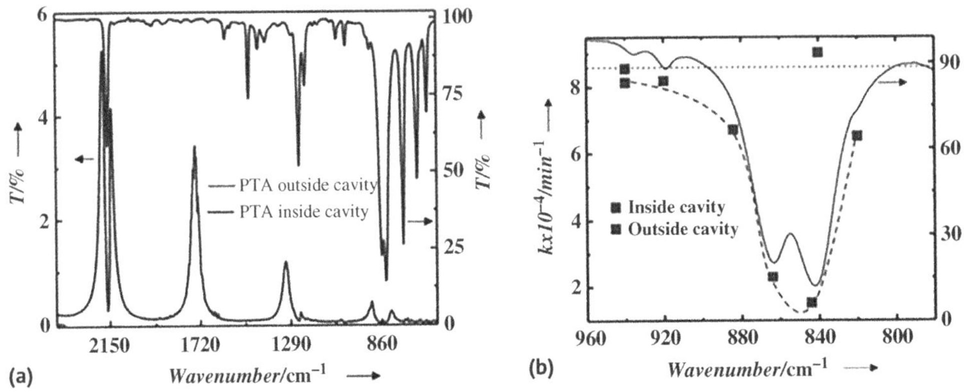

Figure 12.6 (a) IR transmission spectrum of the cavity including 1-phenyl-2-trimethylsilylacetylene (PTA) (bottom). IR transmission spectrum of bare PTA (top). Rabi splitting is observed around 850 cm^{-1} in PTA in the cavity. (b) IR transmission spectrum of the Si—C mode of PTA. The reaction rate depends on cavity tuning for reactions in (squares guided with dotted line) and outside (squares without guided line) the cavity. The reaction rate is reduced when the cavity mode is adjusted to the absorption peak of the Si—C modes. *Source*: Reproduced from Ref. [9] with permission from Wiley-VCH.

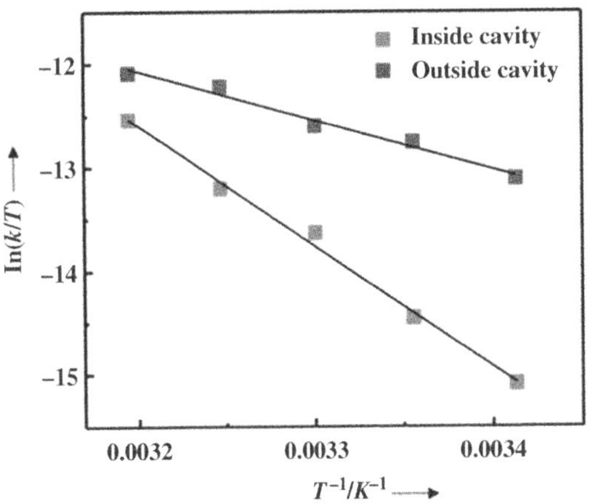

Figure 12.7 Eyring plot of reaction under VSC (diamonds) and noncavity conditions (squares). The gradient and intercept of fitted lines give $\Delta H^{\ddagger}$ and $\Delta S^{\ddagger}$, as shown in Eq. 12.1. *Source*: Reproduced from Ref. [9] with permission from Wiley-VCH.

VSC has also been found to impact various organic reactions. Prins cyclization is an important synthetic reaction for producing cyclic molecules. For this reaction, the hydroxyl group of an alcohol attacks the carbonyl carbon of an aldehyde or ketone, leading to the cyclization reaction. Under the VSC of the C=O stretching vibration, the reaction rate of Prins cyclization was reduced (No. 3, Table 12.1) [12]. Furthermore, under VSC, $\Delta H^{\ddagger}$ increased by approximately 10 kJ mol^{-1}, whereas $\Delta S^{\ddagger}$ remained almost unchanged. The rate-limiting step in Prins cyclization is carbonyl group protonation and oxygen attack on the carbonyl carbon. The modulation of Prins cyclization under VSC was extended to the site-selective reactions. 4-Acetylbenzaldehyde, which contains both ketone and aldehyde groups, can participate in Prins cyclization through either functional group (No. 4, Table 12.1) [13]. Due to the higher reactivity of the aldehyde group, 60% of the reacted carbonyl groups (aldehyde and ketone combined) originated from the aldehyde in the absence of VSC. However, under VSC of the ketone group, the reactivity of the ketone was inhibited, resulting in an increase in the proportion of reacted aldehyde group to 88%. This result indicates that suppressing the reactivity of ketone groups under VSC can promote the site-selective reactions of aldehyde groups. It is noted that this site-selective reaction of Prins cyclization was realized by the FP cavity with autotuning functions. Generally, the refractive indices of the solution are changed as the reaction progresses. The change of refractive indices of solution shifts the cavity mode, leading to the detuning of cavity mode to the target vibrational mode. The piezoelectric actuators are installed in the FP cavity, which is controlled by the computer program. The contraction/expansion of piezoelectric actuators slightly changes the distance of mirrors and cancels the shift of cavity mode caused by the change of refractive indices.

Although both Prins cyclization and deprotection of alkynyl silane were disactivated by VSC, the acceleration of reactions by VSC has also been reported. The hydrolysis of cyanate ions was accelerated under the VSC of the O—H stretching mode of water molecules (No. 5, Table 12.1) [14]. In this hydrolysis reaction, water molecules act as both a reactant and a solvent. VSC might influence both the cleavage of the O—H bond and the hydration properties. Owing to the potential impact of VSC on intermolecular interactions, understanding the effect of the VSC of the O—H stretching mode of water molecules is more complicated.

The effect of VSC was also observed in enantioselective reactions. Aldol reaction catalyzed by L-proline gives chiral organic molecules. The direct asymmetric aldol reactions of 4-nitrobenzaldehyde in acetone catalyzed by were carried out under VSC of C=O stretch (No. 6, Table 12.1) [15]. After the reaction, the reaction mixture was extracted from the cavity, and conversion yields and enantiomeric excess were quantified using chiral high-performance liquid chromatography and nuclear magnetic resonance and under VSC of C=O stretch, conversion yields can increase up to 90% without affecting enantiomeric excess.

12.3.1.2 Organic Reactions Under Cooperative VSC

As explained in Section 12.2, Rabi splitting energy increases in proportion to the square root of the number of molecules. Consequently, it is challenging to achieve a VSC state in a dilute solution containing a small number of molecules. Therefore, cooperative VSC has been investigated, where multiple types of molecules with similar vibrational energy simultaneously couple to the cavity mode. Particularly, when the solvent and reactants have similar vibrational energies, it is possible to create a cooperative VSC state at low concentration of reactants because large amount of solvent molecules are present. The cooperative VSC effect has been also reported for the solvolysis reaction of *p*-nitrophenyl acetate (PNPA). When the C=O stretching modes of ethyl acetate as the solvent and PNPA as the reactant were in a cooperative VSC state, the rate of the solvolysis reaction was increased. Under the VSC of the C=O stretching modes, both $\Delta H^{\ddagger}$ and $\Delta S^{\ddagger}$ decreased (No. 7, Table 12.1) [16]. The decrease in $\Delta S^{\ddagger}$ might be attributable to a change in solvation if VSC can influence intermolecular interactions. Both the reactivity of the carbonyl group in PNPA and the solvation properties of ethyl acetate could influence the reaction kinetics. Therefore, in the future, it is necessary to obtain an understanding of the effect of VSC on solvation properties.

The effect of cooperative VSC was also examined in the reaction of *p*-nitrophenyl benzoate (or its derivatives) with isopropyl acetate. A cooperative VSC state was attained by coupling the C=O stretching modes of both the *p*-nitrophenyl benzoate derivative (the reactant) and isopropyl acetate (the solvent) to the cavity mode. The cooperative VSC increased the reaction rates of *p*-nitrophenyl benzoate derivatives; however, they did not match the Hammett relation. In the Hammett relation, the reaction rate is interpreted based on the substituent constant, which reflects the electron-donating and electron-withdrawing properties. The observed deviation from the Hammett relation may originate from the impact of VSC on both intermolecular and intramolecular interactions (No. 8, Table 12.1) [17].

The effect of solvents on ester solvolysis reactions under cooperative VSC was investigated by coupling the C=O stretching bands of both the reactant and solvent molecules to a cavity mode. Three reactants, *p*-nitrophenylacetate, 3-methyl-para-nitrophenylbenzoate, and bis-(2, 4-dinitrophenyl) oxalate, were studied in two solvents: ethyl acetate and cyclopentanone. The influence of VSC on solvolysis reaction rates was examined. Both solvents enhanced the reaction rates of *p*-nitrophenylacetate and 3-methyl-*p*-nitrophenylbenzoate under VSC. However, the resonance effect occurred at different temperatures for each solvent. Bis-(2, 4-dinitrophenyl) oxalate did not respond to VSC in either solvent system due to poor overlap of the reactant and solvent C=O vibrational bands. As observed in other reports of chemical reactions under VSC, the effect of VSC is maximized when the cavity mode is adjusted to the absorption peak. This resonance effect was observed at different temperatures for different solvents (ethyl acetate and cyclopentanone). The different behavior of solvent molecules under VSC suggests that solvent coordination in the transition state may play a significant role in controlling the reaction mechanism. Cooperative VSC between the solvent and the reactant molecules may better order the vibrational dipoles, similar to the dipole field effect, stabilizing the transition state (No. 9, Table 12.1) [18].

The modulated reaction rates of the alcoholysis of phenyl isocyanate and cyclohexanol were also observed. The NCO stretch of phenyl isocyanate (the reactant), the C=O stretch of cyclohexyl carbamate (the product), and the C—H stretching modes of tetrahydrofuran (THF, the solvent) and cyclohexanol (the reactant) exhibited sufficiently strong absorbance for VSC. The effects of VSC on these three vibrational modes were examined. The reaction rate decreased fivefold for cavities adjusted to the NCO stretch and was reduced by half for cavities tuned to the C=O stretch and C—H strecthing modes (No. 10, Table 12.1) [19].

12.3.2 Enzymatic Reactions Under VSC

VSC has also been utilized in enzymatic reactions. Pepsin, a stomach enzyme that facilitates the digestion of proteins from food intake, showed a deceleration in the reaction rate of peptide hydrolysis when the O—H stretching modes of both water and pepsin were in the cooperative VSC state (No. 11, Table 12.1) [20]. Since the active site of pepsin has hydroxyl groups, the enzymatic activity was changed by the VSC of the O—H stretching mode. It can also be assumed that changes in hydration induced a structural rearrangement of pepsin, thus affecting the hydrolysis activity. The VSC effect has also been demonstrated in the enzymatic reaction of α-chymotrypsin, a pancreatic enzyme responsible for the digestion of proteins (No. 12, Table 12.1) [21]. The reaction rate of ester hydrolysis by α-chymotrypsin was improved when the O—H stretching mode of water as the solvent and the O—H and N—H stretching modes of α-chymotrypsin were in the cooperative VSC state. Since the rate-limiting step of ester hydrolysis involves proton transfer, the VSC of the O—H and N—H stretching modes is considered to enhance proton dissociation.

The activation of enzymatic reactions under VSC was also observed in the hydrolysis of adenosine triphosphate (ATP) by ATPase (No. 13, Table 12.1) [22] and the hydrolysis of sucrose by sucrase (No. 14, Table 12.1) [23]. When the O—H stretch was under VSC, these hydrolysis reactions were significantly accelerated. This suggests that intermolecular forces, such as hydrogen bonding and hydrophobic interactions, are influenced by VSC, resulting in changes to the observed reaction kinetics.

12.3.3 Symmetry of Molecular Vibrations

The symmetry of vibrational modes can influence the impact of VSC on the energy landscape. For example, the equilibrium of the charge-transfer complex of 1,3,5-trimethylbenzene and iodine was altered under VSC, influencing the Gibbs free energy (ΔG°) [47]. Molecular vibrations can be categorized into two types of symmetry: A' (—CH_3 bend and aliphatic C—H stretch) and E' (ring C═C stretch and aromatic C—H stretch). ΔG° increased under the VSC of the vibrations with symmetry A', while ΔG° decreased under the VSC of the vibrations with symmetry E' (No. 15, Table 12.1). The result indicates that the symmetry of vibrational modes plays a role in the deactivation and activation of reactions under VSC.

Indeed, different stereoselectivity was observed for the ring-opening reaction of cyclobutene derivatives under the VSC of the C═O stretching vibration and under the VSC of the C—H vibration. For cyclobutene-*cis*-3,4-dimethylcarboxylate, the conrotatory ring-opening reaction was facilitated and *cis*, *trans*-butadiene derivatives were promoted under the VSC of the C═O stretching vibration [48]. On the other hand, *trans*-butadiene derivatives were obtained when the C—H vibration was subjected to VSC (No. 16, Table 12.1). Thus, depending on the symmetry of the target vibrational modes, the stereoselectivity of the ring-opening reaction could be modulated using VSC.

12.3.4 Interpretation of Vibrational Strong Coupling

One concern in this field is the reproducibility of experimental results [49, 50]. Evaluating reaction kinetics in an optical cavity can easily include experimental errors. For instance, in some cavity setups, the solution is introduced into the optical cavity from an inlet, and the reaction under VSC occurs there. If an excess amount of solution remains in the inlet or outlet (outside the cavity), it can diffuse into the cavity during the reaction, leading to errors in assessing the effect of VSC. Another potential error is the time lag between the true start of the reaction and the moment when the VSC state is achieved. Initially, the solution is prepared and then injected into the cavity. Subsequently, the cavity mode is tuned to achieve VSC. This tuning process requires a certain amount of time, but the exact duration depends on the cavity setup. Since chemical reactions begin when the reaction solution is prepared, there can be a delay between the actual start of the reactions and the start of the VSC state. Significant variations in this time lag can obscure the effects of VSC, making them difficult to observe. As each research group develops its own cavity system, the susceptibility to these experimental errors also depends on the cavity design and the choice of chemical reactions [51]. In this context, it is important to use the average values and standard deviations from repeated experiments to estimate the reaction kinetics.

Including experimental errors would make it difficult to qualitatively understand the effect of VSC. Indeed, the effect of VSC on chemical reactions is not entirely understood yet. Theoretical frameworks are proposed to simulate the effect of VSC [52, 53]. The formation of polaritonic states generates new energy levels nearby the transition states. The modulation of reaction kinetics is discussed based on the transition theory in polaritonic states and the DS. The interactions among the transition states (upper polariton, lower polariton and DS) influence the reaction kinetics [54–56]. In contrast, the cage effect (cavity friction) and modulated vibrational energy transfer are proposed to explain the decreased reaction kinetics [57–59]. The cage effect induces the recrossing in the barrier region and reduces the transmission coefficient. Because of the reduction of the transmission coefficient, a decrease in reaction kinetics can be experimentally observed. This effect is not captured by the transition state theory. From the viewpoint of chemistry, VSC may affect the vibrational energy transfer, resulting in the change of reactivity of specific functional groups [60–63]. The enhanced vibrational energy transfer among strongly coupled molecules was experimentally observed [64], although this is not the case for chemical reactions. The condensation of vibro-polariton is theoretically predicted, which influence kinetics of the chemical reactions [65]. The fields await the development of theoretical framework to understand the effect of VSC. The interpretation of VSC is invariable also for self-assembly or other phenomena influenced by VSC.

12.3.5 Self-assembly and Crystallization Under VSC

VSC can be applied not only to enzymatic and organic reactions but also to self-assembly processes [66]. When metal ions and organic ligands are spontaneously assembled, they form well-defined framework structures, so-called metal–organic frameworks (MOFs) [67, 68]. MOFs are of significant interest because of their functional properties, including adsorption, separation, and heterogeneous catalysis. The same combination of metal ions and organic ligands in solution can yield MOFs with various crystal structures, leading to heterogeneity in their porous characteristics, which can affect performance. Generally, the selectivity of crystal structure relies on the interactions of the metal ions and organic ligands, which are influenced by solvation. For example, when zinc ions and 2-methylimidazole are spontaneously assembled in aqueous solution, two types of MOFs (ZIF-L and ZIF-8) are produced. Under the VSC of the O—H stretching mode of water (the solvent), ZIF-8 was selectively formed (Figure 12.8). The result suggests that the VSC of solvent molecules can influence the self-assembly process (No. 1, Table 12.2).

The effect of VSC on self-assembly has also been investigated in organic polymers. Poly(para-phenylenethynylene) (PPE) in *o*-dichlorobenzene is assembled into fiber-like structures that form gels driven by CH–π interactions. In contrast, when a solution of PPE was introduced into an FP cavity and the aromatic ring C=C stretching vibrations of both *o*-dichlorobenzene and PPE were under VSC, flake-like structures were observed (No. 2, Table 12.2). The result suggests that VSC may influence the van der Waals interactions between the solvent and polymer [25]. The self-assembly behavior of organic molecules is also changed under VSC. The self-assembly of 1,3-bis(phenylethynyl)benzene results in flake-like structures due to π–π interactions. On the other hand, when the C—C stretching mode of the benzene ring was under VSC, spherical assemblies were formed (No. 3, Table 12.2). The difference in the self-assembly

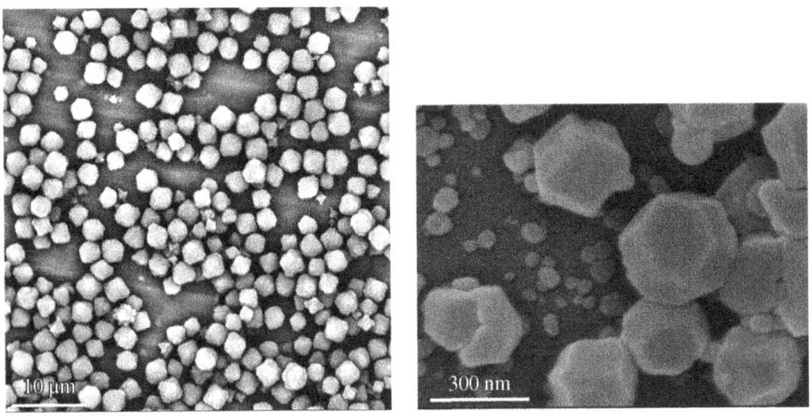

(a) (b)

Figure 12.8 Scanning electron microscope images of crystals obtained (a) in solution (without an optical cavity) and (b) in the FP cavity. Crosshair-star-shaped crystals of ZIF-L and cuboctahedral-shaped crystals of ZIF-8 are observed in (a). Only ZIF-8 is obtained, with no ZIF-L detected in (b).

behavior of 1,3-bis(phenylethynyl)benzene can be due to the modulation of the π–π interactions by VSC [26]. Noted that the VSC of the C═C stretching vibration of 1,4-bis(phenylethynyl)benzene does not affect the assembly structures, suggesting that symmetry of molecular vibration plays a role in the effect of VSC on assembly process.

DNA origami is a technology that folds DNA to create nanoscale structures. The formation of these nanoscale shapes depends on the hydrogen bonding of each base pair. Under VSC of pyridine bases in DNA molecules, the melting temperatures of DNA were reduced, which promotes the formation of origami structures at lower temperatures. These results suggest that VSCs modulate the DNA energy landscape and enhance intermolecular interactions (No. 4, Table 12.2) [27]. The effect of VSC on the supramolecular polymerization of porphyrins was also investigated. The polymerization temperature of chiral zinc porphyrin decreased by about 10°C under VSC of C—H stretch, revealing that VSC modulates solute–solvent interactions, favoring different states of supramolecular polymerization (No. 5, Table 12.2) [28].

12.4 Summary

The field of cavity strong coupling was initiated by physicists. Because of this background, there was little integration between molecular chemistry and cavity strong coupling until the 1990s. In early investigations, cavity strong coupling was observed by placing inorganic materials in a cavity structure [5]. Later, the cavity strong coupling of organic molecules was also observed [6], with an important turning point being the report of the structural isomerization of spiropyran under cavity strong coupling. These investigations focused on the strong coupling of exciton and electromagnetic modes in a cavity vacuum field [59]. However, in 2015, the strong coupling of molecular vibrations and cavity vacuum fields, so-called VSC, was observed [8]. The effect of VSC on chemical reactions was first demonstrated for the deprotection reaction of silyl organic molecules [9]. In the subsequent six years, VSC effects have been reported for various chemical reactions, including solvolysis reactions, cyclization reactions, enzymatic reactions, self-assembly, and crystallization. In solvolysis reactions, cyclization reactions, enzymatic reactions, VSC decelerates or accelerates the chemical reactions. Although this behavior can be rationalized based on the intricacy of reaction mechanisms, further elucidation of the origin of these different effects is required. In particular, reaction acceleration is often observed when the solvent molecules are in the VSC state. The manipulation of self-assembly under VSC suggests that the intermolecular interactions are varied by VSC. If the intermolecular interactions of solvent molecules change under VSC, the VSC effect becomes more complicated in all the chemical reactions, because it cannot be explained by a simple change in bond formation/dissociation energies. More recently, the role of the symmetry of vibrational modes has started to be investigated. The relationship between the VSC effect and the symmetry of vibrational modes is not fully understood. Despite the intense research on the effect of VSC, it is currently difficult to systematically interpret the mechanisms of VSC-induced reactivity changes. Historically, cavity strong coupling has mostly been investigated by physicists. It is hoped that a variety of researchers will enter this field in the future, which will aid in significantly advancing mechanistic interpretation.

References

1 Purcell, E. M. Resonance absorption by nuclear magnetic moments in a solid. *Phys. Rev.* **1946,** *69*(11–12), 681.

2 Holmes, R. J., Forrest, S. R. Strong exciton–photon coupling in organic materials. *Org. Electron.* **2007,** *8*(2–3), 77–93.

3 Ciuti, C., Bastard, G., Carusotto, I. Quantum vacuum properties of the intersubband cavity polariton field. *Phys. Rev. B.* **2005,** *72*(11), 115303.
4 Galego, J., Garcia-Vidal, F. J., Feist, J. Cavity-induced modifications of molecular structure in the strong-coupling regime. *Phys. Rev. X.* **2015,** *5*(4), 041022.
5 Weisbuch, C., Nishioka, M., Ishikawa, A., Arakawa, Y. Observation of the coupled exciton-photon mode splitting in a semiconductor quantum microcavity. *Phys. Rev. Lett.* **1992,** *69*(23), 3314–3317.
6 Lidzey, D. G., Bradley, D. D. C., Skolnick, M. S., Virgili, T., Walker, S., Whittaker, D. M. Strong exciton–photon coupling in an organic semiconductor microcavity. *Nature.* **1998,** *395*(6697), 53–55.
7 Ebbesen, T. W. Hybrid light–matter states in a molecular and material science perspective. *Acc. Chem. Res.* **2016,** *49*(11), 2403–2412.
8 Shalabney, A., George, J., Hutchison, J., Pupillo, G., Genet, C., Ebbesen, T. W. Coherent coupling of molecular resonators with a microcavity mode. *Nat. Commun.* **2015,** *6*, 5981.
9 Thomas, A., George, J., Shalabney, A., Dryzhakov, M., Varma, S. J., Moran, J. et al. Ground-state chemical reactivity under vibrational coupling to the vacuum electromagnetic field. *Angew. Chem. Int. Ed.* **2016,** *55*(38), 11462–11466.
10 Thomas, A., Jayachandran, A., Lethuillier-Karl, L., Vergauwe, R. M. A., Nagarajan, K., Devaux, E. et al. Ground state chemistry under vibrational strong coupling: dependence of thermodynamic parameters on the Rabi splitting energy. *Nanophotonics* **2020,** *9*(2), 249–255.
11 Thomas, A., Lethuillier-Karl, L., Nagarajan, K., Vergauwe, R. M. A., George, J., Chervy, T. et al. Tilting a ground-state reactivity landscape by vibrational strong coupling. *Science* **2019,** *363*(6427), 615–619.
12 Hirai, K., Takeda, R., Hutchison, J. A., Uji-i, H. Modulation of Prins cyclization by vibrational strong coupling. *Angew. Chem. Int. Ed.* **2020,** *59*(13), 5332–5335.
13 Hirai, K., Ishikawa, H., Takahashi, Y., Hutchison, J. A., Uji-i, H. Autotuning of vibrational strong coupling for site-selective reactions. *Chem. Eur.* **2022,** *28*(47), e202201260.
14 Hiura, H., Shalabney, A. Vacuum-field catalysis: accelerated reactions by vibrational ultra strong coupling. *Chem Rxiv.* **2021,** 10.26434/chemrxiv.7234721.v5.
15 Singh, J., Garg, P., Anand, R. V., George, J. Cavity catalysis of an enantioselective reaction under vibrational strong coupling. *Chem. Eur. J.* **2024,** *30*(26), e202400607.
16 Lather, J., Bhatt, P., Thomas, A., Ebbesen, T. W., George, J. Cavity catalysis by cooperative vibrational strong coupling of reactant and solvent molecules. *Angew. Chem. Int. Ed.* **2019,** *58*(31), 10635–10638.
17 Lather, J., Thabassum, A. N. K., Singh, J., George, J. Cavity catalysis: modifying linear free-energy relationship under cooperative vibrational strong coupling. *Chem. Sci.* **2022,** *13*(1), 195–202.
18 Singh, J., Lather, J., George, J. Solvent dependence on cooperative vibrational strong coupling and cavity catalysis. *ChemPhysChem* **2023,** *24*(11), e202300016.
19 Ahn, W., Triana, J., Recabal, F., Herrera, F., Simpkins, B. Modification of ground-state chemical reactivity via light–matter coherence in infrared cavities. *Science* **2023,** *380*(6650), 11165–11168.
20 Vergauwe, R. M. A., Thomas, A., Nagarajan, K., Shalabney, A., George, J., Chervy, T. et al. Modification of enzyme activity by vibrational strong coupling of water. *Angew. Chem. Int. Ed.* **2019,** *58*(43), 15324–15328.
21 Lather, J., George, J. Improving enzyme catalytic efficiency by co-operative vibrational strong coupling of water. *J. Phys. Chem. Lett.* **2021,** *12*(1), 379–384.
22 Gao, F., Guo, J., Si, Q., Wang, L., Zhang, F., Yang, F. Modification of ATP hydrolysis by strong coupling with O–H stretching vibration. *ChemPhotoChem* **2023,** *7*(4), e202200330.
23 Bai, J., Wang, Z., Zhong, C., Hou S., Lian, J., Si, Q. et al. Vibrational coupling with O-H stretching increases catalytic efficiency of sucrase in Fabry-Pérot microcavity. *Biochem. Biophys. Res. Commun.* **2023,** *652*, 31–34.

24 Hirai, K., Ishikawa, H., Chervy, T., Hutchison, J. A., Uji-i, H. Selective crystallization *via* vibrational strong coupling. *Chem. Sci.* **2021,** *12*(36), 11986–11994.

25 Joseph, K., Kushida, S., Smarsly, E., Ihiawakrim, D., Thomas, A., Paravicini-Bagliani, G. L. et al. Supramolecular assembly of conjugated polymers under vibrational strong coupling. *Angew. Chem. Int. Ed.* **2021,** *60*(36), 19665–19670.

26 Sandeep, K., Joseph, K., Gautier, J., Nagarajan, K., Sujith, M., Thomas, K. G. et al. Manipulating the self-assembly of phenyleneethynylenes under vibrational strong coupling. *J. Phys. Chem. Lett.* **2022,** *13*(5), 1209–1214.

27 Zhong, C., Hou, S., Zhao, S., Bai, J., Wang, Z., Gao, F. et al. DNA origami driving DNA origami coassembling by vibrational strong coupling in the dark. *ACS Photonics* **2023,** *10*(5), 1618–1623.

28 Joseph, K., de Waal, B., Jansen, S. A. H., van der Tol, J. J. B., Vantomme, G., Meijer, E. W. Consequences of vibrational strong coupling on supramolecular polymerization of porphyrins. *J. Am. Chem. Soc.* **2024,** *146*(17), 12130–12137.

29 Nagarajan, K., Thomas, A., Ebbesen, T. W. Chemistry under vibrational strong coupling. *J. Am. Chem. Soc.* **2021,** *143*(41), 16877–16889.

30 Dovzhenko, D. S., Ryabchuk, S. V., Rakovich, Y. P., Nabiev, I. R. Light–matter interaction in the strong coupling regime: configurations, conditions, and applications. *Nanoscale.* **2018,** *10*(8), 3589–3605.

31 Hirai, K., Hutchison, J. A., Uji-i, H. Recent progress in vibropolaritonic chemistry. *ChemPlusChem* **2020,** *85*(9), 1981–1988.

32 Hertzog, M., Wang, M., Mony, J., Börjesson, K. Strong light–matter interactions: a new direction within chemistry. *Chem. Soc. Rev.* **2019,** *48*(3), 937–961.

33 Ribeiro, R. F., Martínez-Martínez, L. A., Du, M., Campos-Gonzalez-Angulo, J., Yuen-Zhou, J. Polariton chemistry: controlling molecular dynamics with optical cavities. *Chem. Sci.* **2018,** *9*(30), 6325–6339.

34 Hirai, K., Hutchison, J. A., Uji-i, H. Molecular chemistry in cavity strong coupling. *Chem. Rev.* **2023,** *123*(13), 8099–8126.

35 Simpkins, B. S., Fears, K. P., Dressick, W. J., Spann, B. T., Dunkelberger, A. D., Owrutsky, J. C. Spanning strong to weak normal mode coupling between vibrational and Fabry–Pérot cavity modes through tuning of vibrational absorption strength. *ACS Photonics* **2015,** *2*(10), 1460–1467.

36 George, J., Shalabney, A., Hutchison, J. A., Genet, C., Ebbesen, T. W. Liquid-phase vibrational strong coupling. *J. Phys. Chem. Lett.* **2015,** *6*(6), 1027–1031.

37 Casey, S. R., Sparks, J. R. Vibrational strong coupling of organometallic complexes. *J. Phys. Chem. C.* **2016,** *120*(49), 28138–28143.

38 George, J., Chervy, T., Shalabney, A., Devaux, E., Hiura, H., Genet, C. et al. Multiple Rabi splittings under ultrastrong vibrational coupling. *Phys. Rev. Lett.* **2016,** *117*(15), 153601.

39 Vergauwe, R. M. A., George, J., Chervy, T., Hutchison, J. A., Shalabney, A., Torbeev, V. Y. et al. Quantum strong coupling with protein vibrational modes. *J. Phys. Chem. Lett.* **2016,** *7*(20), 4159–4164.

40 Hertzog, M., Rudquist, P., Hutchison, J. A., George, J., Ebbesen, T. W., Börjesson, K. Voltage-controlled switching of strong light–matter interactions using liquid crystals. *Chem. Eur. J.* **2017,** *23*(72), 18166–18170.

41 Hiura, H., Shalabney, A., George, J. Vibrational ultra strong coupling of water and ice. *ChemRxiv* **2019**. https://doi.org/10.26434/chemrxiv.9808508.v1.

42 Houdré, R., Stanley, R. P., Ilegems, M. Vacuum-field Rabi splitting in the presence of inhomogeneous broadening: resolution of a homogeneous linewidth in an inhomogeneously broadened system. *Phys. Rev. A.* **1996,** *53*(4), 2711–2715.

43 Li, T. E., Subotnik, J. E., Nitzan, A. Cavity molecular dynamics simulations of liquid water under vibrational ultrastrong coupling. *Proc. Natl. Acad. Sci. USA* **2020,** *117*(31), 18324–18331.

44 Long, J. P., Simpkins, B. S. Coherent coupling between a molecular vibration and Fabry-Perot optical cavity to give hybridized states in the strong coupling limit. *ACS Photonics* **2015,** *2*(1), 130–136.

45 Wang, S., Chervy, T., George, J., Hutchison, J. A., Genet, C., Ebbesen, T. W. Quantum yield of polariton emission from hybrid light-matter states. *J. Phys. Chem. Lett.* **2014,** *5*(8), 1433–1439.

46 Ahn, W., Vurgaftman, I., Dunkelberger, A. D., Owrutsky, J. C., Simpkins, B. S. Vibrational strong coupling controlled by spatial distribution of molecules within the optical cavity. *ACS Photonics.* **2018,** *5*(1), 158–166.

47 Pang, Y., Thomas, A., Nagarajan, K., Vergauwe, R. M. A., Joseph, K., Patrahau, B. et al. On the role of symmetry in vibrational strong coupling: The case of charge-transfer complexation. *Angew. Chem. Int. Ed.* **2020,** *59*(26), 10436–10440.

48 Sau, A., Nagarajan, K., Patrahau, B., Lethuillier-Karl, L., Vergauwe, R. M. A., Thomas, A. et al. Modifying Woodward–Hoffmann stereoselectivity under vibrational strong coupling. *Angew. Chem. Int. Ed.* **2021,** *60*(11), 5712–5717.

49 Imperatore, M. V., Asbury, J. B., Giebink, N. C. Reproducibility of cavity-enhanced chemical reaction rates in the vibrational strong coupling regime. *J. Chem. Phys.* **2021,** *154*, 191103.

50 Wiesehan, G. D., Xiong, W. Negligible rate enhancement from reported cooperative vibrational strong coupling catalysis. *J. Chem. Phys.* **2021,** *155*, 241103.

51 Hirai, K., Hutchison, J. A., Uji-i, H. Optical cavity design and functionality for molecular strong coupling. *Chem. Eur. J.* **2024,** *30*, e202303110.

52 Wang, D. S., Yelin, S. F. A. Roadmap toward the theory of vibrational polariton chemistry. *ACS Photonics* **2021,** *8*(10), 2818–2826.

53 Fregoni, J., Garcia-Vidal, F. J., Feist, J. Theoretical challenges in polaritonic chemistry. *ACS Photonics.* **2022,** *9*(4), 1096–1107.

54 Campos-Gonzalez-Angulo, J. A., Ribeiro, R. F., Yuen-Zhou, J. Resonant catalysis of thermally activated chemical reactions with vibrational polaritons. *Nat. Commun.* **2019,** *10*(1), 4685.

55 Yang, P. Y., Cao, J. Quantum effects in chemical reactions under polaritonic vibrational strong coupling. *J. Phys. Chem. Lett.* **2021,** *12*(39), 9531–9538.

56 Du, M., Yuen-Zhou, J. Catalysis by dark states in vibropolaritonic chemistry. *Phys. Rev. Lett.* **2022,** *128*(9), 96001.

57 Li, X., Mandal, A., Huo, P. Cavity frequency-dependent theory for vibrational polariton chemistry. *Nat. Commun.* **2021,** 12, 1315.

58 Li, X., Mandal, A., Huo, P. Theory of mode-selective chemistry through polaritonic vibrational strong coupling. *J. Phys. Chem. Lett.* **2021,** *12*(29), 6974–6982.

59 Philbin, J. P., Wang, Y., Narang, P., Dou, W. Chemical reactions in imperfect cavities: enhancement, suppression, and resonance. *J. Phys. Chem. C.* **2022,** *126*, 14908–14913.

60 Li, T. E., Nitzan, A., Subotnik, J. E. Collective vibrational strong coupling effects on molecular vibrational relaxation and energy transfer: numerical insights via cavity molecular dynamics simulations. *Angew. Chem. Int. Ed.* **2021,** *60*(28), 15533–15540.

61 Wang, D. S., Neuman, T., Yelin, S. F., Flick, J. Cavity-modified unimolecular dissociation reactions via intramolecular vibrational energy redistribution. *J. Phys. Chem. Lett.* **2022,** *13*(15), 3317–3324.

62 Schäfer, C., Flick, J., Ronca, E., Narang, P., Rubio, A. Shining light on the microscopic resonant mechanism responsible for cavity-mediated chemical reactivity. *Nat. Commun.* **2022,** *13*(15), 7817.

63 Wang, D. S., Flick, J., Yelin, S. F. Chemical reactivity under collective vibrational strong coupling. *J. Chem. Phys.* **2022,** *157*, 224304.

64 Xiang, B., Ribeiro, R. F., Du, M., Chen, L., Yang, Z., Wang, J. et al. Intermolecular vibrational energy transfer enabled by microcavity strong light–matter coupling. *Science* **2020,** *368*(6491), 665–667.

65 Pannir-Sivajothi, S., Campos-Gonzalez-Angulo, J. A., Martínez-Martínez, L. A., Sinha, S., Yuen-Zhou, J. Driving chemical reactions with polariton condensates. *Nat. Commun.* **2022,** *13*, 1645.

66 Hirai, K., Uji-i, H. Molecular assembly in optical cavities. *Chem. Asian J.* **2025,** *19*, e202401262.

67 Kitagawa, S., Kitaura, R., Noro, S. Functional porous coordination polymers. *Angew. Chem. Int. Ed.* **2004,** *43*(18), 2334–2375.

68 Furukawa, H., Cordova, K. E., O'Keeffe, M., Yaghi, O. M. The chemistry and applications of metal-organic frameworks. *Science* **2013,** *341*(6149), 1230444.

13

Controlling and Probing Molecular Polaritons

Michael A. Michon and Blake S. Simpkins

Naval Research Laboratory, Chemistry Division, Washington, DC, USA

13.1 Introduction

The goal of this chapter is to accelerate the ability of research groups to contribute quickly, accurately, and substantively, to the field of cavity-modified materials behaviors. The opportunity to tune fundamental chemical and physical properties with an engineered optical environment has inspired images of designer cavity catalysis [1–3], enhanced energy technologies [4], and directed bio-assembly [5]. Although numerous demonstrations of cavity-modified chemistry have been made [6, 7], hesitancy remains due to a lack of mechanistic understanding [8–10], experimentally precarious results [11, 12], and the generally enigmatic character of this phenomenon. In fact, corroborating results [7] have been met with doubt by even close colleagues. However, just as error bars are slowly shaved away, measurement by measurement, to eventually reveal the sought-after truth, we must also continue to evaluate cavity-induced effects, troubleshooting and improving experimental techniques, refining analytical tools, and listening to and addressing legitimate criticism.

The key to achieving this is lowering the barrier of entry into the field. New groups could quickly enter and contribute with accepted best practices, improved and robust experimental platforms, and easily accessible analytical procedures. The following sections are arranged to provide a simple analytical platform for describing cavity-coupled systems, highlights of nonidealities that exist in real systems and guidelines for their magnitudes and avoidance, and finally, a discussion of analytical and experimental techniques that could be adopted.

13.2 Analytical Description of Cavities

13.2.1 Treatment of Lossless Mirrors Bounding an Absorbing Medium

Having an analytical treatment to describe the spectrum of a cavity-coupled system is invaluable in designing and interpreting experimental studies. Fortunately, closed-form expressions for transmission and reflection spectra have already been developed [13–16], and even with simplifying

Polariton Chemistry: Molecules in Cavities, First Edition. Edited by Joel Yuen-Zhou, Noel C. Giebink and Raphael F. Ribeiro.

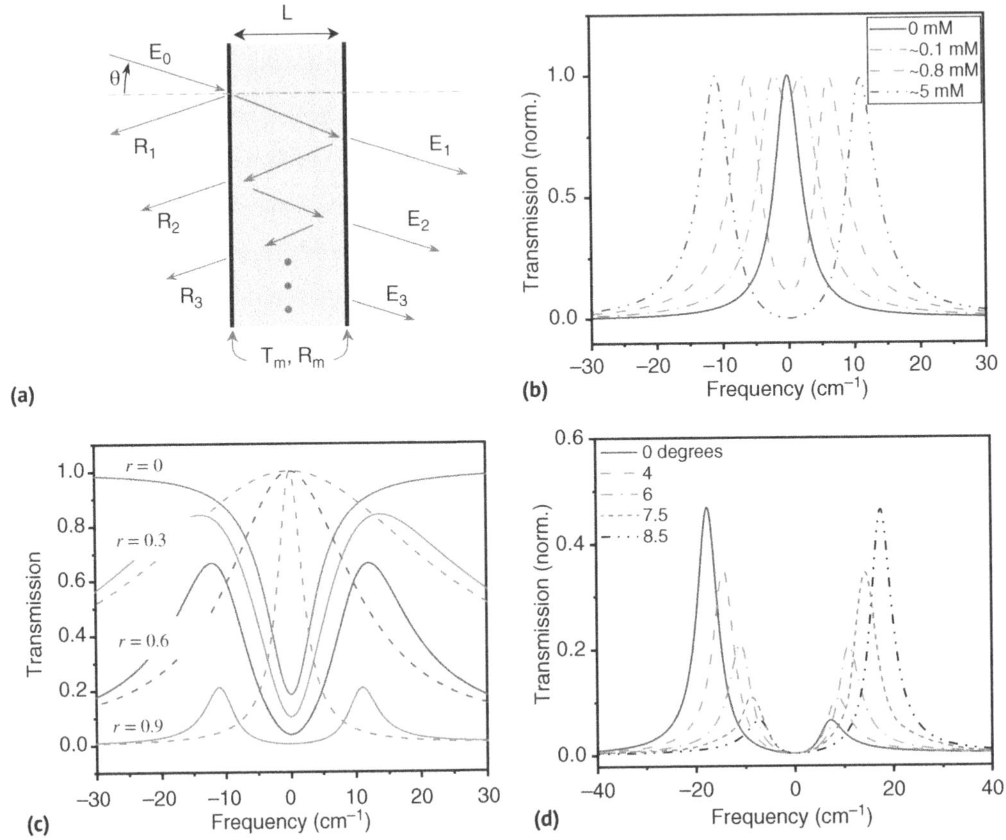

Figure 13.1 Analytical modeling of etalon response. (a) Depicts the propagating ray picture for an incident beam partially reflecting and transmitting upon its encounter with the front and rear partially reflective surfaces. Accounting for the infinite series of transmitted rays results in the analytical expressions presented in the text. Plots depict the cavity transmission spectra as a function of (a) absorber concentration, (b) interface reflectivity, and (c) incident angle.

assumptions, such as lossless mirrors, angle-independent reflectivity, and ignoring the phase shifts that upon optical reflection, these treatments can be informative, predictive, and even allow for the description of time-varying response [17].

Beginning with a lossless slab of material (see Figure 13.1a), an incident field amplitude, E_0, generates a first-pass transmitted amplitude of $E_1 = E_0 t^2 e^{(i\omega t)}$, where t is the transmission amplitude coefficient through the top and bottom slab interfaces. Beams are reflected within the slab with some transmission occurring upon each round trip. In general, the transmission amplitude upon the Nth passage through the film is $E_N = E_0 t^2 r^{2(N-1)} e^{i[\omega t-(N-1)\delta]}$, where reflections are now accounted for with r and the total phase change per round trip within the slab is $\delta = 4\pi nL\nu\cos\theta$, where n is the refractive index, L the cavity length, ν the frequency in cm^{-1}, and θ the angle of incidence. If we define $a = E_0 t^2 e^{i\omega t}$ and $b = r^2 e^{-i\delta}$, then this infinite series of transmitted beams may be summed using a geometric series as:

$$E_{\text{total}} = \sum_N E_N = a + ab + ab^2 + \cdots = \frac{a}{1-b} = \frac{E_0 t^2 e^{i\omega t}}{1 - r^2 e^{-i\delta}}$$

Multiplying this expression with its complex conjugate gives the transmission intensity through a lossless slab, shown in Eq. (13.1), where T_m and R_m are the intensity transmissivity and reflectivity through the slab interface, which act as lossless mirrors, $T_m + R_m = 1$. Absorptive material within the slab can be described through the absorption coefficient, α, which produces an exponential loss of amplitude during each trip through the slab, resulting in Eq. (13.2).

$$T_{\text{lossless}} = \frac{T_m^2}{1 + R_m^2 - 2R_m \cos(4\pi\nu nL \cos\theta)} \tag{13.1}$$

$$T_{\text{abs}} = \frac{T_m^2 e^{-\alpha L}}{1 + R_m^2 e^{-2\alpha L} - 2R_m e^{-\alpha L} \cos(4\pi\nu nL \cos\theta)} \tag{13.2}$$

$$\varepsilon_1 = \varepsilon_\infty + \sum_j \frac{S_j\left(\nu_j{}^2 - \nu^2\right)}{\left(\nu_j{}^2 - \nu^2\right)^2 + \left(\gamma_j \nu\right)^2} \tag{13.3}$$

$$\varepsilon_2 = \sum_j \frac{S_j \gamma_j \nu}{\left(\nu_j{}^2 - \nu^2\right)^2 + \left(\gamma_j \nu\right)^2} \tag{13.4}$$

$$n = \sqrt{\frac{\varepsilon_1 + \sqrt{\varepsilon_1^2 + \varepsilon_2^2}}{2}} \tag{13.5}$$

$$k = \sqrt{\frac{-\varepsilon_1 + \sqrt{\varepsilon_1^2 + \varepsilon_2^2}}{2}} \tag{13.6}$$

$$\alpha = 4\pi\nu k = 4\pi\nu\sqrt{\frac{-\varepsilon_1 + \sqrt{\varepsilon_1^2 + \varepsilon_2^2}}{2}} \tag{13.7}$$

$$A = \frac{\alpha L}{\ln(10)} = \frac{L}{\ln(10)} 4\pi\nu\sqrt{\frac{-\varepsilon_1 + \sqrt{\varepsilon_1^2 + \varepsilon_2^2}}{2}} \tag{13.8}$$

This convenient description of a Fabry–Pérot (FP) cavity allows for the calculation of transmission spectra, as a function of incident angle, as long as the absorption coefficient, α, and interface (i.e. mirror) reflectivity, R, are known. A common approach to modeling the absorption coefficient is to describe each optically active transition as a Lorentzian oscillator. Then, the real and imaginary components of the complex permittivity, ε_1 and ε_2, take the form of a summation over all the transitions one wishes to include, and are shown as Eqs. (13.3 and 13.4,) where ε_∞ is the permittivity at long wavelength (i.e. background refractive index), ν is frequency in units of wavenumbers (cm^{-1}), the ν_j's are the resonant frequencies of the transitions, the S_j's are scaling factors that determine the strength of each transition, and the γ_j's are the linewidths of the Lorentzian forms describing the transitions. If all of these characteristics are known or assumed, one can build a dielectric function defined by Eqs. (13.5–13.7), insert it into Eq. (13.2) and begin modeling. Examples of such calculations are shown in Figure 13.1 and highlight the roles of the various terms. Transmission spectra are modeled for a range of absorber concentrations (i.e. amplitude of the oscillators modeled by Eqs. 13.3–13.4) are shown in Figure 13.1b and show an increasing peak separation (i.e. Rabi splitting, Ω) with increasing absorber concentration [15, 18, 19]. The role of mirror reflectivity is to determine, in part, the linewidth of the cavity mode, which is highlighted in Figure 13.1c. Clearly, higher reflectivity results in narrower cavity and polariton modes, which has immediate relevance for attaining the strong coupling regime, typically defined as a Rabi splitting larger than both the cavity and absorber linewidths [16, 20], but there can be more insight here. The cavity linewidth has implications for energy exchange and lifetime timescales since the relative loss rates between the molecular and cavity photon excitations depend on the cavity linewidth (i.e. lifetime) [21, 22]. The final aspect of the analytical modeling that will be highlighted here is the ability to model angle-dependent transmission response. This allows one to calculate the full dispersion of an FP cavity, which tunes blue as the angle of interrogation increases (Figure 13.1d).

With no knowledge of the amplitude factors, we can reformulate the absorption according to Eq. (13.8), which provides an expression for fitting an experimentally measured absorption spectrum (although one must know or assume the sample thickness and ε_∞) to yield all the pertinent transition characteristics. Specifically, the measured absorption in OD can be expressed as shown in Eq. (13.8), then the curve may be fit with Lorentzian amplitude factors, frequencies, and linewidths as fitting parameters. An example of such a treatment is shown in Figure 13.2, where the absorption coefficient for the phenyl isocyanate is fit with two oscillators. Once the data are fit,

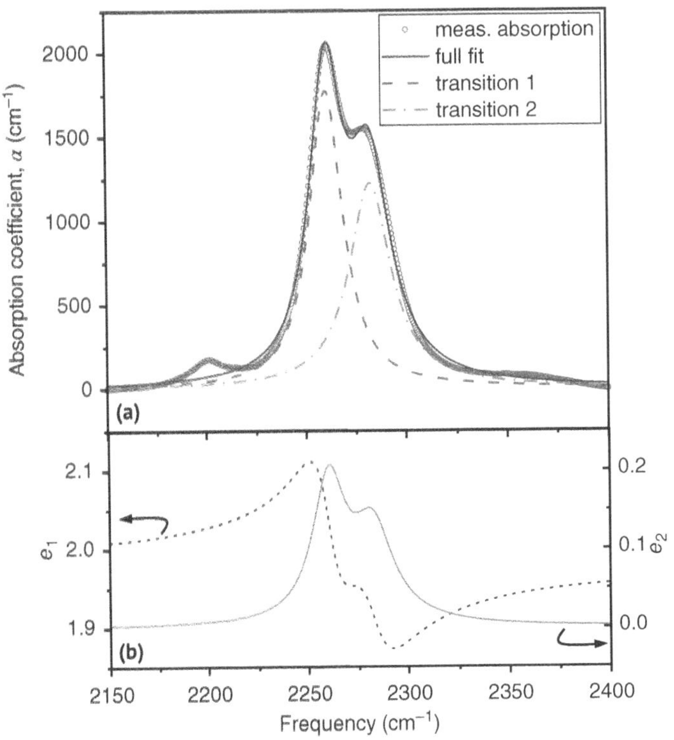

Figure 13.2 Example of Lorentzian model generation. An absorption spectrum for material of interest (a) may be fit with an appropriate number of Lorentzian oscillators to yield amplitude, resonant frequency, and linewidth parameters, which accurately describe the material. These parameters then define the real and imaginary components of the permittivity (b).

Figure 13.2a, the extracted oscillator parameters can be used to model the real and imaginary components of the permittivity (Figure 13.2b).

13.2.2 Semiclassical Coupled Oscillators

Although the classical approaches described in Section 13.2.1 do an excellent job of describing observed spectra, there is no explicit description or dependence on coupling strength (referred to as the Rabi coupling, Ω). On the one hand, this makes the treatments described in Section 13.2.1 equally applicable to weak and strong coupling regimes. On the other hand, if the goal is to explicitly define an interaction energy, one can turn to an approach that utilizes some degree of quantization. Whether one quantizes both the field and the material oscillator (i.e. fully quantum mechanical) or only quantizes the material resonance and leaves the field classical (i.e. semiclassical), the end result is a Hamiltonian that includes three terms, one describing the photon, another for the material excitation, and a third describing their interaction [16]. Applying a coordinate transformation referred to as the rotating wave approximation results in an expression diagonalizable into a 2 × 2 matrix, Eq. (13.9), whose eigenvalues define the polariton energies according to Eq. (13.10) [23]. With such an expression, a fixed material excitation energy, E_{mat}, and an expression describing the dispersive FP cavity resonance, $E_{\mathrm{cav}}(\theta) = E_0\left[1-\left(\frac{\sin^2\theta}{n^2}\right)\right]^{\frac{-1}{2}}$, where E_0 is cavity resonance at normal incidence (i.e. in-plane wave vector, $k_{||} = 0$), θ is the angle of incidence, and n is the background refractive index of the material in the cavity, the angle-dependent dispersion of the normal modes can be generated. A dispersion, modeled to mimic typical experimental conditions, is shown in Figure 13.3a. This dispersion shows an upper polariton (UP) that is relatively flat and nearly isoenergetic to the material resonance at low interrogation angles. At higher angles, the UPs disperse rapidly to higher energy and the lower polariton (LP) is now flat and resides near the material resonance energy. The Rabi splitting of such dispersion would be calculated as the minimum energetic separation between the polaritons, which occurs for this example data near 25°. These dispersing modes have varying photonic and material characters [24, 25], represented by α^2 and β^2. Generally speaking, the closer a polariton is to the material excitation energy, the greater the material character. Therefore, in the example data of Figure 13.3a, the UP is more material like at low angle (i.e. in-plane wave vectors) and more photon like at higher angles. The photonic character is explicitly expressed in Eq. (13.11) and plotted in Figure 13.3b for both polaritons. In addition to the general trends described above, one can see the curves crossing each other at the condition where both polaritons have equal photonic and material content.

$$\begin{bmatrix} E_{\mathrm{cav}}(\theta) & V \\ V & E_{\mathrm{mat}} \end{bmatrix}\begin{bmatrix} \alpha \\ \beta \end{bmatrix} = E_{\mathrm{UP,LP}}\begin{bmatrix} \alpha \\ \beta \end{bmatrix} \tag{13.9}$$

$$E_{\mathrm{UP,\,LP}} = \frac{E_{\mathrm{cav}}+E_{\mathrm{mat}}}{2} \pm \sqrt{\frac{(E_{\mathrm{cav}}+E_{\mathrm{mat}})^2}{4}+\left(\frac{\Omega}{2}\right)^2} \tag{13.10}$$

$$\alpha^2_{\mathrm{UP,LP}} = \frac{\left(\frac{\Omega}{2}\right)^2}{\left(E_{\mathrm{cav}}-E_{\mathrm{UP,LP}}\right)^2+\left(\frac{\Omega}{2}\right)^2} \tag{13.11}$$

$$\beta^2_{\mathrm{UP,LP}} = 1-\alpha^2_{\mathrm{UP,LP}} \tag{13.12}$$

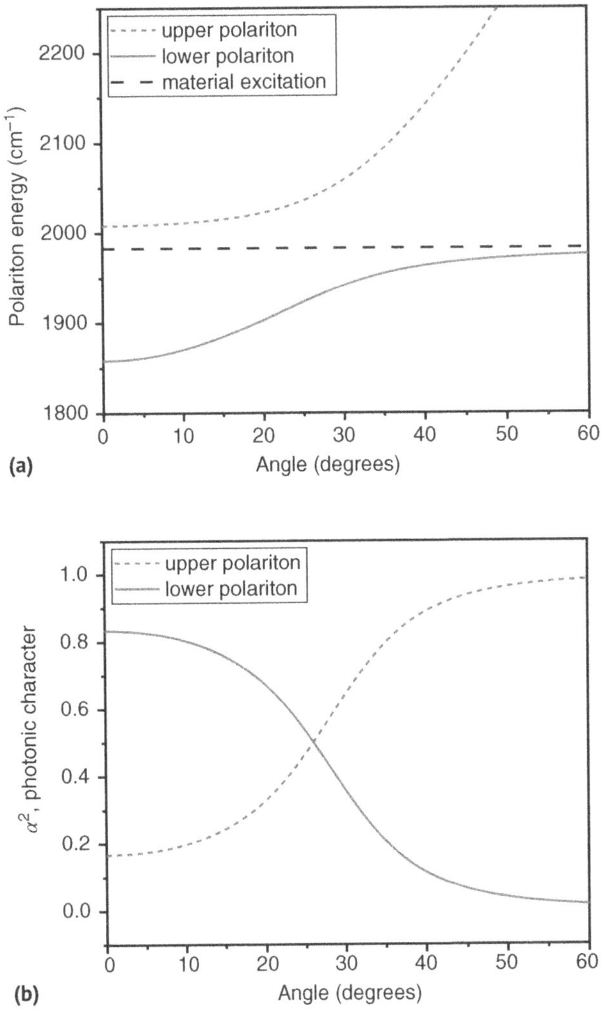

Figure 13.3 Dispersion and photonic character modeled according to Eqs. (13.10 and 13.11). In (a), the UP and LP branches are seen to anticross. A minimum in the branch separation occurs where the bare cavity resonance is coincident with the material excitation (~25° in this example). The Hopfield coefficient describing the fraction of each polariton that is of photonic character is calculated in (b). The material component is simple unity minus this value.

13.3 Nonidealities: That We Must, Nevertheless, Deal with

13.3.1 Details for Dealing with Cavities

Despite the broad utility and robustness of the classical and semiclassical treatments described in Sections 13.2.1 and 13.2.2, there are additional experimental details that should be considered. The first discussed here is the conversion of angle-dependent spectra to a true dispersive relationship. To extract the coupling strength (i.e., Rabi splitting, Ω), one finds the minimum separation between the UP and LP modes in the dispersion response. Angle-tuning of an FP cavity is a convenient method for generating the required dispersion contour. However, it must be remembered that the two polariton modes revealed at any given interrogation angle actually have different internal wavevectors. This means that a single spectrum does not accurately reflect the coherence of the polariton states. Therefore, the dispersion response should be converted from angle-to-wavevector dependence. The in-plane wavevector is $k_{\|} = k_0 \sin\theta = 2\pi\nu\sin\theta$, where ν is the frequency and θ is the angle of incidence. This dependence on both incident angle and frequency is what makes this conversion necessary for accurate determination of the Rabi splitting, since the conversion does not simply stretch or contract the dispersing axis. Instead, it causes a correlated distortion in both the dispersing and energy directions. In practice, we carry out this conversion by first defining an empty array that will eventually house the converted data. Each element is located at $[k_{\|}, \nu]$. Returning to the original data, each data point has a defined ν and θ value, which determines the $[k_{\|}, \nu]$ location where this intensity value should be placed in the destination array. Since the exact $k_{\|}$ likely does exist in the destination array, one can use bidirectional interpolation to place the most accurate value in the most appropriate element location. Correcting the dispersion in this way [26] can alter extracted Rabi splitting values by ~5%.

Next, it is important to identify an outstanding challenge that has plagued several attempts to demonstrate, then verify, cavity-modified chemical reactivity. Accurate methods for monitoring reaction progress in a strong coupling environment are key to establishing this field, and visible absorption spectroscopy is a common method for monitoring product distributions, the formation of reaction intermediates, and overall reaction kinetics. As such, there have been multiple reports of using visible absorbance as a reporter for reaction progress [27–29], as well as reports questioning the robustness of some of these results [11, 12]. These works utilize thin metal mirrors (~10 nm Au), which are reasonably reflective in the infrared (~90%) and modestly transmissive in the visible. Although there is transmission of several percents in the visible, the inevitable formation of weak cavity modes in the visible and significant absorption make quantitative interpretation of absorption spectra problematic. There will be no analytical answer presented here, but an important, and very practical, contribution to the field would be an analytical treatment, which accurately corrects for these unwanted and potentially misleading features. One can imagine a straightforward treatment, such as that used to correct for weak fringing in organic thin films [30], but with the inclusion of metal absorption. On the other hand, the optical properties of metal layers of this dimension may not be well behaved.

One last detail researchers should be aware of is the impact of the passivating coating applied to the mirror surfaces. It is a common practice to coat the metal mirrors with an oxide to insure against material-specific effects (e.g., metal catalysis) when comparing chemical behaviors inside and outside a cavity. If one uses a refractive index shift as a report of reaction progress, which can be problematic in its own right [31], one should note that the cavity is only partially filled with the reaction medium experiencing an index shift, with the remainder filled with an oxide layer exhibiting no index shift. Since the observed shift is due to only a portion of the cavity medium, the extracted modification of the reaction medium will be larger than if one ignores the invariable mirror coatings. For example, the frequency, in units of cm^{-1}, of a resonant mode in an FP cavity is $\nu = \frac{10^4 m}{2nL}$, where m is the mode order, n is the refractive index, and L is the cavity length, measured in cm. The shift in fringe position can then be used as a measure of the evolution of the refractive index due to solution composition variation as a reaction proceeds. The refractive index in the cavity should be described as $n_{\text{true}} = (1 - P)\, n_{\text{rxn}} + P n_{\text{ox}}$, where P is the fraction of the cavity length filled by the passivated layer rather than reacting solution, n_{rxn} is the refractive index of the reacting solution, and n_{ox} the index of the passivating oxide layer. Admittedly, this only amounts to a small correction (~1%) for the systems found in the literature, but this systematic error could become significant for systems where the refractive index difference between the reacting solution and passivation layer is large, the index change between reactants and products is very small, the thickness of the passivating layer is significant (e.g. ~10% of the total cavity), or some combination of these factors. In short, check.

There are other implications for a nonuniform medium filling the cavity. Not only does it amount to a correction that could be applied (described above), but it also reminds us that the field distribution within the cavity is nonuniform. It is nonuniform due to the natural formation of nodes and antinodes, as well as variations in the refractive index. If the passivation layer has a large refractive index, it will garner a greater fraction of the mode, and therefore, have a larger impact on any properties measured via the modes (e.g. fringe position, optical absorption, etc.). These spatial nonuniformities will be discussed below.

13.3.2 Spatially Dependent Response

When recording cavity-modified physics or chemistry and trying to build an understanding of what processes could be modified, it is easy to imagine the functional material within the cavity as a monolith. In other words, place material in a cavity and see that material's response is modified. In truth, there is an inhomogeneity of the cavity mode profile consisting of nodes and antinodes in FP cavities, which immediately yields two important conclusions. First, material located at a modal node will experience a different environment (and potentially a different cavity-induced modification of properties) than material placed at a mode peak. Second, optical probing of a material response will preferentially interrogate material at antinodes supported for the frequency of interrogation while neglecting material located at the nodes generated at the probe frequency. Considering the first point, material at different mode positions *feels* different environments [32]. Although a single-cavity mode extends across the entire cavity, and this delocalization has been cited as responsible for long-range excitation transfer, interaction rates are strongest at antinodes and weak at nodes [33]. In other words, although a single-cavity mode might enable long-range energy transfer via mode delocalization, this transfer is most likely to occur between a material excitation located at or near an antinode and material located some distance away but at another modal antinode [33–35]. This exact concept is used to design cavities for maximal energy transfer [36].

Considering cavity-inhibited chemical reactivity, one might propose that reactivity is slowed most for molecules located at regions of high-field intensity and negligibly at nodal positions. If this were the case, the system as a whole would exhibit a smeared range of rates. However, the means of monitoring the chemical reaction has now become key to determining what would be observed. First, consider electrical or ionic conductivity measured between the two mirrors of an FP cavity with the two mirrors serving as contact electrodes. If the conductivity were modulated due to interaction with the cavity mode, and this modification were greatest at modal antinodes, the system resistance could be coarsely modeled as two resistors in series, $R_{\text{measured}} = R_{\text{node}} + R_{\text{antinode}}$, where R_{node} and R_{antinode} are the resistances of material located at the node and antinode. Intuitively, it is clear that the measures of system resistance will underestimate the modification occurring in the most significantly altered regions. The effects of this spatial variation on measured resistance and its inaccuracy in measuring cavity-modified regions is shown in Figure 13.4. The important points to note are, first, that any increased resistance occurring at the antinodes is systematically measured as a lower value. For example, in Figure 13.4a, see the curve at $f = 10$, indicating a 10-fold increase in resistance at the antinodes. This results in a measured value that has only increased by a factor of 5. On the other hand, for reduced resistance at the antinode, the measurement error becomes much more significant. The overall error, measured as the measured resistance normalized by the resistance of interest (i.e. at the antinode) is $e = \frac{R_{\text{measured}}}{R_{\text{antinode}}} - 1$ (curves in Figure 13.4). One can see that at large increased resistance, the error is relatively small since the relatively large

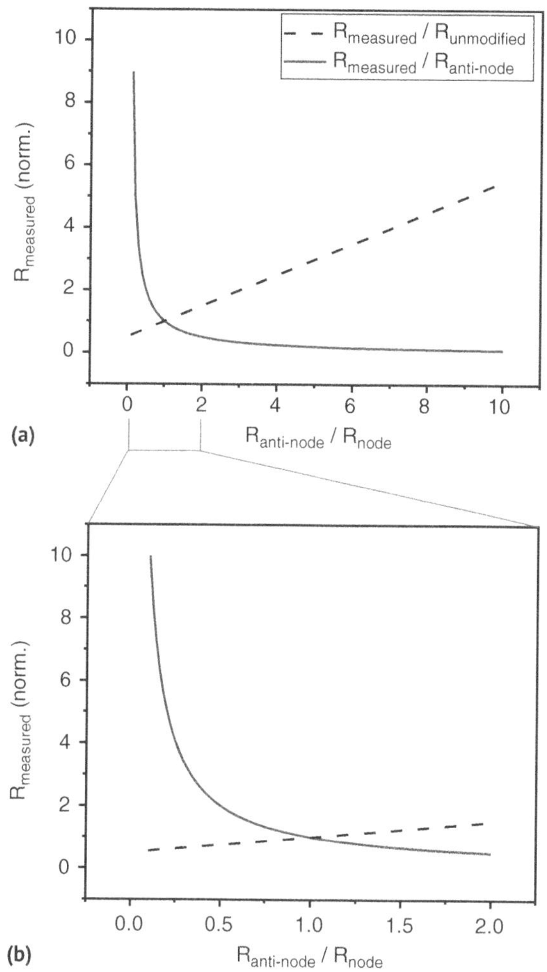

Figure 13.4 These calculated results predict (a) what resistance would be measured as a function of modified behavior at the modal antinode position. The dashed line plots the measured resistance normalized by the unmodified value and shows a linear relationship with a slope less than unity. The curve is essentially a measure of how poorly the measurement captures the behavior at the antinodes. (b) Presents the same data but over a much smaller range to highlight the drastically erroneous measurement that could be made under conditions where the conductivity is reduced at the antinodes.

R_{antinode} value dominates the R_{measured} expression. However, for reduced resistance, this systematic error becomes large (a factor of 10 for a 90% reduction in resistance; see Figure 13.4b for details). Notably, measuring resistance in the perpendicular orientation (current passing in-plane in an FP cavity, modeled as resistors in parallel) would experience the opposite effect (i.e. regions of reduced resistance will dominate the measured response rather than be obfuscated).

Considering the second point identified, which is that the modal structure results in preferential interrogation of certain regions over others [32], we highlight ultrafast spectroscopic results [37]. A demonstration of the effect of molecule location within the cavity will be made by modeling the transient response of a first-order cavity under 1% excitation. However, the spatial distribution of these excited molecules will be confined to different positions within the cavity mode. The magnitude of the resulting differential transmission signal (defined as $-\log^{T}/T_0$) will demonstrate

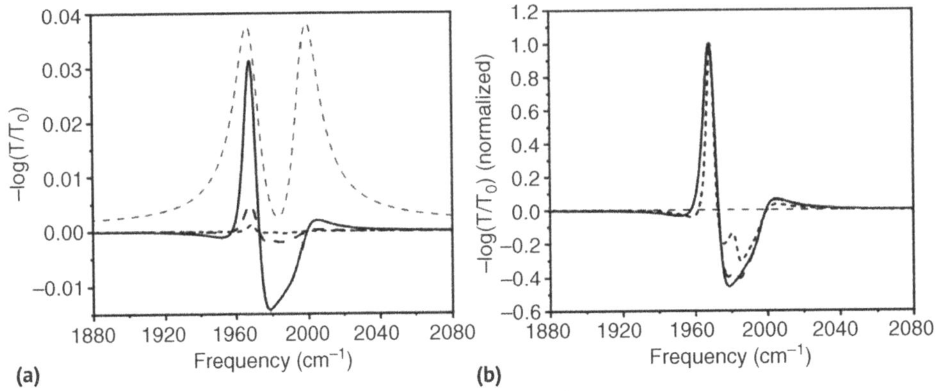

Figure 13.5 Differential transmission spectra are plotted for the same level of molecular excitation but localized to different regions of a first order Fabry-Perot cavity. In (a), it is shown that confining the excitation to a region of high field magnitude (anti-node) results in a much greater signal (solid curve) than when localizing that same excitation to a cavity node (dashed, dotted curves). The spectra are normalized and plotted in (b) to demonstrate that the fundamental spectral signatures are essentially the same and only vary in magnitude.

preferential interrogation of material located at the antinodes. Figure 13.5 plots the transient response of molecular excitation centered at different regions of the cavity. Qualitatively, all curves show a significant peak near the LP frequency and a derivative-like feature near the UP frequency. The origin of these features is well understood, discussed fully in the literature [38, 39], and is not the subject of this work. Instead, the focus here is on signal magnitude. Compare the solid curve, associated with excitation localized to a cavity antinode, to the dashed curve, where the same magnitude of molecular excitation has been localized to a cavity node. The positive peak located near the LP is still evident, but it is greatly diminished in magnitude. Furthermore, confining the same amount of excitation to a narrower region at the cavity node (see the dotted curve) results in nearly complete erasure of the optical signatures associated with the excitation. However, when these data are normalized and replotted (Figure 13.5b), it is evident that the optical signatures of the molecular excitation are essentially the same [37]. They only vary, albeit greatly, in their magnitude. The point here is that if one is using optical interrogation to report on a physical or chemical process within a cavity, the spatial structure of the mode at the frequency of interrogation will cause preferential reporting of the response from some regions (antinodes), while being relatively insensitive to changes occurring in other regions (nodes).

13.3.3 Line Broadening

Measured transmission spectra can be broadened by several experimental realities. Two will be discussed here. First, it is inevitable that there is some degree of nonparallelism between the two mirrors. In a typical vibrational strong coupling (VSC) procedure, mirrors are assembled, and their parallelism is evaluated by examining the number of visible fringes observed across the cavity face. Each visible fringe corresponds to a variation from fringe m to $m + 1$, which is the free spectral range. Since the cavity length, $l = m\frac{\lambda_{res}}{2n}$, where m is the mode order, λ_{res} is the resonance wavelength, and n is the refractive index, the difference in cavity lengths between adjacent modes is therefore $\frac{\lambda_{res}}{2n} \sim \frac{600\text{ nm}}{2*1.5} = 200$ nm for a fringe in the visible. The impact of this magnitude of cavity length variation will be much smaller in the infrared than the visible, but still present. Figure 13.6

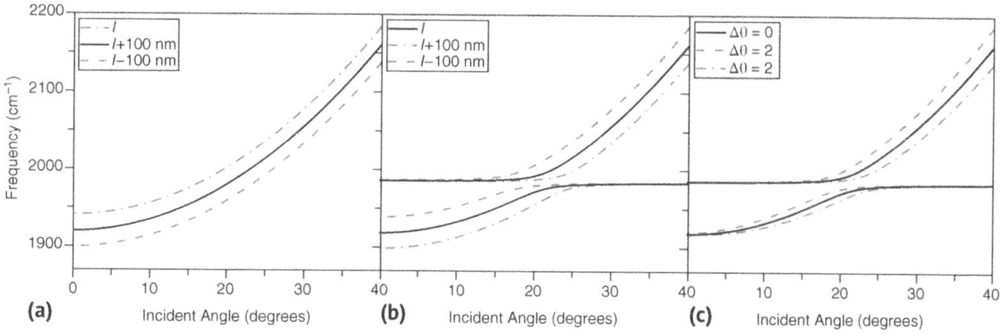

Figure 13.6 Plots showing the effect of mirror nonparallelism on dispersion measurements. First, in (a), an empty cavity dispersion is shown (black curve), along with the same dispersion assuming planarity variations over the interrogated area of ±100 nm (dot-dashed and dashed curves). A cavity exhibiting this degree of nonparallelism would have a dispersion broadening bounded by these curves. In (b), the same cavity is filled with an oscillator, but the anticrossing spectral signature is almost completely obfuscated due to nonparallelism. Finally, in (c), the effect of poor interrogation beam collimation is shown, which results in a range of interrogation angles.

presents calculated results showing the degree of line broadening in the infrared resulting from different degrees of nonparallelism. First, the effect of varying spacing between the mirrors, a result of nonideal parallelism, is shown in Figure 13.6a, b for empty and filled cavities. The black curves represent the response from the nominal cavity separation and the dot-dashed and dashed curves, the response for the cavity when the mirror separation is varied by 100 nm. In a real experiment, subject to these accuracies, the transmission spectra would experience a broadening bounded by the curves. In the case of strong coupling (Figure 13.6b), nearly the entire anticrossing feature is obscured. Another source of experimental broadening comes from the divergence of the probing beam. Conventional Fourier transform infrared (FTIR) spectrometers focus the infrared beam at the point of interrogation. This focus results in divergence angles of several degrees (the exact value depends on aperture settings and instrument specifics). Light impinging on the sample at the nominal angle will report a different resonant frequency for the cavity due to the angular dispersion. This effect is quantified and shown in Figure 13.6c.

In addition to these straightforward causes of measurement broadening, which can be addressed through improved experimental design, there may also be inherent inhomogeneities in the sample. A particularly relevant example is the OH vibrational band of water. This vibration has been the subject of several studies of strong coupling and its effect on chemical processes [2, 27, 28]. However, how to think about and properly describe this severely inhomogeneous vibrational band is an important and open question. Is it akin to "cooperative" coupling [40] since there are distinct vibrational modes contained within the broad envelope? Can one simply treat it as a very broad absorber, or should one think about it as a series of discrete absorbers at different energies? Both give the same modeled optical response but likely have very different chemical implications.

13.4 Current Challenges and Proposed Best Practices

The promise of cavity-modified chemical processes is not matched by its adoption. Although many exciting results exist, such as suppressed and enhanced reactivity, skewed product distribution, and shifted charge exchange equilibria, pursuit, and validation of these processes by a range of groups is not occurring. In fact, there are even reports of irreproducibility. A major obstacle is a reliable, robust, and convenient apparatus for evaluating the effects of cavity coupling on chemical processes. Certainly, a solution housed between two reflective surfaces is conceptually simple; however, the commercial tools available leave quite a lot to be desired in terms of flexibility, repeatability, and fortification against simple user error. It is with these obstacles in mind, and the desire to promote widespread testing and examination of cavity-coupling effects, that this chapter, and the piezo-adjustable reaction cavity design contained within, is put forth.

After describing the most important experimental challenges and the shortcomings of commonly used systems, a new design will be presented. The various functions of this apparatus will be described, and its function validated through repeatability, stability, and reversibility testing. The files needed to fabricate such a tool and a complete parts list are given in the supplemental materials.

13.4.1 Current Challenges

There is, no doubt, a great desire to repeat, verify, and extend the published demonstrations of cavity-modified chemical processes. At the crux of these experiments is assembling a well-controlled cavity and measuring the chemical activity occurring therein. However, a research group aiming to contribute to this burgeoning field is immediately faced with several challenges. In addition to the overall need for a simple and easy-to-use system that junior researchers (graduate or undergraduate students) can fruitfully use to generate contributive data, the most pressing challenges are

1. *Achieving a desired cavity tuning*: These studies require experimental trials carried out with cavity modes tuned to, or controllably detuned from, vibrational modes of interest. One must be able to achieve a targeted cavity resonance repeatedly to generate trustworthy statistics. Furthermore, one must ensure there is no unintended coincidence between higher order cavity modes and molecular vibrational modes, which could make experimental interpretation ambiguous. At odds with these requirements is the fact that commercial fluidic cells use polymer spacers that are only available in a few select thicknesses, which severely limits versatility.
2. *Stability over time*: Any chemical reaction will occur over some time interval (typically minutes to hours). The cavity tuning (at least to within a linewidth) must be stable over this time period. Commercial cells can be susceptible to mechanical relaxation of gaskets, O-rings, fastening screws, etc. Additionally, solvent leakage or evaporation into unintended headspace can alter pressure within the cavity, and therefore the cavity length and resonance frequency.
3. *Cavity parallelism*: The issue of mirror parallelism offers more subtle challenges. Departure from purely parallel mirrors results in a cavity tuning that varies with position (i.e. a cavity mode might be in resonance with a vibration of interest in one location but detuned from that desired mode a few millimeters away). The obvious pitfall is that any observable that samples the entire cavity (e.g. mass spectroscopy of product distribution, electrochemical charge passed, optical transmission that samples a large area), will effectively average the responses from regions of various tunings. Even if the measurement is localized to a region with acceptable tuning variance (e.g., a small spot-size optical measurement or electrochemical current from a small electrode), experimentalists are still subject to diffusion effects, which will mute the cavity-induced response in the same way, though more weakly, as the macroscale measurements described above.

4. *Ability to reliably make electrical contact to access system response*: Electrochemical tools represent an important opportunity for chemical interrogation of cavity-coupled species. Although metallic mirrors, which are commonly used in VSC experiments, can easily serve as a working electrode, commercial apparatuses for fluidic infrared transmission experiments are not easily amenable to making electrical contact to one or both mirror surfaces.
5. *Real-time modulation of cavity tuning*: When imagining potential uses and technologies based on cavity-modified materials, there is an immediate realization of the utility of real-time modulation, allowing slowed or hastened chemical reactivity on command, periodic cycling of a work function for energy control, or even incorporating diagnostic feedback to target a certain final product composition. Current commercial fluidic cells do not allow for real-time tunability.

13.4.2 Measuring Reaction Rates in Optical Cavities

If one hopes to observe any change in material properties or chemistry or under strong coupling, there are three kinds of experiments which are vital to perform: (i) The "out-of-cavity" experiment is a measurement of the system of interest performed outside a cavity. This is a conventional control measurement, which serves to establish a baseline for the system properties being studied. (ii) In the "cavity on-resonance" experiment, the system is measured inside a cavity that is tuned such that the energy of one of its optical modes coincides with that of a bare molecular mode *at normal incidence* – these are the conditions under which modified chemistry has been reported to occur under strong coupling [27, 28]. (iii) In the "cavity off-resonance" experiment, the system is measured inside a cavity whose optical mode is detuned from the bare molecular mode *at normal incidence*. In this negative control, no modification of material properties or chemistry should be observed, and its purpose is to rule-out measurement artifacts or mundane effects caused solely by the presence of the cavity mirrors. It is desirable to fix as many system parameters as possible across these three experiments to eliminate trivial sources of variability. For example, the sample path length should at least be comparable between out-of-cavity, cavity on-resonance, and cavity off-resonance measurements, even if detuning requires it to be changed slightly.

In this section, we will examine the case of chemical reactions under VSC monitored with FTIR spectroscopy. The out-of-cavity control is the same as a *conventional* infrared absorption measurement. Recall the Beer–Lambert law:

$$A \approx \epsilon CL \tag{13.13}$$

The *molar* absorption coefficient ϵ (units of L mol^{-1} cm^{-1}) is obtained from the slope of absorbance A versus concentration C of the absorbing species in a cell of fixed-path length L. It is important to point out that the linear relationship in Eq. (13.13) is only a good approximation for *weak* absorbers and *low* concentrations. Inconveniently, entering the strong coupling regime generally requires *strong* absorbers and *high* concentrations, so one must take care when utilizing Eq. (13.13) and constructing a calibration curve. Rather than correlating the maximum absorbance of a particular absorber $A(\nu_j)$ with its concentration C_j, it is advisable to use a more robust quantity such as *integrated* absorbance, which is always directly proportional to concentration [41, 42]:

$$\int_0^\infty A_j(\nu)\mathrm{d}\nu \propto S_j = s_j C_j \tag{13.14}$$

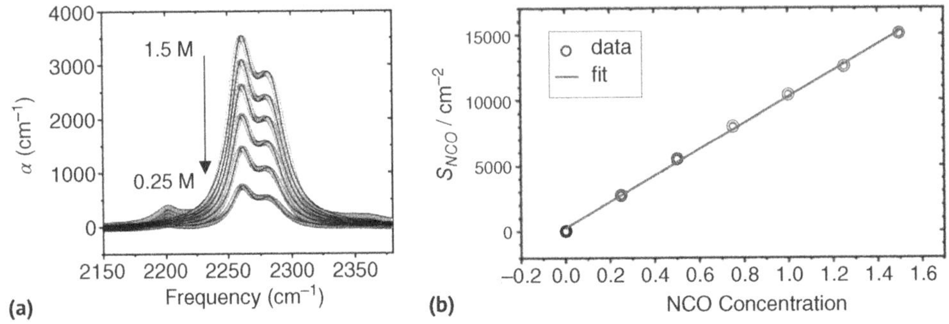

Figure 13.7 (a) Concentration-dependent absorptivity of phenyl isocyanate NCO stretching mode in tetrahydrofuran fit with a pair of Lorentz oscillators. (b) Calibration curve for linearly correlating the integrated absorptivity of the stronger oscillator with the concentration of phenyl isocyanate. *Source*: [7] / American Association for the Advancement of Science.

Here, S_j is the integrated absorptivity (the factor in the numerator of Eqs. (13.3 and 13.4), with units of cm^{-2}), and s_j is the respective molar quantity (units of $L\ mol^{-1}\ cm^{-2}$). To construct a calibration curve, a series of absorbance spectra are measured for known concentrations C_j of the absorbing species j, and s_j is treated as a free parameter in a fit using the Lorentz oscillator model (Figure 13.7a). The best-fit values are then plotted versus concentration and s_j is extracted as the slope of the best-fit line (Figure 13.7b). Whenever possible, any species present in the reaction (including spectators) should be isolated or procured in their pure form and calibrated in the intended reaction solvent. Out-of-cavity kinetic data can then be fit using Eqs. (13.3 and 13.4) (substituting in $S_j = s_j C_j$), allowing for concentration C_j to be extracted at each time point.

Tracking reaction progress within optical cavities requires some additional considerations. Due to the highly reflective broadband mirrors commonly used for VSC, nonzero optical transmittance is limited to the narrow cavity resonances making the conventional infrared absorption measurement described above impossible. Luckily, the analytical description of cavity transmission in Eq. (13.2) is very sensitive to the presence of molecular modes when they overlap in energy with optical modes. Consequently, if the cavity is resonant with a vibrational mode of the reactant, the rate of consumption during the reaction manifests as a time-dependent Rabi contraction. Therefore, reaction kinetics for "cavity on-resonance" systems may be directly extracted from transmission spectra taken at normal incidence via fitting to Eq. (13.2). However, this presents a problem: when the cavity is *detuned* from the molecular mode of interest, the transmission spectrum acquired at normal incidence no longer reports on the molecular concentration – how then can one perform the aforementioned "cavity off-resonance" control measurement? The frequencies of higher order cavity fringes *do* respond to changes in the solution's refractive index taking place over the course of a reaction and this has previously been used to track reaction progress [43–45]. Unfortunately, this method can be an unreliable handle because the positions of these fringes are also susceptible to solvent leakage, evaporation, temperature changes, mechanical stabilization of the cell, and even strong coupling of optical modes in remote spectral regions [31]. A clever alternative approach for measuring reaction rates in off-resonant cavities was developed by Ahn et al. [7]. Their strategy leverages angular dispersion and involves collecting on-resonance transmission spectra at small nonzero angles for cavities, which are detuned at normal incidence. These spectra can then be fit using Eq. (13.2) to extract molecular concentration. In the next section, we will construct a model experiment to computationally validate this method of analysis.

13.4.3 Validating Angle-independent Rate Extraction

The analytical approximation to the transfer matrix (TM) method in Eq. (13.2) describes the transmittance of an optical cavity containing absorbers, but neglects the angle-dependent reflectance of the mirrors. To demonstrate the efficacy of this approximation for analyzing chemical kinetics measured in a cavity at different incident angles, we will simulate the time-dependent optical response of a model system using the full-TM method and show that the approximation gives a result, which is indistinguishable within the margin of experimental error.

The system is defined to emulate the cavity-coupled urethane addition reaction reported by Ahn et al. [7]. This reaction involves the alcoholysis of an isocyanate (R—NCO) by an alcohol (R—OH) in tetrahydrofuran (THF) to form a carbonyl-containing carbanilate (R—C=O):

$$\mathrm{R{-}NCO} + \mathrm{R{-}OH} \xrightarrow{\mathrm{THF}} \mathrm{R{-}C{=}O} \tag{13.15}$$

This bimolecular reaction is first order in each reactant, that is,

$$\frac{\mathrm{d}}{\mathrm{d}t} C_{\mathrm{CO}}\,(\mathrm{t}) = \frac{-\mathrm{d}}{\mathrm{d}t} C_{\mathrm{NCO}}(t) = kC_{\mathrm{NCO}}C_{\mathrm{OH}} \tag{13.16}$$

where k is the second-order rate constant with units of $\mathrm{M}^{-1}\,\mathrm{s}^{-1}$. If we assume an equimolar (3.0 M) starting solution of these reactants, this differential rate equation can be rewritten in terms of a single reactant:

$$\frac{\mathrm{d}}{\mathrm{d}t} C_{\mathrm{NCO}}\,(\mathrm{t}) = -kC_{\mathrm{NCO}}^{2} \text{ when } C_{\mathrm{NCO}}(0) = C_{\mathrm{OH}}(0) \tag{13.17}$$

Integrating both sides yields:

$$C_{\mathrm{NCO}}\,(t) = \frac{1}{C_{\mathrm{NCO}}^{-1}(0) + kt} \implies C_{\mathrm{NCO}}^{-1} = C_{\mathrm{NCO}}^{-1}\,(0) + kt \tag{13.18}$$

The average of six out-of-cavity rate constants and their standard deviation as measured by Ahn et al. [7] is $k_{\mathrm{OUT}} = 2.34\,(\pm 0.2) \times 10^{-5}\,\mathrm{M}^{-1}\,\mathrm{S}^{-1}$. This empirical value will be used as an input to describe the time-dependent complex permittivity of the solution by substituting Eq. (13.18) into Eqs. (13.3 and 13.4). We could define a single-Lorentz oscillator corresponding to the decaying reactant mode coupled to the cavity (e.g. the isocyanate NCO stretch at 2260 cm^{-1}), but in general, it is important to account for any nearby resonances as well, especially if they too vary in time. While not necessary in this case, we will also define the remote product carbonyl at 1730 cm^{-1}, which is produced over the course of the reaction. Given the stoichiometry of these two species in Eq. (13.15), we can rewrite the product concentration $C_{\mathrm{CO}}(t)$ in terms of the reactant concentration $C_{\mathrm{NCO}}(t)$:

$$C_{\mathrm{CO}}\,(t) = C_{\mathrm{NCO}}\,(0) - C_{\mathrm{NCO}}\,(t) \tag{13.19}$$

Finally, the solvent THF contributes a background dielectric constant of $\varepsilon_{\infty} = n_{\infty}^{2} = 1.9802$. The simulated time-dependent out-of-cavity absorption spectra for a 7.4-μm path length cell are presented in Figure 13.8. This dataset represents a series of out-of-cavity control measurements, an analysis of which is expected to return the input reaction rate constant of k_{OUT}.

To accurately model the transmittance under cavity-coupled conditions, we use the TM formalism implemented in the TMM python package [46]. An unpolarized transmittance calculation was run for a cavity defined by two 15-nm thick gold mirrors with the complex refractive obtained by Rakić et al. [47] separated by a 7.4-μm thick solution layer. This configuration places the frequency of the fifth-order cavity mode coincident with ν_{NCO}. The TM-simulated time-dependent

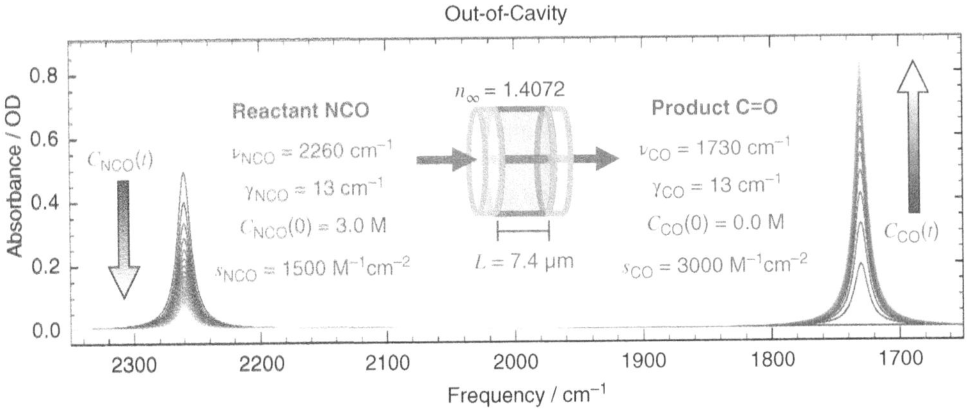

Figure 13.8 Results of a numerical experiment intended to validate the angle independence of reaction rate extraction. The simulated out-of-cavity spectra represent the consumption of the reactant NCO mode (left) and the subsequent generation of the product C=O mode. Parameters utilized in the calculation are provided as well as a schematic depiction of the model system.

cavity transmission spectra for on-resonance (T_{TM}^{ON}) and off-resonance (T_{TM}^{OFF}) conditions are presented in Figure 13.9a, b, respectively, and Figure 13.9b illustrates the transition between these two system geometries. The on-resonance calculation is performed at normal incidence, requiring a specific path length L^{ON} to meet the resonance condition. For the off-resonance calculation, the path length must first be slightly changed to L^{OFF} to detune the cavity mode from the isocyanate mode at normal incidence (top-bold-gray-to-black dashed in Figure 13.9b). Next, the incident angle is changed from 0° until the optical mode disperses back into resonance with the isocyanate mode at some small, nonzero angle (black-dashed-to-bottom-bold gray in Figure 13.9b).

The most notable feature in the time-dependent spectra is the contraction of the Rabi splitting Ω_{ON} and Ω_{OFF} from ~48 to 18 cm^{-1} over the 28-hour interval calculated – as expected, the splitting scales linearly with $\sqrt{C_{NCO}}$ (see Figure 13.10b). It appears that the amplitudes of the UP and LP become more asymmetric as the reactant is converted into the product. This is caused by a passive detuning of the optical mode from the isocyanate due to a change in the solution refractive index during the reaction. This highlights that, to the degree with which the solution dielectric function is adequately described (i.e. Lorentz oscillators describe all relevant optical transitions), any impact on the spectral response due to reactant consumption or product generation will be represented in this model. Changes in the concentration of species whose optically active transitions coincide in frequency with the cavity mode will be the dominant factor defining the spectrum. When L is relaxed in the fit, it can compensate for small variations in the *effective* path length, such as changes in the solution dielectric function not explicitly modeled as well as mechanical instability of the cell.

The TM-simulated datasets T_{TM}^{ON} and T_{TM}^{OFF} are fit using the approximation from Eq. (13.2) giving T_{fit}^{ON} and T_{fit}^{OFF}, respectively, the difference between these two descriptions are plotted at the bottom of Figure 13.9a, c. As evidenced by the small residual, the approximation does an excellent job of reproducing the essential features of cavity transmission for both the on- and off-resonance simulations. The concentrations $C_{NCO}(t)$ extracted from the analytical fits were used to construct the second-order rate plot in Figure 13.10a. Here, the slopes represent the reaction rate constants (with units of 10^{-5} M^{-1} s^{-1}) k_{ON} = 2.37 (O's) and k_{OFF} = 2.33 (X's) while the gray-shaded region represents the uncertainty of ±0.2 in the empirical out-of-cavity rate constant k_{OUT} = 2.34 that was

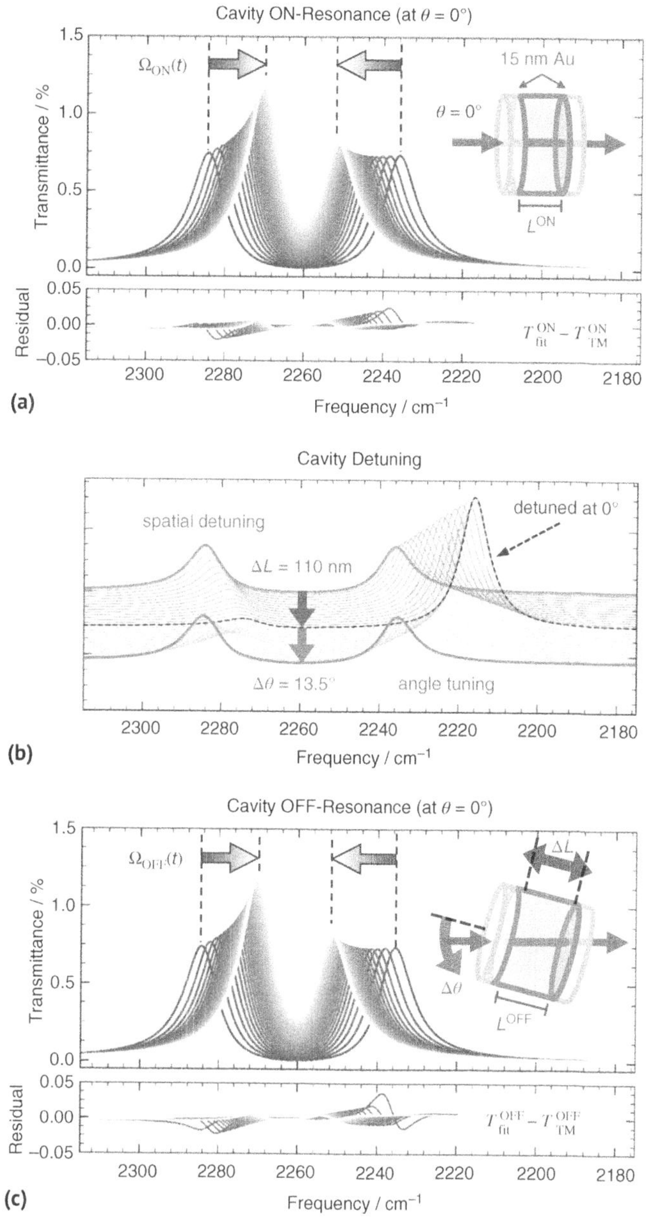

Figure 13.9 Results of a numerical experiment intended to validate the angle-independence of reaction rate extraction. (a) Cavity transmittance spectra calculated at 0° for the reactant NCO mode on-resonance with an optical mode at normal incidence. A schematic depiction of the model system is provided. The bottom subplot shows the difference between the full-TM result and a fit using the analytical approximation in Eq. (13.2). (b) Cavity transmittance spectra calculated for a series of path lengths intended to spatially detune the reactant NCO mode from the optical mode (top-bold gray) and for a series of incident angles intended to bring them back into resonance (bottom-bold gray). The bottom-most spectrum represents the geometry used in the off-resonance control measurement. (c) Cavity transmittance spectra calculated at 13.5° for the reactant NCO mode off-resonance from an optical mode at normal incidence. A schematic depiction of the model system is provided. The bottom subplot shows the difference between the full-TM result and a fit using the analytical approximation in Eq. (13.2). The residuals at the bottom of (a) and (b) are very small suggesting excellent agreement between the full-TM and approximate results.

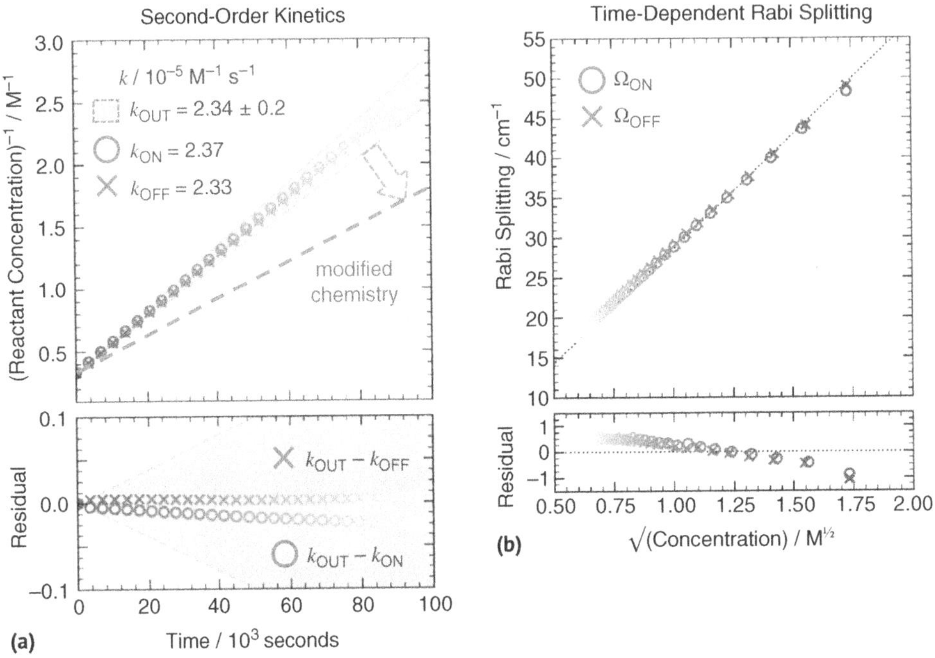

Figure 13.10 Results of a numerical experiment intended to validate the angle independence of reaction rate extraction. (a) Second-order rate plot showing inverse reactant concentration as a function of time. The slopes represent the rate constants k_{ON} and k_{OFF} obtained from simulations of on-resonance (O's) and off-resonance (X's) cavities, respectively. The gray-shaded region represents the uncertainty in the measured out-of-cavity value k_{OUT} used as simulation input. The bottom subplot shows the negligible difference between the simulated on- and off-resonance cavities and a comparison to the empirical out-of-cavity control. (b) The time-dependent Rabi splitting on- (O's) and off-resonance (X's) extracted from Figure 13.9a, c, respectively. Both datasets exhibit a strong linear correlation with $\sqrt{C_{NCO}}$.

used to as input to generate the time-dependent solution dielectric function. Both values obtained from the simulated cavity transmittance fall comfortably within the gray-shaded region bounded by the experimental uncertainty, as is clearly seen from the residuals plotted at the bottom of Figure 13.10a. Thus, we have demonstrated the effectiveness of this approach in accurately determining reaction rates from cavity transmittance spectra, independent of incident angle. The on- and off-resonance simulations are nearly indistinguishable from each other (differing only by ~2%) despite being performed under different system geometries. This is not only expected given the correspondence between angular and spatial dispersion in FP cavities, but also reassuring since our model is purely optical and does not attempt to capture how light–matter interactions might impact the dynamics in principle. This means that if the on-resonant results were to diverge from those of the out-of-cavity and off-resonant controls in a real experiment, it could be argued that some intriguing chemical physics driven by light–matter interactions are at play. The dashed line in Figure 13.10a illustrates the kind of anomalous rate suppression that is frequently reported in chemical reactions under VSC.

13.4.4 Proposed Cavity System Design

The field of VSC-modified chemistry is still relatively young and there exist many limitations with "first-generation" experimental apparatuses, which should be addressed to advance our understanding of cavity-coupled phenomena. The most common etalon systems employed are demountable, fixed-path length microfluidic cells with mirrors consisting of gold film deposited on a transparent substrate and topped with an oxide or polymer-based passivation layer. The mirror separation is often coarsely defined by a polymer spacer, which can be compressed slightly, allowing for some tunability. Computer-controlled nanopositioning systems are commonly available and easily afford the accuracy needed for infrared cavities (~100 nm). Versions of such systems

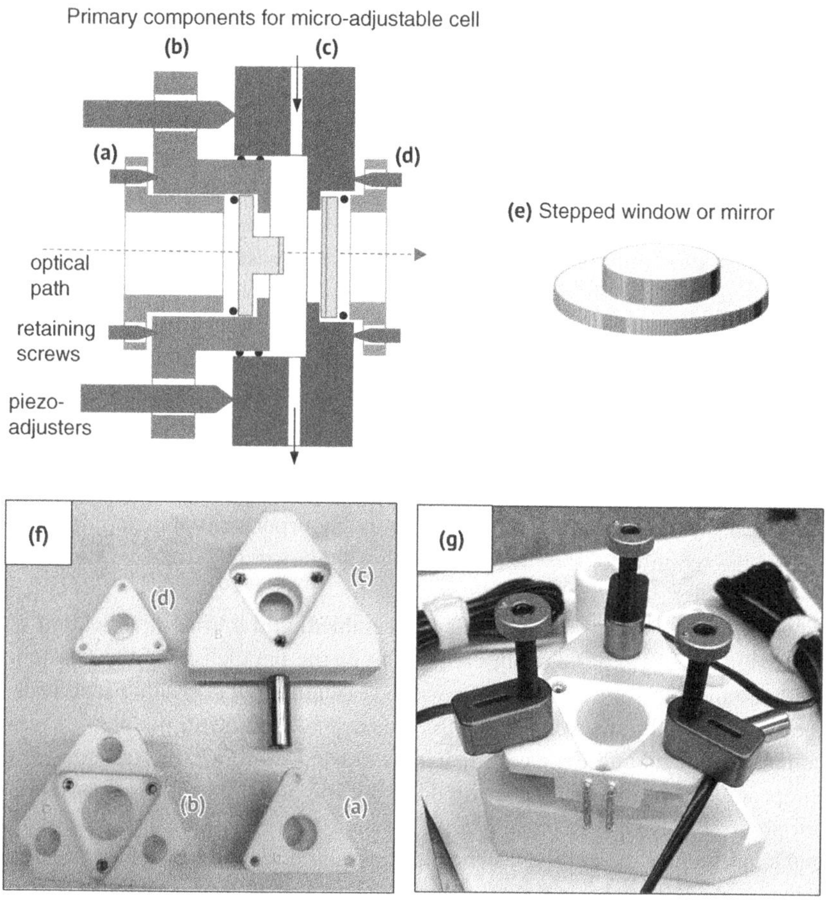

Figure 13.11 Schematic diagram of proposed fluidic test cell. Such a system consists of four parts, which create two half cells. Parts (a) and (b) join together to hold a stepped window or mirror in place. This is half-cell a–b. Retaining screws (dark gray) or a screw fitting (similar to common Harrick cells) provide the retaining force to clamp the window. Parts (c) and (d) retain a simple flat window in a similar fashion. The two half cells are then joined with a–b sliding into c–d. Piezo or stepper motor adjuster provide accurate control of mirror separation. O-rings provide seals on the backside of the windows as well as between the two half cells. A qualitative detail of the stepped window is shown in (e). Such a cell could be 3D printed, and a link to preliminary design files is provided in the references. (f) Shows the 3D printed components, each labeled as (a), (b), (c), (d) and (g) shows the fully assembled configuration.

have been assembled [3, 11, 48, 49] but providing for a sealed liquid environment while maintaining a broad tunability range is challenging. Piezo-actuated cavities have been used to maintain site-selective VSC during chemical reactions while actively compensating for changes in the solution refractive index with a computer-controlled feedback loop, ensuring that the desired coupling condition is sustained for the duration of the measurement [3]. A versatile solution is shown in Figure 13.11 (.stl files suitable for three-dimensional (3D) printing, or as a starting point for modified designs, are given in the supporting information in the arXiv version of this manuscript [50]). This apparatus consists of four main components (labeled a, b, c, and d). One mirror is mounted between components (a) and (b) and another between (c) and (d). One of these mirrors is machined in a stepped geometry (see Figure 13.11e). The second window is a conventional size of 1 in diameter × 2 mm thick. These windows could be used uncoated and serve as transparent boundaries to an infrared transmission measurement or coated with a layered mirror structure to yield an FP cavity. Either mirror, stepped or flat, could be placed in either housing (a–b or c–d) except when electrical contact is desired. The current design utilizes retaining screws to hold (a) and (b) together, thus, holding the window in place. Another option would be to compress (a) into (b) with a large-threaded cover (similar to the conventional Harrick cell design). After installing the windows, the two halves of the system (a–b and c–d) are assembled with a–b sliding into c–d and retaining the two halves to one another with springs (e.g. Six springs of type 1942N4 from McMaster Carr). This brings the windows into proximity of one another. Precise window separation is controlled by piezoelectric actuators (e.g. Newport model 8322NF linear actuator) mounted into part (b) and shown as red in Figure 13.11. Fluidic sealing is maintained by O-rings (shown in black). A single O-ring is placed on the rear of each windows and two are used to seal the interface between the two half cells. Fluidic channels allow solution to be flowed into or out of the call (indicated by black arrows). The design shown in Figure 13.11 is really quite similar to what can be assembled with conventional piezo, stepper, or other precision adjustable optics components; however, there are two important contributions. First is placement of O-rings in relevant locations, making the system compatible with liquid environments. Second is the incorporation of a stepped window, which allows for robust retaining of each window while still allowing micron-scale window separations.

The system described in Figure 13.11 was 3D printed, assembled, and a brief qualification of its function carried out (Figure 13.12). First, rough alignment of the parallelism of the two half cells is carried out by manually adjusting the piezo screw drives (Figure 13.11) to achieve autocollimation of laser reflections from the two windows. After this coarse alignment, the piezo drives were actuated via computer control and the cavity spacing extracted from infrared transmission fringe analysis. The repeatability of the cavity spacing upon cycling of the piezo drives is shown in Figure 13.12a–c for progressively smaller piezo cycling distances. There is some drift representing the limits of cavity length repeatability. These behaviors are quantified in terms of the average displacement and standard deviation of this displacement as a function of number of piezo drive cycles (a single "cycle" is the smallest displacement each drive can accomplish). Several points can be made. First, the displacement (Figure 13.12d) scales linear with the number of cycles for the sparse dataset here. Second, the slope, which represents the distance moved per cycle, is slightly smaller (~10%) for contraction of the cavity than for extension. This may be due to insufficient force produced by the retaining springs used to hold the two half cells together. Third, the standard deviation of the displacement realized per cycle (Figure 13.12e) decreases with increased number of cycles, presumably due to averaging out the slippage events that occasionally occur. Finally, a dispersion contour is presented in Figure 13.12f generated by tuning the cavity length from 10 to 30 μm while recording infrared transmission spectra.

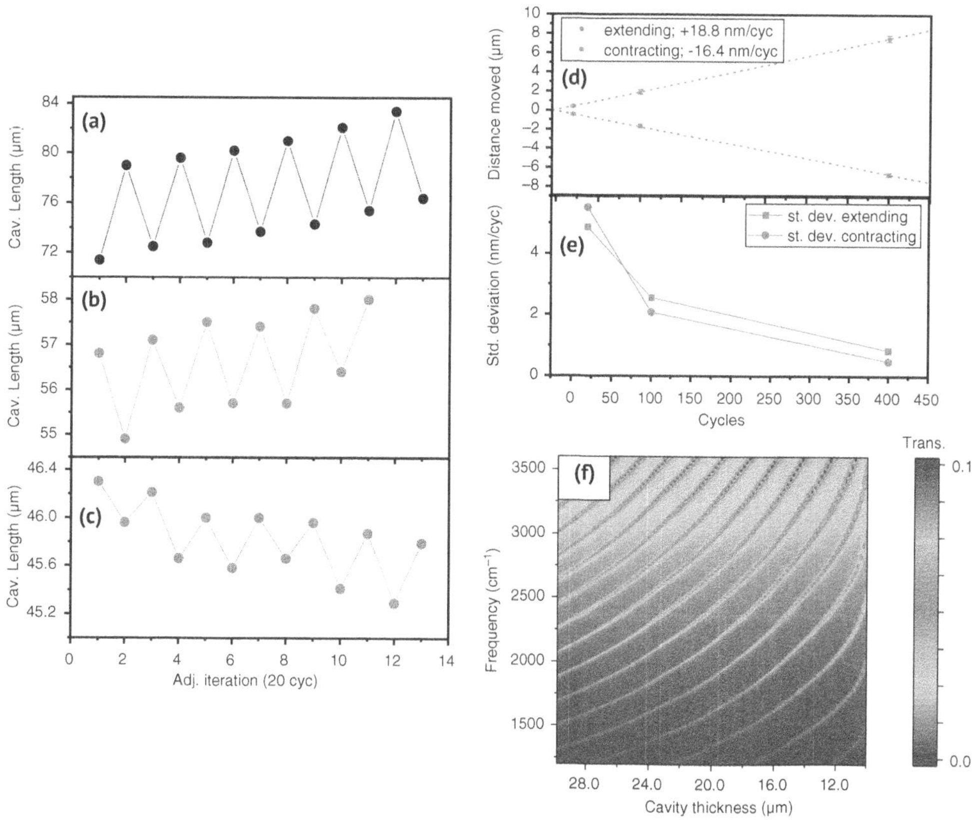

Figure 13.12 Preliminary qualification data for the function of cell design. The reproducibility of positioning, as one actuates the tuning positioners back and forth, is shown in (a), (b), and (c) for movements of ~8, 2, and 0.5 μm, respectively (i.e. 400, 100, and 20 cycles of the piezo adjusters). These responses are used to generate the average displacement and standard deviations of the displacement plotted in (d) and (e) as a function of number of cycles applied. In (f), an example dispersion plot is shown for an air-filled cavity tuned from 10 to 30 μm mirror separation.

A second, improved in some respects, design is illustrated in Figure 13.13. This system is similar in spirit to the model shown in Figure 13.11; however, there are several differences. First, restoring force exerted against the piezo-stepper motors is accomplished by compressed springs (McMaster–Carr part#6547K272; 0.5″ long, OD 0.198″) held by retaining bolts (McMaster-Carr part#10F200KCSFT/X, 10-32, 2″ long hex head screw), which are far easier to implement. Secondly, one mirror is mounted by gluing it in place rather than having both mirrors retained through compression. This necessitates a slightly different sealing mechanism, which can be seen as components (a–c) are bolted together with two square-profile O-rings between them. The fluidic connections are designed to be compatible with ¼-28 standard flat-bottom high-performance liquid chromatography (HPLC) fitting. Finally, the general tolerances have been refined, so the two halves fit much more snugly together. The STL files for these components are also found in the supporting information of the arXiv version of this manuscript [50].

Gold is a convenient choice of mirror coating for a mid-IR cavity, given its simple fabrication and high reflectivity. However, the broadband reflectivity of gold and other metals supports many optical modes, which can inadvertently couple to molecular vibrations outside the region

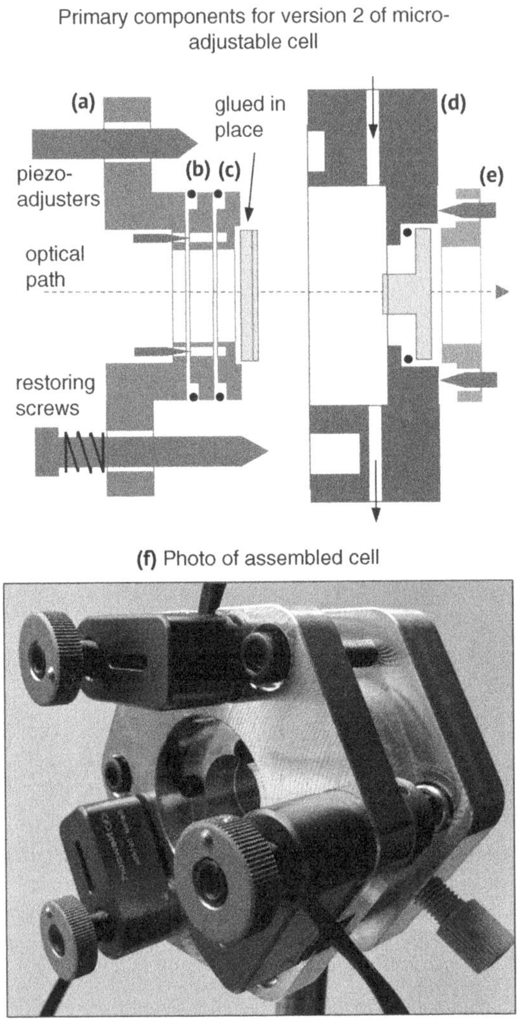

Figure 13.13 Second version of piezo-tunable cell. The schematic at the top shows parts (a)–(c), which are fastened together by three bolts (McMaster-Carr part#91251A113, 4-40 thread, 3/4″ long hex head screw). O-rings fit between each of these to provide the liquid seal. Mirror can be glued in place. This component slides into (d). Restoring screws (gray, McMaster-Carr part#10F200KCSFT/X, 10-32, 2″ long hex head screw) pass through (a), (b), and (c) and screw into (d) and have a spring (McMaster-Carr part#6547K272; 0.5″ long, OD 0.198″) located between the screw head and part (a) to provide restoring force to oppose piezo-stepper motion. Second mirror is fastened between (d) and (e) (McMaster-Carr part#91251A106, 4-40, 1/4″ long hex head screw). The image in (f) shows digital rendering of fully assembled system.

of interest, making it difficult to infer a straightforward mechanism responsible for modified chemistry [7]. As mentioned in Section 13.4.2, the negligible cavity transmission over regions between resonant fringes makes conventional infrared absorption measurements impossible. While the previous section describes a reliable strategy for sidestepping this issue for measurements of chemical kinetics, this is far from ideal and a great deal of valuable spectral information remains inaccessible. This high reflectivity also extends to shorter wavelengths, causing etaloning and reduced transmission in the visible and ultraviolet regimes, where it may be desirable to optically interrogate a cavity-coupled system [12, 27–29]. The resulting spectral artifacts can obfuscate any interesting VSC-induced effects and must be carefully corrected to make a confident determination [12, 30, 40]. Furthermore, the proximity of molecular absorbers to metallic interfaces can induce catalytic [51] or surface enhancement [52, 53] effects. A passivating polymer or oxide layer is deposited on the gold to safeguard against these effects, although this complicates sample fabrication and, under some circumstances, could induce confusing optical effects. A better alternative to metallic mirrors for studying VSC might be distributed Bragg reflectors (DBRs), which consist of alternating layers of dielectric materials with different refractive indices. DBRs can have higher Q-factors than metallic mirrors due to their low loss and high reflectivity making them appealing for VSC applications. However, the effective coupling strength one can achieve with DBRs is limited by a reduction in the number of molecules within a given mode volume due to penetration of the EM field into the dielectric layers [54]. The optical properties of a DBR stack can be tailored by selecting different materials and varying the thickness of each layer. Analytical tools exist that allow for the optimization of these multilayer structures to produce the desired properties [55, 56]. For example, DBRs can be designed to have a narrow reflective band at arbitrary frequencies to enable VSC with a specific molecular vibration while simultaneously providing high, unstructured transmittance in other spectral regions. This enables optical pumping [57], straightforward monitoring of reaction progress in the mid-IR or visible, and prevents unintended coupling between other cavity modes and material excitations.

13.5 Conclusion

This chapter now concludes. In it, we have given guidance regarding the analytical modeling of simple etalon-type cavities, which allows researchers the ability to design and explore experimental regimes quickly on a laptop. Experimental nonidealities that scientists should be aware of, but are often neglected, were described along with estimates of their impact on the experiment. Finally, a preliminary design for a cavity apparatus was presented, which gives greater reliability, experimental repeatability, and convenience of use. We wish for this design to stimulate ongoing experimental apparatus development.

The authors hope this information assists in the validation and understanding of cavity-modified materials behavior by providing tools to ease entry into this field of study. Not only will this lead to repeated and reliable demonstrations, but it will also help participants from a wide variety of scientific disciplines make immediate contributions. Contributions from high-resolution spectroscopists, experts in reaction kinetics, physical chemists, and physicists will help establish rigor, trust, and eventual technological impact in this field.

References

1 Thomas, A., Lethuillier-Karl, L., Nagarajan, K., Vergauwe, R. M. A., George, J., Chervy, T. et al. Tilting a ground-state reactivity landscape by vibrational strong coupling. *Science* **2019**, *363*, 615–619.

2 Hirai, K., Ishikawa, H., Chervy, T., Hutchison, J. A., Uji-I, H. Selective crystallization via vibrational strong coupling. *Chem. Sci.* **2021**, *12*, 11986–11994.

3 Hirai, K., Ishikawa, H., Takahashi, Y., Hutchison, J. A., Uki-I, H. Autotuning of vibrational strong coupling for site-selective reactions. *Chem. Eur. J.* **2022**, *28*, e202201260.

4 Nikolis, V., Mischok, A., Siegmund, B., Kublitski, J., Jia, X., Benduhn, J. et al. Strong light-matter coupling for reduced photon energy losses in organic photovoltaics. *Nat. Commun.* **2019**, *10*, 37062.

5 Zhong, C., Hou, S., Zhao, X., Bai, J., Wang, Z. Gao, F. et al. Driving DNA origami coassebling by vibrational strong coupling in the dark. *ACS Photon* **2023**, *10*, 1618–1623.
6 Garcia-Vidal, F. J., Ciuti, C., Ebbesen, T. W. Manipulating matter by strong coupling to vacuum fields. *Science* **2021**, *373* eabd0336.
7 Ahn, W., Triana, J. F., Recabal, F., Herrera, F., Simpkins, B. S. Modification of ground state chemical reactivity via light-matter coherence in infrared cavities. *Science* **2023**, *380*, 1165.
8 Galego, J., Climent, C., Garcia-Vidal, F. J., Feist, J. Cavity casimir-polder forces and their effects in ground-state chemical reactivity. *Phys. Rev. X.* **2019**, *9*, 021057.
9 Zhdanov, V. P. Vacuum field in a cavity, light-meditated vibrational coupling, and chemical reactivity. *Chem. Phys.* **2020**, *535*, 110767.
10 Vurgaftman, I., Simpkins, B. S., Dunkelberger, A. D., Owrutsky, J. C. Negligible effect of vibrational polaritons on chemical reaction rates via the density of states pathway. *J. Phys. Chem. Lett.* **2020**, *11*, 3557–3562.
11 Imperatore, M. V., Asbury, J. B., Giebink, N. C. Reproducibility of cavity-enhanced chemical reaction rates in the vibrational strong coupling regime. *J. Chem. Phys.* **2021**, *15*, 191103.
12 Wiesehan, G. D., Xiong, W., Negligable rate enhancement from reported cooperative vibrational strong coupling catalysis. *J. Chem. Phys.* **2021**, *155*, 241103.
13 Pedrotti, F. L., Pedrotti, L. S. *Introduction to Optics, Prentice-Hall International*. 2e. Inc., Englewood Cliffs, NJ. 1993.
14 Born, M., Wolf, E. *Principles of Optics: Electromagnetic Theory of Propagation, Interference and Diffraction of Light*. 7e. Cambridge University Press, Cambridge, UK. 1999.
15 Khitrova, G., Gibbs, H. M., Jahnke, F., Kira, M., Koch, S. W. Nonlinear optics of normal-mode-coupling semiconductor micro-cavities. *Rev. Mod. Phys.* **1999**, *71*, 1591–1639.
16 Törmä, P., Barnes, W. L. Strong coupling between surface plasmon polaritons and emitters: a review. *Rep. Prog. Phys.* **2015**, *78*, 013901.
17 Zhu, Y., Gauthier, D. J., Morin, S. E., Wu, Q., Carmichael, H. J., Mossberg, T. W. Vacuum rabi splitting as a feature of linear-dispersion theory: analysis and experimental observations. *Phys. Rev. Lett.* **1990**, *64*, 2499–2502.
18 Houdré, R., Gibernon, J. L., Pellandini, P., Stanley, R. P., Oesterle, U., Weisbuch, C. et al. Saturation of the strong-coupling regime in a semiconductor microcavity: free-carrier bleaching of cavity polaritons. *Phys. Rev. B.* **1995**, *52*, 7810.
19 Agranovich, V. M., Litinskaia, M., Lidzey, D. G. Cavity polaritons in microcavities containing disordered organic semiconductors. *Phys. Rev. B: Condens. Matter Mater. Phys.* **2003**, *67*, 085311.
20 Stanley, R. P., Houdré, R., Weisbuch, C., Oesterle, U., Ilegems, M. Cavity-polariton photoluminescence in semiconductor microcavities: experimental evidence. *Phys. Rev. B.* **1996**, *53*, 10995.
21 Houdré, R., Stanley, R. P., Ilegems, M. Vacuum-field Rabi splitting in the presence of inhomogeneous broadening: resolution of a homogeneous linewidth in an inhomogeneously broadened system. *Phys. Rev. A: At., Mol. Opt. Phys.* **1996**, *53*, 2711–2771.
22 Gibbs, H., Khitrova, G., Koch, S. Exciton–polariton light–semiconductor coupling effects. *Nature. Photon.* **2011**, *5*, 273.
23 Skolnick, M. S., Fisher, T. A., Whittaker, D. M. Strong coupling phenomena in quantum microcavity structure. *Semicond. Sci. Technol.* **1998**, *13*, 645–669.
24 Hopfield, J. J. Theory of the contribution of excitons to the complex dielectric constant of crystals. *Phys. Rev.* **1958**, *112*, 1555–1567.
25 Holmes, R. J., Forrest, S. R. Strong exciton-photon coupling and exciton hybridization in a thermally evaporated polycrystalline film of an organic small molecule. *Phys. Rev. Lett.* **2004**, *93*, 186404.

26 Long, J. P., Simpkins, B. S. Coherent coupling between a molecular vibration and fabry-perot optical cavity to give hybridized states in the strong coupling limit. *ACS Photon* **2015**, *2*, 130–136.
27 Vergauwe, R. M. A., Thomas, A., Nagarajan, K., Shalabney, A., George, J., Chervy, T. et al. Modification of enzyme activity by vibrational strong coupling of water. *Angew. Chem. Int. Ed.* **2019**, *58*, 15324–15328.
28 Lather, J., George, J. Improving enzyme catalytic efficiency by co-operative vibrational strong coupling of water. *J. Phys. Chem. Lett.* **2021**, *12*, 379–384.
29 Pijko, M., Patrahau, B., Joseph, K., Muller, C., Devaux, E., Ebbesen, T. W. et al. Solvent Polarity under vibrational strong coupling. *J. Am. Chem. Soc.* **2023**, *145*, 13215–13222.
30 Saydjari, A., Long, J. P., Dressick, W. J., Simpkins, B. S. Optical ionterference effect corrections for absorbance spectra of layer-by-layer thin films bearing covalently bound dye. *Chem. Phys. Lett.* **2014**, *608*, 328–333.
31 Thomas, P. A., Barnes, W. L. Strong coupling-induced frequency shifts of highly detuned photonic modes in multimode cavities. *J. Chem. Phys.* **2024**, *160*, 204303.
32 Ahn, W., Vurgaftman, I., Dunkelberger, A. D., Owrutsky, J. C., Simpkins, B. S. Vibrational strong coupling controlled by spatial distribution of molecules within the optical cavity. *ACS Photon* **2018**, *5*, 158–166.
33 Coles, D. M., Somaschi, N., Michetti, P., Clark, C., Lagoudakis, P. G., Savvidis, P. G. et al. Polariton-mediated energy transfer between organic dyes in a strongly coupled optical microcavity. *Nat. Mater.* **2014**, *13*, 712–719.
34 Garcia-Vidal, F. J., Feist, J., Long-distance operator for energy transfer. *Science* **2017**, *357*, 1357–1358.
35 Weisbuch, C., Nishioka, M., Ishikawa, A., Arakawa, Y. Observation of the coupled exciton-photon mode splitting in a semiconductor quantum microcavity. *Phys. Rev. Lett.* **1992**, *69*, 3314–3331.
36 Zhong, X., Chervy, T., Zhang, L., Thomas, A., George, J., Genet, C. et al. Energy transfer between spatially separated entangled molecules. *Angewandte. Chemie. Int. Ed.* **2017**, *31*, 9034–9038.
37 Simpkins, B. S., Yang, Z., Dunkelberger, A. D., Vurgaftman, I., Owrutsky, J. C., Xiong, W. Comment on isolating polaritonic 2D-IR transmission spectra. *J. Phys. Chem. Lett.* **2023**, *14*, 983–988.
38 Dunkelberger, A. D., Spann, B. T., Fears, K. P., Simpkins, B. S., Owrutsky, J. C. Modified relaxation dynamics and coherent energy exchange in coupled vibration-cavity polaritons. *Nat. Commun.* **2016**, *7*, 13504.
39 Xiang, B., Ribeiro, R. F., Dunkelberger, A. D., Wang, J., Li, Y., Simpkins, B. S. et al. Two-dimensional infrared spectroscopy of vibrational polaritons. *Proc. Nat. Acad. Sci. U.S.A.* **2018**, *115*, 4845–4850.
40 Singh, J., Lather, J., George, J. Solvent dependence on cooperative vibrational strong coupling and cavity catalysis. *Chem. Phys. Chem.* **2023**, *24*, e202300016.
41 Mayerhofer, T. G., Pipa, A. V., Popp, J. Beer's law-why integrated absorbance depends linearly on concentration. *Chem. Phys. Chem.* **2019**, *20*, 2748–2753.
42 Mayerhofer, T. G., Dabrowska, A., Schwaighofer, A., Lendl, B., Popp, J. Beyond beer's law: why the index of refraction depends (almost) linearly on concentration. *Chem. Phys. Chem.* **2020**, *21*, 707–711.
43 Thomas, A., George, J., Shalabney, A., Dryzhakov, M., Varma, S. J., Moran, J. et al. Ebbesen, ground-state chemical reactivity under vibrational coupling to the vacuum electromagnetic field. *Angew. Chem. Int. Ed.* **2016**, *55*, 11462–11466.
44 Thomas, A., Jayachandran, A., Lethuiller, L., Vergauwe, R. M. A., Nagarajan, K., Devauz, E. et al. Ground state chemistry under vibrational strong coupling: Dependence of thermodynaics parameters on the rabi splitting energy. *Nanophotonics* **2020**, *9*, 249–255.
45 Hirai, K., Takeda, R., Hutchison, J. A., Uji-i, H. Modulation of prins cyclization by vibrational strong coupling. *Angew. Chem. Int. Ed.* **2020**, *59*, 5332–5335.

46 Byrnes, S. J. Transfer-matrix method for optics of thin and thick multilayer films (2022). https://github.com/sbyrnes321/tmm (accessed 2 August 2024).

47 Rakić, A. D., Djurišić, A. B., Elazar, J. M., Majewski, M. L. Optical properties of metallic films for vertical-cavity optoelectronic devices. *Appl. Opt.* **1998**, *37*, 5271–5283.

48 Wright, A. D., Nelson, J. C., Weichman, M. L. Rovibrational polaritons in gas-phase methane. *J. Am. Chem. Soc.* **2023**, *145*, 5982–5987.

49 Yamada, H., Stemo, G., Katsuki, H., Yanagi, H. Development of a spacerless flow-cell cavity for vibrational polaritons. *J. Phys. Chem. B.* **2022**, *126*, 4689–4696.

50 Simpkins, B., Michon, M. ChemRxiv 2023. 10.26434/chemrxiv-2023-c3cqb. (This content is a preprint and has not been peer-reviewed).

51 Liu, L., Corma, A. Metal catalysts for heterogeneous catalysis: From single atoms to nanoclusters and nanoparticles. *Chem. Rev.* **2018**, *118*, 4981–5079.

52 Chance, R. R., Prock, A., Silbey, R. Molecular fluorescence and energy transfer near interfaces. (eds. I. Prigogine and S.A. Rice), *Adv. Chem. Phys.* **1978**, *37*, 1–65.

53 Shalabney, A., George, J., Hiura, H., Hutchison, J. A., Genet, C., Hellwig, P. et al. Enhanced raman scattering from vibro-polariton hybrid states. *Angew. Chem. Int. Ed.* **2015**, *54*, 7971–7975.

54 Simpkins, B. S., Fears, K. P., Dressick, W. J., Spann, B. T., Dunkelberger, A. D., Owrutsky, J. C. Spanning strong to weak normal mode coupling between vibrational and Fabry–Pérot cavity modes through tuning of vibrational absorption strength. *ACS Photon* **2015**, *2*, 1460–1467.

55 Varner, J. F., Wert, D., Matari, A., Nofal, R., Foley, J. J. Accelerating the discovery of multilayer nanostructures with analytic differentiation of the transfer matrix equations. *Phys. Rev. Research* **2020**, *2*, 013018.

56 Varner, J. F., Eldabagh, N., Volta, D., Eldabagh, R., Foley, J. WPTherml: a python package for the design of materials for harnessing heat. *ChemRxiv*. **2019**. (This content is a preprint and has not been peer-reviewed).

57 Fidler, A. P., Chen, L., McKillop, A. M., Weichman, M. L. Ultrafast dynamics of CN radical reactions with chloroform solvent under vibrational strong coupling. arXiv:2307.04875 2023.

Part IV

Frontiers

14

A Comparison of Coulomb and Multipolar Gauge Theories of Cavity Quantum Electrodynamics

Adam Stokes[1] *and Ahsan Nazir*[2]

[1] *School of Mathematics, Statistics, and Physics, Newcastle University, Newcastle upon Tyne, United Kingdom*
[2] *Department of Physics and Astronomy, University of Manchester, Manchester, United Kingdom*

14.1 Introduction

The use of electromagnetic cavities is one of the earliest and most natural methods to manipulate light–matter coupling, which is of importance for both fundamental and applied modern physics. Kaluzny et al. were the first to successfully observe coupling strengths exceeding the losses in the system by using a collection of Rydberg atoms in a high-Q microwave cavity [1]. By now, the vast array of studied cavity quantum electrodynamics (QED) systems also includes optical cavities [2], single atoms [3] and single molecules [4], excitons in semiconductors [5], quantum dots [6] and superconducting qubits (circuit QED) [7]. Among the numerous applications of these systems are high-precision measurements [8], quantum information processing (QIP) [9], and tests of fundamental physics [10]. Recent research has also begun to explore the possibility of using cavities to modify the complex emergent properties of condensed matter systems [11, 12]. This includes the strong light–matter coupling with molecules [4, 13], leading to the possibility of cavity-induced changes in chemical reactions [14–18]. More broadly, cavity-induced modification of charge transport and optical properties in material systems such as organic molecules is of major current interest [19–25].

In describing the increasingly available and diverse ways to access nonstandard light–matter coupling regimes within cavity QED, additional conceptual challenges are encountered. By far, the most commonly employed descriptions are the Coulomb and multipolar gauge theories, and simple generalizations thereof. But while by construction these different descriptions are globally equivalent, they each provide different physical definitions of light (photons) and matter as quantum subsystems. As a result, equality of predictions from different theories for processes in which photons are created, destroyed and exchanged by matter, is only guaranteed under specific conditions fulfilled within standard (weak and Markovian) interaction regimes [26]. There has been significant recent interest in understanding the implications of this fact for the predictions of effective models in which only a finite number of material energy levels and/or photonic modes are retained [26–42], time-dependent interactions

[26, 28, 30, 33, 40], Dicke model superradiance [26, 29, 34, 37], and photodetection theory [33, 40]. The goal of this chapter is to identify the precise physical differences between the light and matter subsystems defined relative to the Coulomb and multipolar gauges in cavity QED. To do so in the simplest possible way, we begin with the example of a perfect parallel plate cavity. We then focus on predictions of the most basic single photon processes, namely, emission, absorption, and photon exchange. We subsequently discuss the extension of these descriptions to include cavity leakage using effective models. Finally, we exemplify the generality of our findings by showing that the same physical differences persist in the general theory of macroscopic QED.

14.2 Perfect Cavity

We begin by briefly reviewing the theory of a perfect parallel plate cavity, which, while obviously idealized, is well-suited to understanding the physical relativity of the light and matter subsystems, and the ensuing implications.

14.2.1 Empty Fabry–Pérot Cavity

We assume perfectly conducting parallel plates are located at $z = 0$ and $z = L$. The canonical fields are [43–45]

$$\mathbf{A}_{\mathrm{T}}(\mathbf{x}) = \sum_{n=0}^{\infty}{}'' \sum_{\sigma=1,2} \int d^2\kappa \frac{1}{\sqrt{\omega_n(\boldsymbol{\kappa})}} \left[\mathbf{f}_{\sigma n}(\boldsymbol{\kappa},\mathbf{x}) a_{\sigma n}(\boldsymbol{\kappa}) + \mathbf{f}^*_{\sigma n}(\boldsymbol{\kappa},\mathbf{x}) a_{\sigma n}(\boldsymbol{\kappa})^\dagger\right], \tag{14.1}$$

$$\Pi(\mathbf{x}) = -i\sum_{n=0}^{\infty}{}'' \sum_{\sigma=1,2} \int d^2\kappa \sqrt{\omega_n(\boldsymbol{\kappa})} \left[\mathbf{f}_{\sigma n}(\boldsymbol{\kappa},\mathbf{x}) a_{\sigma n}(\boldsymbol{\kappa}) - \mathbf{f}^*_{\sigma n}(\boldsymbol{\kappa},\mathbf{x}) a_{\sigma n}(\boldsymbol{\kappa})^\dagger\right], \tag{14.2}$$

where the mode functions are specified as the solution of the boundary problem

$$\hat{\mathbf{z}} \times \mathbf{f}_{\sigma n}(\boldsymbol{\kappa},\mathbf{x})|_{z=0,L} = \mathbf{0}, \tag{14.3}$$

$$\nabla \cdot \mathbf{f}_{\sigma n}(\boldsymbol{\kappa},\mathbf{x}) = 0, \tag{14.4}$$

$$\nabla \times [\nabla \times \mathbf{f}_{\sigma n}(\boldsymbol{\kappa},\mathbf{x})] - \omega_n(\boldsymbol{\kappa})^2 \mathbf{f}_{\sigma n}(\boldsymbol{\kappa},\mathbf{x}) = \mathbf{0} \tag{14.5}$$

with

$$\boldsymbol{\kappa} = (k_x, k_y, 0) \tag{14.6}$$

$$\omega_n(\boldsymbol{\kappa})^2 = \boldsymbol{\kappa}^2 + k_{nz}^2, \qquad k_{nz} = \frac{n\pi}{L}. \tag{14.7}$$

The double-prime on the summations over n in Eqs. (14.1) and (14.2) indicates that a factor of $1/\sqrt{2}$ should be inserted for the $n = 0$ term. Explicitly, the mode functions are

$$\mathbf{f}_{1n}(\boldsymbol{\kappa},\mathbf{x}) = \frac{e^{i\boldsymbol{\kappa}\cdot\mathbf{x}}}{2\pi\sqrt{L}} \hat{\boldsymbol{\kappa}} \times \hat{\mathbf{z}} \sin k_{nz} z, \tag{14.8}$$

$$\mathbf{f}_{2n}(\boldsymbol{\kappa},\mathbf{x}) = \frac{e^{i\boldsymbol{\kappa}\cdot\mathbf{x}}}{2\pi\omega_n(\boldsymbol{\kappa})\sqrt{L}} [i\hat{\boldsymbol{\kappa}} k_{nz} \sin k_{nz} z - \hat{\mathbf{z}}\kappa \cos k_{nz} z]. \tag{14.9}$$

The operators $a_{\sigma n}(\boldsymbol{\kappa})$ obey the usual bosonic commutation relations

$$[a_{\sigma n}(\boldsymbol{\kappa}), a_{\sigma' n'}(\boldsymbol{\kappa}')^\dagger] = \delta_{\sigma\sigma'}\delta_{nn'}\delta(\boldsymbol{\kappa} - \boldsymbol{\kappa}). \tag{14.10}$$

In the absence of charged sources, the electromagnetic fields are entirely transverse, $\mathbf{E} = \mathbf{E}_\mathrm{T} = -\dot{\mathbf{A}}_\mathrm{T} = -\Pi$ and $\mathbf{B} = \nabla \times \mathbf{A}_\mathrm{T}$. The total energy of the transverse (photonic) field inside the cavity is

$$H_\mathrm{ph} = \sum_{n=0}^{\infty} \sum_{\sigma} \int d^2\kappa\, \omega_n(\boldsymbol{\kappa}) \left(a_{\sigma n}(\boldsymbol{\kappa})^\dagger a_{\sigma n}(\boldsymbol{\kappa}) + \frac{1}{2} \right). \tag{14.11}$$

The canonical commutation relation is

$$\begin{aligned} &[A_{\mathrm{T}i}(\mathbf{x}), \Pi_j(\mathbf{x}')] \\ &= i \sum_{n=0}^{\infty}{}' \sum_{\sigma} \int d^2\kappa \left[f_{\sigma ni}(\boldsymbol{\kappa}, \mathbf{x}) f^*_{\sigma nj}(\boldsymbol{\kappa}, \mathbf{x}') + f^*_{\sigma ni}(\boldsymbol{\kappa}, \mathbf{x}) f_{\sigma nj}(\boldsymbol{\kappa}, \mathbf{x}') \right] \\ &= i \left[\delta^\mathrm{T}_{ij}(\mathbf{x} - \mathbf{x}') - \nabla_i \nabla'_j \int_{\bar{v}} d^3s \frac{\xi(\mathbf{s}, \mathbf{x}')}{4\pi|\mathbf{x} - \mathbf{s}|} \right] \end{aligned} \tag{14.12}$$

where the prime on the summation indicates that a factor of 1/2 must be inserted for the $n = 0$ term. The first equality in Eq. (14.12) follows from Eqs. (14.1), (14.2), and (14.10). The second equality in Eq. (14.12) is as found in Ref. [46] and warrants some further remarks. We have used $\bar{v}$ to denote the volume external to the cavity and the distribution $\xi(\mathbf{s}, \mathbf{x})$ vanishes for $\mathbf{s}$ inside the cavity. It specifies the image points of internal cavity points $\mathbf{x}$ reflected in the boundaries. In the simplest case of a single conducting plate at $z = 0$ for example, we have [46]

$$\xi(\mathbf{s}, \mathbf{x}) = -\delta(\mathbf{s} - \check{\mathbf{x}}) \tag{14.13}$$

where $\check{\mathbf{x}} = (x, y, -z)$ is the reflection of $\mathbf{x}$ in the $z = 0$ plane. For two plates at $z = 0$ and $z = L$ the definition of ξ is more involved, requiring an infinite sum over images. For this reason, it is more straightforward to specify ξ by considering the presence of matter inside the cavity and deducing ξ by using it to define an image charge distribution. Using the resulting definition of ξ together with Eqs. (14.8) and (14.9), Eq. (14.12) can be verified by direct calculation, as is demonstrated in Appendix A.

We note finally that the first equality in Eq. (14.12) implies that the commutator is invariant under a simultaneous exchange of Cartesian indices and spatial arguments; $i \leftrightarrow j$ and $\mathbf{x} \leftrightarrow \mathbf{x}'$. This leaves both $\delta^\mathrm{T}_{ij}(\mathbf{x}-\mathbf{x}')$ and $\nabla_i \nabla'_j$ unchanged, whereas $\xi(\mathbf{s}, \mathbf{x}')/(4\pi|\mathbf{x}-\mathbf{s}|)$ is replaced with $\xi(\mathbf{s}, \mathbf{x})/(4\pi|\mathbf{x}'-\mathbf{s}|)$ on the right-hand side of Eq. (14.12). The most convenient form depends on the calculation being performed.

14.2.2 Perfect Cavity Containing Matter

We now consider a neutral hydrogen-like atom consisting of charges $\pm q$ inside the cavity. The nucleus $-q$ is assumed to be clamped at position $\mathbf{R}$ with components $(0, 0, R_z)$, $0 < R_z < L$, while the position of the dynamical charge $+q$ is denoted $\mathbf{r}$. The densities of charge and current are, therefore, defined as

$$\rho(\mathbf{x}) = q[\delta(\mathbf{x} - \mathbf{r}) - \delta(\mathbf{x} - \mathbf{R})], \tag{14.14}$$

$$\mathbf{J}(\mathbf{x}) = \frac{q}{2}[\dot{\mathbf{r}}\delta(\mathbf{x} - \mathbf{r}) + \delta(\mathbf{x} - \mathbf{r})\dot{\mathbf{r}}], \tag{14.15}$$

respectively.

14.2.2.1 Coulomb Gauge

Let us begin by choosing the Coulomb gauge so that $\Pi = -\mathbf{E}_\mathrm{T}$. We know from the free theory given in Section 14.2.1 that by construction $\hat{\mathbf{z}} \times \Pi$ vanishes at the boundaries, but since matter is present the electric field $\mathbf{E}$ no longer coincides with its transverse part $\mathbf{E}_\mathrm{T}$. Importantly then, in the Coulomb gauge the presence of the mirrors is *not* accounted for only by the mode functions of the canonical fields. If we require the *total* electric field to satisfy the cavity boundary conditions, then we must introduce an image distribution $\tilde{\rho}$ supported outside the cavity, giving a total electrostatic potential [46]

$$\begin{aligned}\phi(\mathbf{x}) &= \int_{v} d^3x' \frac{\rho(\mathbf{x}')}{4\pi|\mathbf{x}-\mathbf{x}'|} + \int_{\bar{v}} d^3s \frac{\tilde{\rho}(\mathbf{s})}{4\pi|\mathbf{x}-\mathbf{s}|} \\ &\equiv \phi^{\mathrm{real}}(\mathbf{x}) + \phi^{\mathrm{image}}(\mathbf{x}).\end{aligned} \tag{14.16}$$

The total electrostatic field is then $\mathbf{E}_\mathrm{L} = -\nabla\phi$ and Gauss' law reads $\nabla \cdot \mathbf{E}_\mathrm{L} = \rho + \tilde{\rho}$. Writing the image charge density as

$$\tilde{\rho}(\mathbf{s}) = \int_{v} d^3x \, \xi(\mathbf{s}, \mathbf{x})\rho(\mathbf{x}), \tag{14.17}$$

provides us with an alternative definition of the distribution ξ appearing in Eq. (14.12), whose verification requires us to construct ξ explicitly.

For the two-plate problem, the required ξ can be deduced from the image distribution $\tilde{\rho}$, which can be determined by inspection. Consider the field at the boundary $z = 0$ produced by the charge q at $\mathbf{r}$. To balance this field, a charge $-q$ must be placed at $(r_x, r_y, -r_z)$. Similarly, a charge $-q$ must be placed at $(r_x, r_y, 2L - r_z)$ to balance the field at $z = L$. However, the first of these image charges causes an unbalanced field at the $z = L$ plate while the second causes an imbalance at $z = 0$. Further charges therefore have to be placed at the appropriate locations to balance the fields, but it is clear that any image used to balance the field at the $z = 0$ plate will imbalance it at the $z = L$ plate and vice versa. An infinite proliferation of charges is required to balance the field at both plates [47]. For the (two-charge) atomic system, one deduces that the total distribution of charges such that $\phi = -\nabla^{-2}(\rho + \tilde{\rho})$ must be

$$\rho(\mathbf{x}) + \tilde{\rho}(\mathbf{x}) = q \sum_{n=-\infty}^{\infty} [\delta(\mathbf{x} - \mathbf{r}_n) - \delta(\mathbf{x} - \check{\mathbf{r}}_n) - \delta(\mathbf{x} - \mathbf{R}_n) + \delta(\mathbf{x} - \check{\mathbf{R}}_n)] \tag{14.18}$$

where $\mathbf{y}_n = \mathbf{y} + 2nL\hat{\mathbf{z}}$ and $\check{\mathbf{y}}_n = \check{\mathbf{y}} + 2nL\hat{\mathbf{z}}$ in which $\mathbf{y} = \mathbf{r}, \mathbf{R}$, and as before, $\check{\mathbf{y}} = (y_x, y_y, -y_z)$. It follows that the image density and distribution ξ are given by

$$\tilde{\rho}(\mathbf{x}) = q \sum_{n\neq 0} [\delta(\mathbf{x} - \mathbf{r}_n) - \delta(\mathbf{x} - \mathbf{R}_n)] + q \sum_{n} [\delta(\mathbf{x} - \check{\mathbf{R}}_n) - \delta(\mathbf{x} - \check{\mathbf{r}}_n)], \tag{14.19}$$

$$\xi(\mathbf{s}, \mathbf{x}) = \sum_{n\neq 0} \delta(\mathbf{s} - \mathbf{x}_n) - \sum_{n} \delta(\mathbf{s} - \check{\mathbf{x}}_n). \tag{14.20}$$

In practice, since the images further away from the plates have a smaller effect, a suitable truncation of the summation may be performed. As noted at the end of Section 14.2.1, it remains to verify that Eq. (14.12) holds. That it does so is demonstrated in Appendix A.

The electrostatic energy inside the cavity can be written

$$H_{\rm long} = \frac{1}{2}\int_v d^3x\,\mathbf{E}_{\rm L}(\mathbf{x})^2 = \frac{1}{2}\int_v d^3x\rho(\mathbf{x})\phi(\mathbf{x}) \tag{14.21}$$

where we have used integration by parts with $\mathbf{E}_{\rm L} = -\nabla\phi$ and $-\nabla^2\phi = \rho+\tilde{\rho}$, noting that $\tilde{\rho}$ vanishes inside v. Using Eq. (14.16) in Eq. (14.21) we obtain

$$H_{\rm long} = V(\mathbf{r}) + V^{\rm self} + V^{\rm image}, \tag{14.22}$$

$$V(\mathbf{r}) = -\frac{q^2}{4\pi r}, \tag{14.23}$$

$$V^{\rm image} = \frac{1}{2}\int_v d^3x \int_{\tilde{v}} d^3s \frac{\rho(\mathbf{x})\tilde{\rho}(\mathbf{s})}{4\pi|\mathbf{x}-\mathbf{s}|}. \tag{14.24}$$

Here $V^{\rm self}$ contains the infinite Coulomb self-energies of each real charge, the term $V(\mathbf{r})$ is the usual Coulomb binding energy of the two real charges, and the term $V^{\rm image}$ describes the Coulomb interaction between the real charges and the images.

We have now specified the photonic and the electrostatic field energies in the Coulomb gauge. The kinetic energy of the charges, meanwhile, is $m\dot{\mathbf{r}}^2/2$ where in the Coulomb gauge $m\dot{\mathbf{r}} = \mathbf{p} - q\mathbf{A}_{\rm T}(\mathbf{r})$. The total energy (Hamiltonian) is the sum of these three energies. Ignoring $V^{\rm self}$ this is

$$H_0 = \frac{1}{2m}[\mathbf{p} - q\mathbf{A}_{\rm T}(\mathbf{r})]^2 + V(\mathbf{r}) + V^{\rm image} + H_{\rm ph}, \tag{14.25}$$

which can be partitioned into an unperturbed part h and interaction V^0, in the standard way as

$$H_0 = h + V^0, \tag{14.26}$$

$$h = H_m + H_{\rm ph}, \tag{14.27}$$

$$H_m = \frac{\mathbf{p}^2}{2m} + V(\mathbf{r}), \tag{14.28}$$

$$V^0 = -\frac{q}{m}\mathbf{p}\cdot\mathbf{A}_{\rm T}(\mathbf{r}) + \frac{q^2}{2m}\mathbf{A}_{\rm T}(\mathbf{r})^2 + V^{\rm image}. \tag{14.29}$$

In the EDA, we assume that $|\mathbf{r} - \mathbf{R}|$ is small compared to the relevant resonant wavelengths, such that upon expanding $\mathbf{A}_{\rm T}(\mathbf{r}) = \mathbf{A}_{\rm T}(\mathbf{R} + [\mathbf{r} - \mathbf{R}])$, we need retain only the first term, namely $\mathbf{A}_{\rm T}(\mathbf{R})$. This is equivalent to retaining only the leading contribution, $q\delta(\mathbf{x}-\mathbf{R})$, when expanding the density of the dynamical charge $\rho_{\mathbf{r}}(\mathbf{x}) = q\delta(\mathbf{x} - \mathbf{r})$. When the same expansion is performed in the total charge density, the leading term is $\rho(\mathbf{x}) = -q\mathbf{d}\cdot\nabla\delta(\mathbf{x} - \mathbf{R})$ and the image charge interaction corresponding to this density, written $V^{\rm image} \approx V^{\rm image,dip}$, is a sum of dipole–dipole interactions of the real dipole with each image dipole. In the simplest case of a dipole near a single plate, for example, Eq. (14.13) gives

$$V^{\rm image,dip} = -\frac{1}{2}\frac{d_i\check{d}_j}{4\pi u^3}(\delta_{ij} - 3\hat{u}_i\hat{u}_j) \tag{14.30}$$

where $\mathbf{u} = \mathbf{R} - \check{\mathbf{R}}$. In general, the total Hamiltonian H_0 in the EDA is

$$H_0 \approx \frac{1}{2m}[\mathbf{p} - q\mathbf{A}_{\rm T}(\mathbf{R})]^2 + V(\mathbf{r}) + V^{\rm image,dip} + H_{\rm ph}. \tag{14.31}$$

14.2.2.2 Multipolar Gauge

The Power-Zienau-Woolley (PZW) transformation is defined by

$$R = \exp\left[-i\int_{v} d^3x\,\mathbf{P}(\mathbf{x})\cdot\mathbf{A}_{\mathrm{T}}(\mathbf{x})\right], \tag{14.32}$$

$$\mathbf{P}(\mathbf{x}) = \int_0^1 d\lambda\,\mathbf{d}\delta(\mathbf{x}-\mathbf{R}-\lambda(\mathbf{r}-\mathbf{R})) \tag{14.33}$$

where $\mathbf{d} = q(\mathbf{r}-\mathbf{R})$. By definition, we have $-\nabla\cdot\mathbf{P} = \rho$. Unlike in free space, the longitudinal polarization differs from the longitudinal electric field because its divergence equals (minus) the real charge density alone. Equation (14.32) can also be written as a local phase transformation $R = e^{iq\chi(\mathbf{r})}$ where

$$\chi(\mathbf{r}) = -\int_0^1 d\lambda(\mathbf{r}-\mathbf{R})\cdot\mathbf{A}_{\mathrm{T}}(\mathbf{R}+\lambda(\mathbf{r}-\mathbf{R})). \tag{14.34}$$

The transformation of the photon operator $a_{\sigma n}(\boldsymbol{\kappa})$ is found to be

$$Ra_{\sigma n}(\boldsymbol{\kappa})R^\dagger = a_{\sigma n}(\boldsymbol{\kappa}) + \left[\frac{i}{\sqrt{\omega_n(\boldsymbol{\kappa})}}\int_{v} d^3x\mathbf{P}(\mathbf{x})\cdot\mathbf{f}_{\sigma n}(\boldsymbol{\kappa},\mathbf{x})^*\right]' \tag{14.35}$$

where the notation $[\cdot]'$ indicates that a factor of $1/\sqrt{2}$ must be inserted for the case $n = 0$. It follows that

$$RH_{\mathrm{ph}}R^\dagger = H_{\mathrm{ph}} + \int_{v} d^3x\mathbf{P}(\mathbf{x})\cdot\Pi(\mathbf{x}) + {\sum_n}'\sum_\sigma\int d^2\kappa\left|\int_{v} d^3x\,\mathbf{f}_{\sigma n}(\boldsymbol{\kappa},\mathbf{x})\cdot\mathbf{P}(\mathbf{x})\right|^2. \tag{14.36}$$

The second term on the right-hand side is the expected linear multipolar polarization–photon interaction term. The final term can be computed using Eq. (14.12). One finds that

$${\sum_n}'\sum_\sigma\int d^2\kappa\left|\int_{v} d^3x\,\mathbf{f}_{\sigma n}(\boldsymbol{\kappa},\mathbf{x})\cdot\mathbf{P}(\mathbf{x})\right|^2 = \epsilon^{\mathrm{pol}} - V^{\mathrm{image}} \tag{14.37}$$

where

$$\epsilon^{\mathrm{pol}} := \frac{1}{2}\int_{v} d^3xd^3x'\delta^{\mathrm{T}}_{ij}(\mathbf{x}-\mathbf{x}')\Pi(\mathbf{x})P_j(\mathbf{x}') \tag{14.38}$$

is the expected polarization self-energy term.

Finally, the kinetic momentum transforms, as in the case of free space

$$R[\mathbf{p}-q\mathbf{A}_{\mathrm{T}}(\mathbf{r})]R^\dagger = \mathbf{p}-q\mathbf{A}(\mathbf{r}) \tag{14.39}$$

where

$$\mathbf{A}(\mathbf{x}) = -\int_0^1 d\lambda\,\lambda(\mathbf{x}-\mathbf{R})\times\mathbf{B}(\mathbf{R}+\lambda(\mathbf{x}-\mathbf{R})) \tag{14.40}$$

is the usual multipolar (Poincaré gauge) potential for a charge distribution centered at **R**. Collecting all terms, the multipolar Hamiltonian is found to be

$$H_1 = RH_0R^\dagger = \frac{1}{2m}[\mathbf{p} - q\mathbf{A}(\mathbf{r})]^2 + V(\mathbf{r}) + \epsilon^{\text{pol}} + \int_v d^3x\mathbf{P}(\mathbf{x}) \cdot \Pi(\mathbf{x}) + H_{\text{ph}}, \tag{14.41}$$

which can be partitioned into unperturbed and interaction parts as $H_1 = h + V^1$ where h is defined as in Eq. (14.27). Importantly, there is no image charge interaction in the multipolar gauge theory. The boundary conditions are satisfied entirely by the use of the appropriate mode functions for the transverse field. This occurs for the same reason that inter-atomic electrostatic interactions are canceled in this gauge. Specifically, the multipolar polarization field is localized inside the atom, so there is no overlap between the polarization of the real atom with the polarizations of any of the image atoms.

In the EDA $\mathbf{A}(\mathbf{r}) \approx \mathbf{0}$ and $\mathbf{P}(\mathbf{x}) \approx \mathbf{d}\delta(\mathbf{x} - \mathbf{R})$ so that H_1 becomes

$$H_1 = \frac{\mathbf{p}^2}{2m} + V(\mathbf{r}) + \epsilon^{\text{dip}} + \mathbf{d} \cdot \Pi(\mathbf{R}) + H_{\text{ph}}, \tag{14.42}$$

where

$$\epsilon^{\text{dip}} = \frac{1}{2}\mathbf{d} \cdot \delta^{\text{T}}(\mathbf{0}) \cdot \mathbf{d} \tag{14.43}$$

is the well-known divergent dipolar self-energy.

It is instructive to determine the physical meaning of the canonical momenta in the multipolar and dipole gauges. For the material canonical momentum **p**, we have immediately that, as in free space $\mathbf{p} = m\dot{\mathbf{r}} + q\mathbf{A}(\mathbf{r})$ and in the EDA $\mathbf{p} = m\dot{\mathbf{r}}$. This canonical momentum is much more localized than its Coulomb gauge counterpart [26]. For the field canonical momentum, in the Coulomb gauge we have $\Pi = -\mathbf{E}_\text{T}$, so in the multipolar gauge we have $R\Pi R^\dagger = -\mathbf{E}_\text{T}$ where

$$R\Pi R^\dagger = \Pi + \mathbf{P}_\text{T} + \mathbf{E}_\text{L}^{\text{image}}, \tag{14.44}$$

in which $\mathbf{E}_\text{L}^{\text{image}} = -\nabla\phi^{\text{image}}$. It follows using $\mathbf{E} = \mathbf{E}_\text{T} + \mathbf{E}_\text{L}^{\text{real}} + \mathbf{E}_\text{L}^{\text{image}}$ that

$$\Pi = -\mathbf{E} + \mathbf{E}_\text{L}^{\text{real}} - \mathbf{P}_\text{T} = -\mathbf{E} - \mathbf{P}, \tag{14.45}$$

where $\mathbf{P} = \mathbf{P}_\text{T} + \mathbf{P}_\text{L}$ with $\mathbf{P}_\text{L} = -\mathbf{E}_\text{L}^{\text{real}}$. As in the case of free space, the multipolar canonical momentum is minus the Maxwell displacement field, but here the electric field implicitly includes the description of the parallel plates. For all points outside the atom, we have $\mathbf{P} = \mathbf{0}$, and therefore $\mathbf{P}_\text{T} = \mathbf{E}_\text{L}^{\text{real}}$, so at all such points

$$\Pi = -\mathbf{E}. \tag{14.46}$$

This means, in particular, that the multipolar-gauge canonical momentum equals the *total* electric field at the boundaries. This again explains why there is no image charge electrostatic interaction in the multipolar gauge; the boundary conditions for the *total* electric field are taken care of entirely via use of the appropriate mode functions within the canonical fields [46]. We note that in the EDA all points outside the atom just means all points other than the dipole's position **R**.

14.3 Gauge Relativity

We now turn our attention to understanding the physical differences between the light and matter quantum subsystems defined relative to the Coulomb and multipolar gauges.

14.3.1 Relativity and Invariance

The canonical operators **r** and **p** belong to a material algebra that we denote $\mathcal{A}_m$, whereas $\mathbf{A}_\mathrm{T}$ and Π belong to a photonic algebra that we denote $\mathcal{A}_\mathrm{ph}$. Repeated application of the $a_{\sigma n}(\boldsymbol{\kappa})^\dagger$ to the vacuum vector $|0\rangle$ yields photon number vectors (eigenvectors of H_ph) that span a Hilbert space denoted $\mathcal{H}_\mathrm{ph}$ and that can be used to write down an arbitrary density operator for the photonic reservoir. The space of such density operators is denoted $\mathcal{D}_\mathrm{ph}$. Similarly, via $H_m|n\rangle = \omega_n|n\rangle$ the bare material, Hamiltonian H_m defines discrete energy vectors $\{|n\rangle\}$ that span a Hilbert space denoted $\mathcal{H}_m$ and that can be used to write down arbitrary density operators comprising a space denoted $\mathcal{D}_m$. The Hilbert space, operator algebra, and density operator space of the composite theory are $\mathcal{H} = \mathcal{H}_m \otimes \mathcal{H}_\mathrm{ph}$, $\mathcal{A} = \mathcal{A}_m \otimes \mathcal{A}_\mathrm{ph}$, and $\mathcal{D} = \mathcal{D}_m \otimes \mathcal{D}_\mathrm{ph}$, respectively. We say that the pair $(\mathcal{D}, \mathcal{A})$ defines the *composite quantum system* and that the pairs $(\mathcal{D}_m, \mathcal{A}_m)$ and $(\mathcal{D}_\mathrm{ph}, \mathcal{A}_\mathrm{ph})$ define the canonical quantum subsystems called "matter" and "light" (or "photons") respectively.

To properly understand the physical content of the theory and the essential uniqueness of its predictions, it is imperative to draw a distinction between *physical systems* and *mathematical systems*, which we also refer to as simply *quantum systems*. The latter are defined using vectors and operators. In contrast, we define a physical system as a pair $(\{\mathcal{S}\}, \{\mathcal{O}\})$, which denotes a collection of measurable *observables* $\{\mathcal{O}\}$ and associated *states* $\{\mathcal{S}\}$. An eigenvector $|o\rangle$ of an operator O that represents an observable $\mathcal{O}$, represents a state $\mathcal{S}$ in which the observable $\mathcal{O}$ possesses the eigenvalue o with certainty. Any state can be represented by a linear combination of eigenvectors that form a basis for the Hilbert space and any mixed state can be represented by a linear combination of outer products of the vectors that represent pure states. In other words, according to the postulates of quantum theory, physical states and observables are respectively *represented* by density operators and Hermitian operators over a Hilbert space. However, since quantum theory also specifies that all predictions are obtained from the Hilbert space inner-product, these representations are not unique. Given any one representation, a unitary transformation (a symmetry transformation of the inner-product) can be used to generate another one. This results in a form of relativity when assigning physical meaning to operators and vectors in quantum theory.

By way of example, let us suppose that at time t the physical state $\mathcal{S}_t$ of a matter–photon composite is represented in the Coulomb gauge by a density operator $\rho^0(t)$ and that an arbitrary observable $\mathcal{O}$ is represented in the Coulomb gauge by the operator O^0. In the multipolar gauge, the same state and observable must be represented by the density operator $\rho^1(t) = R\rho^0(t)R^\dagger$ and the operator $O^1 = RO^0R^\dagger$ respectively, because then physical predictions are gauge-invariant, by which we mean that they are the same when calculated in either gauge. Specifically, by virtue of the unitarity of gauge-fixing transformations, viz., $RR^\dagger = I = R^\dagger R$, we have

$$\mathrm{tr}[O^0\rho^0(t)] = \langle \mathcal{O} \rangle_{\mathcal{S}_t} = \mathrm{tr}[O^1\rho^1(t)]. \tag{14.47}$$

Note that this equation holds for any state, for any observable, and for any light–matter system. Gauge-invariance is seen to be a special case of the general quantum-theoretic invariance of predictions under unitary transformations.

However, since the PZW gauge-fixing transformation is not of product form, $R \neq R_m \otimes R_\mathrm{ph}$, it does not leave invariant the quantum subsystems "light" and "matter." If a "material" operator of the form $O_m \otimes I_{\rm ph}$ represents an observable $\mathcal{O}$ in the Coulomb gauge, then the same observable is represented by the operator $RO_m \otimes I_{\rm ph}R^\dagger$ in the multipolar gauge, and this operator is not generally "material," i.e., it does not have the form $O'_m \otimes I_{\rm ph}$. The same operator $O_m \otimes I_{\rm ph}$, meanwhile, must represent two different observables $\mathcal{O}$ and $\mathcal{O}'$ in the two gauges connected by R. Similarly, it is not the case that $\rho_m^0(t) := {\rm tr}_{\rm ph}\rho^0(t)$ is equivalent to $\rho_m^1(t) := {\rm tr}_{\rm ph}\rho^1(t)$. These two reduced density operators represent the states of two different physical subsystems. It follows that the pair $(\mathcal{D}_m, \mathcal{A}_m)$ represents a *different physical (sub)system*, that is, a different collection of states and observables, in each different gauge. The same is true of the photonic (light) quantum subsystem $(\mathcal{D}_{\rm ph}, \mathcal{A}_{\rm ph})$. The canonical quantum subsystems are therefore said to be gauge-relative [26].

Gauge-invariance is a fundamental feature of QED that holds because gauge-fixing transformations are unitary. Subsystem gauge relativity is also fundamental and holds because unitary gauge-fixing transformations are not of product form. Both features arise for arbitrary material systems in arbitrary photonic environments.

14.3.2 Gauge Nonrelativistic Predictions

Linear space (vector space) *relativity* occurs when the assignment of physical meaning to elements of the linear space depends on the chosen frame (basis). In special relativity, for example, any two inertial frames A and B in spacetime M are connected by a Lorentz transformation Λ. Given two events x_1 and x_2 in M, the spatial (temporal) intervals between them found using frames A and B are generally distinct; $\Delta x_A \neq \Delta x_B$ $(\Delta t_A \neq \Delta t_B)$. The intervals Δx_A and Δt_A $(\Delta x_B$ and $\Delta t_B)$ are respectively measured using a ruler and a clock at rest in frame A (B). Each frame yields a set of unique and meaningful physical predictions, but the two sets of predictions refer to different experiments. It is evident that "space" and "time" can only be defined relative to an inertial frame.

When the relative velocity of A and B is sufficiently small, the associated space and time intervals are approximately the same; $\Delta x_A \approx \Delta x_B$ and $\Delta t_A \approx \Delta t_B$. The relativity of space and time, i.e., the mixing effect of Λ, can then be ignored completely, such that space and time emerge as approximately absolute (frame-independent) concepts, as is assumed from the outset in Galilean relativity. In QED, the assignment of physical meaning to the vectors and operators that define the quantum subsystems, conventionally called "light" and "matter," depends on the chosen (gauge) frame. This relativity can be understood as a consequence of the "mixing" effect of gauge-fixing transformations, which are of nonproduct form. It can often be ignored, however, and the regime in which this is valid is thereby called gauge nonrelativistic [26]. In the context of cavity QED, we review some basic gauge nonrelativistic predictions below.

14.3.2.1 Processes Described by On-energy-shell Matrix Elements

The most basic gauge nonrelativistic predictions in QED take the form of on-energy-shell matrix elements that describe *processes* between unperturbed atom-photon states. Two elementary examples are the spontaneous decay and the level shifts of a single dipole within a photonic reservoir. Let us suppose that this composite system is enclosed in a volume v with periodic boundaries, such that the photonic canonical operators can be written

$$\mathbf{A}_{\rm T}(\mathbf{x}) = \sum_{\mathbf{k}\sigma} \frac{1}{\sqrt{\omega}} \mathbf{f}_{\mathbf{k}\sigma}(\mathbf{x}) a_{\mathbf{k}\sigma} + \text{H.c.}, \tag{14.48}$$

$$\Pi(\mathbf{x}) = -i \sum_{\mathbf{k}\sigma} \sqrt{\omega}\, \mathbf{f}_{\mathbf{k}\sigma}(\mathbf{x}) a_{\mathbf{k}\sigma} + \text{H.c.}, \tag{14.49}$$

where as usual, we have

$$[a_{\mathbf{k}\sigma}, a^\dagger_{\mathbf{k}'\sigma'}] = \delta_{\sigma\sigma'}\delta_{\mathbf{k}\mathbf{k}'}. \tag{14.50}$$

The parallel plate cavity considered above is included as a special case with the allowed wave numbers in the x and y directions being $k_i = 2\pi n_i/l$ with $n_i \in \mathbb{Z}$, $i = x, y$, and $v = l^2L$. Note that factors of $1/(2\pi)$ should be replaced by $1/l$ in Eqs. (14.8) and (14.9) [45] and that the condition on the summation over n_z in these equations continues to apply in Eqs. (14.48) and (14.49). The case of free space is also included in Eqs. (14.48) and (14.49) by taking the limit $v \to \infty$ and letting $\mathbf{f}_{\mathbf{k}\sigma}(\mathbf{x}) = \mathbf{e}_{\mathbf{k}\sigma}e^{i\mathbf{k}\cdot\mathbf{x}}/\sqrt{2v}$ where $\mathbf{e}_{\mathbf{k}1}$ and $\mathbf{e}_{\mathbf{k}2}$ are mutually orthogonal unit polarization vectors orthogonal to $\mathbf{k}$.

Let us now consider the process of spontaneous emission. The vector $|e, 0\rangle$ is an eigenvector of the free Hamiltonian $h = H_m + H_{\text{ph}} = H_\alpha - V^\alpha$, $\alpha = 0, 1$. It represents a state with an excited dipole and no photons. One can consider the decay of such a state into a state represented by a vector of the form $|n, 1_{\mathbf{k}\sigma}\rangle$ with $n < e$ and with one photon in the mode $\mathbf{k}\sigma$. This process is called spontaneous emission of a real photon. However, an unperturbed vector $|v\rangle$ (an eigenvector of h) represents two distinct states S_v^0 and S_v^1 in the Coulomb and multipolar gauges respectively. In general, then, the emission process $|i\rangle = |e, 0\rangle \to |f\rangle = |n, 1_{\mathbf{k}\sigma}\rangle$ is not physically unique; it constitutes two distinct physical processes, $S_i^0 \to S_f^0$ and $S_i^1 \to S_f^1$, in the two different gauges. If, however, the interaction is assumed to vanish at the initial and final times, then $|e, 0\rangle$ and $|n, 1_{\mathbf{k}\sigma}\rangle$ do represent unique physical states S_i and S_f. If the interaction is further assumed to be switched on and off adiabatically over an infinite period of time, then the process $S_i \to S_f$ will conserve the unperturbed energy $h = H_m + H_{\text{ph}}$, which equals H before and after the interaction. Under these conditions, the rate of dipolar decay $e \to n$ accompanied by photon emission into any mode can be defined as

$$\Gamma_{en} = 2\pi \sum_{\mathbf{k}\sigma} |\langle e, 0|V^\alpha|n, 1_{\mathbf{k}\sigma}\rangle|^2 \delta(\omega_{en} - \omega) \tag{14.51}$$

where $\omega_{en} = \epsilon_e - \epsilon_n$ and as usual $\omega = |\mathbf{k}|$. The matrix element $\langle e, 0|V^\alpha|n, 1_{\mathbf{k}\sigma}\rangle$ is α-dependent but when it is evaluated on-energy-shell as imposed by the δ-function in Eq. (14.51), all α-dependence drops out.

In free space with $v \to \infty$ Eq. (14.51) yields the well-known result $\Gamma_{en} = \omega_{en}^3|\mathbf{d}_{en}|^2/(3\pi)$ where $\mathbf{d}_{en} = q\langle e|\mathbf{r}|n\rangle$ is the transition dipole moment between the e'th and n'th levels. In the case of a perfect parallel plate cavity, the spontaneous emission rate is instead calculated using the mode functions in Eqs. (14.8) and (14.9). Letting $\omega_{en} = \omega$ and $\mathbf{d}_{ne} = \mathbf{d}$ the rates for a dipole oriented parallel and orthogonal to the plates are found to be [44, 45, 48]

$$\frac{\Gamma_z}{\Gamma_0} = \sum_{n=0}^{[2L/\lambda]} \frac{3\lambda}{4L}\left[1 + \left(\frac{n\lambda}{2L}\right)^2\right]\sin^2\left(\frac{n\pi R_z}{L}\right), \tag{14.52}$$

$$\frac{\Gamma_\rho}{\Gamma_0} = \frac{3\lambda}{2L}\left(\frac{1}{2} + \sum_{n=1}^{[2L/\lambda]}\left[1 - \left(\frac{n\lambda}{2L}\right)^2\right]\right)\cos^2\left(\frac{n\pi R_z}{L}\right) \tag{14.53}$$

where Γ_0 is the corresponding free space emission rate, $\lambda = 2\pi/\omega$, and $[2L/\lambda]$ denotes the integer part of $2L/\lambda$. These rates are plotted in Figure 14.1. Emission for a dipole parallel to the plates is suppressed for $\lambda > 2L$ since then only $n = 0$ modes contribute, but these are restricted to the z direction. For a dipole located in a field node, this rate exhibits jumps at half-odd integer values of L/λ whereat the modes of even n do not contribute. The rate for a dipole oriented orthogonal to the plates is enhanced as L/λ decreases because the $n = 0$ mode scales inversely with L.

In the limit that the plate at $z = L$ is taken to infinity, the rates become those of a dipole near a single plate and read

$$\frac{\Gamma_z}{\Gamma_0} = 3\left(\frac{1}{3} + \frac{\sin\eta}{\eta^3} - \frac{\cos\eta}{\eta^2}\right), \tag{14.54}$$

$$\frac{\Gamma_\rho}{\Gamma_0} = \frac{3}{2}\left(\frac{2}{3} + \frac{\sin\eta}{\eta^3} - \frac{\cos\eta}{\eta^2} - \frac{\sin\eta}{\eta}\right) \tag{14.55}$$

where $\eta = 2\omega R_z$. These rates are also plotted in Figure 14.1. The rate for a parallel-oriented dipole is suppressed near the plate, because its field is canceled by that of its image. The rate is doubled if the dipole is oriented normal to the mirror because it then comprises a Dicke pair with its image [45]. Such selection rules are increasingly weakened as the distance to the mirror increases and for large distances emission becomes that of free space, which is isotropic.

A second type of process occurring in the state represented by $|e, 0\rangle$ is (virtual) photon emission and reabsorption, comprising an energy-conserving overall process that ends back in the same state.

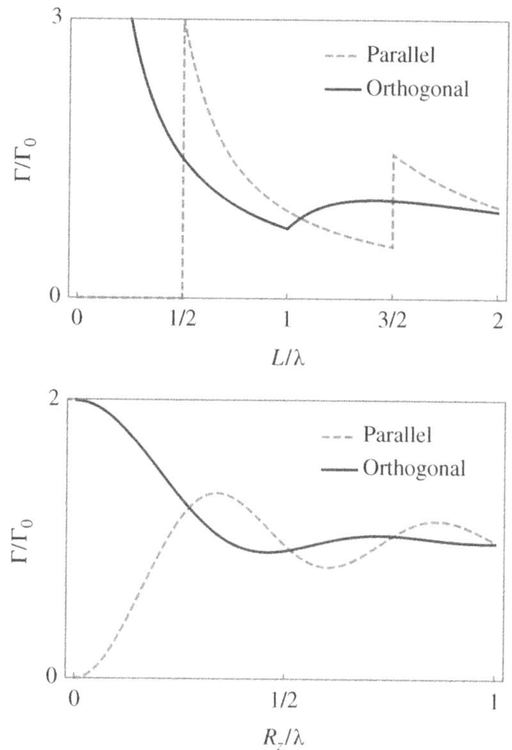

Figure 14.1 Top: The normalized emission rate for a dipole at $\mathbf{R} = (0, 0, L/2)$ with transition frequency ω is plotted as a function of L/λ where $\lambda = 2\pi/\omega$ and L is the plate separation. The cases of a dipole oriented parallel and orthogonal to the plates are shown. Bottom: The parallel/orthogonal oriented dipole normalized emission rates are plotted with R_z/λ for a dipole at $\mathbf{R} = (0, 0, R_z)$ with a single plate at $z = 0$.

These processes result in the (Lamb) shift of the bare dipole energy ϵ_e, which is calculated using second order perturbation theory reads

$$\Delta_e = \langle e,0|V_2^\alpha|e,0\rangle + \sum_{n\neq e}\sum_{\mathbf{k}\sigma} \frac{|\langle e,0|V_1^\alpha|n,1_{\mathbf{k}\sigma}\rangle|^2}{\omega_{en}-\omega} \tag{14.56}$$

where V_1^α and V_2^α are the parts of $V^\alpha = V_1^\alpha + V_2^\alpha$ that are linear and quadratic in the charge q respectively. The shift Δ_e is similarly α-independent despite the α-dependence of the matrix elements of the interaction Hamiltonian appearing therein [26, 49]. Note that $V^{\text{image,dip}}$ and $q^2\mathbf{A}_\mathrm{T}(\mathbf{0})^2/(2m)$ are contained within the second order Coulomb gauge interaction V_2^0 whereas V_2^1 includes only the self-energy term ϵ^{dip}.

The above rates and shifts are well-known and were established early on in the field of cavity QED, but here we have additionally demonstrated that they exemplify the elimination of gauge-relativity within on-energy-shell QED matrix elements. This general property relies upon the following conditions [26, 49–51]: (i) the Hamiltonians of different gauges, such as H^1 and H^0, are unitarily equivalent, (ii) perturbation theory is employed with respect to the same unperturbed Hamiltonian in both cases, that is, $H^0 - V^0 = h = H^1 - V^1$, and (iii) the matrix element is evaluated subject to constraint that the initial and final unperturbed energies are equal. Since these are the conditions that can be said to define a scattering process, the corresponding predictions may be called S-matrix elements.

14.3.2.2 Born–Markov Secular Master Equation

The master equations $\dot\rho^\alpha(t) = i\mathrm{tr}_{\text{ph}}[\rho^\alpha(t), H^\alpha]$ describing the dynamics of the "material" quantum subsystem are not equivalent for the two different values $\alpha = 0$ and $\alpha = 1$. But under enough simplifying assumptions and approximations, it is possible to show that the master equation is in fact unique. The Born–Markov and secular approximations with second order perturbation theory mimic the assumptions that define the S-matrix and are therefore said to define the gauge nonrelativistic regime [26]. In Appendix B, we choose the standard definition of the unperturbed Hamiltonian, namely, $h = H_m + H_{\text{ph}}$, and we show directly that the resulting Born–Markov secular master equation for the dipole considered in Section 14.3.2.1 is indeed unique. The final result is

$$\dot\rho(t) = i[H_m + H_{\text{LS}}, \rho(t)] + \mathcal{D}[\rho(t)] \tag{14.57}$$

where ρ now refers to the reduced material density operator,

$$\mathcal{D}(\rho) := \sum_{\substack{n,m,p,q\\ \omega_{nm}=\omega_{pq}>0}} \gamma_{nmpq}\left(L_{nm}\rho L_{pq}^\dagger - \frac{1}{2}\left\{L_{pq}^\dagger L_{nm}, \rho\right\}\right) \tag{14.58}$$

with $L_{nm} := |m\rangle\langle n|$, and

$$H_m + H_{\text{LS}} = \sum_n (\epsilon_n + \Delta_n)|n\rangle\langle n|. \tag{14.59}$$

The decay rates in Eq. (14.58) are

$$\gamma_{nmpq} = 2\pi \sum_{\mathbf{k}\sigma} \langle n,0|V^\alpha|m,1_{\mathbf{k}\sigma}\rangle\langle p,1_{\mathbf{k}\sigma}|V^\alpha|q,0\rangle\delta(\omega_{nm}-\omega_k). \tag{14.60}$$

It may often be the case that the transition spectrum is nondegenerate over all relevant levels, that is, that there are essentially no dipole transitions for which $\omega_{nm} = \omega_{pq}$ but $|n\rangle \neq |p\rangle$ and $|m\rangle \neq |q\rangle$

within the energy range of interest. In this case, the constraint $\omega_{nm} = \omega_{pq} > 0$ that occurs in Eq. (14.57) implies that the replacements $p \to n$ and $q \to m$ can be made within the summand, and the sum over p and q is removed. The rates become γ_{nmnm} which are nothing but the Γ_{nm} in Eq. (14.51).

The γ_{nmpq} are α-independent because they are evaluated on-energy shell. The level shift Δ_n in Eq. (14.59) is given by Eq. (14.56) and so is also α-independent. Explicit expressions for $\mathcal{D}(\rho)$ and $H_{\rm LS}$ are given in Appendix B. The reduced material density operator found by solving the master equation subject to any chosen initial conditions is α-independent, $\rho^\alpha(t) = \rho(t)$, and so in this case the α label can be dropped as in Eq. (14.57).

For sufficiently large coupling strengths or sufficiently non-Markovian environmental interactions, the Born–Markov and secular approximations generally breakdown and then the gauge relativity inherent within the definitions of the quantum subsystems can no longer be ignored. The master equations $\dot\rho^0_m(t) = i{\rm tr}_{\rm ph}[\rho^0(t), H_0]$ and $\dot\rho^1_m(t) = i{\rm tr}_{\rm ph}[\rho^1(t), H_1]$ are no longer the same, and it must be recognized that the density operators $\rho^0_m(t)$ and $\rho^1_m(t)$ represent two distinct physical states. A discussion of the physical differences between the definitions of the canonical quantum subsystems defined relative to the Coulomb and multipolar gauges can be found in Ref. [26] and is further discussed below.

14.3.3 A Case Study in Gauge Relativity and Gauge Ambiguities: Dipolar Photon Emission and Detection

To exemplify the importance of the gauge relativity of the light and matter quantum subsystems, we now consider in detail the fundamental phenomena of photon emission and detection inside the cavity.

14.3.3.1 Gauge Relativity of Vacuum and Source Fields

In Section 14.2.2, we rotated the Coulomb gauge Hamiltonian to obtain the multipolar Hamiltonian expressed in terms of the same set of canonical operators $\mathbf{y} = \{\mathbf{r}, \mathbf{p}, \mathbf{A}_{\rm T}, \Pi\}$ (equivalently $\{\mathbf{r}, \mathbf{p}, a_{\mathbf{k}\sigma}, a^\dagger_{\mathbf{k}\sigma}\}$). Since $\mathbf{p}$ and Π (equivalently $\mathbf{p}$ and $\{a_{\mathbf{k}\sigma}, a^\dagger_{\mathbf{k}\sigma}\}$) do not commute with R, these operators have different physical meanings in the two gauges connected by R [26, 50]. To understand these differences, it is most convenient to introduce canonical operators associated with each gauge. Specifically, since $H_1(\mathbf{y}) = RH_0(\mathbf{y})R^\dagger = H_0(R\mathbf{y}R^\dagger)$, we introduce two distinct sets related by $\mathbf{y}^1 = R^\dagger \mathbf{y}^0 R$, such that $H_1(\mathbf{y}^1) = H_0(\mathbf{y}^0) = H$ is a unique Hamiltonian operator. Its distinct Coulomb and multipolar functional forms result from its expression in terms of the distinct sets $\mathbf{y}^0$ and $\mathbf{y}^1$ [26, 49, 52]. A Heisenberg picture operator is defined for any (Schrödinger picture) operator O by $O(t) = e^{iHt}Oe^{-iHt}$, and so $\mathbf{y}^0(t) = R(t)\mathbf{y}^1(t)R(t)^\dagger$ for all times t. Solving the Heisenberg equations $\dot{\mathbf{y}} = -i[\mathbf{y}, H_\alpha(\mathbf{y})]$ using the Hamiltonian $H_\alpha(\mathbf{y})$ yields the solution $\mathbf{y}(t) \equiv \mathbf{y}^\alpha(t)$, which is different depending on the value of α. We note that since $\mathbf{r}$ and $\mathbf{A}_{\rm T}$ commute with R, they are the same for both α, as are their derivatives in time. For these operators, the label α can therefore be omitted.

Let us begin by considering photons defined relative to the Coulomb ($\alpha = 0$) and multipolar ($\alpha = 1$) gauges. For a dipole at $\mathbf{R}$, integration of the Heisenberg equation $\dot a^\alpha_{\mathbf{k}\sigma} = -i[a^\alpha_{\mathbf{k}\sigma}, H_\alpha(\mathbf{y}^\alpha)]$ yields a solution that is the sum of free (vacuum) and source parts as

$$a^\alpha_{\mathbf{k}\sigma}(t) = a^\alpha_{\mathbf{k}\sigma,\rm vac}(\mathrm{t}) + \mathrm{a}^\alpha_{\mathbf{k}\sigma,s}(t), \tag{14.61}$$

$$a^\alpha_{\mathbf{k}\sigma,\rm vac}(\mathrm{t}) = \mathrm{a}^\alpha_{\mathbf{k}\sigma}(0)e^{-i\omega t}, \tag{14.62}$$

with

$$a^{0}_{\mathbf{k}\sigma,s}(t) = \frac{i}{\sqrt{\omega}}\int_{0}^{t} dt' e^{i\omega(t'-t)}\mathbf{f}_{\mathbf{k}\sigma}(\mathbf{R})^{*}\cdot\dot{\mathbf{d}}(t') \tag{14.63}$$

and

$$a^{1}_{\mathbf{k}\sigma,s}(t) = \sqrt{\omega}\int_{0}^{t} dt' e^{i\omega(t'-t)}\mathbf{f}_{\mathbf{k}\sigma}(\mathbf{R})^{*}\cdot\mathbf{d}(t'). \tag{14.64}$$

Since both $\mathbf{d}$ and $\dot{\mathbf{d}}$ are gauge-invariant, both source fields are gauge invariant, but that they are different immediately demonstrates the gauge-relativity of the photon concept.

A key difference between the Coulomb and multipolar gauge modes is the scaling of associated quantities with frequency. The two source operators in Eqs. (14.63) and (14.64) both exhibit divergences but at opposite extremes; $\omega \to 0$ in the case of the Coulomb gauge, and $\omega \to \infty$ in the case of the dipole gauge. The infrared divergence of the Coulomb gauge operator occurs because it implicitly includes a static contribution characterized by singular behavior at $\omega = 0$. The ultraviolet divergence of the dipole gauge operator occurs because the dipolar transverse polarization has been taken as that of a point localized dipole. It should be noted that the latter assumption is not necessary. Softening of this point-like localization can be achieved via the introduction of a form factor in $\mathbf{k}$-space, which effectively removes the contribution of frequencies above a chosen cut-off. Indeed, it is usually argued that such a procedure is necessary for consistency with the nonrelativistic treatment and EDA, and it results in a so-called regularized dipole gauge [26, 50, 53, 54].

To relate the two source operators, we integrate Eq. (14.63) by parts, giving

$$a^{0}_{\mathbf{k}\sigma,s}(t) - a^{1}_{\mathbf{k}\sigma,s}(t) = \frac{i}{\sqrt{\omega}}\mathbf{f}^{*}_{\mathbf{k}\sigma}(\mathbf{R})\cdot\left[\mathbf{d}(t) - \mathbf{d}(0)e^{-i\omega t}\right]. \tag{14.65}$$

The difference is seen to be a temporal boundary term. To understand its physical significance, we calculate the electric source field defined relative to each gauge. Since in the multipolar gauge $\Pi^{1} = -\mathbf{E}$ for all $\mathbf{x} \neq \mathbf{R}$, substituting Eq. (14.64) into the negative of Eq. (14.49) immediately yields the desired multipolar electric source field as

$$-\Pi^{1}_{s}(t,\mathbf{x}) = \mathbf{E}^{1}_{s}(t,\mathbf{x}) = \int_{0}^{t} dt'\, \mathbf{G}(\mathbf{x},\mathbf{R},t-t')\cdot\mathbf{d}(t'), \tag{14.66}$$

where G is defined as

$$G_{ij}(\mathbf{x},\mathbf{x}',t) := i\sum_{\mathbf{k}\sigma}\omega f_{\mathbf{k}\sigma i}(\mathbf{x})f_{\mathbf{k}\sigma j}(\mathbf{x}')^{*}e^{-i\omega t} + \text{c.c.} \tag{14.67}$$

in which c.c stands for complex conjugate. In general, the green's function includes both retarded and advanced parts, but the limits of integration in Eq. (14.66) imply that only the retarded part contributes.

In the Coulomb gauge, the use of Eq. (14.61) in Eq. (14.49) gives minus the transverse electric source field $\Pi^{0}_{s} = -\mathbf{E}^{0}_{\mathrm{T},s} = -\mathbf{E}^{0}_{s} + \mathbf{E}_{\mathrm{L}}$, which can be calculated in two ways. Either one substitutes Eq. (14.63) into Eq. (14.49) and calculates the field directly, or one uses Eq. (14.65) in Eq. (14.49) to obtain

$$-\Pi^{0}_{s}(t,\mathbf{x}) = \mathbf{E}^{0}_{\mathrm{T},s}(t,\mathbf{x}) = \mathbf{E}^{1}_{s}(t,\mathbf{x}) + \mathbf{K}(\mathbf{x},\mathbf{R},0)\cdot\mathbf{d}(t) - \mathbf{K}(\mathbf{x},\mathbf{R},t)\cdot\mathbf{d}(0) \tag{14.68}$$

where for $t \geq 0$,

$$K_{ij}(\mathbf{x},\mathbf{x}',t) := -\sum_{\mathbf{k}\sigma} f_{\mathbf{k}\sigma i}(\mathbf{x}) f^*_{\mathbf{k}\sigma j}(\mathbf{x}')e^{-i\omega t} + \text{c.c}, \tag{14.69}$$

$$\frac{\partial}{\partial t}\mathbf{K}(\mathbf{x},\mathbf{x}',t) = \mathbf{G}(\mathbf{x},\mathbf{x}',t). \tag{14.70}$$

Note that $-iK_{ij}(\mathbf{x},\mathbf{x}',0) = [A_{\mathrm{T}i}(\mathbf{x}),\Pi_j(\mathbf{x}')]$, which in the case of a parallel plate cavity is given by Eq. (14.12) and in free space is simply $i\delta^{\mathrm{T}}_{ij}(\mathbf{x}-\mathbf{x}')$. The total electric source field in the Coulomb gauge, $\mathbf{E}^0_s$, is found by adding $\mathbf{E}_{\mathrm{L}}$ to $\mathbf{E}^0_{\mathrm{T},s}(t,\mathbf{x})$ in Eq. (14.68). To find the functions $\mathbf{K}$ and $\mathbf{G}$, we first introduce the elementary retarded green's function $G_\square$ for the wave operator and its antiderivative for $t > 0$;

$$K_\square(\mathbf{x},\mathbf{x}',t) = -\frac{\theta(|\mathbf{x}-\mathbf{x}'|-t)}{4\pi|\mathbf{x}-\mathbf{x}'|}, \tag{14.71}$$

$$G_\square(\mathbf{x},\mathbf{x}',t) = \frac{\delta(t-|\mathbf{x}-\mathbf{x}'|)}{4\pi|\mathbf{x}-\mathbf{x}'|} = \frac{\partial}{\partial t}K_\square(\mathbf{x},\mathbf{x}',t) \tag{14.72}$$

where θ denotes the Heaviside step function. It is also convenient to introduce the dyadic differential operator $\mathcal{D}$ such that $\mathcal{D}\cdot\mathbf{v}(\mathbf{x}) = \nabla\times[\nabla\times\mathbf{v}(\mathbf{x})]$ for any (twice differentiable) vector field $\mathbf{v}$. In components $\mathcal{D}_{ij} = -\delta_{ij}\nabla^2 + \nabla_i\nabla_j$.

We begin by considering the case of free space, for which the functions $\mathbf{K}$ and $\mathbf{G}$ are found to be

$$\mathbf{K}(\mathbf{x},\mathbf{x}',t) = \mathcal{D}K_\square(\mathbf{x},\mathbf{x}',t), \tag{14.73}$$

$$\mathbf{G}(\mathbf{x},\mathbf{x}',t) = \mathcal{D}G_\square(\mathbf{x},\mathbf{x}',t). \tag{14.74}$$

Note that $-iK_{ij}(\mathbf{x},\mathbf{x}',0) = i\delta^{\mathrm{T}}_{ij}(\mathbf{x}-\mathbf{x}')$, which is the expected result for the equal-time commutation relation $[A_{\mathrm{T}i}(\mathbf{x}),\Pi_j(\mathbf{x}')]$ in free-space.

Substituting Eq. (14.74) into Eq. (14.66) yields the multipolar gauge electric source-field, which is nothing but the well-known causal electric source-field of a localized dipole (for $t_r \neq 0$) [55, 56]

$$\mathbf{E}^1_s(t,\mathbf{x}) = \theta(t_r)\nabla\times\left[\nabla\times\frac{\mathbf{d}(t_r)}{4\pi|\mathbf{x}-\mathbf{R}|}\right] \tag{14.75}$$

where $t_r = t - |\mathbf{x}-\mathbf{R}|$ is the retarded time at which the field at $(t,\mathbf{x})$ must depend on a localized source at $\mathbf{R}$. From Eq. (14.68), we obtain the Coulomb-gauge transverse electric source field (for $t_r \neq 0$)

$$\mathbf{E}^0_{\mathrm{T},s}(t,\mathbf{x}) = \mathbf{E}^1_s(t,\mathbf{x}) - \mathbf{E}_{\mathrm{L}}(t,\mathbf{x}) + \theta(-t_r)\mathbf{E}_{\mathrm{L}}(0,\mathbf{x}) \tag{14.76}$$

where $\mathbf{E}_{\mathrm{L}}(t,\mathbf{x}) = \mathbf{P}_{\mathrm{T}}(t,\mathbf{x}) = \mathbf{d}(t)\cdot\delta^{\mathrm{T}}(\mathbf{x}-\mathbf{R})$. The total Coulomb-gauge electric source field is therefore [26, 52]

$$\mathbf{E}^0_s(t,\mathbf{x}) = \mathbf{E}^1_s(t,\mathbf{x}) + \theta(-t_r)\mathbf{E}_{\mathrm{L}}(0,\mathbf{x}). \tag{14.77}$$

Equation (14.76) verifies that the source part of the photonic momentum in the Coulomb gauge is not the causal electric field $\mathbf{E}^1_s$ in Eq. (14.75); rather, it is the difference between this field and the electrostatic field $\mathbf{E}_{\mathrm{L}}$, plus a static precursor term $\theta(-t_r)\mathbf{E}_{\mathrm{L}}(0,\mathbf{x})$. More subtly, Eq. (14.77) demonstrates that the partitioning of a given field into vacuum and source parts is generally gauge relative [26]. The static precursor persists within the Coulomb gauge *total* electric source field. Since $a^0_{\mathbf{k}\sigma,\mathrm{vac}}(t) = Ra^1_{\mathbf{k}\sigma,\mathrm{vac}}(t)R^\dagger$, the two vacuum components of the electric field $\mathbf{E}^0_{\mathrm{vac}}$ and $\mathbf{E}^1_{\mathrm{vac}}$ differ by the negative of this term as $\mathbf{E}^0_{\mathrm{vac}}(t,\mathbf{x}) = -\Pi^0_{\mathrm{vac}}(t,\mathbf{x}) = -R\Pi^1_{\mathrm{vac}}(t,\mathbf{x})R^\dagger = -\Pi^1_{\mathrm{vac}}(t,\mathbf{x}) - \theta(-t_r)\mathbf{P}_{\mathrm{T}}(0,\mathbf{x}) = \mathbf{E}^1_{\mathrm{vac}}(t,\mathbf{x}) - \theta(-t_r)\mathbf{E}_{\mathrm{L}}(0,\mathbf{x})$. It follows using Eq. (14.77) that the total field is, as required, unique; $\mathbf{E}^0_{\mathrm{vac}} + \mathbf{E}^0_s = \mathbf{E} = \mathbf{E}^1_{\mathrm{vac}} + \mathbf{E}^1_s$ [26, 52].

In the Coulomb gauge, each of $\Pi^0 = -\mathbf{E}_\mathrm{T}$, $\mathbf{E}_s^0$, and $\mathbf{E}_\mathrm{vac}^0$ receive nonlocal static contributions when expressed in terms of the unique, local, and primitive field $\mathbf{E}$. This is due to the implicit inclusion of the static field within the definition of the corresponding dipole subsystem. In the dipole gauge, the dipole is defined as a bare mechanical object and $\Pi^1 = -\mathbf{E}$ for all $\mathbf{x} \neq \mathbf{R}$. Neither Π^1, $-\mathbf{E}_s^1$, nor $\mathbf{E}_\mathrm{vac}^1$ receive static contributions when expressed in terms of $\mathbf{E}$.

Calculating the electric field within the parallel plate cavity proceeds in a similar manner. The free-space source-field of the dipole is superposed with the source field of each image dipole. The functions $\mathbf{K}$ and $\mathbf{G}$ are found in Appendix A to be

$$K_{ij}(\mathbf{x},\mathbf{x}',t) = \mathcal{D}_{ij}K_\square(\mathbf{x},\mathbf{x}',t) + \sum_{n\neq 0} \mathcal{D}_{ij}K_\square(\mathbf{x},\mathbf{x}'_n,t) - \sum_{n=-\infty}^{\infty} \sigma_{jk}\mathcal{D}_{ki}K_\square(\mathbf{x},\check{\mathbf{x}}'_n,t) \tag{14.78}$$

and

$$G_{ij}(\mathbf{x},\mathbf{x}',t) = \mathcal{D}_{ij}G_\square(\mathbf{x},\mathbf{x}',t) + \sum_{n\neq 0} \mathcal{D}_{ij}G_\square(\mathbf{x},\mathbf{x}'_n,t) - \sum_{n=-\infty}^{\infty} \sigma_{jk}\mathcal{D}_{ki}G_\square(\mathbf{x},\check{\mathbf{x}}'_n,t) \tag{14.79}$$

where $\sigma = \mathrm{diag}(1,1,-1)$, $\mathbf{x}_n = \mathbf{x} + 2nL\hat{\mathbf{z}}$, and $\check{\mathbf{x}}_n = \check{\mathbf{x}} + 2nL\hat{\mathbf{z}}$ with $\check{\mathbf{x}} = (x,y,-z)$. In each expression, the first term gives the free-space result. Using Eqs. (14.66) and (14.79), the multipolar electric source-field is seen to be a superposition of the causal field of a localized dipole and its images.

The contributions in Eq. (14.68) for $\mathbf{E}_{\mathrm{T},s}^0$ that are additional to $\mathbf{E}_s^1$, can be found using Eq. (14.78). We can straightforwardly show that the $\mathbf{d}(t)$-dependent part is again the electrostatic field by using Eq. (14.12) to obtain

$$-\mathbf{K}(\mathbf{x},\mathbf{R},0)\cdot\mathbf{d}(t) = \mathbf{d}(t)\cdot\delta^\mathrm{T}(\mathbf{x}-\mathbf{R}) - \mathbf{d}(t)\cdot\nabla^\mathrm{R}\nabla\int_{\bar{V}} d^3s\frac{\xi(\mathbf{s},\mathbf{R})}{4\pi|\mathbf{x}-\mathbf{s}|}. \tag{14.80}$$

The electrostatic field of the dipole is $\mathbf{E}_\mathrm{L}^\mathrm{real}(t,\mathbf{x}) = \mathbf{P}_\mathrm{T}(t,\mathbf{x}) = \mathbf{d}(t)\cdot\delta^\mathrm{T}(\mathbf{x}-\mathbf{R})$, which is the first term on the right-hand side of Eq. (14.80). The second term on the right-hand side of Eq. (14.80) is the image electrostatic field $\mathbf{E}_\mathrm{L}^\mathrm{image} = -\nabla\phi^\mathrm{image}$, as can be verified easily using Eqs. (14.16) and (14.17), and the dipolar charge density $\rho(\mathbf{x}) = -\mathbf{d}\cdot\nabla\delta(\mathbf{x}-\mathbf{R})$. Thus, as in free space, the contribution in Eq. (14.80) is the electrostatic field, which in this case reads $\mathbf{E}_\mathrm{L}(t,\mathbf{x}) = \mathbf{E}_\mathrm{L}^\mathrm{real}(t,\mathbf{x}) + \mathbf{E}_\mathrm{L}^\mathrm{image}(t,\mathbf{x})$. An explicit expression can be found either by using Eq. (14.20) in Eq. (14.80) or by using Eq. (14.78) directly. The remaining term in Eq. (14.68) is similarly found using Eq. (14.78) and is again seen to be a static precursor that superposes the static precursor fields of the real dipole and each image.

14.3.3.2 Elimination of the Difference in Free Space Source Fields: Rotating-wave and Markovian Approximations

In Section 14.3.2.1, it was seen that subsystem gauge-relativity has no effect in the calculation of S-matrix elements for which strict bare energy conservation applies. In Section 14.3.2.2, this was also found to be the case in deriving a master equation for the dipole using quantum optical approximations that likewise ensure bare energy-conservation. By imposing the same approximations when deriving the source fields above, the effect of subsystem gauge-relativity is similarly found to disappear. To see this, let us assume the (weak-coupling) perturbative solution

$$\mathbf{d}(t) = \sum_{ab} \mathbf{d}_{ab}|a\rangle\langle b|e^{i\omega_{ab}t}, \tag{14.81}$$

which will be suitable for the evaluation of the fields to first order in q. Immediately, use of Eq. (14.81) combined with an imposition of resonance $\omega = \omega_{ba}$ can be seen to eliminate the difference between the multipolar and Coulomb gauge source operators in Eqs. (14.63) and (14.64).

In calculating the source fields in free space, the resonance condition is routinely brought about by invoking rotating-wave and Markovian approximations after the use of Eq. (14.81) [57]. We begin by substituting Eq. (14.81) into Eq. (14.66) to obtain

$$\Pi_s^1(t,\mathbf{x}) = -\frac{i}{2}\int_0^t dt' \int \frac{d^3k}{(2\pi)^3}\sum_\sigma\sum_{ab}\omega\mathbf{e}_{\mathbf{k}\sigma}\mathbf{e}_{\mathbf{k}\sigma}\cdot\mathbf{d}_{ab} \\ \times|a\rangle\langle b|e^{i(\omega_{ab}+\omega)t'}e^{-i\omega t}e^{i\mathbf{k}\cdot(\mathbf{x}-\mathbf{R})} + \text{H.c.} \tag{14.82}$$

This gives the field as the sum of its photon annihilation part $\Pi_s^{1,(+)}(t,\mathbf{x})$, and the Hermitian conjugate (H.c.), which is the photon creation part $\Pi_s^{1,(-)}(t,\mathbf{x})$. Performing the angular integration gives

$$\Pi_s^1(t,\mathbf{x}) = \frac{-i}{4\pi^2}\int_0^t dt' \int_0^\infty d\omega \sum_{ab} \mathcal{D}\cdot\mathbf{d}_{ab}\frac{\sin(\omega|\mathbf{x}-\mathbf{R}|)}{|\mathbf{x}-\mathbf{R}|} \\ \times|a\rangle\langle b|e^{i(\omega_{ab}+\omega)t'}e^{-i\omega t} + \text{H.c.} \tag{14.83}$$

At this stage one can combine the positive and negative frequency fields such that the frequency integral gives a sum of advanced and retarded green's functions for the wave operator. The limits of the time integral ensure that only the retarded green's function contributes and one obtains the causal first order (in q) source field

$$\Pi_s^1(t,\mathbf{x}) = -\theta(t_r)\sum_{ab}\mathcal{D}\cdot\frac{\mathbf{d}_{ab}}{4\pi|\mathbf{x}-\mathbf{R}|}e^{i\omega_{ab}t_r}|a\rangle\langle b|. \tag{14.84}$$

While we could have obtained this result immediately by simply inserting Eq. (14.81) into the exact expression for the electric source field, Eq. (14.75) [58–60], the route we have opted for allows us to show that the separate creation and annihilation parts in Eq. (14.83) are not themselves causal fields. They are so only within the rotating-wave and Markovian approximations [61, 62]. To see this, let us assume that when $a > b$ the exponential functions in Eq. (14.83) are rapidly oscillating, such that these terms can be ignored (rotating-wave approximation). Assuming subsequently that the time integrals are negligible unless $\omega = \omega_{ba} > 0$, allows us to push the lower limit of the frequency integral to $-\infty$ (Markovian approximation). We thereby obtain

$$\Pi_s^{1,(+)}(t,\mathbf{x}) = -\theta(t_r)\sum_{a<b}\mathcal{D}\cdot\frac{\mathbf{d}_{ab}}{4\pi|\mathbf{x}-\mathbf{R}|}e^{i\omega_{ab}t_r}|a\rangle\langle b|, \tag{14.85}$$

$$\Pi_s^{1,(-)}(t,\mathbf{x}) = -\theta(t_r)\sum_{a>b}\mathcal{D}\cdot\frac{\mathbf{d}_{ab}}{4\pi|\mathbf{x}-\mathbf{R}|}e^{i\omega_{ab}t_r}|a\rangle\langle b|. \tag{14.86}$$

Despite the additional approximations made, when combined these fields give back the exact first order total source field in Eq. (14.84). The rotating wave approximation (RWA) has the effect of associating only downward dipolar transitions with the positive frequency part and only upward transitions with the negative frequency part. This in turn enables the Markovian extension of each of these fields to include both signs of the frequency ω, as is required for causal signal propagation [63]. It is noteworthy that in the multipolar gauge these extra approximations do not alter the final result, Eq. (14.84).

The rotating-wave and Markovian approximations do alter the Coulomb gauge first-order source-field, because by ensuring resonance, they reduce it to the multipolar one. To see this, we now use Eq. (14.81) within the source field $\Pi_s^{0,(+)}(t,\mathbf{x})$ to obtain

$$\Pi_s^0(t,\mathbf{x}) = -\frac{i}{2}\int_0^t \mathrm{d}t' \int \frac{d^3k}{(2\pi)^3} \sum_\sigma \sum_{ab} \omega_{ba} \mathbf{e}_{\mathbf{k}\sigma}\mathbf{e}_{\mathbf{k}\sigma} \cdot \mathbf{d}_{ab} \times |a\rangle\langle b| e^{i(\omega_{ab}+\omega)t'} e^{-i\omega t} e^{i\mathbf{k}\cdot(\mathbf{x}-\mathbf{R})} + \text{H.c.} \tag{14.87}$$

which differs from Eq. (14.82) by a factor of ω_{ba}/ω. Computing the angular integral as before gives

$$\Pi_s^0(t,\mathbf{x}) = \frac{-i}{4\pi^2}\int_0^t \mathrm{d}t' \int_0^\infty \mathrm{d}\omega \sum_{ab} \frac{\omega_{ba}}{\omega} D \cdot \mathbf{d}_{ab} \frac{\sin(\omega|\mathbf{x}-\mathbf{R}|)}{|\mathbf{x}-\mathbf{R}|} \times |a\rangle\langle b| e^{i(\omega_{ab}+\omega)t'} e^{-i\omega t} + \text{H.c.}, \tag{14.88}$$

which as expected, is singular at $\omega = 0$. If, however, we now perform the rotating-wave approximation by restricting the summation to values $a < b$, and we again assume that the time integrals are zero unless $\omega = \omega_{ba}$, then we have ensured resonance. The additional factor of ω_{ba}/ω that is present in Eq. (14.88) as compared with Eq. (14.83) evaluates to unity and by again extending the lower limit of the frequency integral to $-\infty$ we obtain $\Pi_s^{0,(+)}(t,\mathbf{x}) = \Pi_s^{1,(+)}(t,\mathbf{x})$ where $\Pi_s^{1,(+)}(t,\mathbf{x})$ is given by Eq. (14.85). The rotating-wave and Markovian approximations have ensured bare-energy conservation. In so doing, they have removed the static singularity in the Coulomb gauge source field, reducing it to (minus) the same well-known first order total electric source-field that automatically occurs in the multipolar gauge (Eq. 14.84).

14.3.3.3 Gauge Ambiguity

It is important to recognize that it is always possible to express operators in terms of manifestly gauge-invariant and local degrees of freedom, namely, $\mathbf{r}$, $\dot{\mathbf{r}}$ (equivalently ρ and $\mathbf{J}$), $\mathbf{E}$ and $\mathbf{B}$, so there is no ambiguity regarding the physical meaning of any given operator. As already noted, for example, $a^1_{\mathbf{k}\sigma,s}$ and $a^0_{\mathbf{k}\sigma,s}$ are each uniquely defined functions of $\mathbf{d}$ and $\dot{\mathbf{d}}$, similarly, $\mathbf{E}^0_s$ and $\mathbf{E}^1_s$ are merely two different functions of $\mathbf{d}$. Above we have determined the precise physical differences between these operators and verified that such differences cancel out in the total field $\mathbf{E}$. However, the implications of gauge-relativity cannot be avoided simply because particle positions and velocities, $\mathbf{r}$ and $\dot{\mathbf{r}}$, and the total fields, $\mathbf{E}$ and $\mathbf{B}$, are not gauge relative. In the case of spontaneous emission, for example, one considers an initial vector $|e,0\rangle$, which represents two different physical states S_0 and S_1 for $\alpha = 0$ and $\alpha = 1$ respectively. Even if one considers observables such as the total electric energy density $\mathbf{E}^2/2$, averages found using the vector $|0,e\rangle$ are gauge-relative; that is, the predictions $\langle \mathbf{E}^2/2\rangle_{S_0}$ and $\langle \mathbf{E}^2/2\rangle_{S_1}$ are physically distinct. It should also be noted that quadratic functionals of the complete fields receive infinite (vacuum) contributions which must be removed (or renormalized). This makes a separation of fields into vacuum and source parts typically unavoidable, yet such a separation is necessarily gauge relative.

A *gauge ambiguity* arises if (and only if) it is unclear which physical degrees of freedom a considered operational procedure grants access to. For example, does a given preparation procedure result in the physical state S_0 or S_1? Does a given photodetection procedure measure the gauge-invariant physical observable $n^0 = \sum_{\mathbf{k}\sigma} a^{0\dagger}_{\mathbf{k}\sigma} a^0_{\mathbf{k}\sigma}$ or the gauge-invariant physical observable $n^1 = \sum_{\mathbf{k}\sigma} a^{1\dagger}_{\mathbf{k}\sigma} a^1_{\mathbf{k}\sigma}$? Determining when and how the differences between the averages of such observables are significant is an important task. We will confront this task below in the simplest settings of photon emission and absorption in the weak-coupling regime.

14.3.3.4 Spontaneous Emission

In spontaneous emission, two regimes are immediately amenable for weak couplings. These are (i) the Markovian regime $1/\omega \ll t \ll 1/\Gamma$ and (ii) the long-time limit $\Gamma t \gg 1$ after which the dipole has certainly decayed. In case (i) the condition $t \ll 1/\Gamma$ and weak-coupling render the perturbative solution for $\mathbf{d}(t)$ in Eq. (14.81) suitable. Use in Eqs. (14.63) and (14.64) gives the following average photon numbers for the mode $\mathbf{k}\sigma$ in the state represented by $|e,0\rangle$

$$\langle n^0_{\mathbf{k}\sigma}(t)\rangle = \sum_a \frac{\omega^2_{en}}{\omega}|\mathbf{f}_{\mathbf{k}\sigma}(\mathbf{R})\cdot\mathbf{d}_{ea}|^2F_t(\omega-\omega_{ea}), \tag{14.89}$$

$$\langle n^1_{\mathbf{k}\sigma}(t)\rangle = \sum_a \omega|\mathbf{f}_{\mathbf{k}\sigma}(\mathbf{R})\cdot\mathbf{d}_{ea}|^2F_t(\omega-\omega_{ea}) \tag{14.90}$$

where

$$F_t(\omega) := \frac{\sin^2[\omega t/2]}{(\omega/2)^2} = |\zeta_t(\omega)|^2, \tag{14.91}$$

in which

$$\zeta_t(\omega) = i\int_0^t dt' e^{i\omega t'} = \frac{e^{i\omega t}-1}{\omega}. \tag{14.92}$$

The two predictions differ by a factor of ω^2_{ea}/ω^2 within the summands, which are therefore only the same at resonance $\omega = \omega_{ea}$. Indeed, to impose such a resonance condition would be to impose bare-energy conservation, which eliminates subsystem gauge-relativity (see Section 14.3.2.1). In free space, the difference between the averages in Eqs. (14.89) and (14.90) would likely be difficult to probe experimentally, because summing over all modes and noting that $t \gg 1/\omega_{ea}$ allows us to replace $\partial_t F_t(\omega-\omega_{ea})$ with $2\pi\delta(\omega-\omega_{ea})$, we see that the resonance condition is satisfied. We thereby obtain

$$\frac{\mathrm{d}}{\mathrm{d}t}\sum_{\mathbf{k}\sigma} n^0_{\mathbf{k}\sigma}(t)\rangle = \frac{\mathrm{d}}{\mathrm{d}t}\sum_{\mathbf{k}\sigma} n^1_{\mathbf{k}\sigma}(t)\rangle = 2\pi\sum_{\mathbf{k}\sigma}\sum_a \omega|\mathbf{f}_{\mathbf{k}\sigma}(\mathbf{R})\cdot\mathbf{d}_{ea}|^2\delta(\omega-\omega_{ea}). \tag{14.93}$$

In free space and the continuum limit $v \to \infty$, this rate is easily computable as $\sum_{a<e}\Gamma_{ea} = \Gamma$ where $\Gamma_{ea} = \omega^3_{ea}|\mathbf{d}_{ea}|^2/(3\pi)$ (Einstein A-coefficient for the $e \to a$ transition), so that Γ is the total rate out of the level e. The Markovian and perturbative approximations that result in the equality of the rates obtained from distinct photon numbers are nothing but those entering Fermi's golden rule.

The difference between the photon number predictions is expected to be much more significant within a finite cavity and may even be possible to probe in the steady state. In case (ii) ($\Gamma t \gg 1$) the interaction remains weak but excited dipolar states decay exponentially, suggesting the ansatz [64]

$$\mathbf{d}(t) = \sum_{a,b}\mathbf{d}_{ab}|a\rangle\langle b|e^{(i\omega_{ab}-\Gamma_{ab}/2)t}. \tag{14.94}$$

In the simplest example of a two-level dipole for which $\{|n\rangle\} = \{|0\rangle,|1\rangle\}$, the solution in Eq. (14.94) is consistent with the optical Bloch equations, i.e., the two-level dipole Born–Markov secular master equation. Using Eq. (14.94) in Eq. (14.64) gives

$$a^1_{\mathbf{k}\sigma,s}(t) = e^{-i\omega t}\sqrt{\omega}\sum_{a,b}\mathbf{f}_{\mathbf{k}\sigma}(\mathbf{R})^*\cdot\mathbf{d}_{ab}|a\rangle\langle b|i\zeta_t(\omega+\omega_{ab}+i\Gamma_{ab}/2). \tag{14.95}$$

This yields the steady-state average photon number

$$\langle n^1_{\mathbf{k}\sigma}(\infty)\rangle = \sum_a |\mathbf{f}_{\mathbf{k}\sigma}(\mathbf{R})\cdot\mathbf{d}_{ea}|^2 \frac{\omega}{(\omega-\omega_{ea})^2+(\Gamma_{ea}/2)^2}. \tag{14.96}$$

The use of Eq. (14.94) in the boundary term of Eq. (14.65) shows that the contribution from $\mathbf{d}(t)$ vanishes in the steady state, but the contribution from $\mathbf{d}(0)$ remains and is singular at $\omega = 0$. This results in a different steady state prediction. Specifically, using Eq. (14.94) in Eq. (14.63) and retaining terms up to order q yields

$$a^0_{\mathbf{k}\sigma,s}(t) = \frac{e^{-i\omega t}}{\sqrt{\omega}} \sum_{a,b} \omega_{ab}\mathbf{f}_{\mathbf{k}\sigma}(\mathbf{R})^* \cdot \mathbf{d}_{ab}|a\rangle\langle b| i\zeta_t(\omega+\omega_{ab}+i\Gamma_{ab}/2). \tag{14.97}$$

The two source fields in Eqs. (14.95) and (14.97) differ by a factor of ω_{ab}/ω within the summand, such that the resulting photon number predictions in the state $|e,0\rangle$ again differ by a factor of ω^2_{ea}/ω^2 as

$$\langle n^0_{\mathbf{k}\sigma}(\infty)\rangle = \sum_a |\mathbf{f}_{\mathbf{k}\sigma}(\mathbf{R})\cdot\mathbf{d}_{ea}|^2 \frac{\omega^2_{ea}/\omega}{(\omega-\omega_{ea})^2+(\Gamma_{ea}/2)^2} \tag{14.98}$$

which again possesses an additional pole at $\omega = 0$. We can gain further insight by returning to Eqs. (14.63) and (14.64), and considering their different dependence on mode frequency. The Coulomb gauge source operator $a^0_{\mathbf{k}\sigma,s}$ is singular at $\omega = 0$ unlike the multipolar operator $a^1_{\mathbf{k}\sigma,s}$. As we saw in Section 14.3.3.1, this singular boundary term Eq. (14.65) is what leads to the additional static fields within the source component of Π as compared to its dipole gauge counterpart, which is just the causal source field of a localized dipole. Thus, $a^1_{\mathbf{k}\sigma,s}$ is nothing but the operator left when the singular (static) part of $a^0_{\mathbf{k}\sigma,s}$ is separated off and removed.

When using Eq. (14.81) and Eq. (14.94), this same separation is readily achieved by means of the identities [65]

$$\frac{1}{\sqrt{\omega}}\frac{1}{\omega-\omega_{ab}} = \frac{\sqrt{\omega}}{\omega_{ab}}\left[\frac{1}{\omega-\omega_{ab}} - \frac{1}{\omega}\right] \tag{14.99}$$

and

$$\frac{1}{\sqrt{\omega}}\frac{1}{\omega-\omega_{ab}+i\Gamma_{ab}/2} = \frac{\sqrt{\omega}}{\omega_{ab}}\left[\frac{1}{\omega-\omega_{ab}+i\Gamma_{ab}/2} - \frac{1}{\omega}\right] + O(q^2) \tag{14.100}$$

respectively. In the case of the ansatz in Eq. (14.94), for example, substituting Eq. (14.100) into Eq. (14.97) and again retaining terms to the same first order in q as occurs in Eq. (14.95), yields

$$a^0_{\mathbf{k}\sigma,s}(t) = a^1_{\mathbf{k}\sigma,s}(t) + i\frac{e^{-i\omega t}}{\sqrt{\omega}} \sum_{a,b} \mathbf{f}_{\mathbf{k}\sigma}(\mathbf{R})^* \cdot \mathbf{d}_{ab}|a\rangle\langle b|[e^{(i(\omega+\omega_{ab})-\Gamma_{ab}/2)t} - 1]. \tag{14.101}$$

where $a^1_{\mathbf{k}\sigma,s}(t)$ is given by Eq. (14.95). This difference, singular at $\omega = 0$, is simply the boundary term in Eq. (14.65) with the dipole moment given by Eq. (14.94).

The steady-state spectra of the mode continuum limit $v \to \infty$ are defined by summing over polarizations and integrating over all directions as

$$S^\alpha(\omega) = \frac{v}{(2\pi)^3} \int \mathrm{d}\Omega \sum_\sigma n^\alpha_{\mathbf{k}\sigma}(\infty)\rangle, \tag{14.102}$$

such that the difference in photon number predictions actually persists even within free space spectra. Lamb initially noted that only a dipole gauge calculation was in agreement with his experiments measuring emissions from atomic hydrogen [66]. Power and Zienau explained this by separating off and removing the static precursor contributions from the Coulomb gauge source operators using Eq. (14.100) [65]. They noted that this removal is precisely what is achieved by what is now known as multipolar QED or Power-Zienau-Woolley theory. They argued that the dipole gauge spectrum [essentially Eq. (14.96)], which results solely from the Fourier components of the causal source field, is what would be revealed by a Fourier analysis of the actually emitted wave train of the radiation field [65]. Milonni et al. also deemed this result to be the correct one and noted that it is obtained using the Coulomb gauge photon number operator if and only if the boundary contribution coming from $\mathbf{d}(0)$ is ignored in Eq. (14.65) [67].

14.3.3.5 Absorption

We now consider the process of photon absorption. In order to make contact with standard textbook developments of (Glauber) photodetection theory, it is most convenient to use the interaction picture rather than the Heisenberg picture used above. Letting $H = h + V$ and O respectively denote the Hamiltonian and an arbitrary operator in the Schrödinger picture, the corresponding interaction picture operator is defined by $O_I(t) := e^{iht}Oe^{-iht}$. In the case of a dipole-cavity system, a natural unperturbed Hamiltonian is the usual sum of bare Hamiltonians $h = H_m + H_{ph}$. The probability amplitude that the system, which at time $t = t_i$ is in a state represented by unperturbed vector $|i\rangle$, has by time $t_f > t_i$ evolved into a state represented by unperturbed vector $|f\rangle$, can be expressed as $c_{fi} = \langle f|U_I(t_f, t_i)|i\rangle$. The interaction picture evolution $U_I(t_f, t_i) = e^{iht_f}U(t_f, t_i)e^{-iht_i}$ satisfies $U_I(t, t) = I$ and $i\partial_t U_I(t, t_i) = V_I(t)U_I(t, t_i)$. Direct integration yields the solution $U_I(t_f, t_i) = I - i\int_{t_i}^{t_f} \mathrm{d}t V_I(t)U_I(t, t_i)$ which can be used to recursively generate the Dyson series

$$U_I(t_f, t_i) = I - i\int_{t_i}^{t_f} \mathrm{d}t V_I(t) - \int_{t_i}^{t_f} \mathrm{d}t \int_{t_i}^{t'} \mathrm{d}t' V_I(t)V_I(t') + \ldots \tag{14.103}$$

Setting $t_i = 0$ and $t_f = t$, assuming $|f\rangle \neq |i\rangle$, and truncating the Dyson series at first order gives

$$c_{fi} = -i\int_0^t \mathrm{d}t' \,\langle f|V_I(t')|i\rangle = -\langle f|V|i\rangle\, \zeta_t(\omega_{fi}) \tag{14.104}$$

where $\omega_{fi} = \omega_f - \omega_i$ is the energy difference between the initial and final unperturbed states. Instead, choosing $t_i = -t/2 = -t_f$ gives

$$c_{fi} = -\langle f|V|i\rangle i\xi_t(\omega_{fi}) \tag{14.105}$$

where

$$\xi_t(\omega) := \frac{\sin(\omega t/2)}{\omega/2} = 2\mathrm{Im}\zeta_{t/2}(\omega) = 2\partial_t F_{t/2}(\omega). \tag{14.106}$$

Since $\xi_t(\omega)^2 = F_t(\omega) = |\zeta_t(\omega)|^2$, the probabilities $|c_{fi}|^2$ obtained from Eqs. (14.104) and (14.105) are identical. For $t \gg 1/\omega$ the function ξ_t satisfies $\xi_t(\omega) \to 2\pi\delta(\omega)$ and for finite t is usually interpreted as expressing the conservation of bare energy $[\omega = 0]$ to within an uncertainty of $1/t$ [68].

We will focus on comparing the results obtained when using the Coulomb and multipolar-gauge interaction Hamiltonians. To describe one-photon absorption from a mode $\mathbf{k}\lambda$, we let the initial and final states be represented by the unperturbed vectors $|i\rangle = |n_{\mathbf{k}\lambda}; g\rangle$ and $|f\rangle = |(n-1)_{\mathbf{k}\lambda}, e\rangle$ where $|e\rangle$ represents an arbitrary final excited level of the dipole. The probability amplitude for this process is found using Eq. (14.104)

$$c_e(n,t) = -\langle f|V|i\rangle \zeta_t(\omega_{eg} - \omega). \tag{14.107}$$

In the Coulomb and multipolar gauges the amplitudes are therefore found, respectively, to be

$$c_e^\alpha(n,t) = i\sqrt{n}\Lambda^\alpha_{eg,\mathbf{k}\lambda}(t) \tag{14.108}$$

where

$$\Lambda^0_{eg,\mathbf{k}\lambda}(t) = \mathbf{d}_{eg}\cdot\mathbf{f}_{\mathbf{k}\lambda}(\mathbf{R})\frac{\omega_{eg}}{\sqrt{\omega}}\zeta_t(\omega_{eg}-\omega), \tag{14.109}$$

$$\Lambda^1_{eg,\mathbf{k}\lambda}(t) = \mathbf{d}_{eg}\cdot\mathbf{f}_{\mathbf{k}\lambda}(\mathbf{R})\sqrt{\omega}\zeta_t(\omega_{eg}-\omega). \tag{14.110}$$

We observe the familiar difference between the Coulomb and dipole gauge predictions of a factor of ω_{eg}/ω, which is unity only at resonance. When Eq. (14.129) is used instead, the function ζ_t should be replaced with $i\xi_t$.

For times $t \gg 1/(\omega_{eg}-\omega)$, we may replace $F_t(\omega_{eg}-\omega) = |\zeta_t(\omega_{eg}-\omega)|^2$ by $2\pi t\delta(\omega_{eg}-\omega)$ such that the associated probabilities become identical, $|c_e^0(t,n)|^2 = |c_e^1(t,n)|^2$, and proportional to t. The corresponding rate is therefore time-independent and, in free space, is straightforwardly relatable to the Einstein B-coefficient [49]. This equality of rates is again seen to be the result of the (Markovian) assumptions defining Fermi's golden rule.

More generally, we observe, as expected, that c_e^0 is infrared divergent as $1/\sqrt{\omega}$ whereas c_e^1 is ultraviolet divergent as $\sqrt{\omega}$. We can relate the absorption amplitudes by using Eq. (14.99), giving [cf. Eq. (14.101)]

$$\Lambda^0_{eg,\mathbf{k}\lambda}(t) = \Lambda^1_{eg,\mathbf{k}\lambda}(t) + \frac{1}{\sqrt{\omega}}\mathbf{f}_{\mathbf{k}\lambda}(\mathbf{R})\cdot\mathbf{d}_{eg}(e^{i(\omega_{eg}-\omega)t}-1). \tag{14.111}$$

We see again that if the singular contribution $c_e^1(n,t) - c_e^0(n,t)$ resulting from the implicit inclusion of the static field within the Coulomb gauge dipole is separated off and removed, then one is left with the absorption amplitude $c_e^1(n,t)$ of a localized bare dipole.

The one-photon absorption process can easily be generalized to the case of arbitrary single-mode states $|i\rangle = \sum_n \psi_n|n_{\mathbf{k}\lambda}\rangle$ and $|f\rangle = \sum_n \phi_n|n_{\mathbf{k}\lambda}\rangle$. In this case, both normally and anti-normally ordered products of field operators will generally contribute to the probabilities. However, for optical frequencies, one typically invokes the RWA, which constitutes retaining only the absorptive combination. For a bilinear interaction Hamiltonian the amplitude

$$C_e(t) = \sum_{n,n'} \phi^*_{n'}\psi_n\langle n'_{\mathbf{k}\lambda}, e|V|n_{\mathbf{k}\lambda}, g\rangle\zeta_t(\omega_{eg}+(n'-n)\omega) \tag{14.112}$$

is nonzero only if $n' - n = \pm 1$. Within the RWA one neglects the contribution from the case $n' - n = 1$ corresponding to a large energy denominator $\omega_{eg} + \omega$ and rapidly oscillating function $\zeta_t(\omega_{eg}+\omega)$, giving

$$\begin{aligned} C_e(t) &= \sum_n \phi^*_{n-1}\psi_n\,\langle(n-1)_{\mathbf{k}\lambda}, e|V|n_{\mathbf{k}\lambda}, g\rangle\,\zeta_t(\omega_{eg}-\omega), \\ &= \sum_m \phi^*_{n-1}\psi_n c_e(n,t) \end{aligned} \tag{14.113}$$

where $c_e(n, t)$ is the one-photon amplitude for absorption out of the mode $\mathbf{k}\lambda$ (Eq. 14.107). If we instead assume an initial vector $|i\rangle = \sum_n \psi_n |n_{\mathbf{k}\lambda}\rangle$, but sum over all final field states, which is equivalent to leaving the final field state unspecified, then within the RWA, we obtain a detection probability

$$P_e(t) = |c_e(\langle n_{\mathbf{k}\lambda}\rangle_i, t)|^2 \tag{14.114}$$

where $\langle n_{\mathbf{k}\lambda}\rangle_i = \sum_n n|\psi_n|^2$ is the average number of photons in the initial state.

In both Eqs. (14.113) and (14.114), as in Eq. (14.107), a static contribution, singular at $\omega = 0$, is present within the Coulomb gauge prediction and absent from the dipole gauge prediction. These differences vanish for sufficiently long times, as the function F_t becomes increasingly sharp at zero.

We note that the descriptions of stimulated emission in the Coulomb and dipole gauges can also be related to each other along the same lines as the absorption and spontaneous emission processes we have considered. Only one or two modes are involved in these processes, demonstrating that when used judiciously, a theory of the dipole's response to the field that is restricted to only a few modes is sufficient. This can even remain true when the field under consideration is from an identical dipole at $\mathbf{0}$ undergoing resonant emission that can be viewed as completely independent of the detector dipole. Let us consider a simple periodic cavity. In the long-time limit, the rate of absorption of a single photon with momentum $\mathbf{k}$ found from Eq. (14.108) using Fermi's golden rule is $w^1 = \pi I \sum_\lambda |\mathbf{d} \cdot \mathbf{e}_{\mathbf{k}\lambda}|^2 \rho_e$ where ρ_e is the density of final detector states and $I = \omega_{eg}/v$ is the intensity of the incident photon. Since v is arbitrary, the intensity I may be taken as an arbitrary single-photon intensity [69].

In the process of spontaneous emission by a second dipole at $\mathbf{0}$ only a single photon with frequency ω_{eg} is emitted. The intensity incident upon the detector dipole at $\mathbf{R}$ is that of emission in the direction $\hat{\mathbf{R}}$. Thus, the detection rate is $w = \pi I \sum_\lambda |\mathbf{d}_{eg} \cdot \mathbf{e}_{\hat{\mathbf{R}}\lambda}|^2 \rho_e = \pi I \mathbf{d}_{eg}^* \cdot \gamma \cdot \mathbf{d}_{eg} \rho_e$ in which $\gamma_{ij} = \delta_{ij} - \hat{R}_i \hat{R}_j$ and I denotes the intensity of spontaneous emission from the dipole at $\mathbf{0}$. This can be found from the spontaneous emission rate in the direction $\hat{\mathbf{R}}$ which is similarly found using Eq. (14.104) and Fermi's golden rule to be $\Gamma = \omega_{eg}^3 \mathbf{d}_{eg}^* \cdot \gamma \cdot \mathbf{d}_{eg}/(8\pi^2)$ in which the density of final states is that of photonic modes $\rho_{\text{ph}}(\omega) = v\omega^2/(2\pi)^3$. The intensity (power per unit area) of emission incident upon the detector is $I = \omega_{eg}\Gamma/R^2$ where $\omega_{eg}\Gamma$ is the power of spontaneous emission and R^2 is the appropriate cross-sectional area. The rate of detection is therefore

$$w = \frac{\rho_e}{8\pi}\left(\frac{\omega_{eg}^2}{R}\mathbf{d}_{eg}^* \cdot \gamma \cdot \mathbf{d}_{eg}\right)^2. \tag{14.115}$$

Note that only two-modes (those with $\mathbf{k} = \omega_{eg}\hat{\mathbf{R}}$) and first-order perturbation theory have been used to derive this expression, which is nevertheless identical to the far-field limit of the rate of resonant energy transfer obtained from a complete (unrestricted) model and second-order perturbation theory (cf. Section 14.3.3.8) [49, 69, 70].

14.3.3.6 Photodetection

We can generalize the description of absorption in Section 14.3.3.5 to describe photodetection. We do so, following Glauber [71], by again using second order perturbation theory (Eq. 14.104) and the RWA, but taking an arbitrary generally mixed initial state of the field, represented by $\rho = \sum_i \Pi |i\rangle\langle i|$. Glauber also supposed that given absorption resulting in excitation to the level e, the actual probability of a registered detection event is $\mathcal{D}_e$. Glauber employed the dipole–gauge interaction

Hamiltonian [71]. We give the final detection probability for both the dipole and Coulomb gauge interactions

$$P^{\alpha}(t) = \int_{0}^{t} dt' dt'' S_{ij}^{\alpha}(t'' - t') g_{ij}^{\alpha}(t', t'') \tag{14.116}$$

where

$$g_{ij}^{0}(t, t') = \mathrm{tr}\left[\rho A_{\mathrm{T}i}^{(-)}(t, \mathbf{R}) A_{\mathrm{T}j}^{(+)}(t', \mathbf{R})\right], \tag{14.117}$$

$$g_{ij}^{1}(t, t') = \mathrm{tr}\left[\rho \Pi_{i}^{(-)}(t, \mathbf{R}) \Pi_{j}^{(+)}(t', \mathbf{R})\right] \tag{14.118}$$

and

$$S_{ij}^{0}(t) = \sum_{e} D_{e} \omega_{eg}^{2} d_{eg,i}^{*} d_{eg,j} e^{i\omega_{eg} t}, \tag{14.119}$$

$$S_{ij}^{1}(t) = \sum_{e} D_{e} d_{eg,i}^{*} d_{eg,j} e^{i\omega_{eg} t}. \tag{14.120}$$

Computing the temporal integrals in Eq. (14.116) yields

$$P^{\alpha}(t) = \sum_{e} D_{e} \sum_{\mathbf{k}\lambda, \mathbf{k}'\lambda'} \mathrm{tr}(\rho a_{\mathbf{k}\lambda}^{\dagger} a_{\mathbf{k}'\lambda'}) \Lambda_{eg,\mathbf{k}\lambda}^{\alpha}(t)^{*} \Lambda_{eg,\mathbf{k}'\lambda'}^{\alpha}(t). \tag{14.121}$$

The implicit static contributions to the Coulomb gauge prediction $P^{0}(t)$ that are characterized by additional terms with poles at $\omega = 0$, can be separated-off using Eq. (14.111). When these terms are removed, one is left with the dipole gauge prediction $P^{1}(t)$. Thus, the same basic physical difference observed in the description of elementary absorption persists within photodetection theory.

Since the probabilities D_e within the definition of the detector sensitivity are arbitrary (they cannot be determined without a detailed theory of the detection experiment), one typically posits a form for the sensitivity function depending on the detection scheme envisioned. Two limiting cases of opposite extreme are of immediate interest: (i) Broadband detection whereby $S_{ij}(t) = s_{ij}\delta(t)$ and (ii) narrowband detection whereby $S_{ij}(t) = s_{ij}e^{i\Omega t}$ for some $\Omega > 0$. In each case, the probabilities can be related to the absorption amplitude obtained in Section 14.3.3.5.

In case (ii), if we let $\Omega = \omega_{eg}$ and $s_{ij} = d_{eg,i}^{*} d_{eg,j}$ a consideration of the single-mode initial field state $|i\rangle = \sum_n \psi_n |n_{\mathbf{k}\lambda}\rangle$ essentially leads back to Eq. (14.114) for absorption found in Section 14.3.3.5. Glauber argues [71] that case (i) is more basic, because narrowband detectors are usually realized by placing frequency filtering devices in front of broadband detectors. Glauber's broadband detection rate is

$$w^{1}(t) = \frac{dP^{1}(t)}{dt} = s_{ij}^{1} g_{ij}^{1}(t, t) \tag{14.122}$$

which if $s_{ij}^{1} \propto \delta_{ij}$ is proportional to the average Glauber intensity

$$I_{G}(t) = \langle \Pi^{(-)}(t, \mathbf{R}) \cdot \Pi^{(+)}(t, \mathbf{R}) \rangle \tag{14.123}$$

where the average is taken in the initial state ρ and the field operators are taken in the interaction picture. We note that this form for the intensity is obtained if and only if the detector is defined relative to the dipole gauge, because it derives directly from the correlation function appearing in the absorption probability.

The broadband limit of a completely flat (frequency-independent) spectral response, $s_{ij}(\omega)$, defined as the Fourier transform of $S_{ij}(t)$, cannot be strictly satisfied, because $s_{ij}(\omega) = 0$ for $\omega < 0$. The idealization makes sense if the field correlation function itself is suitably band limited. Specifically, it is reasonable provided that the width $\Delta\omega$ of the sensitivity function $s_{ij}(\omega)$ satisfies $\Delta\omega \gg 1/t$ and $\Delta\omega \gg \delta\omega$, where $\delta\omega$ is the bandwidth of $\int_{-\infty}^{\infty} dt dt' g_{ij}(t,t')e^{i\omega(t'-t)}$ [71]. For example, the broadband limit is not sensible when moving beyond the RWA to describe virtual absorption from the vacuum. These processes occur over short timescales and incur broad field correlations.

14.3.3.7 Source Intensity

In the foregoing discussion, we have considered absorption and detection by specifying initial states of the photonic field rather than specifying the source that produces photons. In the presence of a source, the field $\Pi(t, \mathbf{x})$ is a superposition of the vacuum, detector, and external source fields such as those produced by a dipole emitter; $\Pi = \Pi_{\rm vac} + \Pi_d + \Pi_s$. The Glauber intensity operator is therefore a sum of products of creation and annihilation components of each of these three parts. Normal ordering resulting from the RWA implies that $\Pi_{\rm vac}$ does not contribute in the photon vacuum, which is a natural state to take within the weak-coupling regime, because the source of photons has now been explicitly included. Similarly, the contribution that depends only on the detector's own reaction field does not contribute in the RWA for a ground state detector, and it is by definition independent of the source. The terms describing interference between the source and detector fields do not contribute in second order perturbation theory [i.e., weak-coupling, cf. Eqs. (14.81), (14.94), and (14.104)], provided the source is also in a sharp bare state, such that the initial state of the system is $|e_s, g_d, 0\rangle$.

Under these assumptions, one obtains an intensity that depends exclusively on the source as $I_{G,s}(t) = \langle \Pi_s^{(-)}(t,\mathbf{x}) \cdot \Pi_s^{(+)}(t,\mathbf{x})\rangle$. In the dipole gauge $\Pi_s^1(t,\mathbf{x}) = -\mathbf{E}_s^1(t,\mathbf{x})$ provided that the source and detector are spatially disjoint, so one recovers the electric source-field intensity $I_{G,s}(t) = \langle \mathbf{E}_s^{1,(-)}(t,\mathbf{x}) \cdot \mathbf{E}_s^{1,(+)}(t,\mathbf{x})\rangle$. In sufficiently strong coupling or non-Markovian regimes, the assumptions leading to this result will generally no longer be valid and the total field, including its detector and vacuum parts, would need to be considered more carefully.

It should also be noted that the Glauber intensity is not the intensity of emission. The radiated power of free-space spontaneous emission of a dipole into levels $n < e$ from the initial vector $|e, 0\rangle$, when defined using $I_{G,s}$ is to order q^2 found to be

$$\int d\Omega x^2 I_{G,s} = \frac{1}{2}\sum_{n<e} \omega_{en}\Gamma_{en}, \tag{14.124}$$

which is only half the correct result [62]. The latter is obtained by replacing $I_{G,s}$ with the average outward energy flux $\langle \hat{\mathbf{x}} \cdot \mathbf{S}\rangle$, where $\mathbf{S}$ is the Poynting vector, which is not normally ordered. This average involves not only a product of order q source fields but also products of the second order source fields with the vacuum (zeroth order) fields [59, 60, 62, 72, 73].

Importantly, the Glauber intensity obtained above is defined using the local electric source field which is equal to the source component of the photonic momentum in the dipole gauge. As was shown in Section 14.3.3.1, in the Coulomb gauge the field correlation function from which the intensity derives is defined in terms of the transverse potential $\mathbf{A}_{\rm T}$. The source component of the photonic momentum Π meanwhile is not equal to the source component of the electric field defined relative to the Coulomb gauge, and moreover, the latter is not equal to the local and causal electric field of a localized dipole.

The intensity is nevertheless often defined using the Coulomb-gauge transverse electric source field $\mathbf{E}_{\mathrm{T},s}^{0,(+)}(t,\mathbf{x})$. When it is restricted to a subset of modes (is band-limited), the Glauber type intensity defined in this way will receive source-dependent static contributions that do not generally vanish in the far-field. Contributions of this type are discussed further in the context of source-detector energy transfer in Section 14.3.3.8.

Here, we determine the differences between Glauber-type detection intensities $I_{G,s}(t) = \langle \Pi_s^{(-)}(t,\mathbf{x}) \cdot \Pi_s^{(+)}(t,\mathbf{x})\rangle$ that result when the fields $\Pi^{(\pm)}$ are defined relative to the Coulomb and dipole gauges, which is a topic of current interest [33]. The analysis of the fields in Section 14.3.3.1 can be carried over immediately to the photon creation and annihilation parts defined by

$$\Pi_s^{\alpha,(+)}(t,\mathbf{x}) \;\; := -i\sum_{\mathbf{k}\sigma}\sqrt{\omega}\mathbf{f}_{\mathbf{k}\sigma}(\mathbf{x})a_{\mathbf{k}\sigma,s}^{\alpha}(t), \tag{14.125}$$

which are such that $\Pi_s^{1,(+)}(t,\mathbf{x}) = -\mathbf{E}_s^{1,(+)}(t,\mathbf{x})$ and $\Pi_s^{0,(+)}(t,\mathbf{x}) = -\mathbf{E}_{\mathrm{T},s}^{0,(+)}(t,\mathbf{x})$. We can again make use of Equation (14.65) to obtain analogously to Eq. (14.68)

$$\mathbf{E}_{\mathrm{T},s}^{0,(+)}(t,\mathbf{x}) = \mathbf{E}_s^{1,(+)}(t,\mathbf{x}) + \mathbf{K}^{(+)}(\mathbf{x},\mathbf{R},0)\cdot\mathbf{d}(t) - \mathbf{K}^{(+)}(\mathbf{x},\mathbf{R},t)\cdot\mathbf{d}(0) \tag{14.126}$$

where

$$K_{ij}^{(+)}(\mathbf{x},\mathbf{x}',t) := -\sum_{\mathbf{k}\sigma} f_{\mathbf{k}\sigma i}(\mathbf{x})f_{\mathbf{k}\sigma j}^{*}(\mathbf{x}')e^{-i\omega t}. \tag{14.127}$$

Similarly, when using Eqs. (14.81) and (14.94) for the dipole moment operator, the identities (14.99) and (14.100) can be used, as in the case of the total fields, to separate the Coulomb-gauge source field into a singular component plus the multipolar source field. Returning to Eq. (14.126), in the case of free space $\mathbf{f}_{\mathbf{k}\sigma}(\mathbf{x}) = \mathbf{e}_{\mathbf{k}\sigma}e^{i\mathbf{k}\cdot\mathbf{x}}/\sqrt{2v}$ with $v \to \infty$ we obtain

$$\begin{aligned}\mathbf{E}_{\mathrm{T},s}^{0,(+)}(t,\mathbf{x}) \;\; &:= \;\; i\sum_{\mathbf{k}\sigma}\sqrt{\omega}\mathbf{f}_{\mathbf{k}\sigma}(\mathbf{x})a_{\mathbf{k}\sigma,s}^{0}(t)\\ &= \;\; \mathbf{E}_s^{1,(+)}(t,\mathbf{x}) - \frac{1}{2}\left[\mathbf{E}_{\mathrm{L}}(t,\mathbf{x}) - \mathbf{E}_{\mathrm{L}}(0,\mathbf{x})\theta(-t_r)\right]\\ &\quad -\frac{i}{\pi}\mathbf{E}_{\mathrm{L}}(0,\mathbf{x})\operatorname{arctanh}\left(\frac{t}{|\mathbf{x}-\mathbf{R}|}\right).\end{aligned} \tag{14.128}$$

The last line is anti-Hermitian, so it does not contribute to the total field and could therefore be removed through a redefinition of the transverse electric field's positive and negative frequency parts.

The two intensities generally only coincide when the additional terms in Eq. (14.126) can be ignored. As well as the intensity of the difference $\mathbf{E}_{\mathrm{T},s}^{0} - \mathbf{E}_s^{1}$ resulting entirely from the temporal boundary terms in Eq. (14.128) (the singular part of the Coulomb-gauge source field), there are also contributions describing interference between these terms and the field $\mathbf{E}_s^1$. These contributions will generally possess a longer range character due to the radiative part of $\mathbf{E}_s^1$.

In Eq. (14.128) the additional terms depend only on $\mathbf{E}_{\mathrm{L}}$, which can be ignored in the far field. Inside a finite cavity, however, it is not clear that such contributions can generally be neglected even for a single dipole. If a transition energy ω_{eg} of the dipole is resonant with the n'th mode of the parallel plate cavity, then the cavity length can be expressed as $L = n\pi/\sqrt{\omega_{eg}^2 - \kappa^2}$. For a point $\mathbf{x} = (0,0,aL)$ inside the cavity $(0 < a < 1)$ we have $\omega_{eg}x = an\pi\omega_{eg}/\sqrt{\omega_{eg}^2 - \kappa^2}$ such that if $\kappa \approx 0$ (emission in the z-direction) we obtain $\omega_{eg}x = an\pi$, which is of order unity for small n. As discussed in Section 14.3.2.1, some of the most interesting cavity QED effects occur precisely when the cavity length is of the same order as resonant wavelengths, that is, when $\omega_{eg}x \sim 1$ for points $\mathbf{x}$ inside the cavity. In this regime, the static, near, intermediate, and far fields of the source and its images are of similar magnitudes. This point is exemplified clearly by the calculation in Appendix C in which we re-derive Eqs. (14.54) and (14.55) for the emission rate of a dipole close to a single plate, by using the electric source fields of the image dipole.

It is clear that, in general, the source fields defined relative to different gauges may be markedly different within a cavity. However, in Section 14.3.3.2, weak-coupling conditions under which such differences disappear were identified. Unless the source and detector dipoles have markedly dissimilar characteristics, a source-field interaction outside of the weak-coupling regime would entail a similar detector-field coupling and vice versa. In this case, moving beyond the weak-coupling solutions in Eq. (14.81) and Eq. (14.94), but retaining the pure source intensity $I_{G,s}$ derived under the assumption of a weak detector-field interaction, would not be consistent. Restricting one's attention to Eq. (14.81) or Eq. (14.94), the difference between the intensities defined using $\mathbf{E}^0_{\mathrm{T},s}$ and $\mathbf{E}^1_s$ is easily found. It should be noted, however, that for the case of free space at least, it was found in Section 14.3.3.2 that, if the rotating-wave and Markovian approximations are invoked, then the differences between source fields, and between the positive and negative frequency parts, are eliminated.

14.3.3.8 Energy Transfer

When emission and absorption can be considered independent processes, the descriptions in Sections 14.3.3.4 and 14.3.3.5 are suitable. Emission results in a certain final state of the field during which one ignores the detector (Section 14.3.3.4). This state can be taken as the initial state of the field for the description of absorption while ignoring the source (Section 14.3.3.4). Section 14.3.3.5 further considers Glauber-type detection intensities that depend only on the source. But approaches using a first order truncation of the Dyson series cannot in general describe energy transfer from source to detector as a single (second order) coherent process.

In describing energy transfer, the initial and final vectors are $|i\rangle = |e_s, g_d, 0\rangle$ and $|f\rangle = |g_s, e_d, 0\rangle$, where g_μ and e_μ denote ground and excited levels of the material system $\mu = s, d$. It is convenient to choose $t_i = -t/2 = -t_f$ when truncating the Dyson series (14.103) at second order, which yields [68]

$$c_{fi} = -i\,\langle f|V|i\rangle\,\xi_t(\omega_{fi}) - i\int_{-\infty}^{\infty}\frac{d\omega}{2\pi}\,\langle f|W(\omega)|i\rangle\,\xi_t(\omega-\omega_f)\xi_t(\omega-\omega_i) \tag{14.129}$$

where

$$W(\omega) = \sum_n \frac{V|n\rangle\langle n|V}{\omega-\omega_n+i\eta} \tag{14.130}$$

in which the limit $\eta \to 0^+$ is to be taken. To arrive at Eq. (14.129), we have eliminated the time ordering within the temporal integrals of Eq. (14.103) by inserting a step function $\theta(t'-t'')$, and we have then used the identity [68]

$$e^{-i\omega t}\theta(t) = i\int_{-\infty}^{\infty}\frac{d\omega'}{2\pi}\,\frac{e^{-i\omega' t}}{\omega'-\omega+i\eta}. \tag{14.131}$$

Provided $\langle f|W(\omega)|i\rangle$ varies slowly with ω near to ω_i over the interval of width $1/t$, we may let $\langle f|W(\omega)|i\rangle \approx \langle f|W(\omega_i)|i\rangle$. This allows us to compute the integral over ω in Eq. (14.132) immediately, with the result

$$c_{fi} = -i\langle f|[V + W(\omega_i)]|i\rangle\xi_t(\omega_{fi}). \tag{14.132}$$

The term $\langle f|[V + W(\omega_i)]|i\rangle$ is called a T-matrix element while the multiplying function $\xi_t(\omega_{fi})$ describes the conservation of bare energy to within $1/t$. For sufficiently long times, the T-matrix element is evaluated on-energy-shell, $\omega_{fi} = 0$.

In describing resonant energy transfer, the energy shell condition reads $\omega_{s,eg} = \omega_{d,eg} = \omega_{eg}$, where $\omega_{\mu,eg}$ is the transition energy between the initial and final states of the μ'th dipole. In the dipole gauge, the bilinear interaction Hamiltonian reads $V^1 = \sum_{\mu=s,d} \mathbf{d}_\mu \cdot \Pi(\mathbf{R}_\mu)$, where $\mathbf{R}_\mu$ is the position of the μ'th dipole and $\mathbf{d}_\mu$ its dipole moment operator. The self-contributions comprising ϵ^{dip} do not contribute at order q^2. The matrix element $\langle f|V|i\rangle$ vanishes and the remaining contribution is

$$\begin{aligned}&\langle f|W(\omega_i)|i\rangle \\ &= \sum_{\mathbf{k}\lambda} \omega \left[\frac{\mathbf{f}^*_{\mathbf{k}\lambda}(\mathbf{R}_s)\cdot\mathbf{d}^*_{s,eg}\mathbf{f}_{\mathbf{k}\lambda}(\mathbf{R}_d)\cdot\mathbf{d}_{d,eg}}{\omega_{eg}-\omega} - \frac{\mathbf{f}_{\mathbf{k}\lambda}(\mathbf{R}_s)\cdot\mathbf{d}^*_{s,eg}\mathbf{f}^*_{\mathbf{k}\lambda}(\mathbf{R}_d)\cdot\mathbf{d}_{d,eg}}{\omega_{eg}+\omega}\right].\end{aligned} \tag{14.133}$$

where $\mathbf{d}_{\mu,eg}$ is the transition dipole moment of the μ'th dipole.

In the Coulomb gauge, the bilinear dipole-field coupling is $-\sum_{\mu=s,d}(q/m_\mu)\mathbf{p}_\mu \cdot \mathbf{A}_\text{T}(\mathbf{R}_\mu)$, which yields

$$\begin{aligned}&\langle f|W(\omega_i)|i\rangle \\ &= \sum_{\mathbf{k}\lambda} \frac{\omega_{eg}^2}{\omega} \left[\frac{\mathbf{f}^*_{\mathbf{k}\lambda}(\mathbf{R}_s)\cdot\mathbf{d}^*_{s,eg}\mathbf{f}_{\mathbf{k}\lambda}(\mathbf{R}_d)\cdot\mathbf{d}_{d,eg}}{\omega_{eg}-\omega} - \frac{\mathbf{f}_{\mathbf{k}\lambda}(\mathbf{R}_s)\cdot\mathbf{d}^*_{s,eg}\mathbf{f}^*_{\mathbf{k}\lambda}(\mathbf{R}_d)\cdot\mathbf{d}_{d,eg}}{\omega_{eg}+\omega}\right].\end{aligned} \tag{14.134}$$

This result differs to that in Eq. (14.133) by a factor of ω_{eg}^2/ω^2 and thereby possesses an additional pole at $\omega = 0$.

In Eq. (14.132), the T-matrix element is evaluated on shell (i.e., at resonance) for sufficiently long times, such that the Coulomb and dipole gauge total amplitudes should become identical. Note, however, that while naively evaluating the prefactor ω_{eg}^2/ω in Eq. (14.134) at resonance does indeed yield the dipole gauge prediction in Eq. (14.133), this procedure does not constitute an evaluation of the matrix element on shell. It yields an incorrect prediction in the Coulomb gauge, wherein the first order part $\langle f|V|i\rangle$ of the T-matrix element is necessarily nonzero. The sum over modes within $\langle f|W(\omega_i)|i\rangle$ corresponds to intermediate photon exchange that is not constrained by bare energy conservation. In the Coulomb gauge, it includes a static component that exactly cancels the first order static term $\langle f|V|i\rangle$. We now demonstrate this in the simplest case of free space by again using the same procedure of separating off the singularity at $\omega = 0$. We begin by noting that in the dipole gauge only Eq. (14.133) contributes and can be written

$$\langle f|W(\omega_i)|i\rangle = \frac{1}{2v}\sum_{\mathbf{k}} d^*_{s,eg,i}d_{d,eg,j}(\delta_{ij} - \hat{k}_i\hat{k}_j)e^{i\mathbf{k}\cdot\mathbf{R}}\omega\left[\frac{1}{\omega_{eg}-\omega} - \frac{1}{\omega_{eg}+\omega}\right]. \tag{14.135}$$

In the Coulomb gauge, the direct static dipole–dipole interaction between s and d yields the nonzero first order nonlocal contribution

$$\begin{aligned}\langle f|V|i\rangle &= \mathbf{d}^*_{s,eg}\cdot\delta^\text{L}(\mathbf{R})\cdot\mathbf{d}_{d,eg} = -\mathbf{d}^*_{s,eg}\cdot\delta^\text{T}(\mathbf{R})\cdot\mathbf{d}_{d,eg} \\ &= \frac{d^*_{s,eg,i}d_{d,eg,j}}{4\pi R^3}(\delta_{ij} - 3\hat{R}_i\hat{R}_j) =: \mathcal{V}^\text{dip}\end{aligned} \tag{14.136}$$

where $\mathbf{R} = \mathbf{R}_d - \mathbf{R}_s$. This matrix element is nothing but the matrix element of the electrostatic overlap energy between the nonlocal dipoles s and d defined relative to the Coulomb gauge, namely [74]

$$\mathcal{V}^\text{dip} = \langle f|\int d^3x\, \mathbf{E}_{\text{L}s}(\mathbf{x})\cdot\mathbf{E}_{\text{L}d}(\mathbf{x})|i\rangle \tag{14.137}$$

where $\mathbf{E}_{L\mu} = -\nabla\phi_\mu$ with ϕ_μ the Coulomb potential generated by the dipolar density $\rho_\mu(\mathbf{x}) = -\mathbf{d}_\mu \cdot \nabla\delta(\mathbf{x}-\mathbf{R}_\mu)$.

The accompanying second order term is found to be given by Eq. (14.135) but with the prefactor of ω on the second line replaced by ω_{eg}^2/ω. It therefore possesses an additional pole at $\omega = 0$, which can be separated-off via repeated application of Eq. (14.99) to give

$$\langle f|W(\omega_i)|i\rangle = \frac{1}{2v}\sum_{\mathbf{k}} d^*_{s,eg,i} d_{d,eg,j}(\delta_{ij} - \hat{k}_i\hat{k}_j)e^{i\mathbf{k}\cdot\mathbf{R}}\omega\left[\frac{1}{\omega_{eg}-\omega} - \frac{1}{\omega_{eg}+\omega} + \frac{2}{\omega}\right]. \tag{14.138}$$

The contribution of the final term singular at $\omega = 0$ is $\mathbf{d}^*_{s,eg} \cdot \delta^{\mathrm{T}}(\mathbf{R}) \cdot \mathbf{d}_{d,eg}$. We see therefore that the sum of Eqs. (14.138) and (14.136) yields the dipole-gauge matrix element in Eq. (14.135). Thus, while in the dipole gauge the contribution of the photonic interaction gives the final retarded amplitude, which we express as $c_{fi} = c_{fi}^{\mathrm{ph},1}$, in the Coulomb gauge the final result is the sum of static and photonic terms, as $c_{fi} = \mathcal{V}^{\mathrm{dip}} + c_{fi}^{\mathrm{ph},0}$ where $c_{fi}^{\mathrm{ph},0} = c_{fi} - \mathcal{V}^{\mathrm{dip}}$.

The unique on-shell matrix element c_{fi} is well-known. In the mode continuum limit the poles within the integrand can be handled using the principal value identity $\lim_{\eta\to 0^+}(x \pm i\eta)^{-1} = \mp i\pi\delta(x) + \mathcal{P}/x$. One obtains straightforwardly

$$\langle f|W(\omega_i)|i\rangle = \mathcal{R} + i\mathcal{I} \tag{14.139}$$

where

$$\mathcal{R} = \frac{3\Gamma_{ij}}{4}\left[\beta_{ij}\left(\frac{\cos\eta}{\eta^3} + \frac{\sin\eta}{\eta^2}\right) - \gamma_{ij}\frac{\cos\eta}{\eta}\right], \tag{14.140}$$

$$\mathcal{I} = \frac{3\Gamma_{ij}}{4}\left[\beta_{ij}\left(\frac{\sin\eta}{\eta^3} - \frac{\cos\eta}{\eta^2}\right) - \gamma_{ij}\frac{\sin\eta}{\eta}\right] \tag{14.141}$$

in which $\Gamma_{ij} = \omega_{eg}^3 d^*_{s,eg,i} d_{d,eg,j}/(3\pi)$, $\beta_{ij} = \delta_{ij} - 3\hat{R}_i\hat{R}_j$, $\gamma_{ij} = \delta_{ij} - \hat{R}_i\hat{R}_j$, and $\eta = \omega_{eg}R$. Another way to evaluate the integrals over $\mathbf{k}$ in Eq. (14.135) is to instead force convergence using a suitable damping factor and then employ the method of residues. This method gives the real part $\mathcal{R}$ alone (the principal value) [49, 74, 75] and suffices to observe the cancellation of the (also real) contribution $\mathcal{V}^{\mathrm{dip}}$ from the direct static interaction within Eq. (14.138), which possesses real part $\mathcal{R} - \mathcal{V}^{\mathrm{dip}}$.

In the far-field all but the radiative terms that decay as $1/R$ in Eq. (14.139) are negligible, as is $\mathcal{V}^{\mathrm{dip}}$. This implies in particular that $c_{fi} = c_{fi}^{\mathrm{ph},1} = c_{fi}^{\mathrm{ph},0}$ and for identical dipoles the Fermi golden rule-rate obtained from this amplitude is precisely Eq. (14.115), derived in Section 14.3.3.5. Note, however, that if the number of modes is restricted in the derivation above, then the partition of c_{fi} into near ($\sim R^{-3}$), intermediate ($\sim R^{-2}$), and far-field ($\sim R^{-1}$) parts dissolves. On the other hand, since the static term $-\mathcal{V}^{\mathrm{dip}}$ within $c_{fi}^{\mathrm{ph},0}$ is characterized entirely by a pole at $\omega = 0$ the partition, $c_{fi} = \mathcal{V}^{\mathrm{dip}} + c_{fi}^{\mathrm{ph},0}$ can still be made. As is evident from Eq. (14.138), on the level of individual modes, the static part of $c_{fi}^{\mathrm{ph},0}$ does not vanish in the far field.

The matrix element in Eq. (14.139) can be written succinctly as

$$\langle f|W(\omega_i)|i\rangle = -\mathbf{d}^*_{s,eg} \cdot \mathbf{G}(\mathbf{R}_s, \mathbf{R}_d, \omega_{eg}) \cdot \mathbf{d}_{d,eg} \tag{14.142}$$

where

$$\mathbf{G}(\mathbf{x}, \mathbf{x}', \omega) = \mathcal{D}g(\mathbf{x}, \mathbf{x}', \omega), \tag{14.143}$$

$$g(\mathbf{x}, \mathbf{x}', \omega) = \int dt\, G_\square(\mathbf{x}, \mathbf{x}', t)e^{i\omega t} = \frac{e^{i\omega|\mathbf{x}-\mathbf{x}'|}}{4\pi|\mathbf{x}-\mathbf{x}'|} \tag{14.144}$$

in which $G_{\square}$ is defined in Eq. (14.72), such that g is the retarded green's function for the Helmholtz equation. This in turn implies that the matrix element can also be written as the source part of the second order (in q) dipole-field interaction

$$\langle f|W(\omega_i)|i\rangle = \langle g_s, e_d|\mathbf{d}_d(t)\cdot\Pi_s(t,\mathbf{R}_d)|e_s, g_d\rangle \tag{14.145}$$

where $\Pi_s(t,\mathbf{R}_d)$ is defined by Eq. (14.84) and $\mathbf{d}_d(t)$ is given by Eq. (14.81) [75]. This shows that the detector responds to the (causal) multipolar source-field. In the case of the parallel plate cavity, one expects, based on Eqs. (14.66) and (14.79), that a detector dipole will respond additionally to the various source and detector image dipoles.

Although we have considered the nonlocal (i.e., joint) probability that the source, the detector, and the field will each be found in specified final states, if instead, one specifies the final detector state only, summing over all final source and field states, then one obtains the "local" (i.e., nonjoint) probability of detector excitation independent of the source and field.

14.3.3.9 Summary

In view of the fundamental physical differences between the Coulomb and multipolar gauge quantum subsystems, we have compared various Coulomb and dipole gauge subsystem predictions, focusing specifically on spontaneous emission, absorption and detection, and energy transfer. We have shown that:

1. Differences between Coulomb and dipole gauge subsystem predictions will generally disappear when bare energy conservation is imposed.
2. In the Coulomb gauge, each photonic mode provides additional static contributions to probability amplitudes, characterized by singular behavior at zero frequency.
3. These static contributions can be separated-off at the level of individual photonic modes via Eq. (14.65), Eq. (14.99), or Eq. (14.100). What remains are the subsystem predictions of the dipole gauge.
4. In particular, predictions for single-photon absorption can be related to the predictions of photodetection theory (Section 14.3.3.6) in the narrowband and broadband limits, showing that the same differences occur in this theory.
5. When considering energy transfer as a single coherent process (Section 14.3.3.8), the static contribution implicit within the Coulomb gauge light–matter interaction is exactly canceled by the contribution of the direct electrostatic interaction, only if the on-energy-shell condition is imposed. The static contribution to detector excitation from the individual modes within the Coulomb gauge is nonvanishing even in the far-field of the source.

Our analysis has employed approximations and has been restricted to specific physical situations, but we may expect the qualitative conclusions drawn to be general. We have focused on understanding the nature of individual photonic modes within the two gauges. The total static contribution of the Coulomb gauge light–matter interaction that results from summing the static contributions of all modes, can be expected to become negligible in the far-field [see, for example, Eqs. (14.68), (14.76), (14.77), (14.126), (14.128), (14.138)], but this is not true of the individual modal contributions themselves (see Sections 14.3.3.4 and 14.3.3.5).

We remark finally that there are myriad other atom-photon processes that might be considered using the same method of analysis as we have used, such as stimulated emission, Raman and Raleigh scattering, and dispersive forces, to name only a few. In each case, individual modes of the

Coulomb gauge will receive implicit static contributions. As in the specific cases we have considered, it should always be possible to separate off such contributions in such a way that what remains is the corresponding prediction found in the dipole gauge.

14.4 Imperfect Cavities

We now confront the task of describing more realistic imperfect (leaky) cavities, about which there is, of course, a wealth of literature. Our focus, however, will be as above, on discerning the precise physical differences between the light and matter subsystems defined relative to conventionally used gauges. We begin by briefly providing a reasonable semi-phenomenological extension of the perfect cavity theory given above in terms of global modes that describe the inside and outside of the cavity. We discuss simplified one-dimensional fully phenomenological descriptions of a high-Q cavity consisting of an interaction between internal and external modes treated as separate reservoirs. We briefly remark on truncations of material energy levels and on the number of internal cavity modes. We finally discuss subsystem gauge-relativity in macroscopic QED, focusing on generalized Coulomb and multipolar gauges. We show rigorously, that the same basic physical differences identified in Section 14.3.3 persist in this theory.

14.4.1 Phenomenological Descriptions

14.4.1.1 Empty Imperfect Parallel Plate Cavity

We begin by reviewing a description of an imperfect empty cavity given by Dutra [43] for the case that the plate at $z = L$ is no longer a perfect conductor. The canonical fields are given by [43, 76]

$$\mathbf{A}_{\mathrm{T}}(\mathbf{x}) = \sum_{\sigma=1,2} \int' d^3k \frac{1}{\sqrt{\omega}} \left[\mathbf{f}_\sigma(\mathbf{k},\mathbf{x})a_\sigma(\mathbf{k}) + \mathbf{f}^*_\sigma(\mathbf{k},\mathbf{x})a_\sigma(\mathbf{k})^\dagger\right], \tag{14.146}$$

$$\Pi(\mathbf{x}) = -i \sum_{\sigma=1,2} \int' d^3k \sqrt{\omega} \left[\mathbf{f}_\sigma(\mathbf{k},\mathbf{x})a_\sigma(\mathbf{k}) - \mathbf{f}^*_\sigma(\mathbf{k},\mathbf{x})a_\sigma(\mathbf{k})^\dagger\right], \tag{14.147}$$

where a prime is used to indicate that integration is over the half-space of positive k_z due to the perfect mirror at $z = 0$. Inside the cavity, $0 < z < L$, the mode functions are defined as

$$\mathbf{f}^<_1(\mathbf{k},\mathbf{x}) = \frac{1}{\sqrt{4\pi^3}} \mathcal{L}_1(\mathbf{k}) e^{i\kappa\cdot\mathbf{x}} \hat{\kappa} \times \hat{\mathbf{z}} \sin k_z z, \tag{14.148}$$

$$\mathbf{f}^<_2(\mathbf{k},\mathbf{x}) = \frac{1}{\sqrt{4\pi^3}} \mathcal{L}_2(\mathbf{k}) e^{i\kappa\cdot\mathbf{x}} \left[i\hat{\kappa} \cos\theta \sin k_z z \hat{\mathbf{z}} \sin\theta \cos k_z z\right], \tag{14.149}$$

and outside the cavity, $z > L$, they are

$$\mathbf{f}^>_1(\mathbf{k},\mathbf{x}) = \frac{1}{\sqrt{4\pi^3}} \mathcal{L}_1(\mathbf{k}) e^{i\kappa\cdot\mathbf{x}} \hat{\kappa} \times \hat{\mathbf{z}} \left[\sin k_z z + 2i \frac{r_1(\theta)}{t_1(\theta)} \sin\delta \sin(k_z z - \delta)\right], \tag{14.150}$$

$$\begin{aligned}\mathbf{f}^>_2(\mathbf{k},\mathbf{x}) = \frac{1}{\sqrt{4\pi^3}} \mathcal{L}_2(\mathbf{k}) e^{i\kappa\cdot\mathbf{x}} \Bigg\{ & i\hat{\kappa} k_z \cos\theta \left[\sin k_z z - 2i \frac{r_2(\theta)}{t_2(\theta)} \sin\delta \sin(k_z z - \delta)\right] \\ & -\hat{\mathbf{z}}\kappa \sin\theta \left[\cos k_z z - 2i \frac{r_2(\theta)}{t_2(\theta)} \sin\delta \cos(k_z z - \delta)\right]\Bigg\}.\end{aligned} \tag{14.151}$$

Here θ is the angle (of incidence) between $\mathbf{k}$ and $\hat{\mathbf{z}}$ and

$$\mathcal{L}_\sigma(\mathbf{k}) = \frac{t_\sigma(\theta)}{1 - m_\sigma r_\sigma(\theta)e^{2i\delta}}, \tag{14.152}$$

$$\delta = \omega L \cos\theta = k_z L, \tag{14.153}$$

$$|r_\sigma|^2 + |t_\sigma|^2 = 1, \tag{14.154}$$

$$r_\sigma t_\sigma^* + t_\sigma r_\sigma^* = 0, \tag{14.155}$$

$$m_1 = -1 \text{ and } m_2 = 1. \tag{14.156}$$

The numbers $r_\sigma(\theta)$ and $t_\sigma(\theta)$ are reflection and transmission coefficients. For generality, they are allowed to depend on the angle of incidence θ and on the polarization σ. Equation (14.154) implies that the imperfect mirror incurs no absorption losses. The photon operators satisfy

$$[a_\sigma(\mathbf{k}), a_{\sigma'}(\mathbf{k}')^\dagger] = \delta_{\sigma\sigma'}\delta(\mathbf{k} - \mathbf{k}') \tag{14.157}$$

and the energy of the system is

$$H_{\text{ph}} = \int' d^3k\, \omega \left(a_\sigma(\mathbf{k})^\dagger a_\sigma(\mathbf{k}) + \frac{1}{2}\right). \tag{14.158}$$

In the perfect cavity limit, $t_\sigma \to 0$ and $r_\sigma \to 1$, the cavity resonances become infinitely sharp, that is (see Appendix D) [43]

$$\lim_{t_\sigma \to 0} |\mathcal{L}_\sigma(\mathbf{k})|^2 = \frac{\pi}{L} \sum_{n=-\infty}^{\infty} \delta(k_z - k_{nz}) \tag{14.159}$$

where $k_{nz} = n\pi/L$. One can define discrete bosonic operators satisfying Eq. (14.10) by $a_{\sigma n}(\boldsymbol{\kappa}) = \lim_{t_\sigma \to 0} \sqrt{\pi/L}\, b_\sigma(\boldsymbol{\kappa} + k_{nz}\hat{\mathbf{z}})$ where $a_\sigma(\mathbf{k}) = \mathcal{L}_\sigma(\mathbf{k})^* b_\sigma(\mathbf{k})$ [43]. Substituting these expressions into Eqs. (14.146) and (14.147) and taking the perfect cavity limit, the expansions in Eqs. (14.1) and (14.2) are recovered for $n \neq 0$.

The perfect cavity limit must be taken judiciously to recover the expected result for the $n = 0$ ($k_z = 0$) term describing propagation parallel to the cavity plates in Eq. (14.159). Assuming the lower limit $-\epsilon \to 0^-$ of the k_z integrals in Eqs. (14.146) and (14.147) avoids evaluation of $\pi\delta(k_z)/L$ at the boundary. The mode functions in Eqs. (14.149) and (14.151) must then be multiplied by $1/\sqrt{2}$ for $k_z = 0$, which amounts to imposing the same condition as occurs in Eqs. (14.1) and (14.2) for the $n = 0$ terms of the mode summations therein. If the strict lower limit of 0 is instead retained for the k_z integrals in Eqs. (14.146) and (14.147), then an extra factor of 1/2 is introduced by the δ-function $\delta(k_z)$ at the boundary in Eq. (14.159). In this case, the mode functions in Eqs. (14.149) and (14.151) must be multiplied by $\sqrt{2}$ for $k_z = 0$ to recover the mode expansions in Eqs. (14.1) and (14.2).

14.4.1.2 Imperfect Cavity Containing Matter

A naive means by which the description above may be extended to include an atomic emitter within the cavity is to replace the operators and mode functions from Section 14.2 (perfect cavity case) with their imperfect cavity counterparts from Section 14.4.1.1. The Coulomb and multipolar gauge Hamiltonians will be formally identical to their perfect cavity counterparts given in Eqs. (14.25) and (14.41) respectively, but the photonic canonical operators and photon Hamiltonians therein will be given by their expressions in Section 14.4.1.1 [Eqs. (14.48), (14.49), and (14.158)] rather than the corresponding expressions in Section 14.2 [Eqs. (14.1), (14.2), and (14.11)]. For example, the dipole gauge Hamiltonian reads

$$H_1 = H_m + \epsilon^{\rm dip} + H_{\rm ph} + \mathbf{d} \cdot \Pi(\mathbf{R}) \tag{14.160}$$

where Π and $H_{\rm ph}$ are given by (14.49) and (14.158) respectively, $H_m = \mathbf{p}^2/(2m) + V(\mathbf{r})$ is the usual bare atomic Hamiltonian, and $\epsilon^{\rm dip}$ is given in Eq. (14.43).

The naive replacement has a different effect in the Coulomb and multipolar gauges. One therefore obtains nonequivalent theories depending on the gauge in which the replacement is made. In the Coulomb gauge $\Pi = -\mathbf{E}_{\rm T}$, so the replacement only effects the transverse electric field while the separated-off electrostatic interactions are left unchanged. In this case the naive replacement specifies a kind of filtering boundary at $z = L$, that allows leakage of transverse-electric energy but not of electrostatic energy. In contrast, in the multipolar gauge $\Pi = -\mathbf{E}$ at the imperfect boundary $z = L$ and there are no electrostatic image interactions. Thus, the naive replacement provides a description of a leakage in terms of the total (local) electric field.

Let us denote the nonequivalent Hamiltonians resulting from the naive replacement of mode functions performed in the Coulomb and multipolar gauges by $\mathcal{H}_0$ and $\mathcal{H}_1$ respectively. The S-matrix elements calculated using these distinct theories do not generally coincide for a given definition of the unperturbed Hamiltonian. If the latter does not include $V^{\rm image}$ then the rate of spontaneous emission is the same in both descriptions because $V^{\rm image}$ does not directly contribute. The result for a transition $n \to m$ is as ever

$$\Gamma_{nm} = 2\pi \int' d^3k \sum_\sigma \omega |\mathbf{d}^{nm} \cdot \mathbf{f}_\sigma(\mathbf{k}, \mathbf{R})|^2 \delta(\omega_{nm} - \omega). \tag{14.161}$$

For level shifts, which can be understood as the result of a dispersive interaction of a dipole with itself [77], distinct results are obtained because $V^{\rm image}$ does directly contribute. To determine the difference, we return to Eq. (14.37). The term $\epsilon^{\rm pol}$ therein appears in the multipolar but not the Coulomb gauge Hamiltonian, whereas the term $V^{\rm image}$ appears in the Coulomb-gauge but not the multipolar Hamiltonian. If the mode functions are replaced with their imperfect cavity counterparts, then Eq. (14.37) no longer holds, so a Coulomb gauge theory in which $V^{\rm image}$ is left the same (or simply ignored) will no longer be equivalent to the multipolar theory of the imperfect cavity.

To obtain a Coulomb gauge theory equivalent to $\mathcal{H}_1$ the perfect cavity image interaction $V^{\rm image}$ must be replaced by an imperfect electrostatic interaction defined using the imperfect cavity mode functions by [cf. Eq. (14.37)]

$$V^{\rm image,imp} = \epsilon^{\rm pol} - \sum_\sigma \int' d^3k \left| \int_v d^3x\, \mathbf{f}_\sigma(\mathbf{k}, \mathbf{x}) \cdot \mathbf{P}(\mathbf{x}) \right|^2. \tag{14.162}$$

With this replacement, the transformation of the Coulomb gauge (multipolar) Hamiltonian into the multipolar (Coulomb gauge) Hamiltonian proceeds as in Section 14.2 for the perfect cavity. The Coulomb gauge theory is found from the multipolar theory as $\tilde{\mathcal{H}}_0 = R^\dagger \mathcal{H}_1 R \neq \mathcal{H}_0$ where

$$R = \exp\left[-i \int_v d^3x \mathbf{P}(\mathbf{x}) \cdot \mathbf{A}_{\rm T}(\mathbf{x}) \right]. \tag{14.163}$$

in which $\mathbf{A}_{\rm T}$ is given by Eq. (14.146). This modification amounts to describing within the Coulomb gauge, a plate that is imperfect with respect to the total electric field.

Alternatively, one might consider replacing ϵ^{pol} in the multipolar Hamiltonian $\mathcal{H}_1$ as

$$\epsilon^{\text{pol,imp}} = V^{\text{image}} + \sum_{\sigma}\int' d^3k \left| \int_v d^3x\, \mathbf{f}_\sigma(\mathbf{k},\mathbf{x}) \cdot \mathbf{P}(\mathbf{x}) \right|^2 . \tag{14.164}$$

One then obtains a multipolar theory equivalent to the naive Coulomb-gauge theory as $\tilde{\mathcal{H}}_1 = R\mathcal{H}_0 R^\dagger \neq \mathcal{H}_1$. This amounts to describing within the multipolar gauge, a plate that is imperfect with respect to the transverse electric field only.

It should be noted that V^{image} and associated level shift phenomena are especially important when dealing with multiple material systems (atoms and molecules) inside a cavity. In this case, V^{image} constitutes a nontrivial interaction term between distinct material systems that can strongly affect energy exchange and dispersive interactions. The two distinct theories constructed above in which image electrostatic interactions are treated differently, can be expected to yield significantly different predictions relating to these phenomena.

Beyond scattering theory, the subsystems within any one of these theories remain gauge-relative and the implications of this fact that were discussed in Section 14.2 continue to apply. We note further that since the mode expansions in Eqs. (14.146) and (14.147) are of the form in Eqs. (14.48) and (14.49) all of the expressions from Section 14.3.3 that are written in terms of the mode functions $\mathbf{f}_{\mathbf{k}\sigma}(\mathbf{x})$ continue to apply. We note in particular, that using the mode functions in Eqs. (14.148–14.151) together with Eqs. (14.61–14.70) and (14.125–14.127), one can compute electric source fields and related intensities at any point $\mathbf{x}$, inside or outside of the cavity. For external points far from the cavity, the $\mathbf{K}$-dependent terms in Eqs. (14.68) and (14.126) should be negligible, such that each electric source field will produce the same results. We note, however, that the dynamics of the dipole moment operator appearing within the source fields will generally depend on which of the two nonequivalent models derived above is employed.

14.4.1.3 Phenomenological Descriptions

A common approach to model cavity leakage is to instead assume a linear coupling between internal modes and external modes treated as a separate bosonic reservoir (Gardiner-Collet type models) within a one-dimensional model of the form [78]

$$H_{\text{G-C}} = \sum_n \omega_n a_n^\dagger a_n + \int_0^\infty d\omega\, \omega b^\dagger(\omega)b(\omega) + \left[\int_0^\infty d\omega \sum_{n=1}^\infty g_n(\omega) a_n b^\dagger(\omega) + \text{H.c.}\right] \tag{14.165}$$

where a_n and $b(\omega)$ are discrete and continuous bosonic annihilation operators that define the cavity and external modes, respectively. Such a description can be expected to hold only for certain high Q cavities with $|r|$ very close to unity and $|t|$ very close to zero [43].

If supplemented with suitable mode expansions for canonical fields, the above model may be augmented to include a dipole at position R with canonical operators r, p such that $[r,p] = i$, and bare Hamiltonian $p^2/(2m) + V(r)$. An interaction term $V^0 = -qpA(R)/m + q^2A(R)^2/(2m)$ gives a Coulomb gauge coupling (without image charge effects) or an interaction $V^1 = d\Pi_<(R_z) + \epsilon^{\text{dip}}$ gives a dipole gauge coupling. Unsurprisingly, the corresponding total Hamiltonians H_0 and H_1 defined in this way can produce markedly different predictions beyond standard weak interaction regimes [26, 79].

Gauge freedom is ultimately vectorial in origin. One-dimensional theories cannot accommodate the decomposition of fields into transverse and longitudinal parts, and so they cannot reproduce important properties of a three-dimensional description which distinguish the physical meanings of the canonical operators in distinct gauges. There are, however, two main reasons for considering such simple models. On the one hand, they enable a determination of when the different coupling forms used naively will produce significantly different predictions. On the other hand, owing to their simplicity, they may enable one to compare how well approximations such as material level truncations and restrictions on the numbers of modes perform when assuming different coupling forms. It is important to distinguish between these two aspects. An approximation $\tilde{H}_0$ of H_0 may be more accurate at reproducing the predictions of H_0 than an approximation $\tilde{H}_1$ of H_1 is at reproducing the predictions of H_1, but it does not follow that $\tilde{H}_0$ is more physically "correct" than $\tilde{H}_1$, because the models H_0 and H_1 are not themselves physically equivalent.

Truncating the number of material energy levels necessarily breaks gauge-invariance as it is defined in nontruncated theories [26, 27, 35]. When considering a single mode, it is typically a better approximation in the dipole-gauge because the coupling does not scale with the material transition energies, unlike in the Coulomb gauge [26, 27, 30, 32, 35]. The optimal gauge for material truncation will largely depend on the number of modes considered, the particular mode under consideration, the prediction sought, and the physical regime considered [26].

When restricting the number of photonic modes, it is important to draw a distinction between a restriction within the light–matter interaction Hamiltonian (at the positions of the sources), from a restriction of fields at an arbitrary point $\mathbf{x}$ away from the sources. The point $\mathbf{x}$ could, for example, be the position of a detector that is not itself explicitly modeled. In Sections 14.3.3.5 and 14.3.3.6, it was seen that mode-restricted absorption and detection may naturally arise due to the physical situation considered. The point $\mathbf{x}$ could also be a point at a cavity boundary. As summarized in points 2–5 of Section 14.3.3.9, in addition to the modes of the causal and local fields that are manifest within the dipole gauge, the individual modes of the Coulomb gauge include highly nonlocal static contributions, which do not vanish at points $\mathbf{x}$ arbitrarily far from the source.

Restricting the number of photonic modes within the light–matter interaction Hamiltonian should be understood as an approximation intended to isolate and retain only those modes that are dominant therein. For a single dipole and a perfect cavity the approximation can be implemented while preserving the unitarity of gauge-fixing transformations and without therefore breaking gauge-invariance [26]. This is not necessarily the case when considering multiple dipoles. Of course, nonequivalent mode restricted theories will also be obtained if the restriction is implemented within already nonequivalent theories, such as those resulting from material level truncation, or those resulting from different models of leakage (see Section 14.4.1.2). In situations where mode-restricted theories are not equivalent, one typically expects the restriction within the light–matter coupling to perform better in the Coulomb gauge with coupling scaling as $1/\sqrt{\omega}$, than in the dipole gauge, with coupling scaling as $\sqrt{\omega}$. We reiterate that it does not follow, however, that the Coulomb gauge photonic momentum $-\mathbf{E}_\mathrm{T}$ is the physical observable (rather than, say, the dipole gauge momentum $-\mathbf{D}_\mathrm{T}$) by which the cavity must couple to external modes.

The more severe ultraviolet divergences that occur with the scaling $\sqrt{\omega}$ are not unexpected in either nonrelativistic or relativistic QED. They are dealt with via renormalization or appropriate physical regularization. In the dipole gauge, the point localization of the dipole can be softened via a form factor in Fourier space that cuts-off high-frequency modes. Such regularization can be viewed as necessary for consistency with a nonrelativistic treatment and the EDA. Regularized dipole gauge descriptions have been considered in some detail [50, 53, 54]. Within such a theory it should always be possible to restrict the number of modes within the light–matter interaction to some finite number that land within the envelope of the overall frequency cut-off employed. In such cases, even a restriction to a single mode may yet yield simple toy models sufficient to capture key physical features [74].

Finally, we note that while we have focused here on the Coulomb and dipole gauges, other gauges may be noteworthy for interesting and advantageous properties exhibited by the subsystems therein. In particular, for simple harmonic or two-level dipoles, a so-called Jaynes-Cummings (JC) gauge can be chosen that symmetrically mixes the Coulomb and dipole gauge linear couplings. This results in a Hamiltonian without counter-rotating terms within the linear interaction Hamiltonian. Provided that any other quadratic nonresonant interactions can be ignored or removed by local subsystem unitary transformations, this Hamiltonian enjoys the property that the unperturbed ground state coincides with the true ground state. Physically, the dipole subsystem defined relative to this gauge can be interpreted as implicitly including the ground state virtual photons dressing the bare dipole (as defined relative to the dipole gauge). One expects the degree of localization of this dipole to be between that of the fully localized dipole of the dipole gauge and the fully electrostatically dressed dipole of the Coulomb gauge. The coupling in the JC gauge is neither ultraviolet divergent (in contrast to the dipole gauge coupling) nor infrared divergent (in contrast to the Coulomb gauge coupling). The JC gauge has therefore been advocated as providing the most physically suitable description of photodetection, that is, as the gauge relative to which one obtains the most satisfactory definition of *measurable* photons [26, 80].

14.4.2 Subsystem Gauge Relativity in Macroscopic QED

We now consider the description of atoms and photons in dielectric media. For simplicity, we again consider a single guest hydrogen-like atom within a nondispersive nonabsorbing medium with real permittivity $\epsilon(\mathbf{x})$, noting however, that the same basic conclusions are found to hold for the more general case of multiple guest atoms in an absorbing and dispersive medium [81–83]. The most useful and natural variant of the Coulomb gauge to consider is the generalized Coulomb gauge such that $\nabla \cdot \epsilon\mathbf{A} = 0$. The corresponding generalized multipolar gauge can be obtained from this theory via a PZW gauge-fixing transformation. The most straightforward way to arrive at a correct canonical formalism is to write down the total energy of the system, then make suitable ansatzë for the canonical operators and their canonical commutation relations, and finally verify that the theory so constructed does indeed yield the correct Maxwell–Lorentz dynamical equations. The latter read

$$\dot{\mathbf{D}}_M = \nabla \times \mathbf{B} - \mathbf{J}, \tag{14.166}$$

$$m\ddot{\mathbf{r}} = q\mathbf{E}(\mathbf{r}) + \frac{q}{2}\left[\dot{\mathbf{r}} \times \mathbf{B}(\mathbf{r}) - \mathbf{B}(\mathbf{r}) \times \dot{\mathbf{r}}\right] \tag{14.167}$$

where $\mathbf{D}_M = \epsilon\mathbf{E}$. The homogeneous equations $\nabla \cdot \mathbf{B} = 0$ and $\dot{\mathbf{B}} = -\nabla \times \mathbf{E}$ are satisfied identically through the introduction of potentials as $\mathbf{B} = \nabla \times \mathbf{A}$ and $\mathbf{E} = -\nabla A_0 - \dot{\mathbf{A}}$. The Gauss law constraint $\nabla \cdot \mathbf{D}_M = \rho$ will be imposed via suitable definition of the constrained part of $\mathbf{D}_M$. In the above equations, the material fields ρ and $\mathbf{J}$ are those of the guest atom (Eqs. 14.14 and 14.15).

In order to construct the canonical theory, we will begin by generalizing the Helmholtz decomposition of vector fields, which reads

$$\mathbf{V} = \mathbf{V}_\mathrm{T} + \mathbf{V}_\mathrm{L}, \tag{14.168}$$

$$\nabla \cdot \mathbf{V}_\mathrm{T} = 0, \tag{14.169}$$

$$\nabla \times \mathbf{V}_\mathrm{L} = \mathbf{0}. \tag{14.170}$$

The addition and subtraction of any transverse field $\mathbf{W}_\mathrm{T}$ provides a repartitioning into a transverse field and a nonlongitudinal field as

$$\mathbf{V} = \tilde{\mathbf{V}}_\mathrm{T} + [\mathbf{V}_\mathrm{L} - \mathbf{W}_\mathrm{T}], \tag{14.171}$$

$$\tilde{\mathbf{V}}_\mathrm{T} = \mathbf{V}_\mathrm{T} + \mathbf{W}_\mathrm{T}. \tag{14.172}$$

An example is the decomposition of the electric field as $\mathbf{E} = -\Pi - \mathbf{P}$, where $\Pi = -\mathbf{E}_\mathrm{T} - \mathbf{P}_\mathrm{T}$ is a transverse photonic momentum and $\mathbf{P} = \mathbf{P}_\mathrm{L} + \mathbf{P}_\mathrm{T} = -\mathbf{E}_\mathrm{L} + \mathbf{P}_\mathrm{T}$ is a generally nonlongitudinal polarization satisfying Gauss law; $-\nabla \cdot \mathbf{P} = \rho$. In the generalized Coulomb gauge $\rho = \nabla \cdot \mathbf{D}_M = -\nabla \cdot (\epsilon \nabla \phi)$ where $\phi = A_0|_{\mathrm{gCG}}$ is the scalar potential (gCG stands for generalized Coulomb gauge), and $\epsilon \nabla \phi$ is not generally longitudinal, unlike $\nabla \phi$. The field $\mathbf{D}_M$ can therefore be partitioned as $\mathbf{D}_M = -\Pi - \epsilon \nabla \phi$ where Π is transverse. Similarly, the current can be partitioned as $\mathbf{J} = \tilde{\mathbf{J}}_\mathrm{T} + \epsilon \nabla \dot{\phi}$ where $\tilde{\mathbf{J}}_\mathrm{T} = \mathbf{J} - \epsilon \nabla \dot{\phi}$. Local charge conservation, $\nabla \cdot \mathbf{J} = -\dot{\rho}$, implies that $\tilde{\mathbf{J}}_\mathrm{T}$ is transverse.

We now suppose that the fields $\mathbf{A}$ and Π can be expanded in terms of generally complex mode functions $\mathbf{f}_k(\mathbf{x})$ as [82]

$$\mathbf{A}(\mathbf{x}) = \sum_k q_k \mathbf{f}_k(\mathbf{x}) = \sum_k q_k^\dagger \mathbf{f}_k^*(\mathbf{x}), \tag{14.173}$$

$$\Pi(\mathbf{x}) = \sum_k p_k \epsilon(\mathbf{x}) \mathbf{f}_k^*(\mathbf{x}) = \sum_k p_k^\dagger \epsilon(\mathbf{x}) \mathbf{f}_k(\mathbf{x}) \tag{14.174}$$

where q_k and p_k are generally non-Hermitian and such that $[q_k, p_{k'}] = i\delta_{kk'}$. The mode functions satisfy [cf. Eqs. (14.4) and (14.5)]

$$\nabla \cdot (\epsilon \mathbf{f}_k) = 0, \tag{14.175}$$

$$\nabla \times (\nabla \times \mathbf{f}_k) - \omega_k^2 \epsilon \mathbf{f}_k = \mathbf{0}, \tag{14.176}$$

in which the ω_k^2 are real (eigenvalues). The $\mathbf{f}_k(\mathbf{x})$ are orthonormal with respect to the weighted inner-product defined by the measure $d\mu_\epsilon(\mathbf{x}) = d^3x\epsilon(\mathbf{x})$, that is,

$$\int d\mu_\epsilon(\mathbf{x}) \mathbf{f}_k(\mathbf{x}) \cdot \mathbf{f}_{k'}^*(\mathbf{x}) = \delta_{kk'}. \tag{14.177}$$

The Maxwell field canonical commutation relation calculated using Eqs. (14.173) and (14.174) reads

$$\begin{aligned}[A_i(\mathbf{x}), \Pi_j(\mathbf{x}')] &= i \sum_k f_{ki}(\mathbf{x}) f_{kj}^*(\mathbf{x}') \epsilon(\mathbf{x}') = i \sum_k f_{ki}^*(\mathbf{x}) f_{kj}(\mathbf{x}') \epsilon(\mathbf{x}') \\ &=: i\delta_{\epsilon ij}^\mathrm{T}(\mathbf{x}, \mathbf{x}'),\end{aligned} \tag{14.178}$$

which defines the distribution $\delta_\epsilon^\mathrm{T}(\mathbf{x}, \mathbf{x}')$. It follows from Eq. (14.178) that $\delta_\epsilon^\mathrm{T}(\mathbf{x}, \mathbf{x}')$ is real, and in general, is transverse with respect to the second argument only.

Consider now the arbitrary hermitian generalized transverse field operator $\mathbf{V} = \sum_k v_k \mathbf{f}_k + \mathrm{H.c.}$ where the v_k are arbitrary operators. Using Eqs. (14.177) and (14.178), one deduces that

$$\int d^3x' \delta_\epsilon^\mathrm{T}(\mathbf{x}, \mathbf{x}') \cdot \mathbf{V}(\mathbf{x}') = \mathbf{V}(\mathbf{x}). \tag{14.179}$$

Noting that if $\mathbf{V}_\mathrm{T}(\mathbf{x})$ is transverse, then $\mathbf{V}_\mathrm{T}(\mathbf{x})/\epsilon(\mathbf{x})$ is generalized transverse, it also follows that

$$\int d^3x' \mathbf{V}_\mathrm{T}(\mathbf{x}') \cdot \delta_\epsilon^\mathrm{T}(\mathbf{x}', \mathbf{x}) = \mathbf{V}_\mathrm{T}(\mathbf{x}). \tag{14.180}$$

Finally, since $\epsilon(\mathbf{x}')\delta^{\mathrm{T}}_{\epsilon}(\mathbf{x}',\mathbf{x})$ is transverse with respect to $\mathbf{x}'$, it follows that

$$\int d^3x'\epsilon(\mathbf{x}')\mathbf{V}_{\mathrm{L}}(\mathbf{x}')\cdot\delta^{\mathrm{T}}_{\epsilon}(\mathbf{x}',\mathbf{x}) = 0. \tag{14.181}$$

Eqs. (14.180), (14.181), and (14.15) imply, in particular, that

$$\tilde{J}_{\mathrm{T}i}(\mathbf{x}) = \int d^3x' J_j(\mathbf{x}')\delta^{\mathrm{T}}_{\epsilon ji}(\mathbf{x}',\mathbf{x}) = \frac{q}{2}\left[\dot{r}_j\delta^{\mathrm{T}}_{\epsilon ji}(\mathbf{r},\mathbf{x}) + \delta^{\mathrm{T}}_{\epsilon ji}(\mathbf{r},\mathbf{x})\dot{r}_j\right]. \tag{14.182}$$

For completeness, we note before continuing that one can define [82]

$$U_{kk'} = \int d^3x\epsilon(\mathbf{x})\mathbf{f}_k(\mathbf{x})\cdot\mathbf{f}_{k'}(\mathbf{x}) \tag{14.183}$$

for which it is easily verified that

$$\mathbf{f}_k = \sum_{k'}\mathbf{f}^*_{k'}U_{k'k}, \tag{14.184}$$

$$q_k = \sum_{k'} q^{\dagger}_{k'}U^*_{kk'}, \tag{14.185}$$

$$\sum_{k''} U^*_{kk''}U_{k'k''} = \delta_{kk'}. \tag{14.186}$$

This enables one to define bosonic operators a_k with $[a_k, a^{\dagger}_{k'}] = \delta_{kk'}$ and $[a_k, a_{k'}] = 0$, by

$$q_k = \frac{1}{\sqrt{2\omega_k}}\left(a_k + \sum_{k'} U^*_{kk'}a^{\dagger}_{k'}\right), \tag{14.187}$$

$$p_k = i\sqrt{\frac{\omega_k}{2}}\left(a^{\dagger}_k - \sum_{k'} U_{kk'}a_{k'}\right), \tag{14.188}$$

such that the required algebraic properties of the q_k and p_k are satisfied. Substituting these expressions into Eqs. (14.173) and (14.174) yields bosonic mode expansions

$$\mathbf{A}(\mathbf{x}) = \sum_k \frac{1}{\sqrt{2\omega_k}}\mathbf{f}_k(\mathbf{x})a_k + \mathrm{H.c.}, \tag{14.189}$$

$$\Pi(\mathbf{x}) = -\,i\epsilon(\mathbf{x})\sum_k\sqrt{\frac{\omega_k}{2}}\mathbf{f}_k(\mathbf{x})a_k + \mathrm{H.c.}, \tag{14.190}$$

of the type we have used in previous sections.

We are now in a position to specify the Hamiltonian and demonstrate that it produces the correct dynamical Eqs. (14.166) and (14.167). The total energy of the system is

$$E = \frac{1}{2}m\dot{\mathbf{r}}^2 + \frac{1}{2}\int d^3x\left[\frac{\mathbf{D}_M(\mathbf{x})^2}{\epsilon(\mathbf{x})} + \mathbf{B}(\mathbf{x})^2\right]. \tag{14.191}$$

For the canonical momenta of the generalized Coulomb gauge, we make the following ansatzë

$$\mathbf{p} = m\dot{\mathbf{r}} + q\mathbf{A}(\mathbf{r}), \tag{14.192}$$

$$\Pi = -\mathbf{D}_M - \epsilon\nabla\phi \equiv \epsilon\dot{\mathbf{A}}, \tag{14.193}$$

which when substituted into Eq. (14.191) yields the generalized Coulomb gauge Hamiltonian

$$H_0 = \frac{1}{2m}[\mathbf{p} - q\mathbf{A}(\mathbf{r})]^2 + V(\mathbf{r}) + \frac{1}{2}\int d^3x\left[\frac{\Pi(\mathbf{x})^2}{\epsilon(\mathbf{x})} + \mathbf{B}(\mathbf{x})^2\right] \tag{14.194}$$

where

$$V(\mathbf{r}) = \frac{1}{2}\int d^3x \rho(\mathbf{x})\phi(\mathbf{x}) \tag{14.195}$$

is the Coulomb energy of the guest charges within the medium. It can be verified immediately that the Hamiltonian and canonical commutation relations yield Eqs. (14.192) and (14.193), demonstrating that these ansatzë are self-consistent. It is also straightforward to show that $-\dot{\Pi}(\mathbf{x}) = i[\Pi(\mathbf{x}), H_0] = \nabla\times\mathbf{B}(\mathbf{x})-\tilde{\mathbf{J}}_{\rm T}(\mathbf{x})$ with $\tilde{\mathbf{J}}_{\rm T}$ as given by the right-hand-side of Eq. (14.182). From this and Eq. (14.193) the medium-assisted Maxwell Ampere law (14.166) follows immediately. The Lorentz-force law (14.167) can be verified via similar calculations.

We have therefore successfully constructed the correct generalized Coulomb gauge theory. The quantum subsystems in this gauge are directly analogous to those of the Coulomb gauge in the absence of a medium. According to Eq. (14.193), the photonic momentum is nonlocally connected to the local field $\mathbf{D}_M = \epsilon\mathbf{E}$, while the canonical momentum of the dynamical charge differs from the local mechanical momentum by $-q\mathbf{A}(\mathbf{r})$. In the long wavelength approximation, i.e., the electric dipole approximation, this difference is approximated as $-q\mathbf{A}(\mathbf{R})$.

Passage to the generalized multipolar gauge is achieved via the PZW transformation

$$R = \exp\left[-i\int d^3x\mathbf{P}(\mathbf{x})\cdot\mathbf{A}(\mathbf{x})\right] = \exp\left[-i\int d^3x\tilde{\mathbf{P}}_{\rm T}(\mathbf{x})\cdot\mathbf{A}(\mathbf{x})\right] \tag{14.196}$$

where $\tilde{\mathbf{P}}_{\rm T} = \mathbf{P}(\mathbf{x}) - \epsilon(\mathbf{x})\nabla\phi(\mathbf{x})$ is transverse, because $\mathbf{P}$ defined in Eq. (14.33) satisfies $-\nabla\cdot\mathbf{P} = \rho$. Equations (14.180) and (14.181) imply [cf. Eq. (14.182)]

$$\tilde{\mathbf{P}}_{\rm T}(\mathbf{x}) = \int_0^1 d\lambda\, \mathbf{d}\cdot\delta^{\rm T}_{\epsilon}(\mathbf{R}+\lambda(\mathbf{r}-\mathbf{R}),\mathbf{x}). \tag{14.197}$$

Equation (14.196) can also be written $R = e^{iq\chi(\mathbf{r})}$ where

$$\chi(\mathbf{r}) = -\int_0^1 d\lambda(\mathbf{r}-\mathbf{R})\cdot\mathbf{A}(\mathbf{R}+\lambda(\mathbf{r}-\mathbf{R})). \tag{14.198}$$

The canonical momenta of the generalized Coulomb gauge transform as $R\Pi R^\dagger = \Pi + \tilde{\mathbf{P}}_{\rm T}$ and $R\mathbf{p}R^\dagger = \mathbf{p} - \nabla_{\mathbf{r}}\chi(\mathbf{r})$. The multipolar Hamiltonian $H_1 = RH_0R^\dagger$ is therefore

$$H_1 = \frac{1}{2m}[\mathbf{p} - q\mathbf{A}_1(\mathbf{r})]^2 + V(\mathbf{r}) + \frac{1}{2}\int d^3x\left[\frac{[\Pi(\mathbf{x})+\tilde{\mathbf{P}}_{\rm T}(\mathbf{x})]^2}{\epsilon(\mathbf{x})} + \mathbf{B}(\mathbf{x})^2\right] \tag{14.199}$$

where $\mathbf{A}_1(\mathbf{x}) = \mathbf{A}(\mathbf{x}) + \nabla\chi(\mathbf{x})$. The physical meaning of the canonical momenta of this gauge can be deduced using Eqs. (14.192) and (14.193) or using the Hamiltonian. In the multipolar gauge, the operator $R\Pi R^\dagger$ must represent the observable on the right-hand side of Eq. (14.193), so that, $\Pi + \tilde{\mathbf{P}}_{\rm T} = \epsilon\dot{\mathbf{A}}$. This equation can also be obtained using the multipolar Hamiltonian; $\epsilon\dot{\mathbf{A}} = -i\epsilon[\mathbf{A}, H] = \Pi + \tilde{\mathbf{P}}_{\rm T}$. Since $\mathbf{D}_M = -\epsilon(\dot{\mathbf{A}} + \nabla\phi)$ and $\tilde{\mathbf{P}}_{\rm T} = \mathbf{P} - \epsilon\nabla\phi$, it follows that the multipolar gauge photonic canonical momentum is minus the *total* displacement field, that is, $\Pi = -\mathbf{D}_M - \mathbf{P} =: -\mathbf{D}$. At all points $\mathbf{x}$ outside of the guest atom, this field equals minus the local material displacement field; $-\mathbf{D}_M = -\epsilon\mathbf{E}$. As in the case of free space, this lies in sharp contrast to the corresponding Coulomb gauge momentum $\Pi = \epsilon\dot{\mathbf{A}}$, which is not locally connected to either of the local fields $\mathbf{E}_M$ or $\mathbf{D}_M$. Similarly, the material canonical momentum of the multipolar gauge is $\mathbf{p} = m\dot{\mathbf{r}} + q\mathbf{A}_1(\mathbf{r})$.

In the EDA the multipolar polarization within the PZW transformation becomes

$$\mathbf{P}(\mathbf{x}) \approx \mathbf{d}\delta(\mathbf{x}-\mathbf{R}), \tag{14.200}$$

$$\tilde{\mathbf{P}}_{\mathrm{T}}(\mathbf{x}) \approx \mathbf{d} \cdot \delta_{\epsilon}^{\mathrm{T}}(\mathbf{R},\mathbf{x}). \tag{14.201}$$

yielding $R = e^{-iq\mathbf{r}\cdot\mathbf{A}(\mathbf{R})}$. The dipole gauge material canonical momentum is the localized mechanical momentum; $\mathbf{p} = m\dot{\mathbf{r}}$, and the dipole gauge photonic momentum differs from the local field $\mathbf{D}_M$ only at the position $\mathbf{R}$ of the point guest dipole; $\Pi = -\mathbf{D}_M - \mathbf{d}\delta(\mathbf{x}-\mathbf{R})$.

In summary, the same basic physical differences between the (generalized) Coulomb and multipolar gauges identified and discussed throughout previous sections persist when there is a background material medium. The photonic momentum of the multipolar gauge is the total displacement field equal to the medium's local displacement field everywhere outside of the guest atoms. The field Π propagates outward causally from the guest atoms. In particular, when more than one atom is considered and the EDA is used, the field Π mediates all inter-dipole interactions. In the Coulomb gauge, the photonic momentum differs from the local material displacement by a much less localized medium-assisted electrostatic field (Eq. 14.193). Guest atoms defined relative to this gauge concurrently implicitly include the electrostatic field, and this incurs the presence of an explicit electrostatic interaction Hamiltonian coupling different guest atoms. The individual modes of the generalized Coulomb gauge receive static contributions of the same kind as we have discussed for empty cavities throughout the previous sections.

14.5 Conclusions

We have compared the Coulomb and multipolar gauge theories of atoms contained within simple electromagnetic cavities. We have identified the basic physical differences between the light and matter subsystems defined relative to these two gauges, which can be understood as stemming from different partitions of the electric field $\mathbf{E} = -\Pi - \mathbf{P}$, into a "material" part $-\mathbf{P}$ and a "photonic" part $-\Pi$. The interactions between real atoms and between real atoms and their images or with background media, are dependent on the adopted definition of atom. In the dipole gauge $\mathbf{P}$ is point localized, so the local electric source field of a dipole is obtained directly from the photonic canonical momentum $-\Pi$. In contrast, within the Coulomb gauge $\mathbf{P} = -\mathbf{E}_{\mathrm{L}}$, so $\Pi = -\mathbf{E} + \mathbf{E}_{\mathrm{L}}$ is nonlocal receiving additional static contributions. As a result, image electrostatic interactions are explicitly present in the Coulomb gauge and absent in the dipole gauge.

The differences in the subsystems defined relative to the two gauges result in different dynamical evolution of subsystem operators. This results, in particular, in different partitions of the *same observable* field $\mathbf{E} = \mathbf{E}_0 + \mathbf{E}_s$ into source, $\mathbf{E}_s$, and vacuum, $\mathbf{E}_0$, parts. In the Coulomb gauge $\mathbf{E}_s$ receives nonlocal *static precursor* contributions in addition to the well-known local and causal field of the dipole that is automatically obtained in the dipole gauge. Within a parallel plate cavity, this basic difference occurs not only for the field produced directly by a real dipole inside the cavity, but also for all of the source fields resulting from this dipole's (infinity of) images.

We have exemplified the gauge nonrelativistic property of bare energy-conserving processes over infinite times (S-matrix elements) and noted that this generally requires that all explicit electrostatic interactions are retained in the Coulomb gauge. More broadly, we have noted that the relativity of the light and matter subsystems is least important in standard weak-coupling and Markovian regimes. Nevertheless, by focusing on elementary one-photon processes, including emission, absorption/detection, and energy transfer, we have determined, at the level of *individual modes*, how the relativity of the subsystems affects basic atom-photon predictions. We have found quite generally, that predictions pertaining to one and two dipoles defined relative to the Coulomb gauge contain additional static contributions characterized by singular behavior at zero mode frequency. When these contributions are removed, the corresponding dipole gauge prediction is what is left over. This insight should prove valuable when looking ahead to the description of nonstandard regimes featuring greater coupling strengths and non-Markovian environments in which subsystem relativity becomes increasingly important.

We have extended our consideration to imperfect cavities, showing first that naive models obtained by replacing the perfect cavity canonical fields with their imperfect cavity counterparts generally result in nonequivalent Coulomb and multipolar theories. The reason is that, due to subsystem relativity, the replacement constitutes a different physical assumption about cavity leakage when carried out in the two different gauges, and in fact, either model can be obtained in either gauge. We have further briefly discussed the effect of truncating the number of material levels and photonic modes to obtain toy models. Finally, we have shown that the basic physical differences between the subsystems defined relative to the two gauges persist when one assumes the presence of a background medium. While the generalized dipole gauge photonic momentum within a dielectric medium is the local displacement field, the generalized Coulomb gauge momentum differs from either the displacement or the medium electric field and is nonlocal. The same differences in basic atom photon processes identified in the case of empty cavities can, therefore, be expected to persist when a medium is present.

References

1 Kaluzny, Y., Goy, P., Gross, M., Raimond, J. M., Haroche, S. Observation of self-induced Rabi oscillations in two-level atoms excited inside a resonant cavity: the ringing regime of superradiance. *Phys. Rev. Lett.* **1983**, *51*(13), 1175–1178. https://doi.org/10.1103/PhysRevLett.51.1175.

2 Thompson, R. J., Rempe, G., Kimble, H. J. Observation of normal-mode splitting for an atom in an optical cavity. *Phys. Rev. Lett.* **1992**, *68*(8), 1132–1135. https://doi.org/10.1103/PhysRevLett.68.1132.

3 Meschede, D., Walther, H., Müller, G. One-atom maser. *Phys. Rev. Lett.* **1985**, *54*(6), 551–554. https://doi.org/10.1103/PhysRevLett.54.551.

4 Chikkaraddy, R., De Nijs, B., Benz, F., Barrow, S. J., Scherman, O. A., Rosta, E. et al. Single-molecule strong coupling at room temperature in plasmonic nanocavities. *Nature* **2016**, *535*(7610), 127–130.

5 Weisbuch, C., Nishioka, M., Ishikawa, A., Arakawa, Y. Observation of the coupled exciton-photon mode splitting in a semiconductor quantum microcavity. *Phys. Rev. Lett.* **1992**, *69*(23), 3314–3317. https://doi.org/10.1103/PhysRevLett.69.3314.

6 Lodahl, P., Mahmoodian, S., Stobbe, S. Interfacing single photons and single quantum dots with photonic nanostructures. *Rev. Mod. Phys.* **2015**, *87*(2), 347–400. https://doi.org/10.1103/RevModPhys.87.347.

7 Blais, A., Grimsmo, A. L., Girvin, S. M., Wallraff, A. Circuit quantum electrodynamics. *Rev. Mod. Phys.* **2021**, *93*(2), 025005. https://doi.org/10.1103/RevModPhys.93.025005.

8 Degen, C. L., Reinhard, F., Cappellaro, P. Quantum sensing. *Rev. Mod. Phys.* **2017**, *89*(3), 035002. https://doi.org/10.1103/RevModPhys.89.035002.

9 Meher, N., Sivakumar, S. A review on quantum information processing in cavities. *Eur. Phys. J. Plus.* **2022**, *137*(8), 985. https://doi.org/10.1140/epjp/s13360-022-03172-x.

10 Haroche, S. Nobel lecture: controlling photons in a box and exploring the quantum to classical boundary. *Rev. Mod. Phys.* **2013**, *85*(3), 1083–1102. https://doi.org/10.1103/RevModPhys.85.1083.

11 Ebbesen, T. W. Hybrid light–matter states in a molecular and material science perspective. *Acc. Chem. Res.* **2016**, *49*(11), 2403–2412. https://doi.org/10.1021/acs.accounts.6b00295.

12 Garcia-Vidal, F. J., Ciuti, C., Ebbesen, T. W. Manipulating matter by strong coupling to vacuum fields. *Science* **2021**, *373*(6551), eabd0336. https://doi.org/10.1126/science.abd0336.

13 Santhosh, K., Bitton, O., Chuntonov, L., Haran, G. Vacuum Rabi splitting in a plasmonic cavity at the single quantum emitter limit. *Nat. Commun.* **2016**, *7*(1), ncomms11823.

14 Feist, J., Galego, J., Garcia-Vidal, F. J. Polaritonic chemistry with organic molecules. *ACS Photonics* **2018**, *5*(1), 205–216. https://doi.org/10.1021/acsphotonics.7b00680.

15 Ruggenthaler, M., Tancogne-Dejean, N., Flick, J., Appel, H., Rubio, A. From a quantum-electrodynamical light–matter description to novel spectroscopies. *Nat. Rev. Chem.* **2018**, *2*(3), 1–16.

16 Ribeiro, R. F., Martínez-Martínez, L. A., Du, M., Campos-Gonzalez-Angulo, J., Yuen-Zhou, J. Polariton chemistry: controlling molecular dynamics with optical cavities. *Chem. Sci.* **2018**, *9*(30), 6325–6339.

17 Flick, J., Rivera, N., Narang, P. Strong light-matter coupling in quantum chemistry and quantum photonics. *Nanophotonics* **2018**, *7*(9), 1479–1501. https://doi.org/10.1515/nanoph-2018-0067/html.

18 Reitz, M., Sommer, C., Genes, C. Cooperative quantum phenomena in light-matter platforms. *PRX Quantum* **2022**, *3*(1), 010201. https://doi.org/10.1103/PRXQuantum.3.010201.

19 Orgiu, E., George, J., Hutchison, J. A., Devaux, E., Dayen, J. F., Doudin, B. Conductivity in organic semiconductors hybridized with the vacuum field. *Nat. Mater.* **2015**, *14*(11), 1123–1129.

20 Schachenmayer, J., Genes, C., Tignone, E., Pupillo, G. Cavity-enhanced transport of excitons. *Phys. Rev. Lett.* **2015**, *114*(19), 196403. https://doi.org/10.1103/PhysRevLett.114.196403.

21 Hagenmüller, D., Schachenmayer, J., Schütz, S., Genes, C., Pupillo, G. Cavity-enhanced transport of charge. *Phys. Rev. Lett.* **2017**, *119*(22), 223601. https://doi.org/10.1103/PhysRevLett.119.223601.

22 Hagenmüller, D., Schütz, S., Schachenmayer, J., Genes, C., Pupillo, G. Cavity-assisted mesoscopic transport of fermions: coherent and dissipative dynamics. *Phys. Rev. B* **2018**, *97*(20), 205303. https://doi.org/10.1103/PhysRevB.97.205303.

23 Rokaj, V., Ruggenthaler, M., Eich, F. G., Rubio, A. Free electron gas in cavity quantum electrodynamics. *Phys. Rev. Res.* **2022**, *4*(1), 013012. https://doi.org/10.1103/PhysRevResearch.4.013012.

24 Eckhardt, C. J., Passetti, G., Othman, M., Karrasch, C., Cavaliere, F., Sentef, M. A. et al. Quantum Floquet engineering with an exactly solvable tight-binding chain in a cavity. *Commun. Phys.* **2022**, *5*(1), 122. https://doi.org/10.1038/s42005-022-00880-9.

25 Amelio, I., Korosec, L., Carusotto, I., Mazza, G. Optical dressing of the electronic response of two-dimensional semiconductors in quantum and classical descriptions of cavity electrodynamics. *Phys. Rev. B* **2021**, *104*(23), 235120. https://doi.org/10.1103/PhysRevB.104.235120.

26 Stokes, A., Nazir, A. Implications of gauge freedom for nonrelativistic quantum electrodynamics. *Rev. Mod. Phys.* **2022**, *94*(4), 045003. https://doi.org/10.1103/RevModPhys.94.045003.

27 Stokes, A., Nazir, A. Gauge ambiguities imply Jaynes-Cummings physics remains valid in ultrastrong coupling QED. *Nat. Commun.* **2019**, *10*(1), 499. https://doi.org/10.1038/s41467-018-08101-0.

28 Stokes, A., Nazir, A. Ultrastrong time-dependent light-matter interactions are gauge relative. *Phys. Rev. Res.* **2021**, *3*(1), 013116. https://doi.org/10.1103/PhysRevResearch.3.013116.

29 Stokes, A, Nazir, A. Uniqueness of the phase transition in many-dipole cavity QED systems. *arXiv* **2020b**. https://link.aps.org/doi/10.1103/PhysRevLett.125.143603.

30 Di Stefano, O., Settineri, A., Macrì, V., Garziano, L., Stassi, R., Savasta, S. et al. Resolution of gauge ambiguities in ultrastrong-coupling cavity quantum electrodynamics. *Nat. Phys.* **2019**, *15*, 803–808.

31 Roth, M., Hassler, F., DiVincenzo, D. P. Optimal gauge for the multimode Rabi model in circuit QED. *Phys. Rev. Res.* **2019**, *1*(3), 033128. https://doi.org/10.1103/PhysRevResearch.1.033128.

32 De Bernardis, D., Pilar, P., Jaako, T., De Liberato, S., Rabl, P. Breakdown of gauge invariance in ultrastrong-coupling cavity QED. *Phys. Rev. A* **2018b**, *98*(5), 053819. https://doi.org/10.1103/PhysRevA.98.053819.

33 Settineri, A., Di Stefano, O., Zueco, D., Hughes, S., Savasta, S., Nori, F. Gauge freedom, quantum measurements, and time-dependent interactions in cavity QED. *Phys. Rev Res.* **2021**, *3*(2), 023079. https://doi.org/10.1103/PhysRevResearch.3.023079.

34 De Bernardis, D., Jaako, T., Rabl, P. Cavity quantum electrodynamics in the nonperturbative regime. *Phys. Rev. A* **2018a**, *97*(4), 043820. https://doi.org/10.1103/PhysRevA.97.043820.

35 Stokes, A., Nazir, A. Gauge non-invariance due to material truncation in ultrastrong-coupling quantum electrodynamics. *Nat. Phys.* **2024**, *20*(3), 376–378. https://doi.org/10.1038/s41567-023-02155-8.

36 Taylor, M. A., Mandal, A., Zhou, W., Huo, P. Resolution of gauge ambiguities in molecular cavity quantum electrodynamics. *Phys. Rev. Lett.* **2020**, *125*(12), 123602. https://doi.org/10.1103/PhysRevLett.125.123602.

37 Garziano, L., Settineri, A., Di Stefano, O., Savasta, S., Nori, F. Gauge invariance of the Dicke and Hopfield models. *Phys. Rev. A* **2020**, *102*(2), 023718. https://doi.org/10.1103/PhysRevA.102.023718.

38 Ashida, Y., İmamoğlu, A., Demler, E. Cavity quantum electrodynamics at arbitrary light-matter coupling strengths. *Phys. Rev. Lett.* **2021**, *126*(15), 153603. https://doi.org/10.1103/PhysRevLett.126.153603.

39 Li, J., Golez, D., Mazza, G., Millis, A.J., Georges, A., Eckstein, M. Electromagnetic coupling in tight-binding models for strongly correlated light and matter. *Phys. Rev. B* **2020**, *101*(20), 205140. https://doi.org/10.1103/PhysRevB.101.205140.

40 Gustin, C., Franke, S., Hughes, S. Gauge-invariant theory of truncated quantum light-matter interactions in arbitrary media. *Phys. Rev. A* **2023**, *107*(1), 013722. https://doi.org/10.1103/PhysRevA.107.013722.

41 Akbari, K., Salmon, W., Nori, F., Hughes, S. Generalized Dicke model and gauge-invariant master equations for two atoms in ultrastrongly-coupled cavity quantum electrodynamics. *Phys. Rev. Res.* **2023**, *5*(3), 033002. https://doi.org/10.1103/PhysRevResearch.5.033002.

42 Arwas, G., Manucharyan, V. E. and Ciuti, C. Metrics and properties of optimal gauges in multimode cavity QED. *Phys. Rev. A* **2023**, *108*(2), 023714. https://doi.org/10.1103/PhysRevA.108.023714.

43 Dutra, S. M. *Cavity Quantum Electrodynamics: The Strange Theory of Light in a Box*. Wiley-VCH, 2004, p. 408. https://doi.org/10.1002/0471713465.ch6.

44 Barton, G. Quantum electrodynamics of spinless particles between conducting plates. *Proc. R. Soc. Lond. A Math. Phys. Sci.* **1970**, *320*(1541), 251–275. https://doi.org/10.1098/rspa.1970.0208.

45 Hinds, E. A. Cavity quantum electrodynamics. In: B. David, B. Benjamin (Eds). *Advances in Atomic, Molecular, and Optical Physics*, vol. 28. Academic Press, 1990, pp. 237–289. https://doi.org/10.1016/S1049-250X(08)60201-3.

46 Power, E. A., Thirunamachandran, T. Quantum electrodynamics in a cavity. *Phys. Rev. A* **1982**, *25*(5), 2473–2484. https://doi.org/10.1103/PhysRevA.25.2473.

47 Vukics, A., Domokos, P. Adequacy of the Dicke model in cavity QED: a counter-no-go statement. *Phys. Rev. A* **2012**, *86*(5), 053807. https://doi.org/10.1103/PhysRevA.86.053807.

48 Philpott, M. R. Fluorescence from molecules between mirrors. *Chem. Phys. Lett.* **1973**, *19*(3), 435–439.

49 Craig, D. P., Thirunamachandran, T. *Molecular Quantum Electrodynamics: An Introduction to Radiation-molecule Interactions*. Courier Corporation, 1998.

50 Cohen-Tannoudji, C., Dupont-Roc, J., Grynberg, G. *Photons and Atoms: Introduction to Quantum Electrodynamics*. Wiley, 1989.

51 Woolley, R. G. Gauge invariance in non-relativistic electrodynamics. *Proc. R. Soc. Lond. A* **2000**, *456*(2000), 1803–1819. https://doi.org/10.1098/rspa.2000.0587.

52 Power, E. A., Thirunamachandran, T. Time dependence of operators in minimal and multipolar nonrelativistic quantum electrodynamics II. Analysis of the functional forms of operators in the two frameworks. *Phys. Rev. A* **1999**, *60*, 4936–4942.

53 Vukics, A., Grießer, T., Domokos, P. Fundamental limitation of ultrastrong coupling between light and atoms. *Phys. Rev. A* **2015**, *92*(4), 043835. https://doi.org/10.1103/PhysRevA.92.043835.

54 Brooke, P. G., Marzlin, K. P., Cresser, J. D., Sanders, B. C. Super- and subradiant emission of two-level systems in the near-Dicke limit. *Phys. Rev. A* **2008**, *77*(3), 033844. https://doi.org/10.1103/PhysRevA.77.033844.

55 Jackson, J. D. *Classical Electrodynamics*, 3rd edn. John Wiley & Sons: New York, NY, 1998.

56 Zangwill, A. *Modern Electrodynamics*. Cambridge University Press: Cambridge, 2012.

57 Carmichael, H. J. *Statistical Methods in Quantum Optics 1*. Springer: Berlin, Heidelberg, 1999. https://doi.org/10.1007/978-3-662-03875-8.

58 Power, E. A., Thirunamachandran, T. Quantum electrodynamics with nonrelativistic sources. III. Intermolecular interactions. *Phys. Rev. A* **1983**, *28*(5), 2671–2675. https://doi.org/10.1103/PhysRevA.28.2671.

59 Salam, A. Molecular quantum electrodynamics in the Heisenberg picture: a field theoretic viewpoint. *Int. Rev. Phys. Chem.* **2008**, *27*(3), 405–448. https://doi.org/10.1080/01442350802045206.

60 Power, E. A. Perturbation of the electromagnetic vacuum by atoms and molecules. *Phys. Scr.* **1988**, *1988*(T21), 110. https://dx.doi.org/10.1088/0031-8949/1988/T21/020.

61 Milonni, P. W., James, D. F. V., Fearn, H. Photodetection and causality in quantum optics. *Phys. Rev. A* **1995**, *52*, 1525–1537.

62 Stokes, A. Vacuum source-field correlations and advanced waves in quantum optics. *Quantum* **2018**, *2*, 46. https://doi.org/10.22331/q-2018-01-18-46.

63 Fulling, S. A. *Aspects of Quantum Field Theory in Curved Spacetime*. London Mathematical Society Student Texts. Cambridge University Press: Cambridge, 1989.

64 Power, E. A., Thirunamachandran, T. Time dependence of operators in minimal and multipolar nonrelativistic quantum electrodynamics. I. Electromagnetic fields in the neighborhood of an atom. *Phys. Rev. A* **1999**, *60*(6), 4927–4935. https://doi.org/10.1103/PhysRevA.60.4927.

65 Power, E. A., Zienau, S. Coulomb gauge in non-relativistic quantum electro-dynamics and the shape of spectral lines. *Philos. Trans. R. Soc. Lond. Ser. A Math. Phys. Sci.* **1959**, *251*(999), 427–454. https://doi.org/10.1098/rsta.1959.0008.

66 Lamb, W. E. Fine structure of the hydrogen atom. III. *Phys. Rev.* **1952**, *85*(2), 259–276. https://doi.org/10.1103/PhysRev.85.259.

67 Milonni, P. W., Cook, R. J., Ackerhalt, J. R. Natural line shape. *Phys. Rev. A.* **1989**, *40*(7), 3764–3768. https://link.aps.org/doi/10.1103/PhysRevA.40.3764.

68 Cohen-Tannoudji, C., Dunpont-Roc, J., Grynberg, G. *Atom-Photon Interactions: Basic Process and Applications: Basic Processes and Applications*, New edn. Wiley-VCH: New York, 2010.

69 Andrews, D. L. A unified theory of radiative and radiationless molecular energy transfer. *Chem. Phys.* **1989**, *135*(2), 195–201. https://doi.org/10.1016/0301-0104(89)87019-3.

70 Salam, A. Resonant transfer of excitation between two molecules using Maxwell fields. *J. Chem. Phys.* **2005**, *122*(4), 044113. https://doi.org/10.1063/1.1827596.

71 Glauber, R. J. *Quantum Theory of Optical Coherence: Selected Papers and Lectures*. Wiley VCH: Weinheim, 2007.

72 Power, E. A., Thirunamachandran, T. Quantum electrodynamics with nonrelativistic sources. IV. Poynting vector, energy densities, and other quadratic operators of the electromagnetic field. *Phys. Rev. A* **1992**, *45*(1), 54–63. https://doi.org/10.1103/PhysRevA.45.54.

73 Stokes, A. Quantum optical dipole radiation fields. *Eur. J. Phys.* **2016**, *37*(3), 034001. https://doi.org/10.1088/0143-0807/37/3/034001.

74 Stokes, A., Riley, H., Nazir, A. The gauge-relativity of quantum light, matter, and information. *Open Syst. Inf. Dyn.* **2023**, *30*(03), 2350016. https://doi.org/10.1142/S1230161223500166.

75 Craig, D. P., Thirunamachandran, T. Third-body mediation of resonance coupling between identical molecules. *Chem. Phys.* **1989**, *135*(1), 37–48. https://doi.org/10.1016/0301-0104(89)87004-1.

76 Dutra, S. M., Nienhuis, G. Quantized mode of a leaky cavity. *Phys. Rev. A* **2000**, *62*(6), 063805. https://doi.org/10.1103/PhysRevA.62.063805.

77 Power, E. A., Thirunamachandran, T. Zero-point energy differences and many-body dispersion forces. *Phys. Rev. A* **1994**, *50*(5), 3929–3939. https://doi.org/10.1103/PhysRevA.50.3929.

78 Gardiner, C. W., Collett, M. J. Input and output in damped quantum systems: quantum stochastic differential equations and the master equation. *Phys. Rev. A* **1985**, *31*(6), 3761–3774. https://doi.org/10.1103/PhysRevA.31.3761.

79 Salmon, W., Gustin, C., Settineri, A., Di Stefano, O., Zueco, D., Savasta, S. et al. Gauge-independent emission spectra and quantum correlations in the ultrastrong coupling regime of open system cavity-QED. *Nanophotonics* **2022**, *11*(8), 1573–1590.

80 Drummond, P. D. Unifying the p.A and d.E interactions in photodetector theory. *Phys. Rev. A* **1987**, *35*(10), 4253–4262. https://doi.org/10.1103/PhysRevA.35.4253.

81 Wubs, M., Suttorp, L. G., Lagendijk, A. Multipole interaction between atoms and their photonic environment. *Phys. Rev. A* **2003**, *68*(1), 013822. https://doi.org/10.1103/PhysRevA.68.013822.

82 Glauber, R. J., Lewenstein, M. Quantum optics of dielectric media. *Phys. Rev. A* **1991**, *43*(1), 467–491. https://doi.org/10.1103/PhysRevA.43.467.

83 Knoll, L., Scheel, S., Welsch, D. G. QED in dispersing and absorbing media. In: Peřina, J. ed. *Coherence and Statistics of Photons and Atoms*. John Wiley & Sons, 2001.

Appendix A: Computation of K and G, and the Field Canonical Commutation Relation in the Case of a Perfect Parallel Plate Cavity

Using Eqs. (14.8) and (14.9) in Eq. (14.69) for $i = j = z$ gives

$$\begin{aligned}
-K_{zz}(\mathbf{x},\mathbf{x}',t) &= \sum_{n=0}^{\infty}{}' \sum_{\sigma} \int d^2\kappa \left[f_{\sigma n z}(\boldsymbol{\kappa},\mathbf{x}) f^*_{\sigma n z}(\boldsymbol{\kappa},\mathbf{x}') e^{-i\omega_n(\kappa)t} + f^*_{\sigma n z}(\boldsymbol{\kappa},\mathbf{x}) f_{\sigma n z}(\boldsymbol{\kappa},\mathbf{x}') e^{i\omega_n(\kappa)t} \right] \\
&= \sum_{n=0}^{\infty}{}' \int d^2\kappa \left(\frac{\kappa}{\omega_n(\kappa)}\right)^2 \frac{\cos k_{nz}z \cos k_{nz}z'}{4\pi^2 L} \left[e^{i\boldsymbol{\kappa}\cdot(\mathbf{x}-\mathbf{x}')} e^{-i\omega_n(\kappa)t} + e^{-i\boldsymbol{\kappa}\cdot(\mathbf{x}-\mathbf{x}')} e^{i\omega_n(\kappa)t} \right] \\
&= \sum_{n=0}^{\infty}{}' \int d^2\kappa \left(\frac{\kappa}{\omega_n(\kappa)}\right)^2 \frac{e^{i\boldsymbol{\kappa}\cdot(\mathbf{x}'-\mathbf{x})}}{8\pi^2 L} \cos(\omega_n(\kappa)t) \left[e^{-ik_{nz}(z+z')} + e^{ik_{nz}(z'-z)} + \mathrm{c.c} \right] \\
&= \sum_{n=-\infty}^{\infty} \int d^2\kappa \left(\frac{\kappa}{\omega_n(\kappa)}\right)^2 \frac{e^{i\boldsymbol{\kappa}\cdot(\mathbf{x}'-\mathbf{x})}}{8\pi^2 L} \cos(\omega_n(\kappa)t) \left[e^{-ik_{nz}(z+z')} + e^{ik_{nz}(z'-z)} \right]
\end{aligned} \tag{A.201}$$

To obtain the final equality, one notes that if $f(-n) = f(n)^*$ then

$$\sum_{n=0}^{\infty}{}' f(n) + \mathrm{c.c.} = f(0) + \left[\sum_{n=1}^{\infty} f(n) + \mathrm{c.c.} \right] = \sum_{n=-\infty}^{\infty} f(n). \tag{A.202}$$

Since $k_{nz} = n\pi/L$, Eq. (A.201) can be written

$$-K_{zz}(\mathbf{x},\mathbf{x}',t) = \sum_{n=-\infty}^{\infty} \int \frac{d^3k}{(2\pi)^3} \left(\frac{\kappa}{\omega}\right)^2 e^{i\boldsymbol{\kappa}\cdot(\mathbf{x}'-\mathbf{x})} \cos(\omega t) \left[e^{-ik_z(z+z')} + e^{ik_z(z'-z)} \right] \delta\left(\frac{Lk_z}{\pi} - n \right) \tag{A.203}$$

where $\omega = |\mathbf{k}|$. Via the Dirac comb formula

$$\sum_{n=-\infty}^{\infty} e^{2ink_z L} = \sum_{n=-\infty}^{\infty} \delta\left(\frac{Lk_z}{\pi} - n \right) \tag{A.204}$$

Eq. (A.203) can be written

$$-K_{zz}(\mathbf{x},\mathbf{x}',t) = \sum_{n=-\infty}^{\infty} \int \frac{d^3k}{(2\pi)^3} \left(\frac{\kappa}{\omega}\right)^2 e^{i\boldsymbol{\kappa}\cdot(\mathbf{x}'-\mathbf{x})} \cos(\omega t) \left[e^{-ik_z(z+z')} + e^{ik_z(z'-z)} \right] e^{2ink_z L} \tag{A.205}$$

$$\begin{aligned}
&= \sum_{n=-\infty}^{\infty} \int \frac{d^3k}{(2\pi)^3} \left(\frac{\kappa}{\omega}\right)^2 e^{i\mathbf{k}\cdot(\check{\mathbf{x}}_n'-\mathbf{x})} \cos(\omega t) \\
&\quad + \sum_{n\neq 0} \int \frac{d^3k}{(2\pi)^3} \left(\frac{\kappa}{\omega}\right)^2 e^{i\mathbf{k}\cdot(\mathbf{x}_n'-\mathbf{x})} \cos(\omega t) \\
&\quad + \int \frac{d^3k}{(2\pi)^3} \left(\frac{\kappa}{\omega}\right)^2 e^{i\mathbf{k}\cdot(\mathbf{x}'-\mathbf{x})} \cos(\omega t)
\end{aligned} \tag{A.206}$$

where $\mathbf{x}_n = \mathbf{x} + 2nL\hat{\mathbf{z}}$ and $\check{\mathbf{x}}_n = \check{\mathbf{x}} + 2nL\hat{\mathbf{z}}$ with $\check{\mathbf{x}} = (x, y, -z)$. The three terms on the right-hand side are computed separately. The last of these is the only contribution that occurs in free space and produces the field of the real dipole within the cavity. We compute this term as

$$
\begin{aligned}
&\int \frac{d^3k}{(2\pi)^3}\left(\frac{\kappa}{\omega}\right)^2 e^{i\mathbf{k}\cdot(\mathbf{x}'-\mathbf{x})}\cos(\omega t)\\
&= \left(-\nabla^2+\nabla_z\nabla_z\right)\int \frac{d^3k}{(2\pi)^3}\frac{1}{\omega^2}e^{i\mathbf{k}\cdot(\mathbf{x}'-\mathbf{x})}\cos(\omega t)\\
&= \left(-\nabla^2+\nabla_z\nabla_z\right)\frac{1}{4\pi|\mathbf{x}-\mathbf{x}'|}\int_0^\infty d\omega\,\frac{2}{\pi\omega}\sin(\omega|\mathbf{x}-\mathbf{x}'|)\cos(\omega t)\\
&= \left(-\nabla^2+\nabla_z\nabla_z\right)\frac{\theta(|\mathbf{x}-\mathbf{x}'|-t)}{4\pi|\mathbf{x}-\mathbf{x}'|}
\end{aligned}
\tag{A.207}
$$

The second and first terms on the right-hand side of Eq. (A.206) are respectively computed similarly as

$$
\sum_{n\neq 0}\int \frac{d^3k}{(2\pi)^3}\left(\frac{\kappa}{\omega}\right)^2 e^{i\mathbf{k}\cdot(\mathbf{x}_n'-\mathbf{x})}\cos(\omega t) = \sum_{n\neq 0}\left(-\nabla^2+\nabla_z\nabla_z\right)\frac{\theta(|\mathbf{x}-\mathbf{x}_n'|-t)}{4\pi|\mathbf{x}-\mathbf{x}_n'|}
\tag{A.208}
$$

and

$$
\sum_{n=-\infty}^{\infty}\int \frac{d^3k}{(2\pi)^3}\left(\frac{\kappa}{\omega}\right)^2 e^{i\mathbf{k}\cdot(\check{\mathbf{x}}_n'-\mathbf{x})}\cos(\omega t) = -\sum_{n=-\infty}^{\infty}\sigma_{zz}\left(-\nabla^2+\nabla_z\nabla_z\right)\frac{\theta(|\mathbf{x}-\check{\mathbf{x}}_n'|-t)}{4\pi|\mathbf{x}-\check{\mathbf{x}}_n'|}
\tag{A.209}
$$

where $\sigma = \mathrm{diag}(1,1,-1)$. The function K_{zz} is the sum of all three terms. Similar calculations yield the remaining components of $\mathbf{K}$, which is found finally as

$$
\begin{aligned}
-K_{ij}(\mathbf{x},\mathbf{x}',t) &= \left(-\delta_{ij}\nabla^2+\nabla_i\nabla_j\right)\frac{\theta(|\mathbf{x}-\mathbf{x}'|-t)}{4\pi|\mathbf{x}-\mathbf{x}'|}+\sum_{n\neq 0}\left(-\delta_{ij}\nabla^2+\nabla_i\nabla_j\right)\frac{\theta(|\mathbf{x}-\mathbf{x}_n'|-t)}{4\pi|\mathbf{x}-\mathbf{x}_n'|}\\
&\quad -\sum_{n=-\infty}^{\infty}\sigma_{jk}\left(-\delta_{ki}\nabla^2+\nabla_k\nabla_i\right)\frac{\theta(|\mathbf{x}-\check{\mathbf{x}}_n'|-t)}{4\pi|\mathbf{x}-\check{\mathbf{x}}_n'|}.
\end{aligned}
\tag{A.210}
$$

The green's function $G_{ij} = \partial K_{ij}/\partial t$ is therefore

$$
\begin{aligned}
G_{ij}(\mathbf{x},\mathbf{x}',t) &= \left(-\delta_{ij}\nabla^2+\nabla_i\nabla_j\right)\frac{\delta(t-|\mathbf{x}-\mathbf{x}'|)}{4\pi|\mathbf{x}-\mathbf{x}'|}+\sum_{n\neq 0}\left(-\delta_{ij}\nabla^2+\nabla_i\nabla_j\right)\frac{\delta(t-|\mathbf{x}-\mathbf{x}_n'|)}{4\pi|\mathbf{x}-\mathbf{x}_n'|}\\
&\quad -\sum_{n=-\infty}^{\infty}\sigma_{jk}\left(-\delta_{ki}\nabla^2+\nabla_k\nabla_i\right)\frac{\delta(t-|\mathbf{x}-\check{\mathbf{x}}_n'|)}{4\pi|\mathbf{x}-\check{\mathbf{x}}_n'|}.
\end{aligned}
\tag{A.211}
$$

The first term is the only contribution for a dipole in free space, while the remaining terms describe the image dipoles due to the plates.

Finally, we verify Eq. (14.12) giving the field canonical commutation relation. We substitute Eq. (14.20) into the right-hand side of Eq. (14.12) to obtain

$$
\begin{aligned}
&i\delta^{\mathrm{T}}_{ij}(\mathbf{x}-\mathbf{x}') - i\nabla_i\nabla_j'\sum_{n\neq 0}\frac{1}{4\pi|\mathbf{x}-\mathbf{x}_n'|}+i\nabla_i\nabla_j'\sum_{n}\frac{1}{4\pi|\mathbf{x}-\check{\mathbf{x}}_n'|}\\
&= i\delta^{\mathrm{T}}_{ij}(\mathbf{x}-\mathbf{x}') - i\sum_{n\neq 0}\delta^{\mathrm{L}}_{ij}(\mathbf{x}-\mathbf{x}_n') + i\sum_{n}\sigma_{jk}\delta^{\mathrm{L}}_{ki}(\mathbf{x}-\check{\mathbf{x}}_n')\\
&= i\delta^{\mathrm{T}}_{ij}(\mathbf{x}-\mathbf{x}') + i\sum_{n\neq 0}\delta^{\mathrm{T}}_{ij}(\mathbf{x}-\mathbf{x}_n') - i\sum_{n}\sigma_{jk}\delta^{\mathrm{T}}_{ki}(\mathbf{x}-\check{\mathbf{x}}_n'),
\end{aligned}
\tag{A.212}
$$

By noting that $\delta^{\mathrm{T}}_{ij}(\mathbf{x}) = \left(-\delta_{ij}\nabla^2+\nabla_i\nabla_j\right)[1/(4\pi|\mathbf{x}|)]$ and using Eq. (A.210) this is immediately seen to equal $-iK_{ij}(\mathbf{x},\mathbf{x}',0)$, which by definition is the left-hand side of Eq. (14.12).

Appendix B: Born–Markov-secular Master Equation for the Dipole in a Fabry–Pérot Cavity

We consider the standard definition of h, namely

$$h = h_m + H_{\rm ph} = \frac{\mathbf{p}^2}{2m} + V(\mathbf{r}) + H_{\rm ph} \tag{B.213}$$

The transition energies of h_m are denoted $\omega_{nm} = \epsilon_n - \epsilon_m$ and the eigenvectors are denoted $|n\rangle$ where $H_m|n\rangle = \epsilon_n|n\rangle$. It follows that (in the EDA)

$$V^0 = \frac{q^2}{2m}\mathbf{A}_{\rm T}(\mathbf{R})^2 - \frac{q}{m}\mathbf{p}\cdot\mathbf{A}_{\rm T}(\mathbf{R}) + V^{\rm image,dip} \tag{B.214}$$

$$V^1 = \mathbf{d}\cdot\Pi(\mathbf{R}) + \epsilon^{\rm dip} \tag{B.215}$$

The second-order vacuum (zero temperature) Born–Markov-secular master equation for these two cases is [4]

$$\dot{\rho}(t) = i[H_m + H^{\alpha}_{\rm LS}, \rho(t)] + \mathcal{D}^{\alpha}[\rho(t)] \tag{B.216}$$

where we now use $\rho(t)$ to denote the reduced (material) system density operator. The level-shift Hamiltonians and dissipators are defined by

$$H^0_{\rm LS} = \frac{q^2}{2m}\langle\mathbf{A}_{\rm T}(\mathbf{R})^2\rangle_0 + \bar{V}^{\rm image} + \sum_{\omega}\sum_{ij} S^0_{ij}(\omega)A_i^{0\dagger}(\omega)A^0_j(\omega) \tag{B.217}$$

$$H^1_{\rm LS} = \bar{\epsilon}^{\rm dip} + \sum_{\omega}\sum_{ij} S^1_{ij}(\omega)A_i^{1\dagger}(\omega)A^1_j(\omega), \tag{B.218}$$

$$\mathcal{D}^{\alpha}(\rho) = \sum_{\omega>0}\sum_{ij}\gamma^{\alpha}_{ij}(\omega)\left(A^{\alpha}_j(\omega)\rho A_i^{\alpha\dagger}(\omega) - \frac{1}{2}\left\{A_i^{\alpha\dagger}(\omega)A^{\alpha}_j(\omega),\rho\right\}\right). \tag{B.219}$$

Here, $\bar{\epsilon}^{\rm dip}$ and $\bar{V}^{\rm image}$ are used to denote the parts of $\epsilon^{\rm dip}$ and $V^{\rm image}$ that are diagonal in the eigenbasis of h_m. The off-diagonal contributions are neglected as part of the secular approximation. Note that the vacuum expectation of $\mathbf{A}_{\rm T}^2$ in Eq. (B.217) is independent of the matter system, so it does not contribute to the commutator in the master equation. We retain it within the definition of the level shift Hamiltonian only to make contact with the standard QED second-order Lamb-shift matrix element.

The eigenoperators A^{α}_i are defined using the material operators within the linear (in q) interaction Hamiltonians, namely, $-q\mathbf{p}\cdot\mathbf{A}_{\rm T}(\mathbf{R})/m$ in the Coulomb gauge and $\mathbf{d}\cdot\Pi(\mathbf{R})$ in the dipole gauge, as

$$A^1_i(\omega) = \sum_{\substack{n,m\\ \omega_{nm}=\omega}} |m\rangle\langle m|d_i|n\rangle\langle n| = \sum_{\substack{n,m\\ \omega_{nm}=\omega}} d_i^{mn}|m\rangle\langle n|, \tag{B.220}$$

$$A^0_i(\omega) = -\frac{q}{m}\sum_{\substack{n,m\\ \omega_{nm}=\omega}} |m\rangle\langle m|\Pi|n\rangle\langle n| = -\frac{q}{m}\sum_{\substack{n,m\\ \omega_{nm}=\omega}} p_i^{mn}|m\rangle\langle n| = i\sum_{\substack{n,m\\ \omega_{nm}=\omega}} \omega_{nm}d_i^{mn}|m\rangle\langle n| \tag{B.221}$$

Similarly, the coefficients $\gamma^{\alpha}_{ij}(\omega_{nm})$ and $S^{\alpha}_{ij}(\omega_{nm})$ are Hermitian and anti-Hermitian parts of the vacuum bath correlation functions defined using the photonic fields within the linear interaction Hamiltonians;

$$\frac{1}{2}\gamma_{ij}^{\alpha} + iS_{ij}^{\alpha} = \Gamma_{ij}^{\alpha}, \tag{B.222}$$

$$\Gamma_{ij}^{1}(\omega) := \int_0^\infty ds e^{i\omega s}\langle \Pi_{\text{vac}i}^{(+)}(s,\mathbf{R})\Pi_{\text{vac}j}^{(-)}(0,\mathbf{R})\rangle_0, \tag{B.223}$$

$$\Gamma_{ij}^{0}(\omega) := \int_0^\infty ds e^{i\omega s}\langle A_{\text{Tvac}i}^{(+)}(s,\mathbf{R})A_{\text{Tvac}j}^{(-)}(0,\mathbf{R})\rangle_0. \tag{B.224}$$

The subscript vac indicates that the free field components are to be used (this is the result of second order perturbation theory in the interaction picture w.r.t h);

$$\Pi_{\text{vac}}^{(+)}(t,\mathbf{x}) = -i\sum_{\sigma=1,2}\sum_{l}{}'' \int d^2\kappa\sqrt{\omega_l}\mathbf{f}_{\sigma l}(\boldsymbol{\kappa},\mathbf{x})a_{\sigma l}(0,\boldsymbol{\kappa})e^{-i\omega_l t}, \tag{B.225}$$

$$\mathbf{A}_{\text{Tvac}}^{(+)}(t,\mathbf{x}) = \sum_{\sigma=1,2}\sum_{l}{}'' \int d^2\kappa\frac{1}{\sqrt{\omega_l}}\mathbf{f}_{\sigma l}(\boldsymbol{\kappa},\mathbf{x})a_{\sigma l}(0,\boldsymbol{\kappa})e^{-i\omega_l t}. \tag{B.226}$$

By computing the master equation coefficients, and using Eqs. (A.220) and (A.221), we will show that $\mathcal{D}^0(\rho) = \mathcal{D}^1(\rho)$ and that $H_{\text{LS}}^0 = H_{\text{LS}}^1$. The mechanism by which these equalities occur can already be seen by inspecting the above definitions. Quadratic combinations of the Coulomb and dipole-gauge eigenoperators differ by a factor of ω_{nm}^2 whereas the integrand/summand of the Coulomb and dipole-gauge master equation coefficients differ by a factor of $1/\omega_l^2$. The Born–Markov-secular approximations enforce an evaluation of the coefficients on resonance $\omega_l = \omega_{nm}$ (the coefficients become on-energy-shell S-matrix elements) such that the factor ω_{nm}^2/ω_l^2 by which each master equation term would otherwise differ, becomes unity.

For the level shift Hamiltonian, the proof is a little bit more subtle, so we begin with the dissipator. We find by direct calculation that

$$\gamma_{ij}^1(\omega_{nm}) = 2\pi\sum_{\sigma=1,2}\sum_{l}{}' \int d^2\kappa\,\omega_l f_{\sigma l,i}(\boldsymbol{\kappa},\mathbf{R})f_{\sigma l,j}(\boldsymbol{\kappa},\mathbf{R})^*\delta(\omega_{nm}-\omega_l), \tag{B.227}$$

$$\gamma_{ij}^0(\omega_{nm})\omega_{nm}^2 = 2\pi\sum_{\sigma=1,2}\sum_{l}{}' \int d^2\kappa\,\frac{\omega_{nm}^2}{\omega_l}f_{\sigma l,i}(\boldsymbol{\kappa},\mathbf{R})f_{\sigma l,j}(\boldsymbol{\kappa},\mathbf{R})^*\delta(\omega_{nm}-\omega_l) = \gamma_{ij}^1(\omega_{nm}) \tag{B.228}$$

We denote this unique rate by γ_{nmpq}. The master equation dissipator is seen to be unique and can be written as in free space

$$\mathcal{D}(\rho) = \sum_{\substack{n,m,p,q\\ \omega_{nm}=\omega_{pq}>0}} \gamma_{nmpq}\left(L_{nm}\rho L_{pq}^\dagger - \frac{1}{2}\{L_{pq}^\dagger L_{nm},\rho\}\right) \tag{B.229}$$

where $L_{nm} = |m\rangle\langle n|$.

It remains to consider the Lamb-shift Hamiltonians. We begin by noting that

$$\epsilon^{\text{dip}} = V^{\text{image,dip}} + \sum_{nmp}\sum_{l}{}'\sum_{\sigma}\int d^2\kappa\,\mathbf{d}^{np}\cdot\mathbf{f}_{\sigma l}(\boldsymbol{\kappa},\mathbf{R})\,\mathbf{d}^{pm}\cdot\mathbf{f}_{\sigma l}^*(\boldsymbol{\kappa},\mathbf{R})|n\rangle\langle m| \tag{B.230}$$

and therefore

$$\bar{\epsilon}^{\text{dip}} = \bar{V}^{\text{image,dip}} + \sum_{nm}\sum_{l}{}'\sum_{\sigma}\int d^2\kappa|\mathbf{d}^{nm}\cdot\mathbf{f}_{\sigma l}(\boldsymbol{\kappa},\mathbf{R})|^2|n\rangle\langle n|. \tag{B.231}$$

Next, we compute $S^1_{ij}(\omega)$, finding that

$$\begin{aligned}\sum_\omega\sum_{ij} S^1_{ij}(\omega)A^{1\dagger}_i(\omega)A^1_j(\omega) &= \sum_{nm}\sum_{ij} S^1_{ij}(\omega_{nm})d^{nm}_i d^{mn}_j|n\rangle\langle n| \\ &= -\sum_{nm}\sum_l{}'\sum_\sigma \int d^2\kappa |\mathbf{d}^{nm}\cdot\mathbf{f}_{\sigma l}(\boldsymbol{\kappa},\mathbf{R})|^2\frac{\omega_l}{\omega_{mn}+\omega_l}|n\rangle\langle n|,\end{aligned} \tag{B.232}$$

where the integral takes the principal value. Using the identity

$$\frac{\omega_l}{\omega_{mn}+\omega_l} = 1-\frac{\omega_{mn}}{\omega_l}+\frac{\omega^2_{nm}}{\omega_l}\frac{1}{\omega_{mn}+\omega_l} \tag{B.233}$$

and adding Eqs. (A.231) and (A.232) gives

$$\begin{aligned}H^1_{\rm LS} = &-\sum_{nm}\sum_l{}'\sum_\sigma\int d^2\kappa\,\frac{\omega^2_{nm}}{\omega_l}|\mathbf{d}^{nm}\cdot\mathbf{f}_{\sigma l}(\boldsymbol{\kappa},\mathbf{R})|^2\frac{1}{\omega_{mn}+\omega_l}|n\rangle\langle n| \\ &+\sum_{nm}\sum_l{}'\sum_\sigma\int d^2\kappa\frac{\omega_{mn}}{\omega_l}|\mathbf{d}^{nm}\cdot\mathbf{f}_{\sigma l}(\boldsymbol{\kappa},\mathbf{R})|^2|n\rangle\langle n| + \overline{V}^{\rm image}.\end{aligned} \tag{B.234}$$

Using the (Thomas-Reiche-Kuhn) identity $\sum_m \omega_{mn}d^{nm}_i d^{mn}_j = q^2\delta_{ij}/(2m)$, the first term on the second line is seen to be nothing but $q^2\langle\mathbf{A}_{\rm T}(\mathbf{R})^2\rangle_0/(2m)$. Therefore

$$H^1_{\rm LS} = -\sum_{nm}\sum_l{}'\sum_\sigma\int d^2\kappa\,\frac{\omega^2_{nm}}{\omega_l}|\mathbf{d}^{nm}\cdot\mathbf{f}_{\sigma l}(\boldsymbol{\kappa},\mathbf{R})|^2\frac{1}{\omega_{mn}+\omega_l}|n\rangle\left\langle n| + \frac{q^2}{2m}\langle\mathbf{A}_{\rm T}(\mathbf{R})^2\right\rangle_0 + \overline{V}^{\rm image}. \tag{B.235}$$

As in $H^0_{\rm LS}$, the vacuum expectation of the $\mathbf{A}^2_{\rm T}$-term is constant in the matter Hilbert space and so does not contribute to the master equation itself. In order to have demonstrated that $H^1_{\rm LS} = H^0_{\rm LS}$, it only remains for us to show that the first term in Eq. (B.235) is equal to

$$\sum_\omega\sum_{ij} S^0_{ij}(\omega)A^{0\dagger}_i(\omega)A^0_j(\omega) = \sum_{nm}\sum_{ij} S^0_{ij}(\omega_{nm})\omega^2_{nm}d^{nm}_i d^{mn}_j|n\rangle\langle n|. \tag{B.236}$$

Direct calculation of $\omega^2_{nm}S^0_{ij}(\omega_{nm})$ shows that it does indeed differ from $S^1_{ij}(\omega_{nm})$ in Eq. (B.232) by the expected factor of ω^2_{nm}/ω^2_l;

$$\sum_{ij} S^0_{ij}(\omega_{nm})\omega^2_{nm}d^{nm}_i d^{mn}_j = -\sum_l{}'\sum_\sigma\int d^2\kappa\,\frac{\omega^2_{nm}}{\omega_l}|\mathbf{d}^{nm}\cdot\mathbf{f}_{\sigma l}(\boldsymbol{\kappa},\mathbf{R})|^2\frac{1}{\omega_{mn}+\omega_l}. \tag{B.237}$$

Using this in Eq. (B.236) gives the first term of Eq. (B.235). It follows that $H^1_{\rm LS} = H^0_{\rm LS}$.

Appendix C: Emission Rates of a Dipole Near a Single Plate

Let us consider a dipole close to a single plate at $z = 0$. The image distribution is given by Eq. (14.13), which results from ignoring all but the $n = 0$ term within Eq. (14.20), that is, ignoring everything but the direct reflection of the real dipole in the $z = 0$ plate. The corresponding green's function is obtained by similarly retaining only the $n = 0$ contribution in Eq. (14.79). Using this green's function in Eq. (14.66) gives the electric source field of the image dipole as

$$\mathbf{E}_s^{1,\text{image}}(\text{t}, \mathbf{x}) = -\mathcal{D} \cdot \frac{\check{\mathbf{d}}(\check{t}_r)}{4\pi|\mathbf{x} - \check{\mathbf{R}}|} \tag{C.238}$$

where $\check{t}_r = t - |\mathbf{x} - \check{\mathbf{R}}|$. The dipole moment $-\check{\mathbf{d}}$ is that of the (inverted) image dipole. Using Eq. (14.81) and evaluating the derivatives gives a superposition of well-known classical dipole fields oscillating at the transition frequencies ω_{ab}. At the position of the real dipole, $\mathbf{R} = (0, 0, R_z)$, this field is

$$\mathbf{E}_s^{1,\text{image}}(\text{t}, \mathbf{R}) = \sum_{ab} \frac{\omega_{ab}^3}{4\pi} |a\rangle\langle b| e^{i\omega_{ab}t} e^{-i\eta_{ab}} \check{\mathbf{d}}_{ab} \cdot \left[\beta \left(\frac{1}{\eta_{ab}^3} + \frac{i}{\eta_{ab}^2} \right) - \frac{\gamma}{\eta_{ab}} \right]. \tag{C.239}$$

where $\eta_{ab} = 2\omega_{ab} R_z$, $\beta_{ij} = \delta_{ij} - 3\hat{z}_i\hat{z}_j$ and $\gamma_{ij} = \delta_{ij} - \hat{z}_i\hat{z}_j$. When η_{ab} is of order unity, the near-field $\sim 1/\eta_{ab}^3$, intermediate field $\sim 1/\eta_{ab}^2$, and radiation field $\sim 1/\eta_{ab}$, are of the same order. To show that here the image has a profound effect on the real dipole, we re-derive Eqs. (14.54) and (14.55). We view the matrix element $\langle a|\mathbf{E}_s^{1,\text{image}}(t, \mathbf{R})|e\rangle$ as a classical field given by the real part. Writing $\omega_{ea} = \omega$, $\eta_{ea} = \eta$, and $\mathbf{d}_{ae} = \mathbf{d}$, we find the components of this field parallel and orthogonal to $\hat{\mathbf{z}}$ to be $E_{s,z}^{1,\text{image}}(t, \mathbf{R}) = \mathcal{E}_z^{\text{image}}(\eta)e^{-i\omega t} + \text{c.c}$ and $E_{s,\rho}^{1,\text{image}}(t, \mathbf{R}) = \mathcal{E}_\rho^{\text{image}}(\eta)e^{-i\omega t} + \text{c.c}$ respectively, where

$$\mathcal{E}_z^{\text{image}}(\eta) = \frac{\omega^3 d_z}{2\pi} \left(\frac{1}{\eta^3} - \frac{i}{\eta^2} \right) e^{i\eta}, \tag{C.240}$$

$$\mathcal{E}_\rho^{\text{image}}(\eta) = \frac{\omega^3 d_\rho}{4\pi} \left(\frac{1}{\eta^3} - \frac{i}{\eta^2} - \frac{1}{\eta} \right) e^{i\eta}. \tag{C.241}$$

The total source field is $-\Pi_s^1(t, \mathbf{R}) = \mathcal{E}(\eta)e^{-i\omega t} + \text{c.c.} = [\mathcal{E}^{\text{real}} + \mathcal{E}^{\text{image}}(\eta)]e^{-i\omega t} + \text{c.c.}$ where the real field $-\Pi_s^{1,\text{real}}(t, \mathbf{R}) = \mathcal{E}e^{-i\omega t} + \text{c.c.}$ results from the first term in Eq. (14.79). By Poynting's theorem, the time-averaged power of dipolar radiation is

$$P = \frac{\omega}{2} \text{Im}[\mathbf{d}^* \cdot \mathcal{E}(\eta)] = P_0 + \frac{\omega}{2} \text{Im}[\mathbf{d}^* \cdot \mathcal{E}^{\text{image}}(\eta)] \tag{C.242}$$

where the free space power is $P_0 = \omega \text{Im}[\mathbf{d}^* \cdot \mathcal{E}^{\text{real}}(\eta)]/2 = \omega^4 |\mathbf{d}|^2/(12\pi)$. Letting $P = \omega\Gamma$ and $P_0 = \omega\Gamma_0$, Eqs. (A.240)–(A.242) yield Eqs. (14.54) and (14.55). As seen in Figure 14.1, the plate has a dramatic effect when the wavelength $2\pi/\omega$ is comparable to R_z, which is also where the near, intermediate, and far field components of the source-fields (dipole and image) are all of the same order.

Appendix D: Proof that the Sum of Imperfect Cavity Lorenztians Gives a Dirac Comb in the Perfect Cavity Limit

Consider the function

$$\mathcal{L}(k_z) = \frac{t}{1 - mre^{2ik_zL}} \tag{D.243}$$

where $|t|^2 + |r|^2 = 1$ and $m = \pm 1$. We let $\phi = 2k_zL + \Delta$ where $r = |r|e^{i\Delta}$. When $t \to 0$ and $r \to m$ (perfect cavity limit) $\Delta \to 0$ for $m = 1$ and $\Delta \to \pi$ for $m = -1$. We have

$$\begin{aligned}|\mathcal{L}(k_z)|^2 &= \frac{|t|^2/2}{(1+|r|^2)/2 - |r|m\cos\phi} = \frac{|t|^2([1+|r|^2]/2 + |r|m\cos\phi)/2}{([1+|r|^2]/2 - |r|m\cos\phi)([1+|r|^2]/2 + |r|m\cos\phi)} \\ &= \frac{|t|^2/2}{|r|^2\sin^2\phi + |t|^4/4}([1+|r|^2]/2 + |r|m\cos\phi)\end{aligned} \tag{D.244}$$

From the right-hand side of this expression, in the limit $t \to 0$ (such that $|r| \to 1$) we obtain

$$\lim_{t\to 0} |\mathcal{L}(k_z)|^2 = \pi(1 + m\cos\phi)\delta(\sin\phi) \tag{D.245}$$

where $\phi = 2k_zL$ for $m = 1$ and $\phi = 2k_zL + \pi$ for $m = -1$. The δ-function is simplified using the identity

$$\delta(g(x)) = \sum_{x_0} \frac{\delta(x - x_0)}{|g'(x_0)|} \tag{D.246}$$

where the x_0 are the zeros of $g(x)$. Consider the case $m = 1$. The zeros of $g(\phi) = \sin\phi$ occur at $2k_zL = n\pi$ at which points $|g'(\phi)| = |\cos n\pi| = 1$. Therefore

$$\lim_{t\to 0} |\mathcal{L}(k_z)|^2 = \pi \sum_n (1 + \cos\phi)\delta(\phi - n\pi) \tag{D.247}$$

The function $1 + \cos\phi$ equals $1 + (-1)^n$ at $\phi = n\pi$. Thus,

$$\lim_{t\to 0} |\mathcal{L}(k_z)|^2 = 2\pi \sum_{n\ \text{even}} \delta(\phi - n\pi) = 2\pi \sum_n \delta(2k_zL - 2n\pi) = \frac{\pi}{L}\sum_n \delta\left(k_z - \frac{n\pi}{L}\right). \tag{D.248}$$

The same result is obtained when considering $m = -1$.

15

The Vacuum in Ultrastrong Coupling Cavity Quantum Electrodynamics

Peter Rabl

Physics Department, Technical University of Munich, TUM School of Natural Sciences, Garching, Germany
Walther-Meißner-Institut, Bayerische Akademie der Wissenschaften, Garching, Germany

15.1 Introduction

While the coupling of individual atoms or molecules to a single quantized mode of the electromagnetic field is usually rather weak, light–matter interactions can be collectively enhanced when many emitters are present. As first discussed by Dicke in his famous paper in 1954 [1], collective effects not only increase the polaritonic coupling strength, but they can also lead to qualitatively new quantum radiation phenomena. Based on a simplified model for two-level emitters coupled to a common field mode, he showed that the decay of a dense ensemble of inverted atoms or molecules can be drastically accelerated by the built-up of correlations, an effect known as *superradiance*. It was later shown by Hepp and Lieb [2] and Wang and Hioe [3] that even the ground state of this model exhibits an unexpected behavior and undergoes a phase transition when the collective coupling strength becomes comparable to the bare atomic and photonic energy scales. Beyond this point, the cavity quantum electrodynamics (QED) system enters a new state, which is characterized by a finite polarization of the atoms and a finite expectation value of the electromagnetic field. In view of this property, this phase became known as the *superradiant phase*, although it describes an equilibrium state of the coupled light–matter system and should not be confused with the similarly termed collective radiation phenomenon mentioned above.

Naturally, the prediction of such a superradiant ground state with a nonvanishing photon number was met with skepticism and raised concerns about the validity of the Dicke model for describing light–matter interactions under such ultrastrong coupling conditions. Indeed, soon after its prediction, it was pointed out that the phase transition disappears when a more rigorous model is considered, which includes the diamagnetic A^2-term [4]. However, this analysis was also based on many simplifications and subsequently, it was shown that this *no-go theorem* for the superradiant phase could be violated when the longitudinal components [5] and the full multimode nature of the electromagnetic field were taken into account [6]. Over several decades, various studies appeared that argued both for and against the existence of cavity-induced ground-state instabilities but without providing a concluding answer.

Polariton Chemistry: Molecules in Cavities, First Edition. Edited by Joel Yuen-Zhou, Noel C. Giebink and Raphael F. Ribeiro. Published 2026 by John Wiley & Sons, Inc.

With the emergence of the field of polariton chemistry, questions about the ground state of cavity QED and other vacuum-induced modifications of material properties have regained considerable interest. Although a general theoretical understanding of these effects is still premature, many relevant new insights on this topic have recently been obtained from the analysis of idealized systems of two-level dipoles coupled to a single-cavity mode. After a brief review of the superradiant phases transition in the original Dicke model, the main purpose of this chapter is to summarize the basic effective cavity QED models that can be derived for such settings and be used to predict the resulting many-body ground states at arbitrary coupling strength. While such an analysis indicates that the observation of genuine vacuum-induced phase transitions requires rather extreme experimental conditions, it also reveals astonishing new many-body effects, once this nonperturbative regime of light–matter interaction can be realized.

15.2 The Dicke Model

The Dicke model, introduced by Dicke in 1954, is frequently used both in atomic and solid-state systems to describe phenomena related to the collective coupling of $N \gg 1$ emitters to a single confined radiation mode. Within this model, atoms, molecules, or solid-state emitters are approximated by ideal two-level systems with a ground state $|g\rangle$ and an excited state $|e\rangle$, separated by a transition frequency ω_0. The emitters couple to the cavity mode of frequency ω_c via the electric dipole coupling, which is assumed to be approximately the same for all dipoles. Under these assumptions, the usually considered Hamiltonian for the coupled light–matter system is

$$H_{\rm DM} = \hbar\omega_c a^\dagger a + \hbar\omega_0 S_z + \hbar\, g(a + a^\dagger)S_x \tag{15.1}$$

where a and $a^\dagger$ are bosonic annihilation and creation operators, the $S_k = \frac{1}{2}\sum_{i=1}^{N}\sigma_k^i$ are collective spin operators, and g denotes the coupling strength between a single dipole and a single photon.

15.2.1 Collective Light–Matter Interactions

On resonance, $\omega_c = \omega_0$, and as long as the coupling strength g is sufficiently small, one can apply a rotating-wave approximation (RWA) and replace the interaction term in Eq. (15.1) by $H_{\rm int} \approx \hbar g(S_+a+S_-a^\dagger)/2$, such that only excitation-number conserving processes are retained. Here, $S_+ = (S_-)^\dagger = \sum_{i=1}^{N}\sigma_+^i$ is the spin-raising operator. It creates a single collective matter excitation when acting on the fully polarized state, i.e., $S_+|gg\ldots g\rangle = \sqrt{N}|1\rangle$, where

$$|1\rangle = \frac{1}{\sqrt{N}}\sum_{i=1}^{N}|gg\ldots e_i\ldots g\rangle \tag{15.2}$$

is a properly normalized superposition state with a single excitation equally shared among all dipoles. This state then couples to a single photon in the cavity with an enhanced coupling strength $G = g\sqrt{N}$. Therefore, within the validity of the RWA, the Dicke model exhibits a trivial ground state $|\mathcal{G}\rangle = |0\rangle_c \otimes |g\rangle^{\otimes N}$ with energy $E_{\mathcal{G}} = -N\hbar\omega_0/2$ and two *bright* polaritonic states $|\pm\rangle$ with energies

$$E_\pm = E_{\mathcal{G}} + \hbar\omega_c \pm \frac{G}{2} \tag{15.3}$$

In addition, within the single-excitation subspace, there are $N-1$ *dark* polaritonic states. These states, however, are orthogonal to the superposition in Eq. (15.2) and are therefore completely decoupled from the cavity mode.

Equation (15.3) illustrates the collective enhancement of the coherent interaction between photons and excited matter states, which enables the observation of polaritonic strong-coupling effects in experiments where the single-dipole coupling strength g is still too weak to be resolved. This result also predicts that as one enters the collective ultrastrong coupling regime, $G \gtrsim \omega_c, \omega_0$, the energy of the lower polariton mode eventually lies below the energy of the completely de-excited state $|\mathcal{G}\rangle$. While this behavior already indicates an instability, the required coupling strength lies outside the validity of the RWA, under which this result has been derived.

15.2.2 Superradiant Instability

In the limit $N \to \infty$, but keeping a finite value for the collective coupling strength $G = g\sqrt{N}$, the ground state of the Dicke model can be investigated by using a variational Ansatz of the form

$$|\tilde{\mathcal{G}}\rangle = |\alpha\rangle_c \otimes \prod_{i=1}^{N} |\theta\rangle_i \tag{15.4}$$

Here, $|\alpha\rangle_c$ is a coherent state with complex amplitude α and $|\theta\rangle = \cos(\theta)|g\rangle - \sin(\theta)|e\rangle$. For this state, the resulting variational energy of the Dicke model is

$$E_{\mathrm{var}}(\alpha, \theta) = \langle\tilde{\mathcal{G}}|H_{\mathrm{DM}}|\tilde{\mathcal{G}}\rangle = \hbar\omega_c|\alpha|^2 - \frac{N}{2}\left[\hbar\omega_0 \cos(\theta) + \hbar\, g(\alpha + \alpha^*)\sin(\theta)\right] \tag{15.5}$$

which can be minimized with respect to α and θ to get a bound for the actual ground state energy, i.e., $E_{\mathcal{G}} \leq \min\{E_{\mathrm{var}}(\alpha, \theta)|\alpha, \theta\}$. For a given $\alpha = \alpha_R + i\alpha_I$, the energy of each TLS is minimized for $\tan(\theta) = 2\alpha_R g/\omega_0$. After reinserting this result back into Eq. (15.5), one obtains an energy that depends on the field amplitude α only,

$$E_{\mathrm{var}}(\alpha) = \hbar\omega_c|\alpha|^2 - \frac{N}{2}\sqrt{\omega_0^2 + 4g^2\alpha_R^2} \tag{15.6}$$

For small values of g, this energy has a single minimum at $\alpha = 0$, which corresponds to the *normal* phase with a vanishing photon number and a vanishing polarization $\langle S_x\rangle = -N\sin(\theta)/2 = 0$. However, above a critical coupling strength, $G > G_c = \sqrt{\omega_0\omega_c}$, the energy $E_{\mathrm{var}}(\alpha)$ exhibits two minima at finite values of α_R. These minima correspond to the two degenerate *superradiant* ground states with expectation values

$$\frac{\langle a\rangle}{\sqrt{N}} = \pm\frac{G}{2\omega_c}\sqrt{1 - \left(\frac{G_c}{G}\right)^4} \quad \text{and} \quad \frac{\langle S_x\rangle}{N} = \mp\frac{1}{2}\sqrt{1 - \left(\frac{G_c}{G}\right)^4} \tag{15.7}$$

that break the Z_2-symmetry of the Dicke model. Thus, as the collective coupling increases, the ground state of the Dicke model undergoes a second-order quantum phase transition at a critical value of $G = G_c$, beyond which the cavity mode acquires a nonvanishing field expectation value.

While the current prediction of the superradiant phase transition is based on a variational calculation, an exact diagonalization of the Dicke model shown in Figure 15.1 confirms that this analysis becomes more and more accurate, as the number of dipoles increases. The superradiant phase transition has also been observed experimentally in cavity QED systems with cold atoms by Baumann et al. [7], where the Dicke model can be engineered as an effective model with the help of Raman processes that involve, however, additional laser fields. Therefore, the controversy around this topic does not concern the existence of the superradiant phase per se, but whether or not the Dicke model is a valid description for light–matter interactions in an *equilibrium* cavity QED system, i.e., without any external driving fields.

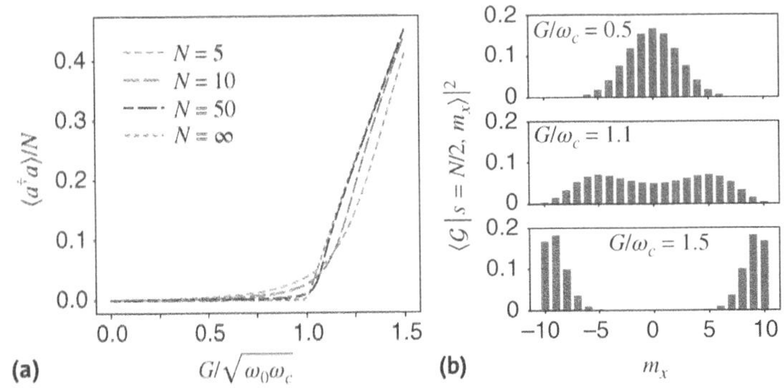

Figure 15.1 Superradiant phase transition. (a) Plot of the average photon number $\langle a^\dagger a\rangle$ in the ground state of the Dicke model $H_{\rm DM}$ given in Eq. (15.1). As the number N of two-level systems increases, a sharp transition between the normal ground state with $\langle a^\dagger a\rangle \approx 0$ and the superradiant phase with $\langle a^\dagger a\rangle \neq 0$ occurs at a critical coupling of $G \simeq \sqrt{\omega_0\omega_c}$. (b) Distribution of the m_x quantum number in the ground state of the Dicke model for three different values of G/ω_c. It illustrates the transition from a single- to a bi-modal distribution of the polarization, which leads to a Z_2-symmetry-breaking transition in the limit $N \to \infty$. For both plots, resonance conditions, $\omega_0 = \omega_c$, have been assumed.

15.3 Effective Models in Cavity QED

In the nonrelativistic limit, light–matter interactions are modeled on a fundamental level in terms of the so-called *minimal-coupling Hamiltonian*, which is directly obtained from a canonical quantization of Maxwell's equations in the Coulomb gauge. For a set of N particles with mass m and charge q, the minimal-coupling Hamiltonian reads

$$H = \sum_{i=1}^{N} \frac{[\vec{p}_i - q\vec{A}(\vec{x}_i)]^2}{2m} + V_C(\{\vec{x}_i\}) + \sum_\lambda \hbar\omega_\lambda a_\lambda^\dagger a_\lambda \tag{15.8}$$

where $\vec{x}_i$ and $\vec{p}_i$ are the position and momentum operators of the ith particle and $V_C(\{\vec{x}_i\})$ denotes the electrostatic Coulomb potential. In Eq. (15.8),

$$\vec{A}(\vec{x}) = \sum_\lambda \sqrt{\frac{\hbar}{2\epsilon_0\omega_\lambda}}\vec{u}_\lambda(\vec{x})\left(a_\lambda + a_\lambda^\dagger\right) \tag{15.9}$$

is the quantized vector potential expressed in terms of a set of orthogonal eigenmodes with bosonic annihilation operators a_λ. The corresponding mode frequencies ω_λ and mode functions $\vec{u}_\lambda(\vec{x})$ satisfy the Helmholtz equation for the boundary conditions imposed by the cavity geometry. Note that these boundaries also modify the Coulomb potential compared to free space, for example, due to the presence of image charges in a metallic mirror. Therefore, in Eq. (15.8), it is implicitly assumed that these modifications are already included in the definition of $V_C(\{\vec{x}_i\})$.

15.3.1 Cavity QED in the Coulomb Gauge

While the minimal-coupling Hamiltonian describes the exact dynamics of a cavity QED system, it involves an infinite number of modes and describes the full complexity of the atomic or molecular constituents. For any practical calculations, additional simplifications are necessary. In a cavity setup, it is often possible to restrict the coupling to a single transverse mode and perform a dipole approximation by neglecting the variation of the vector potential over the extent of the individual dipoles. For the purpose of the following discussion, it is further assumed that the mode profile of this particular cavity mode, $\vec{u}_c(\vec{x}) = 1/\sqrt{V_c}$, is constant across the whole interaction region and that the individual molecular dipoles can be modeled as effective particles of charge q and a one-dimensional binding potential $V(x)$. Under these simplifications, the full Hamiltonian in Eq. (15.8) reduces to

$$\begin{aligned} H_C &= \sum_{i=1}^{N}\left[\frac{(p_i - qA)^2}{2m} + V(x_i)\right] + \hbar\omega_c a^\dagger a + H_{dd} \\ &= \sum_{i=1}^{N} H_{dip}^{(i)} - \frac{q\mathcal{A}_0}{m}\sum_{i=1}^{N}(a + a^\dagger)p_i + \frac{Nq^2\mathcal{A}_0^2}{2m}(a + a^\dagger)^2 + H_{dd} \end{aligned} \tag{15.10}$$

where $A = \mathcal{A}_0(a + a^\dagger)$ is the relevant component of the vector potential with a zero-point amplitude $\mathcal{A}_0$. In Eq. (15.10), $H_{dip}^{(i)} = \frac{p_i^2}{2m} + V(x_i)$ is the bare Hamiltonian of a single dipole and $H_{dd} \sim \sum_{i\neq j}(\dots)x_i x_j$ accounts for the electrostatic dipole–dipole interactions, which depend explicitly on the arrangement of the matter constituents. Note that despite these crude simplifications, H_C preserves the basic structure of the original minimal-coupling Hamiltonian.

To establish a direct connection to the Dicke model from above, the bare Hamiltonian for the individual dipoles can be written in a diagonal form as,

$$H_{dip} = \frac{p^2}{2m} + V(x) = \sum_{n=0}^{\infty} E_n\,|n\rangle\langle n| \tag{15.11}$$

where E_n denotes the energy of the nth molecular eigenstate $|n\rangle$. Then, by performing a two-level approximation that restricts the dynamics of the dipole to the ground state $|g\rangle \equiv |n = 0\rangle$ and the first excited state $|e\rangle \equiv |n = 1\rangle$, with energy difference $E_1 - E_0 = \hbar\omega_0$, one obtains the cavity QED Hamiltonian in the Coulomb gauge,

$$H_{cQED}^{C} = \hbar\omega_c a^\dagger a + \hbar\omega_0 S_z + \hbar\, g_C\,(a + a^\dagger)\,S_y + \frac{\hbar\, N\Omega^2}{4\omega_c}(a + a^\dagger)^2 + H_{dd} \tag{15.12}$$

Here,

$$g_C = \frac{2q\mathcal{A}_0}{\hbar\, m}|p_{eg}| \quad \text{and} \quad \Omega^2 = \frac{2q^2\mathcal{A}_0^2\omega_c}{\hbar\, m} \tag{15.13}$$

determine the strength of the dipole-field coupling, with matrix element $p_{eg} = \langle e|p|g\rangle$, and that of the A^2-term, respectively. Furthermore, within the two-level approximation, the electrostatic interactions can be written as,

$$H_{dd} = \frac{\hbar}{4}\sum_{i\neq j} J_{ij}\sigma_x^i\sigma_x^j, \tag{15.14}$$

where the couplings $J_{ij} \sim d_{eg}^i d_{eg}^j$ are proportional to the transition dipole matrix elements, $d_{eg} = q\langle e|x|g\rangle$.

15.3.1.1 No-go Theorem

By ignoring H_{dd} and considering the coupling to the transverse cavity mode only, one can repeat the variational analysis from Section 15.2.2 to obtain the energy functional

$$E_{var}(\alpha) = \hbar\omega_c|\alpha|^2 - \frac{N}{2}\sqrt{\omega_0^2 + 4g_C^2\alpha_R^2} + \hbar\frac{N\Omega^2}{\omega_c}\alpha_R^2 \tag{15.15}$$

A closer inspection of this expression shows that the existence of a superradiant phase, as characterized by a minimum of this energy at a finite value $\alpha_R \neq 0$, requires that

$$Ng_C^2 \geq \omega_0\omega_c\left(1 + N\frac{\Omega^2}{\omega_c^2}\right) \tag{15.16}$$

which, at first sight, could be reached when $\sqrt{N}g_C$ is large enough. However, the value of g_C and Ω are derived from the same minimal-coupling term and are thus not independent of each other. To see this explicitly, one can use the following general relation between the matrix elements of the momentum and the position operator,

$$\langle n|p|k\rangle = im\frac{(E_n - E_k)}{\hbar}\langle n|x|k\rangle \tag{15.17}$$

together with the Thomas–Reiche–Kuhn sum rule

$$\sum_{n>0}(E_n - E_0)|\langle n|x|0\rangle|^2 = \frac{\hbar^2}{2m} \tag{15.18}$$

When combined, these two general relations state that for an arbitrary potential $V(x)$ the matrix element $|p_{eg}| \leq \sqrt{\hbar m\omega_0/2}$, which appears in the definition of g_C, is bounded by the corresponding matrix element of a harmonic oscillator with frequency ω_0. It follows that

$$|g_C|^2 \leq \frac{2q^2A_0^2\omega_0}{\hbar\, m} = \frac{\omega_0\Omega^2}{\omega_c} \tag{15.19}$$

and the condition for the ground state instability in Eq. (15.16) cannot be satisfied [4]. The additional energy penalty arising from the A^2-term always compensates the energy gain from anti-aligning the dipoles and the field amplitude. This result has become known as the no-go theorem for the superradiant phase transition.

This no-go theorem seemingly proves that the superradiant phase transition is merely an artifact of neglecting the A^2-term in the Dicke model, which is usually a good approximation in the weak-coupling regime, but fails to capture the equilibrium states of ultrastrongly coupled light–matter systems. However, also this derivation has many deficiencies. First, it ignores the role of electrostatic dipole–dipole interactions, which are known to induce phase transitions in the context of ferroelectricity. A second, more subtle issue arises from the two-level approximation, which, as detailed in the following section, is actually invalid in the Coulomb gauge, even for highly nonlinear potentials. Finally, while the variational analysis indicates the absence of a superradiant phase in the limit $N \to \infty$ and $g \to 0$, it does not make predictions about the actual ground state of the cavity QED system, and is expected to fail when the coupling strength g for each individual dipole becomes relevant.

15.3.2 Cavity QED in the Dipole Gauge

In the nonrelativistic limit and under the validity of the dipole approximation, it is often more convenient to study light–matter interactions in the so-called *dipole gauge* or *length gauge* by performing the unitary transformation $H_D = UH_CU^\dagger$. Here,

$$U = e^{-iqA\sum_i x_i/\hbar} \tag{15.20}$$

transforms the momentum and field operators as,

$$Up_iU^\dagger = p_i + qA \quad \text{and} \quad UaU^\dagger = a + i\frac{qA_0}{\hbar}\sum_i x_i \tag{15.21}$$

such that in this new representation the single-mode QED Hamiltonian reads

$$H_{\mathrm{D}} = \sum_{i=1}^{N} H_{\mathrm{dip}}^{(i)} + \hbar\omega_c a^\dagger a + i\omega_c q A_0 (a^\dagger - a) X + \frac{m\Omega^2}{2} X^2 + H_{\mathrm{dd}} \tag{15.22}$$

with $X = \sum_i x_i$ denoting the collective displacement. In the dipole gauge, the A^2-term has been eliminated, but instead an additional all-to-all coupling between the dipoles, the so-called P^2-term appears (see discussion below). In a final step, a two-level truncation is performed to obtain the canonical form of the QED Hamiltonian in the dipole gauge [8, 9],

$$H_{\mathrm{cQED}} = \hbar\omega_c a^\dagger a + \hbar\omega_0 S_z + \hbar g \left(a + a^\dagger\right) S_x + \frac{\hbar g^2}{\omega_c} S_x^2 + H_{\mathrm{dd}} \tag{15.23}$$

where an additional transformation $a \to a e^{i\pi/2}$ has been introduced to match the convention of the Dicke model from above. In Eq. (15.23),

$$g = \frac{2\omega_c q A_0 |\langle g|x|e\rangle|}{\hbar} \tag{15.24}$$

denotes the coupling strength and the identity $2m\Omega^2|\langle g|x|e\rangle|^2 = \hbar g^2/\omega_c$ has been used to highlight the underlying relation between the dipole–field coupling and the P^2-term proportional to S_x^2.

15.3.2.1 Coulomb or Dipole Gauge?

While the structure of $H_{\mathrm{cQED}}^{\mathrm{C}}$ and H_{cQED} is very similar, there are important qualitative differences arising from the different nature of the A^2-term and the P^2-term. This can be best seen when considering only a single dipole, $N = 1$. In this case, $S_x^2 = 1/4$, and Hamiltonian $H_{\mathrm{cQED}}(N = 1) = H_{\mathrm{Rabi}} + \hbar g^2/(4\omega_c)$ reduces to the quantum Rabi model

$$H_{\mathrm{Rabi}} = \hbar\omega_c a^\dagger a + \frac{\hbar\omega_0}{2}\sigma_z + \frac{\hbar g}{2}\left(a + a^\dagger\right)\sigma_x \tag{15.25}$$

Therefore, in contrast to the corresponding model $H_{\mathrm{cQED}}^{\mathrm{C}}(N = 1)$ in the Coulomb gauge, which still contains the A^2-term $\sim \left(a + a^\dagger\right)^2$, in the dipole gauge, there is no immediate constraint on the coupling parameter $g^2/(\omega_0\omega_c)$, which led to the no-go theorem discussed in Section 15.3.1.1. Consequently, the effective cavity QED Hamiltonians derived in different gauges make very different physical predictions, which raises the important question: Which of these models is correct?

Although the nonequivalence of effective models derived in different gauges has been encountered in different fields of research, in the context of ultrastrongly coupled light–matter systems, this problem was only identified and addressed recently [9–11]. Specifically, De Bernardis et al. [9] compared the predictions of $H_{\mathrm{cQED}}^{\mathrm{C}}$ and H_{cQED} to those of an exact model without the two-level approximation. This direct comparison is reproduced for the case $N = 1$ in Figure 15.2 and shows that for highly nonlinear systems, where the two-level approximation is expected to be accurate, H_{cQED} captures the exact spectrum very accurately while $H_{\mathrm{cQED}}^{\mathrm{C}}$ completely fails to predict the ultrastrong-coupling physics. This failure of the two-level approximation in the Coulomb gauge can be traced back to the scaling of the matrix elements of the momentum operator given in Eq. (15.17). These matrix elements scale with the energy difference $\langle n|p|k\rangle \sim (E_n - E_k)$. As a consequence, it is fundamentally not possible to neglect transitions to higher states in a systematic way. This implies that H_{cQED} is indeed the more appropriate model for modeling cavity QED systems in the ultrastrong coupling regime. Of course, this difference disappears once the full Hamiltonians are taken into account, because H_{C} and H_{D} are unitarily equivalent and therefore their spectra must be identical when no truncation of the Hilbert spaces is applied.

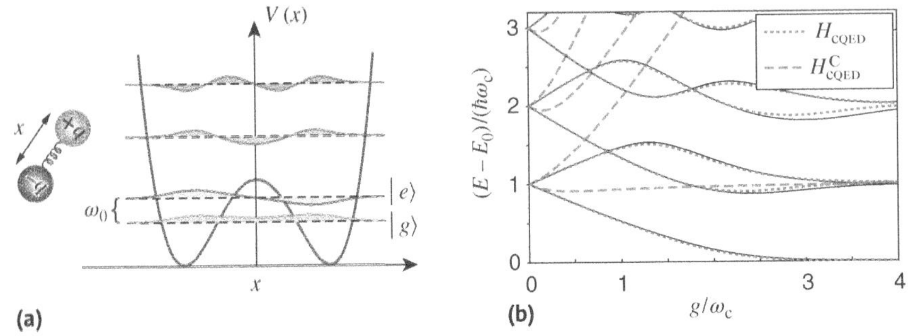

Figure 15.2 Coulomb gauge versus dipole gauge. (a) A molecular dipole is modeled as an effective particle moving in a one-dimensional double-well potential $V(x)$. The filled lines and the dashed black lines indicate the eigenstates and eigenenergies for this potential, respectively. For parameters where the lowest two states are energetically well separated from the higher states, a two-level approximation can be performed. Under such conditions, the plot in (b) compares the spectra of H^C_{cQED} and H_{cQED} with the spectrum of the full multilevel Hamiltonian H_C (solid black lines) for the case of a single dipole. For this plot, a highly nonlinear potential with a level-spacing ratio of $(E_2 - E_1)/(E_1 - E_0) \approx 20$ has been assumed.

15.3.2.2 Physical Quantities

A frequent source of confusion in the field of ultrastrong light–matter interactions arises from a misinterpretation of the predictions obtained from abstract Hamiltonians, such as H_{cQED}, in terms of measurable quantities. In the Coulomb gauge, the relevant component of the quantized electric field is $E = -\dot{A} = i\omega_c A_0(a - a^\dagger)$ and the total electromagnetic energy that is stored in this mode, $H_{\mathrm{EM}} = \hbar\omega_c a^\dagger a$, can be expressed in terms of the photon number operator $a^\dagger a$. Instead, in view of the transformation in Eq. (15.21), the electric field operator in the dipole gauge is

$$E = i\sqrt{\frac{\hbar\omega_c}{2\epsilon_0 V_c}}(a - a^\dagger) - \frac{P}{\epsilon_0} \tag{15.26}$$

where $P = q\sum_i x_i/V_c$ is the polarization density of the matter system. This means that the field operators a and $a^\dagger$ are associated with the displacement field, $D = \epsilon_0 E + P$, rather than the electric field. This relation also explains the appearance of the P^2-term in Eq. (15.22), which originates from the expansion of the electric field energy, $\sim E^2$, in terms of a and $a^\dagger$.

Similarly, after applying the two-level approximation and taking the additional rotation $a \to ae^{i\pi/2}$ into account, the total electromagnetic energy in the dipole gauge,

$$H_{\mathrm{EM}} = \hbar\omega_c d^\dagger d \quad \text{with} \quad d = \left(a + \frac{g}{\omega_c} S_x\right) \tag{15.27}$$

is no longer given by the photon number, but involves operators of the matter system as well. It is therefore instructive to write the total cavity QED Hamiltonian as

$$H_{\mathrm{cQED}} = H_{\mathrm{EM}} + \hbar\omega_0 S_z + H_{\mathrm{dd}} \tag{15.28}$$

such that the three contributions to the total energy, namely the energy of the quantized electromagnetic mode, H_{EM}, the bare energy of the matter system, $\hbar\omega_0 S_z$, and the dipole–dipole interactions, H_{dd}, appear explicitly. Eq. (15.28) also shows that in the limit $\omega_c, \omega_0 \to 0$, the effective model H_{cQED} reproduces the correct electrostatic limit,

$$H_{\mathrm{cQED}}(\omega_c, \omega_0 \to 0) \simeq H_{\mathrm{dd}} \tag{15.29}$$

where dipoles only interact via electrostatic Coulomb forces. This is not the case for the Dicke model or many other cavity QED Hamiltonians, where this important contribution is often omitted or treated inconsistently.

15.4 The Ground States in Cavity QED

Given that Eq. (15.23) represents a simplified, but still physically meaningful model for a dipolar cavity QED system, it can be used to investigate the ground state phases of cavity QED at arbitrary interaction strength. In general, this is still a complicated task, since H_{cQED} contains both strong collective interactions with the cavity mode and short-ranged dipole–dipole interactions described by H_{dd}. This competition makes both analytic and numerical treatments of this model rather challenging. For a qualitative analysis, it thus useful to introduce another simplification and replace the exact dipole–dipole interactions by an equivalent all-to-all coupling,

$$H_{\mathrm{dd}} = \frac{\hbar}{4}\sum_{i\neq j} J_{ij}\sigma_x^i\sigma_x^j \rightarrow \frac{\hbar\bar{J}}{N}S_x^2 \tag{15.30}$$

with a characteristic strength $\bar{J} = \sum_{i\neq j} J_{ij}/N$. Under this simplification, the interaction strength between the dipoles remains the same on average, but the resulting cavity QED Hamiltonian reduces to the extended Dicke model,

$$H_{\mathrm{EDM}} = \hbar\omega_c a^\dagger a + \hbar\omega_0 S_z + \hbar\, g\left(a + a^\dagger\right) S_x + \hbar\left(\frac{\bar{J}}{N} + \frac{g^2}{\omega_c}\right) S_x^2 \tag{15.31}$$

which only contains collective spin operators. Although this model still does not permit exact analytic solutions, it can be analyzed more conveniently in terms of the angular momentum basis states $|s, m_x\rangle$ obeying

$$\vec{S}^2|s, m_x\rangle = s(s+1)|s, m_x\rangle \quad \text{and} \quad S_x|s, m_x\rangle = m_x|s, m_x\rangle \tag{15.32}$$

Here, $s = 0, \ldots, N/2$ is the total spin quantum number and $m_x = -s, -s+1, \ldots, s$ the spin projection quantum number. Since H_{EDM} conserves the total spin, i.e., $\left[H_{\mathrm{EDM}}, \vec{S}^2\right] = 0$, it can be separated into decoupled sectors with at most $2s + 1$ basis states for the matter component.

15.4.1 Boundary-induced Ferroelectricity

To understand the potential vacuum-induced modification of the cavity QED ground state, it is, first of all, necessary to characterize the bare phases of the system in the limit $g \rightarrow 0$. In this case, the extended Dicke model reduces to the Lipkin–Meshkov–Glick (LMG) model [12]

$$H_{\mathrm{LMG}} = \hbar\omega_0 S_z + \frac{\hbar\bar{J}}{N}S_x^2 \tag{15.33}$$

which serves as a basic model to describe ferromagnetic or ferroelectric phase transitions when $\bar{J} < 0$. Specifically, in the limit $S = N/2 \rightarrow \infty$ one can use mean-field theory or a variational Ansatz as above to derive a critical interaction strength,

$$\bar{J}_c = -\omega_0 \tag{15.34}$$

below which the system undergoes a second-order phase transition into an ordered ferroelectric state with $\langle S_x\rangle \neq 0$.

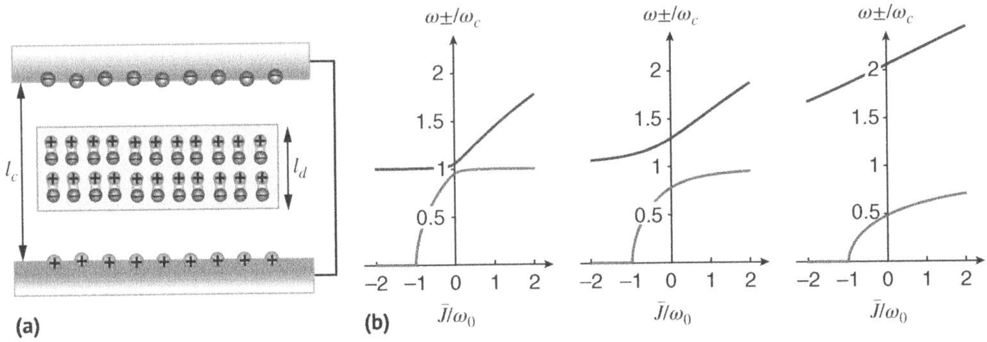

Figure 15.3 Ferroelectricity in a cavity. (a) Sketch of a layer of dipoles of thickness ℓ_d, which is placed between two metallic plates separated by a distance ℓ_c. The induced image charges reduce the average dipole–dipole interaction strength $\bar{J}$ according to Eq. (15.35). (b) The frequencies $\omega_\pm$ for the upper and lower polariton branches in the collective ultrastrong coupling regime, see Eq. (15.38), are plotted as a function of $\bar{J}$ and for $G/\omega_c = 0.1, 0.5, 1.5$ (from left to right) and $\omega_c = \omega_0$. The vanishing of ω_- at $\bar{J} = -\omega_0$ indicates an instability in the normal paraelectric phase and a transition into a ferroelectric phase, which is no longer captured by the Holstein–Primakoff approximation.

As already pointed out below Eq. (15.8), even though for $g = 0$, there is no coupling to the quantized electromagnetic field, the presence of a cavity influences as well the electrostatic interactions, for example, through image charges in the metallic boundaries. This scenario is depicted in Figure 15.3a, where a thin slab of dipoles of thickness ℓ_d is placed between two metallic plates separated by a distance ℓ_c. In the limit of many dipoles, $N \gg 1$, this configuration can be analyzed using macroscopic electrodynamics and results in the following dependence of the average dipole–dipole interaction strength [9],

$$\bar{J} \simeq \bar{J}_0 \left(\frac{2}{3} - \frac{\ell_d}{\ell_c} \right) \tag{15.35}$$

This scaling predicts a transition from an overall repulsive to an overall attractive dipole–dipole interaction, as ℓ_c is reduced. Eq. (15.34) then implies that under appropriate conditions also the phase of the dipolar system can be influenced by the presence of boundaries [13].

Although the highly oversimplified analysis presented here is not able to make quantitative predictions for real materials, it illustrates how the mere presence of boundaries can induce a transition into a polarized state with $\langle S_x \rangle \neq 0$. For any finite g this phase would also be associated with a finite expectation value of $\langle a \rangle \simeq -g\langle S_x \rangle/\omega_c$, similar to the characteristics of the superradiant phase given in Eq. (15.7). However, for small g, the existence of this phase can be solely attributed to the electrostatic interactions and it is unaffected by the coupling to the dynamical cavity mode. For clarity, one should thus speak of a ferroelectric phase.

15.4.2 Collective Ultrastrong Coupling Regime

While in most cavity experiments the coupling strength g between a single dipole and the cavity mode is small, the collective coupling $G = g\sqrt{N}$ can already become comparable to ω_0 and ω_c. Under these conditions, the ground state of a cavity QED system is still to a good approximation given by the decoupled state $|\mathcal{G}\rangle \simeq |0\rangle_c \otimes |g\rangle^{\otimes N}$, and corrections on top of it can be treated as collective bosonic excitations. Formally, this is done by making a Holstein–Primakoff approximation [14],

$$S_- \simeq \sqrt{N} b, \quad S_+ \simeq \sqrt{N} b, \quad S_z = b^\dagger b - \frac{N}{2} \tag{15.36}$$

which replaces the collective spin by a bosonic mode with annihilation operator b. This approximation is valid as long as the total number of excitations is small, $\langle b^\dagger b\rangle \ll N$, and a saturation of the individual two-level dipoles is negligible.

Within the validity of the Holstein–Primakoff approximation, the cavity QED Hamiltonian reduces to that of two coupled harmonic modes,

$$\begin{aligned} H_{\rm HP} &= \hbar\omega_c a^\dagger a + \hbar\omega_0\left(b^\dagger b - \frac{N}{2}\right) \\ &+ \frac{\hbar G}{2}(a+a^\dagger)(b+b^\dagger) + \frac{\hbar}{4}\left(\bar{J} + \frac{G^2}{\omega_c}\right)(b+b^\dagger)^2 \\ &= \hbar\omega_+ p_+^\dagger p_+ + \hbar\omega_- p_-^\dagger p_- + E_G \end{aligned} \tag{15.37}$$

which can be diagonalized and written in terms of the two polaritonic eigenmodes with bosonic mode operators $p_\pm$ and a slightly modified ground-state energy E_G. The corresponding eigenfrequencies are given by

$$\omega_\pm^2 = \frac{1}{2}\left[\omega_c^2 + \tilde{\omega}_0^2 \pm \sqrt{(\omega_c^2 - \tilde{\omega}_0^2)^2 + 4G^2\omega_c\omega_0}\right] \tag{15.38}$$

where $\tilde{\omega}_0^2 = \omega_0(\omega_0 + \bar{J} + G^2/\omega_c)$. The dependence of these frequencies on the dipole–dipole interaction strength $\bar{J}$ is plotted in Figure 15.3b for different values of the collective coupling G. These plots illustrate that while the light–matter coupling has a strong influence on the excited polariton states, it does not affect the point at which the ground state becomes unstable. This point is determined by a vanishing lower polariton frequency,

$$\omega_- = 0 \Leftrightarrow \bar{J} = -\omega_c \tag{15.39}$$

which occurs at the same critical value of $\bar{J}$ as in the decoupled system [9]. This can be understood from the fact that near the phase transition point, the dynamics of the relevant matter excitation slows down, $\omega_- \to 0$. As a result, the unstable dipole mode decouples from the dynamical cavity mode, which interacts with the current density of the matter system.

In summary, the analysis of $H_{\rm EDM}$ in terms of the Holstein–Primakoff approximation confirms the findings obtained in Section 15.3.1.1 about the absence of a ground-state instability that is solely induced by the coupling to a quantized cavity mode. While in the dipole gauge, there is no constraint on the single-dipole coupling strength, for $N \gg 1$ the P^2 term plays a similar role as the A^2 term in the Coulomb gauge and stabilizes the system. This is also consistent with our intuition that the addition of a single photonic degree of freedom should not influence the thermodynamics of a large system [15]. Note, however, that for a fixed value of G, taking the limit $N \to \infty$ also assumes that $g \to 0$. Therefore, in the collective ultrastrong coupling regime, the cavity mode has at most a perturbative effect on each individual dipole.

15.4.3 Nonperturbative Coupling Regime

The physics of a cavity QED system changes drastically once $g \sim \omega_c$, in which case light–matter interactions become nonperturbative at the level of individual dipoles. To analyze this regime, it is more convenient to change to a polaron frame via the unitary transformation $U = e^{\frac{g}{\omega_c} S_x (a^\dagger - a)}$. In this new frame, the transformed cavity QED Hamiltonian $\tilde{H}_{\text{EDM}} = U H_{\text{EDM}} U^\dagger$ reads

$$\tilde{H}_{\text{EDM}} = \hbar\omega_c a^\dagger a + \frac{\hbar \bar{J}}{N} S_x^2 + \tilde{H}_{\text{dip}} \tag{15.40}$$

where $\tilde{H}_{\text{dip}} = \hbar\omega_0 U S_z U^\dagger$. In the electrostatic limit, $\omega_0 \to 0$, the Hamiltonian $\tilde{H}_{\text{EDM}}$ becomes diagonal with the dipoles decoupled from the cavity mode. The corresponding eigenstates are of the form $|n\rangle_c \otimes |s, m_x\rangle$, which in the original frame,

$$|\Psi_{n,s,m_x}\rangle = e^{-\frac{g}{\omega_c} m_x (a^\dagger - a)} |n\rangle_c \otimes |s, m_x\rangle \tag{15.41}$$

correspond to photon number eigenstates that are displaced proportional to the spin projection quantum number m_x. The ground-state spin configuration is determined by the sign of $\bar{J}$, while for $\bar{J} = 0$ all states $|0\rangle_c \otimes |s, m_x\rangle$ are degenerate.

At any finite ω_0, the quantum fluctuations of the dipoles become relevant and the term $\tilde{H}_{\text{dip}}$ couples states with different polarization m_x. In the regime $g \gtrsim \omega_0, \omega_c$, this effect can be accounted for by treating $\tilde{H}_{\text{dip}}$ up to second-order perturbation theory. As discussed in more detail in [9, 16], restricted to the lowest photon number manifold, the result of this perturbation theory is an effective low-energy model for the dipoles only,

$$H_{\text{eff}} = \hbar\omega_0 e^{-\frac{g^2}{2\omega_c^2}} S_z + \frac{\hbar\omega_0^2\omega_c}{2g^2}\left(S_x^2 - \vec{S}^2\right) + \frac{\hbar\bar{J}}{N} S_x^2 \tag{15.42}$$

This effective model highlights two essential cavity-induced modifications of the dipolar ensemble. First, the bare oscillation frequency is reduced by a factor $e^{-\frac{g^2}{2\omega_c^2}}$. This frequency renormalization can be understood from the reduced overlap between two displaced photon states, $\langle\Psi_{0,s,m_x'}|\Psi_{0,s,m_x}\rangle = e^{-\frac{g^2(m_x - m_x')^2}{2\omega_c^2}}$, which are, therefore, coupled less strongly by the term $\sim S_z$. This effect can also be seen as the cavity analog of mass renormalization in QED, where the dipole drags along the cavity field as it oscillates.

While the frequency renormalization is a pure single-particle effect, the second term in Eq. (15.42) arises from the collective coupling of multiple dipoles to the cavity mode. It energetically favors states with maximal total spin s, but minimal m_x. Therefore, opposite to the picture conveyed by the Dicke model, at very large interaction strengths, the cavity induces an effective anti-ferroelectric coupling between the dipoles. This implies that for $\bar{J} = 0$ and large enough g/ω_c the ground state of the cavity QED system minimizes the contribution $\sim S_x^2$ and is given by

$$|G\rangle \simeq |0\rangle_c \otimes |s = N/2, m_x = 0\rangle \tag{15.43}$$

This state is highly entangled and represents a symmetric superposition of all possible dipole configurations consistent with $m_x = 0$. This property also implies that there is no net polarization and despite the large coupling strength, the dipoles are decoupled from the cavity field. One thus speaks of a *collective subradiant phase.* Figure 15.4a shows a plot of the photon number expectation value $\langle a^\dagger a\rangle$ for the exact ground state of the full extended Dicke model for a small number of $N = 10$ dipoles and $\bar{J} = 0$. As expected, for increasing g, the photon number initially increases due to a hybridization between photonic and matter degrees of freedom. However, above a value of about $g/\omega_c \approx 3$ this trend starts to reverse as a result of the formation of the subradiant state [16]. This occurs as a smooth crossover rather than a sharp phase transition.

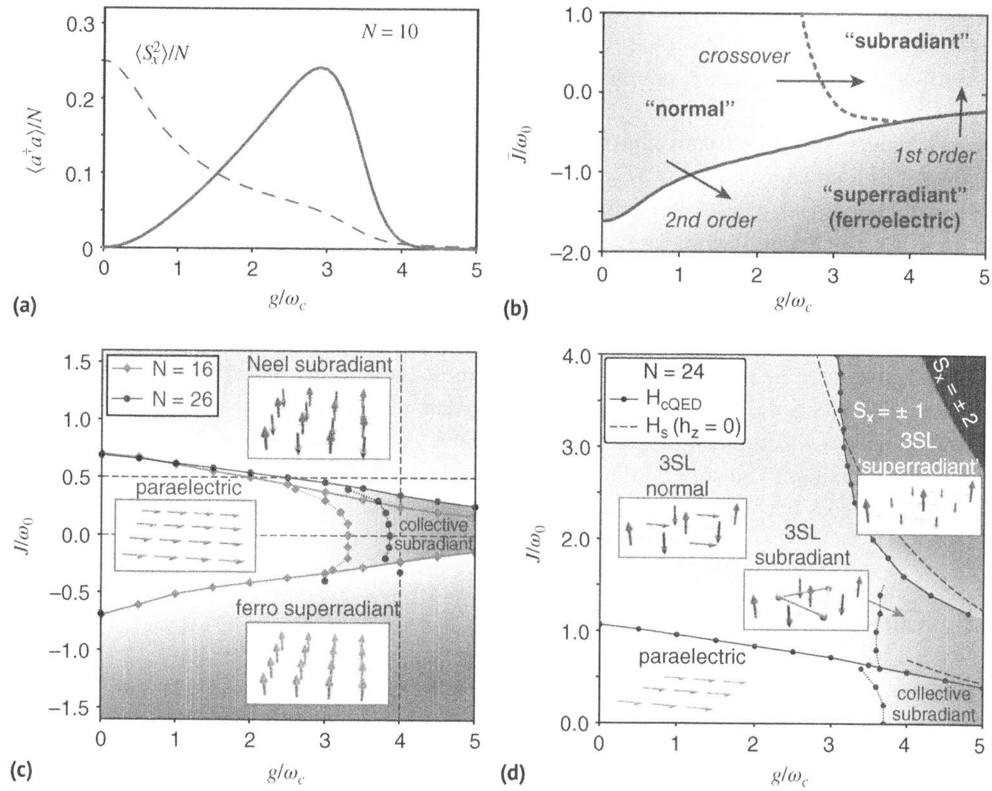

Figure 15.4 Ground-state phases in cavity QED. (a) Plot of the average photon number in the ground state of the extended Dicke model $H_{\rm EDM}$ for $\bar{J} = 0$ and $N = 10$. The dashed line indicates the corresponding values of the polarization fluctuations. (b) Full ground state phase diagram of the extended Dicke model as a function of the dipole–dipole interaction strength $\bar{J}$ and the dipole–field coupling g. The different colors indicate the distinct phases that have been obtained from an exact diagonalization of $H_{\rm EDM}$ for $N = 10$. (c) and (d) Plots of the ground-state phase diagrams of $H_{\rm cQED}$ with nearest-neighbor dipole–dipole interactions obtained in [17]. The plot in (c) shows the case of dipoles arranged on a two-dimensional square lattice and the plot in (d) for a triangular lattice. In the latter case, the frustration effects for $\bar{J} > 0$ give rise to additional three-sublattice (3SL) phases, which crossover into several new 3SL subradiant and superradiant phases. In all plots, $\omega_0 = \omega_c$ [17].

15.4.4 Ground-state Phases in Cavity QED

Based on these analytic results for various limiting cases, Figure 15.4b shows the complete ground-state phase diagram of the extended Dicke model obtained from an exact numerical diagonalization of $H_{\rm EDM}$ for a small number of $N = 10$ dipoles. For $g \to 0$, this model exhibits a transition between the ferroelectric and the normal (paraelectric) phase at a critical value $\bar{J}_c$, which approaches the mean-field prediction given in Eq. (15.34), when the number of dipoles is increased. For the assumed all-to-all dipole–dipole interactions, there is no further transition at positive $\bar{J}$, although the dipoles become gradually more correlated. As g/ω_c increases, one observes a significant shift of the ferroelectric phase transition point toward lower values of $|\bar{J}_c|$. Qualitatively, this effect can be understood from the cavity-induced suppression of the oscillation frequency, $\omega_0 \to \omega_0 e^{-\frac{g^2}{2\omega_c^2}}$, which stabilizes the polarized phase. In contrast, for $\bar{J} \gtrsim 0$ and sufficiently large g/ω_c the paraelectric phase crosses over into the subradiant phase. As discussed above, this phase is characterized by a vanishing photon number $\langle a^\dagger a \rangle \to 0$, but, compared to the normal phase, also by vanishing spin fluctuations, $\langle S_x^2 \rangle \to 0$. This phase is stabilized by dipole–dipole interactions for $\bar{J} > 0$, but also extends into the ferroelectric regime, $\bar{J} < 0$, as predicted by the effective low-energy model in Eq. (15.42).

Going beyond the collective spin approximation, Figure 15.4c, d shows the results of a numerical study of the ground-state phases of H_{cQED} for dipoles on a two-dimensional square lattice with nearest-neighbor dipole–dipole interactions of strength $J_{ij} = J$. This phase diagram, obtained in [17] from an exact numerical diagonalization of H_{cQED} for up to $N \approx 26$ dipoles, confirms that the qualitative picture obtained from the analysis of the extended Dicke model remains valid in the case of short-range dipole–dipole interactions. For the square lattice, the main difference arises from the appearance of a Neel-ordered antiferroelectric phase already at $g = 0$, which at larger couplings is separated by a sharp transition from both the paraelectric and the subradiant phase. On a triangular lattice, the frustrated interactions between neighboring dipoles for $J > 0$ result in additional phases with a three-sublattice order parameter. Interestingly, this feature also gives rise to additional subradiant and superradiant phases for $g/\omega_c \gg 1$. However, a detailed discussion of all these phases goes beyond the scope of the present overview and for further details the reader is referred to [17].

15.5 Conclusions

In summary, the current chapter presents a brief overview about the possible ground states of dipolar cavity QED systems at arbitrary interaction strength. Starting from the historic discussion about the superradiant phase transition in the Dicke model, it was demonstrated that this model is not suited to describe the equilibrium states of strongly coupled light–matter systems. Instead, extended models that include the appropriate A^2 or P^2 terms as well as electrostatic interactions must be used to obtain approximate, but still physically consistent predictions. Such models also permit a cleaner distinction between purely electrostatic boundary effects and genuine vacuum-induced modifications.

The analysis of these basic cavity QED models in different parameter regimes confirms the stability of the normal ground state, even in the collective ultrastrong coupling regime $G \gtrsim \omega_0, \omega_c$, where the excited polaritonic states are already strongly modified. However, by increasing the coupling even further, completely new phases of light and matter with surprising properties can emerge. So far, these phases have been investigated primarily for the case of two-level dipoles, but the underlying mechanism, such as frequency renormalization and a cavity-induced anti-ferroelectric coupling are expected to occur in other settings as well. While the conditions to reach the required coupling strengths are experimentally very challenging, it is in principle possible to access this non-perturbative coupling regime using high-impedance modes [18], as encountered, for example, in cavity QED systems with lump-element LC resonators, nanoplasmonic spheres, or hyperbolic van der Waals materials [19].

References

1 Dicke, R. H. Coherence in spontaneous radiation processes. *Phys. Rev.* **1954,** *93*, 99–110. https://doi.org/10.1103/PhysRev.93.99.

2 Hepp, K., Lieb, E. H. On the superradiant phase transition for molecules in a quantized radiation field: the dicke maser model. *Ann. Phys.* **1973,** *76*(2), 360–404. https://doi.org/10.1016/0003-4916(73)90039-0.

3 Wang Y. K., Hioe. F. T. Phase transition in the Dicke model of superradiance. *Phys. Rev. A.* **1973,** *7*, 831–836. https://doi.org/10.1103/PhysRevA.7.831.

4 Rzażewski, K., Wódkiewicz, K., Żakowicz, W. Phase transitions, two-level atoms, and the A^2 term. *Phys. Rev. Lett.* **1975,** *35*, 432–434. https://doi.org/10.1103/PhysRevLett.35.432.

5 Keeling, J. Coulomb interactions, gauge invariance, and phase transitions of the Dicke model. *J. Condens. Matter Phys.* **2007,** *19*(29), 295213. https://doi.org/10.1088/0953-8984/19/29/295213.

6 Vukics, A., Grießer, T., Domokos, P. Elimination of the a-square problem from cavity QED. *Phys. Rev. Lett.* **2014,** *112*, 073601. https://doi.org/10.1103/PhysRevLett.112.073601.

7 Baumann, K., Guerlin, C., Brennecke, F., Esslinger, T. Dicke quantum phase transition with a superfluid gas in an optical cavity. *Nature* **2010,** *464*(7293), 1301–1306. https://doi.org/10.1038/nature09009.

8 Todorov, Y., Sirtori, C. Few-electron ultrastrong light-matter coupling in a quantum lc circuit. *Phys. Rev. X.* **2014,** *4*, 041031. https://doi.org/10.1103/PhysRevX.4.041031.

9 De Bernardis, D., Jaako, T., Rabl, P. Cavity quantum electrodynamics in the nonperturbative regime. *Phys. Rev. A.* **2018,** *97*, 043820. https://doi.org/10.1103/PhysRevA.97.043820.

10 Stokes, A., Nazir, A. Gauge ambiguities imply Jaynes-Cummings physics remains valid in ultrastrong coupling QED. *Nat. Commun.* **2019,** *10*(1), 499. https://doi.org/10.1038/s41467-018-08101-0.

11 Di Stefano, O., Settineri, A., Macrì, V., Garziano, L., Stassi, R., Savasta, S. et al. Resolution of gauge ambiguities in ultrastrong-coupling cavity quantum electrodynamics. *Nat. Phys.* **2019,** *15*(8), 803–808. https://doi.org/10.1038/s41567-019-0534-4.

12 Lipkin, H. J., Meshkov, N., Glick, A. J. Validity of many-body approximation methods for a solvable model: (I). Exact solutions and perturbation theory. *Nucl. Phys.* **1965,** *62*(2), 188–198. https://doi.org/10.1016/0029-5582(65)90862-X.

13 Batra, I. P., Silverman, B. D. Thermodynamic stability of thin ferroelectric films. *Solid State Commun.* **1972,** *11*, 291–294. https://doi.org/10.1016/0038-1098(72)91180-5.

14 Holstein, T., Primakoff, H. Field dependence of the intrinsic domain magnetization of a ferromagnet. *Phys. Rev.* **1940,** *58*, 1098–1113. https://doi.org/10.1103/PhysRev.58.1098.

15 Pilar, P., De Bernardis, D., Rabl, P. Thermodynamics of ultrastrongly coupled light-matter systems. *Quantum* **2020,** *4*, 335. https://doi.org/10.22331/q-2020-09-28-335.

16 Jaako, T., Xiang, Z. L., Garcia-Ripoll, J. J., Rabl, P. Ultrastrong-coupling phenomena beyond the Dicke model. *Phys. Rev. A.* **2016,** *94*, 033850. https://doi.org/10.1103/PhysRevA.94.033850.

17 Schuler, M., De Bernardis, D., Läuchli, A., Rabl, P. The vacua of dipolar cavity quantum electrodynamics. *SciPost Phys.* **2020,** *9*, 066. https://doi.org/10.21468/SciPostPhys.9.5.066.

18 Sáez-Blázquez, R., De Bernardis, D., Feist, J., Rabl, P. Can we observe nonperturbative vacuum shifts in cavity QED? *Phys. Rev. Lett.* **2023,** *131*, 013602. https://doi.org/10.1103/PhysRevLett.131.013602.

19 Ashida, Y., İmamoğlu, A., Demler, E. Cavity quantum electrodynamics with hyperbolic van der Waals materials. *Phys. Rev. Lett.* **2023,** *130*, 216901. https://doi.org/10.1103/PhysRevLett.130.216901.

Afterword

The idea for this book originated during the COVID-19 pandemic, with the goal of providing a central reference point that would help students and new researchers get involved in this field. Putting together the content of this book has, if anything, reinforced how humbling it is to work in this area, with our own understanding having evolved considerably over the intervening years. The story of polariton chemistry is still being written. Yet, with so many different opportunities for impact communicated in the chapters of this book, there is good reason to believe that the best is yet to come. We hope this volume will incentivize researchers to take new avenues of inquiry and to help solve some of the outstanding questions in the field.

Polariton Chemistry: Molecules in Cavities, First Edition. Edited by Joel Yuen-Zhou, Noel C. Giebink and Raphael F. Ribeiro. Published 2026 by John Wiley & Sons, Inc.

Index

Note: Page numbers in *italics* and **bold** refers to figures and tables respectively.

q

r

s

Printed and bound by CPI Group (UK) Ltd, Croydon, CR0 4YY
17/09/2026
14964257-0004